T0213544

Lecture Notes in Computer Science 10574

Commenced Publication in 1973
Founding and Former Series Editors:
Gerhard Goos, Juris Hartmanis, and Jan van Leeuwen

More information about this series at http://www.springer.com/series/7408

Hervé Panetto · Christophe Debruyne
Walid Gaaloul · Mike Papazoglou
Adrian Paschke · Claudio Agostino Ardagna
Robert Meersman (Eds.)

On the Move to Meaningful Internet Systems

OTM 2017 Conferences

Confederated International Conferences:
CoopIS, C&TC, and ODBASE 2017
Rhodes, Greece, October 23–27, 2017
Proceedings, Part II

 Springer

Editors
Hervé Panetto
University of Lorraine
Nancy
France

Christophe Debruyne
Odisee University College
Brussels
Belgium

Walid Gaaloul
Télécom SudParis
Évry
France

Mike Papazoglou
Tilburg University
Tilburg
The Netherlands

Adrian Paschke
Freie Universität Berlin and Fraunhofer
  FOKUS
Berlin
Germany

Claudio Agostino Ardagna
Università degli Studi di Milano
Crema
Italy

Robert Meersman
TU Graz
Graz
Austria

ISSN 0302-9743   ISSN 1611-3349 (electronic)
Lecture Notes in Computer Science
ISBN 978-3-319-69458-0   ISBN 978-3-319-69459-7 (eBook)
https://doi.org/10.1007/978-3-319-69459-7

Library of Congress Control Number: 2017956721

LNCS Sublibrary: SL2 – Programming and Software Engineering

Printed on acid-free paper

This Springer imprint is published by Springer Nature
The registered company is Springer International Publishing AG
The registered company address is: Gewerbestrasse 11, 6330 Cham, Switzerland

General Co-chairs and Editors' Message for OnTheMove 2017

The OnTheMove 2017 event held October 23–27 in Rhodes, Greece, further consolidated the importance of the series of annual conferences that was started in 2002 in Irvine, California. It then moved to Catania, Sicily in 2003, to Cyprus in 2004 and 2005, Montpellier in 2006, Vilamoura in 2007 and 2009, in 2008 to Monterrey, Mexico, to Heraklion, Crete in 2010 and 2011, Rome 2012, Graz in 2013, Amantea, Italy in 2014 and lastly in Rhodes in 2015 and 2016 as well.

This prime event continues to attract a diverse and relevant selection of today's research worldwide on the scientific concepts underlying new computing paradigms, which of necessity must be distributed, heterogeneous and supporting an environment of resources that are autonomous yet must meaningfully cooperate. Indeed, as such large, complex and networked intelligent information systems become the focus and norm for computing, there continues to be an acute and even increasing need to address the respective software, system, and enterprise issues and discuss them face to face in an integrated forum that covers methodological, semantic, theoretical, and application issues as well. As we all realize, e-mail, the Internet, and even video conferences are not by themselves optimal or even sufficient for effective and efficient scientific exchange.

The OnTheMove (OTM) International Federated Conference series has been created precisely to cover the scientific exchange needs of the communities that work in the broad yet closely connected fundamental technological spectrum of Web-based distributed computing. The OTM program every year covers data and Web semantics, distributed objects, Web services, databases, information systems, enterprise workflow and collaboration, ubiquity, interoperability, mobility, and grid and high-performance computing.

OnTheMove is proud to give meaning to the "federated" aspect in its full title: It aspires to be a primary scientific meeting place where all aspects of research and development of Internet- and intranet-based systems in organizations and for e-business are discussed in a scientifically motivated way, in a forum of interconnected workshops and conferences. This year's 15th edition of the OTM Federated Conferences event therefore once more provided an opportunity for researchers and practitioners to understand, discuss, and publish these developments within the broader context of distributed, ubiquitous computing. To further promote synergy and coherence, the main conferences of OTM 2017 were conceived against a background of their three interlocking global themes:

- Trusted Cloud Computing Infrastructures Emphasizing Security and Privacy
- Technology and Methodology for Data and Knowledge Resources on the (Semantic) Web

– Deployment of Collaborative and Social Computing for and in an Enterprise Context

Originally the federative structure of OTM was formed by the co-location of three related, complementary, and successful main conference series: DOA (Distributed Objects and Applications, held since 1999), covering the relevant infrastructure-enabling technologies, ODBASE (Ontologies, DataBases and Applications of SEmantics, since 2002) covering Web semantics, XML databases and ontologies, and of course CoopIS (Cooperative Information Systems, held since 1993) which studies the application of these technologies in an enterprise context through, e.g., workflow systems and knowledge management. In the 2011 edition security aspects issues, originally started as topics of the IS workshop in OTM 2006, became the focus of DOA as secure virtual infrastructures, further broadened to cover aspects of trust and privacy in so-called Cloud-based systems. As this latter aspect came to dominate agendas in this and overlapping research communities, we decided in 2014 to rename the event as the Cloud and Trusted Computing (C&TC) conference, and originally launched in a workshop format.

These three main conferences specifically seek high-quality, contributions of a more mature nature and encourage researchers to treat their respective topics within a framework that simultaneously incorporates (a) theory, (b) conceptual design and development, (c) methodology and pragmatics, and (d) application in particular case studies and industrial solutions.

As in previous years we again solicited and selected additional quality workshop proposals to complement the more mature and "archival" nature of the main conferences. Our workshops are intended to serve as "incubators" for emergent research results in selected areas related, or becoming related, to the general domain of Web-based distributed computing. This year this difficult and time-consuming job of selecting and coordinating the workshops was brought to a successful end by Ioana Ciuciu, and we were very glad to see that our earlier successful workshops (EI2N, META4eS, FBM) re-appeared in 2017, in some cases in alliance with other older or newly emerging workshops. The Fact Based Modeling (FBM) workshop in 2015 succeeded and expanded the scope of the successful earlier ORM workshop. The Industry Case Studies Program, started in 2011 under the leadership of Hervé Panetto and OMG's Richard Mark Soley, further gained momentum and visibility in its 7th edition this year.

The OTM registration format ("one workshop resp. conference buys all workshops resp. conferences") actively intends to promote synergy between related areas in the field of distributed computing and to stimulate workshop audiences to productively mingle with each other and, optionally, with those of the main conferences. In particular EI2N continues to so create and exploit a visible cross-pollination with CoopIS.

We were very happy to see that in 2017 the number of quality submissions for the OnTheMove Academy (OTMA) noticeably increased. OTMA implements our unique, actively coached and therefore very time- and effort-intensive formula to bring PhD students together, and aims to carry our "vision for the future" in research in the areas covered by OTM. Its 2017 edition was organized and managed by a dedicated team of collaborators and faculty, Peter Spyns, Maria-Esther Vidal, inspired as always by OTMA Dean, Erich Neuhold.

In the OTM Academy, PhD research proposals are submitted by students for peer review; selected submissions and their approaches are to be presented by the students in front of a wider audience at the conference, and are independently and extensively analyzed and discussed in front of this audience by a panel of senior professors. One may readily appreciate the time, effort, and funds invested in this by OnTheMove and especially by the OTMA Faculty.

As the three main conferences and the associated workshops all share the distributed aspects of modern computing systems, they experience the application pull created by the Internet and by the so-called Semantic Web, in particular developments of big data, increased importance of security issues, and the globalization of mobile-based technologies. For ODBASE 2017, the focus somewhat shifted from knowledge bases and methods required for enabling the use of formal semantics in Web-based databases and information systems to applications, especially those within IT-driven communities. For CoopIS 2017, the focus as before was on the interaction of such technologies and methods with business process issues, such as occur in networked organizations and enterprises. These subject areas overlap in a scientifically natural and fascinating fashion and many submissions in fact also covered and exploited the mutual impact among them. For our event C&TC 2017, the primary emphasis was again squarely put on the virtual and security aspects of Web-based computing in the broadest sense. As with the earlier OnTheMove editions, the organizers wanted to stimulate this cross-pollination by a program of engaging keynote speakers from academia and industry and shared by all OTM component events. We are quite proud to list for this year:

- Stephen Mellor, Industrial Internet Consortium, Needham, USA
- Markus Lanthaler, Google, Switzerland

The general downturn in submissions observed in recent years for almost all conferences in computer science and IT has also affected OnTheMove, but this year the harvest again stabilized at a total of 180 submissions for the three main conferences and 40 submissions in total for the workshops. Not only may we indeed again claim success in attracting a representative volume of scientific papers, many from the USA and Asia, but these numbers of course allow the respective Program Committees to again compose a high-quality cross-section of current research in the areas covered by OTM. Acceptance rates vary but the aim was to stay consistently at about one accepted full paper for three submitted, yet as always these rates are subject to professional peer assessment of proper scientific quality.

As usual we separated the proceedings into two volumes with their own titles, one for the main conferences and one for the workshops and posters. But in a different approach to previous years, we decided the latter should appear after the event and thus allow workshop authors to improve their peer-reviewed papers based on the critiques by the Program Committees and on the live interaction at OTM. The resulting additional complexity and effort of editing the proceedings was professionally shouldered by our leading editor, Christophe Debruyne, with the general chairs for the conference volume, and with Ioana Ciuciu and Hervé Panetto for the workshop volume. We are again most grateful to the Springer LNCS team in Heidelberg for their professional support, suggestions, and meticulous collaboration in producing the files and indexes ready for downloading on the USB sticks. It is a pleasure to work with staff that so deeply understands the scientific context at large and the specific logistics of conference proceedings publication.

The reviewing process by the respective OTM Program Committees was performed to professional quality standards: Each paper review in the main conferences was assigned to at least three referees, with arbitrated e-mail discussions in the case of strongly diverging evaluations. It may be worth emphasizing once more that it is an explicit OnTheMove policy that all conference Program Committees and chairs make their selections in a completely sovereign manner, autonomous and independent from any OTM organizational considerations. As in recent years, proceedings in paper form are now only available to be ordered separately.

The general chairs are once more especially grateful to the many people directly or indirectly involved in the set-up of these federated conferences. Not everyone realizes the large number of qualified persons that need to be involved, and the huge amount of work, commitment, and financial risk in the uncertain economic and funding climate of 2017 that is entailed by the organization of an event like OTM. Apart from the persons in their roles mentioned earlier, we therefore wish to thank in particular explicitly our main conference Program Committee chairs:

- CoopIS 2017: Mike Papazoglou, Walid Gaaloul, and Liang Zhang
- ODBASE 2017: Declan O'Sullivan, Joseph Davis, and Satya Sahoo
- C&TC 2017: Adrian Paschke, Hans Weigand, and Nick Bassiliades

And similarly we thank the Program Committee (Co-)chairs of the 2017 ICSP, OTMA and Workshops (in their order of appearance on the website): Peter Spyns, Maria-Esther Vidal, Mario Lezoche, Wided Guédria, Qing Li, Georg Weichhart, Peter Bollen, Hans Mulder, Maurice Nijssen, Anna Fensel, and Ioana Ciuciu. Together with their many Program Committee members, they performed a superb and professional job in managing the difficult yet existential process of peer review and selection of the best papers from the harvest of submissions. We all also owe a significant debt of gratitude to our supremely competent and experienced conference secretariat and technical admin staff in Guadalajara and Dublin, respectively, Daniel Meersman and Christophe Debruyne.

The general conference and workshop co-chairs also thankfully acknowledge the academic freedom, logistic support, and facilities they enjoy from their respective institutions — Technical University of Graz, Austria; Université de Lorraine, Nancy, France; Latrobe University, Melbourne, Australia; and Babes-Bolyai University, Cluj, Romania — without which such a project quite simply would not be feasible. Reader, we do hope that the results of this federated scientific enterprise contribute to your research and your place in the scientific network... and we hope to welcome you at next year's event!

September 2017

Robert Meersman
Hervé Panetto
Christophe Debruyne

Organization

OTM (On The Move) is a federated event involving a series of major international conferences and workshops. These proceedings contain the papers presented at the OTM 2017 Federated conferences, consisting of CoopIS 2017 (Cooperative Information Systems), C&TC 2017 (Cloud and Trusted Computing), and ODBASE 2017 (Ontologies, Databases, and Applications of Semantics).

Executive Committee

General Co-chairs

Robert Meersman	TU Graz, Austria
Tharam Dillon	La Trobe University, Melbourne, Australia
Hervé Panetto	University of Lorraine, France
Ernesto Damiani	Politecnico di Milano, Italy

OnTheMove Academy Dean

Erich Neuhold	University of Vienna, Austria

Industry Case Studies Program Chair

Hervé Panetto	University of Lorraine, France

CoopIS 2017 PC Co-chairs

Mike Papazoglou	European Research Institute in Service Science, Tilburg University, The Netherlands
Walid Gaaloul	TELECOM SudParis, France
Liang Zhang	Fudan University, China

ODBASE 2017 PC Co-chairs

Adrian Paschke	Freie Universität Berlin and Fraunhofer FOKUS, Germany
Nick Bassiliades	Aristotle University of Thessaloniki, Greece
Hans Weigand	Tilburg School of Economics and Management, The Netherlands

C&TC 2017 PC Co-chairs

Claudio Ardagna	Università degli Studi di Milano, Italy
Adrian Belmonte	European Union Agency for Network and Information Security (ENISA), Greece

Konstantinos   Royal Holloway, University of London, UK
  Markantonakis

Local Organization Chair

Stefanos Gritzalis   University of the Aegean, Greece

Publication Chair

Christophe Debruyne   Odisee University College, Belgium

Logistics Team

Daniel Meersman

CoopIS 2017 Program Committee

Aditya Ghose
Akhil Kumar
Alex Norta
Alfredo Cuzzocrea
Aly Megahed
Amal Elgammal
Amel Bouzeghoub
Amel Mammar
Andreas Andreou
Andreas Oberweis
Andreas Opdahl
Antonio Ruiz Cortés
Arturo Molina
Athman Bouguettaya
Barbara Pernici
Barbara Weber
Beatrice Finance
Bruno Defude
Carlo Combi
Cesare Pautasso
Chengzheng Sun
Chihab Hanachi
Chirine Ghedira
Christian Huemer
Claude Godart
Claudia Diamantini
Daniel Florian
Daniela Grigori
David Carlos Romero Díaz

Djamal Benslimane
Djamel Belaid
Elisabettta di Nitto
Epaminondas Kapetanios
Ernesto Exposito
Eva Kühn
Faiez Gargouri
Farouk Toumani
Francois Charoy
Frank Leymann
Frank-Walter Jäkel
Georg Weichhart
George Samaras
Gerald Oster
Giancarlo Guizzardi
Guido Wirtz
Heiko Ludwig
Heinrich Mayr
Hongji Yang
Imen Grida Ben Yahia
Ivona Brandic
Jan Mendling
Jian Yang
Jiang Cao
Jianwen Su
John Miller
Joonsoo Bae
Jörg Niemöller
Jose Luis Garrido

José Palazzo Moreira de Oliveira
Joyce El Haddad
Juan Manuel Murillo Rodríguez
Juan Manuel Vara Mesa
Julius Köpke
Kais Klai
Karim Baina
Khalid Belhajjame
Khalil Drira
Kostas Magoutis
Lakshmish Ramaswamy
Layth Sliman
Leandro Krug Wives
Liang Zhang
Lijie Wen
Lin Liu
Lucinéia Heloisa Thom
Mahmoud Barhamgi
Manfred Jeusfeld
Manfred Reichert
Marcelo Fantinato
Marco Aiello
Maristella Matera
Marouane Kessentini
Martin Gaedke
Martine Collard
Massimo Mecella
Matthias Klusch
Maurizio Lenzerini
Mehdi Ahmed-Nacer
Michael Mrissa
Michael Rosemann
Michele Missikoff
Mike Papazoglou
Mohamed Graiet
Mohamed Jmaiel
Mohamed Mohamed
Mohamed Sellami

Mohammed Ouzzif
Mohand-Said Hacid
Mourad Kmimech
Narjes Bellamine-Ben Saoud
Nizar Messai
Nour Assy
Oktay Turetken
Olivier Perrin
Oscar Pastor
Pablo Villarreal
Paolo Giorgini
Peter Forbrig
Philippe Merle
Richard Chbeir
Rik Eshuis
Salima Benbernou
Sami Bhiri
Sami Yangui
Samir Tata
Sanjay K. Madria
Selmin Nurcan
Shazia Sadiq
Sherif Sakr
Slim Kallel
Sonia Bergamaschi
Sotiris Koussouris
Stefan Jablonski
Tiziana Catarci
Vassilios Andrikopoulos
Wil M.P. van der Aalst
Walid Gaaloul
Willem-Jan van den Heuvel
Yehia Taher
Youcef Baghdadi
Zakaria Maamar
Zhangbing Zhou
Zohra Bellahsene

ODBASE 2017 Program Committee

Adrian Paschke
Alessandra Mileo
Alexander Artikis
Anastasios Gounaris

Anna Fensel
Annika Hinze
Asuncion Gomez Perez
Athanasios Tsadiras

Bernd Neumayr
Charalampos Bratsas
Christian Kop
Christophe Debruyne
Costin Badica
Danh Le Phuoc
Dietrich Rebholz
Dimitris Plexousakis
Dumitru Roman
Efstratios Kontopoulos
Fotios Kokkoras
Georg Rehm
George Vouros
Georgios Meditskos
Gines Moreno
Giorgos Giannopoulos
Giorgos Stamou
Giorgos Stoilos
Gokhan Coskun
Grigoris Antoniou
Grzegorz J. Nalepa
Hans Weigand
Harald Sack

Harry Halpin
Heiko Paulheim
Ioannis Katakis
Irlán Grangel-González
Kalliopi Kravari
Kia Teymourian
Manolis Koubarakis
Marcin Wylot
Markus Luczak-Roesch
Naouel Karam
Nick Bassiliades
Olga Streibel
Oscar Corcho
Ralph Schäfermeier
Rolf Fricke
Ruben Verborgh
Soren Auer
Sotiris Batsakis
Stefania Costantini
Vadim Ermolayev
Vassilios Peristeras
Witold Abramowicz

C&TC 2017 Program Committee

Marco Anisetti
Claudio A. Ardagna
Rasool Asal
Ioannis Askoxylakis
Adrian Belmonte
Michele Bezzi
David Chadwick
Mauro Conti
Ernesto Damiani
Francesco Di Cerbo
Scharam Dustdar
Nabil El Ioini
Stefanos Gritzalis
Marit Hansen
Sotiris Ioannidis
Martin Jaatun

Meiko Jensen
Gwanggil Jeon
George Karabatis
Antonio Mana
Konstantinos Markantonakis
Raja Naeem Akram
Eugenia Nikolouzou
Claus Pahl
Konstantinos Rantos
Damien Sauveron
Stefan Schulte
Julian Schutte
Daniele Sgandurra
Miguel Vargas Martin
Luca Viganò
Christos Xenakis

OnTheMove 2017 Keynotes

Pragmatic Semantics at Web Scale

Markus Lanthaler

Google, Switzerland

Short Bio

Dr. Markus Lanthaler is a software engineer and tech lead at Google where he currently works on YouTube. He received his Ph.D. in Computer Science from the Graz University of Technology in 2014 for his research on Web APIs and Linked Data. Dr. Lanthaler is one of the core designers of JSON-LD and the inventor of Hydra. He has published several scientific articles, is a frequent speaker at conferences, and chairs the Hydra W3C Community Group.

Talk

Despite huge investments, the traditional Semantic Web stack failed to gain widespread adoption and deliver on its promises. The proposed solutions focused almost exclusively on theoretical purity at the expense of their usability. Both academia and industry ignored for a long time the fact that the Web is more a social creation than a technical one. After a long period of disillusionment, we see a renewed interest in the problems the Semantic Web set out to solve and first practical approaches delivering promising results. More than 30% of all websites contain structured information now. Initiatives such as Schema.org allow, e.g., search engines to extract and understand such data, integrate it, and create knowledge graphs to improve their services.

This talk analyzes the problems that hindered the adoption of the Semantic Web, present new, promising technologies and shows how they might be used to build the foundation of the longstanding vision of a Semantic Web of Services.

Evolution of the Industrial Internet of Things: Preparing for Change

Stephen Mellor

Industrial Internet Consortium, Needham, MA 02492, USA

Short Bio

Stephen Mellor is the Chief Technical Officer for the Industrial Internet Consortium, where he directs the standards requirements and technology & security priorities for the Industrial Internet. In that role, he coordinates the activities of the several engineering, architecture, security and testbed working groups and teams. He also co-chairs both the Definitions, Taxonomy and Reference Architecture workgroup and the Use Cases workgroup for the NIST CPS PWG (National Institute for Standards and Technology Cyberphysical System Public Working Group).

He is a well-known technology consultant on methods for the construction of real-time and embedded systems, a signatory to the Agile Manifesto, and adjunct professor at the Australian National University in Canberra, ACT, Australia. Stephen is the author of Structured Development for Real-Time Systems, Object Lifecycles, Executable UML, and MDA Distilled.

Until recently, he was Chief Scientist of the Embedded Software Division at Mentor Graphics, and founder and past president of Project Technology, Inc., before its acquisition. He participated in multiple UML/modeling-related activities at the Object Management Group (OMG), and was a member of the OMG Architecture Board, which is the final technical gateway for all OMG standards. Stephen was the Chairman of the Advisory Board to IEEE Software for ten years and a two-time Guest Editor of the magazine, most recently for an issue on Model-Driven Development.

Talk

The fundamental technological trends presently are more connectivity and more capability to analyze large quantities of data cheaply. But no one knows where those technological trends will take us, so we need to prepare for change.

Prediction is difficult, especially about the future, as several people are reputed to have said. But this keynote will peer ahead into several areas that we can see need attention, such as:

- Security for everything
- Innovation and funding
- Learning, deployment and competitiveness

We need strategies to prepare for evolution in these areas, and we also need to understand longer term trends. Already we see improvements in operational efficiency, and changes in the economy from pay-per-asset to pay-per-use. More changes are likely, towards pay-per-outcome and direct consumer access to "pull" products autonomously.

These changes will fundamentally change the economy and drive technological innovation. The industrial internet is only at the beginning of perhaps forty more years of change.

Contents – Part II

Contents – Part I

Cloud and Trusted Computing (C&TC) 2017

C&TC 2017 PC Co-chairs' Message

Claudio A. Ardagna, Adrian Belmonte,
and Konstantinos Markantonakis

Welcome to the Cloud and Trusted Computing 2017 (C&TC2017), the 7th International Symposium on Secure Virtual Infrastructures, held in Rhodes, Greece, as part of the OnTheMove Federated Conferences & Workshops 2017.

The conference solicited submissions from both academia and industry presenting novel research in the context of cloud and trusted computing. Continuing the successful events of previous years, the C&TC 2017 edition focused on the special theme " *Secure and Trustworthy Big Data Analytics and IoT Integration: From the Periphery to the Cloud.*"

Inside this theme, theoretical and practical approaches for the following areas had been called:

– Trust, security, privacy and risk management
– Data Management
– Computing infrastructures and architectures
– Applications

In this scope, a multitude of specific challenges have been addressed by our authors. These challenges included emergency management, privacy preserving encryption, Big Data analytics, quality of service management in the cloud, and software verification. All submitted papers passed through a rigorous selection process involving at least three reviews. In the end, we decided to accept 5 full papers and 2 short papers, reflecting a selection of the best among the excellent. In addition to the technical program composed of the papers in the proceedings, the workshop included a keynote and a panel.

Organizing a conference like C&TC is a team effort, and many people need to be acknowledged. First, we would like to thank everyone who submitted their contributions to this event for having chosen C&TC to present and discuss their work. Their contributions were the basis for the success of the conference.

Second, we would like to acknowledge the hard work of all our colleagues from the Program Committee, experts in the research domains of the conference, for performing the extremely valuable tasks of reviewing and discussing the many excellent contributions.

Finally, we would like to thank everyone at the OTM organizers team for their exceptional support and, in particular, the OTM Conferences & Workshops General Chairs Robert Meersman, Tharam Dillon, Hervé Panetto, and Ernesto Damiani, and the Publication Chair Christophe Debruyne.

All of these people contributed to the Proceedings of the 7th International Conference on Cloud and Trusted Computing, and all of them deserve our highest gratitude. Thank you!

Property Preserving Encryption in NoSQL Wide Column Stores

Tim Waage[✉] and Lena Wiese

Research Group Knowledge Engineering, Georg August Universität Göttingen,
Institute of Computer Science, Goldschmidtstraße 7, 37077 Göttingen, Germany
{tim.waage,lena.wiese}@informatik.uni-goettingen.de

Abstract. Property preserving encryption (PPE) can enable database systems to process queries over encrypted data. While a lot of research in this area focusses on doing so with SQL databases, NoSQL (Not only SQL) cloud databases are good candidates either. On the one hand, they usually provide enough space to store the typically larger ciphertexts and special indexes of PPE-schemes. On the other hand in contrast to approaches for SQL systems, despite PPE the query expressiveness remains almost unaffected. Thus, in this paper we investigate (i) how PPE can be used in the popular NoSQL sub-category of so-called wide column stores in order to protect sensitive data in the threat model of a persistent honest-but-curious database provider, (ii) what PPE schemes are suited for this task and (iii) what performance levels can be expected.

Keywords: Database security · NoSQL databases · Property preserving encryption · Wide column stores

1 Introduction

In times of the "Web 2.0" [1] traditional SQL-based database services struggle with the changing demands arising in distributed cloud environments. They are not well suited to represent loosely structured data items like they are typical for the Web 2.0. As NoSQL databases [2,3] were designed for meeting those new requirements, they attracted more and more attention in the last years, especially in the sub-category of so-called wide column stores (WCS, see Sect. 3.1). Popular companies developed their own solutions, e.g. Google developed "Bigtable" [4] (used in over 60 Google services) and Facebook developed "Cassandra" [5]. Nowadays it is common to use WCSs on a smaller scale, too. Many cloud database providers offer flexible on-demand services with simple web interfaces for running WCSs remotely in their clusters (e.g. the Google Cloud Platform, Microsoft Azure or Amazon EC2) to exploit the well known benefits of outsourcing databases.

However, storing and processing sensitive data on infrastructures provided by a third party increases the risk of unauthorized disclosure if the infrastructure is compromised by an adversary. Unfortunately, WCSs usually lack security

© Springer International Publishing AG 2017
H. Panetto et al. (Eds.): OTM 2017 Conferences, Part II, LNCS 10574, pp. 3–21, 2017.
https://doi.org/10.1007/978-3-319-69459-7_1

features like access control, which an external front end or a firewall is assumed to take care of. Hence there is a strong need for providing security and privacy guarantees, because there are several ways how sensitive data can be leaked. An adversary can exploit software vulnerabilities, curious or malicious administrators at hosting providers can snoop on private data or attackers with physical access to servers can steal data from disk and memory [6]. Various examples show, that these threats are not only theoretical [7–10].

The straight-forward solution to reduce the damage caused by server compromises is encrypting the data on a trusted client before it gets uploaded to the cloud servers, then process queries by reading it back from the server to the client, decrypt it, and process the query on the client machine. Unfortunately this requires transferring much more data than necessary (typically large fractions of data are read in order to create relatively small data aggregations) and moves a major part of query computation to the client, which defeats the general purpose of (remote) databases.

Existing approaches (see Sect. 2) make use of property-preserving encryption (PPE, see Sect. 3.2) to enable query execution over encrypted data, but the vast majority of existing solutions focusses only on SQL-based technologies and avoids PPE schemes, that rely on index data structures to improve their performance. Hence, this paper makes the following contributions:

- It identifies the requirements for utilizing PPE in the context of WCSs.
- It surveys the practical feasibility of various schemes for order-preserving and searchable encryption with focus on these requirements.
- It shows how PPE can be applied to the data model of WCSs in the honest-but-curious adversary scenario, meaning the database server carries out its tasks as expected, but tries to learn about the data it hosts.
- It implements selected schemes and conducts a practical and comprehensive benchmark using the currently most popular WCSs [11] Cassandra [5] and HBase [12] as underlying platforms. In order to obtain practically relevant results these databases remain unmodified. Hence, we match the conditions that can be found in today's cloud database provider's offers for quantifying the performance loss due to PPE.

2   Related Work

This section surveys related work, limited to approaches that are also designed for the honest-but-curious adversary model, compute over encrypted data and rely on encryption to provide data confidentiality.

Approaches for Relational Databases. One of the first approaches to processing queries over encrypted data is from [13]. Unfortunately in their approach hardware requirements on client side were similar to the ones on the server side. The most popular work on performing queries over encrypted data is "CryptDB" [14] for MySQL and PostgreSQL. It was the first system that could be considered practical, introducing a variety of innovative features, most importantly:

the onion layer model (OLM), which this work uses as well in an improved adaptation (see Sect. 5.1). However, it uses only PPE schemes that are slow for querying, because the authors avoid (client or server side) indexes. Thus, CryptDB does not scale well when datasets reach a certain size. However, it still receives a lot of scientific attention, in favourable [15,16] as well as critic ways [17]. "Monomi" [18] can be considered being an extension of CryptDB, trying to support arbitrary SQL queries with the cost of higher requirements for the client machine. "BlindSeer" [19] addresses efficient sub-linear searches for SQL-queries that can be represented as a monotone boolean formula, consisting of the search conditions: keyword match, range and negation. "DBMask" [20] enforces access control cryptographically, based on attribute based access control and combining broadcast and hierarchical key management. It also uses PPE, but not in an OLM-fashion. The authors of "L-EncDB" [21] propose to use so-called format-preserving encryption to realize fuzzy searches.

Approaches for Non-relational Databases. An approach for executing queries over encrypted triple patterns using SPARQL is presented by [22]. Unfortunately, the number of cryptographic keys in their approach is high and every plaintext triple results in eight ciphertext triples, which leads to much overhead in terms of processing and storage inefficiency. To build a distributed key-value store the recent approach of [23] introduces an additional architectural component called dispatcher, that distributes encrypted data to all the database server nodes evenly. Another very recent approach is "Arx" [24] on top of MongoDB, which introduces two proxy servers and needs to know in advance what operations are to be performed on what fields in order to maintain the required indexes.

Hardware Architectures for Encrypted Databases. "Cipherbase" [25] is an extension of Microsoft's SQL Server with two modified parts: the ODBC driver at the client side and the query processor at the server side, that integrates a secure coprocessor within a so-called "trusted machine", realized utilizing field programmable gate arrays (FPGAs). "TrustedDB" [26] is a similar approach, but with tamper-proof cryptographic coprocessors (SCPUs) instead of FPGAs. Even though hardware approaches like these overcome some limitations of CryptDB-based techniques (in particular regarding the query expressiveness), they rely on expensive trusted hardware and require the database to have the user's decryption keys.

3 Background

3.1 The Data Model of Wide Column Stores

WCSs are inspired by Googles BigTable architecture [4], but there are also publicly available open source databases, that rely on the same or a very similar data model, e.g. Hypertable [27], Accumulo [28] as well as used for practical experiments in this work: Cassandra [5] and HBase [12].

WCSs use tables, rows and columns like traditional relational (SQL-based) databases. However, the fundamental difference is that columns are created for each row instead of being predefined by the table structure. Every row has at least one mandatory column containing its identifier[1]. Except for this column, two rows of the same table can háve completely disjunct sets of columns. The identifier of a row has to be unique for the whole table and cannot be used by another row.

Rows are maintained in lexicographic order by their identifier. As WCSs are distributed systems, ranges of such row identifiers serve as units of distribution. Hence, similar row identifiers (and thus data items that are likely to be semantically related to each other) are always kept physically close together, so that reads of ranges require communication to a minimum number of machines. Because row identifiers are used for coordinating distribution this way, changing them would result in changing the row's physical position within the database (cluster), which is prohibitively expensive and thus not allowed. Thus, a row identifier cannot be changed after the row was inserted.

3.2   Property-Preserving Encryption

The types of PPE relevant for this work are deterministic encryption (DET), order-preserving encryption (OPE) and searchable encryption (SE). Other types of PPE would have no value for supporting encrypted queries in WCSs. In particular, homomorphic encryption is not considered. Apart from its runtime deficits [18,29], query mechanisms of WCS would benefit either only minimal (e.g. only SUM() and AVG() in Cassandra's query language) or not at all (e.g. HBase).

DET. The purpose of DET is enabling the database server to check for equality by mapping identical plaintexts to identical ciphertexts.

OPE. The purpose of OPE is enabling a server to learn about the relative order of data elements without gaining information about their exact values. Therefore it encrypts two elements p_1, p_2 of a domain D in a way that $p_1 \leq p_2 \Rightarrow Enc(p_1) \leq Enc(p_2)$ for all $p \in D$. Thus, its use cases are sorting and range queries over encrypted data. A lot of OPE schemes have been proposed with different strategies to map a plaintext to a ciphertext domain (e.g. [30–36]).

SE. The purpose of SE is enabling a server to search over encrypted data without gaining information about the plaintext data. Most SE schemes use indexes (e.g. [37–44]), which are encrypted in a way, that only a trapdoor allows for comparing the searchword with the ciphertext. However, there are also schemes, that avoid having an index by embedding the trapdoor in a special format into the ciphertext itself (e.g. [45]).

4   Selecting Practical PPE Schemes

Having introduced the key characteristics of the WCS data model and the tasks of the different kinds of PPE, we now move on to identify actually feasible PPE schemes for our architecture presented in Sect. 5.

---

[1] Commonly referred to as "row key". However we use "row identifier" to avoid confusions with cryptographic keys.

4.1   Deterministic Encryption

In contrast to OPE and SE (see below), the only relevant criterion for practicability of DET is the determinism itself. There are plenty of well known DET schemes, that have proven to work well in practice. For this work we use the Advanced Encryption Standard (AES, [46]).

4.2   Order-Preserving Encryption

The criteria for OPE are more complex, due to the working principles of OPE and the WCS data model as described in Sect. 3.1. We focus on five aspects:

I. Ciphertext (im-)mutability. Ciphertexts produced by an OPE scheme are either *mutable,* meaning they may change as more and more input gets encrypted (e.g. in [35], causing re-writes to the database), or *immutable,* meaning they are final (e.g. [34]). Encrypting row identifier columns of WCS tables with OPE requires immutable ciphertexts, as discussed in Sect. 3.1. However, other columns can be encrypted using mutable schemes.

II. Need for additional data structures. If they are not stateless, OPE schemes require additional data structures for storing their state. That can be done using indexes, trees, dictionaries etc., either on client side (or at least a trusted environment), e.g. [35], or on server side, e.g. [33, 47]. In particular maintaining tree structures is expensive for WCSs.

III. Need for additional architectural components. Some OPE schemes require components running co-located to the database server (e.g. [33]), which cannot be considered practical due to the architectural overhead, that is usually not covered by today's cloud database providers.

IV. Input capabilities. Some OPE schemes require detailed knowledge of all plaintexts before encryption (e.g. [32]), which is hard to realize in practical scenarios as databases may grow unpredictably over time. Some schemes even need to encrypt the whole plaintext space in advance [34, 48], instead of encrypting only the desired values on demand.

V. Security. Nearly every OPE scheme comes with its own or no formal security definition (see Table 1). Practical security issues resulting from the database scenario and the efforts for corresponding counter measures (e.g. plaintext shifting [31] or using fake queries to hide the data distribution [49]) have to be taken into account too.

Table 1 shows an overview and brief evaluation of the schemes that we have investigated using the above described criteria. Based on this, we selected to implement the schemes of [31] (modular OPE = *"mOPE"*), [34] (Random Subrange Selection = *"RSS"*) and [35] (Optimal Average-Complexity Ideal-Security = *"OACIS"*) for further experiments (see Sect. 6.2).

To give an idea of why other OPE schemes from Table 1 have been considered impractical, we point out a few of their characteristics that cannot be read from this table: The approaches of [51] and [48] require splitting and partitioning of the

Table 1. Evaluation of the practical feasibility of popular OPE schemes based on the criteria introduced in Sect. 4.2, ordered by date of publication

OPE scheme	I	II	III	IV	V[a]
[30]	+	--	+	-	-- (?[b])
[31] (mOPE)	+	++	+	-	+ (POPF-CCA)
[32]	+	--	+	--	-- (?[c])
[33]	-	--	-	++	+ (IND-OCPA)
[34] (RSS)	+	-	+	+	++ (> IND-OCPA[d])
[48]	+	-	+	-	-- (?[b])
[35] (OACIS)	-	-	+	++	+ (IND-OCPA)
[36]	+	+	+	-	++ (> POPF-CCA[d])

[a] IND-OCPA = indistinguishability under ordered chosen-plaintext attack, POPF-CCA = pseudo random order-preserving function against chosen-ciphertext attack (for both, see [50]).
[b] only informal security analysis provided by the authors.
[c] no security analysis provided by the authors.
[d] ">" = proved by the authors to be better than...

plaintext space, causing much metadata overhead. The scheme of [32] requires detailed knowledge of the plaintext space e.g. the smallest distance between two plaintext values, a requirement that hardly can be met in unpredictably growing datasets. As mentioned before the approach of [33] needs an additional component running co-located to the database server, which often is not possible or does not fit to the offers of cloud database providers and causes network communication overhead.

4.3   Searchable Encryption

An evaluation of practical feasibility of the schemes for SE can be done similarly based on the following criteria:

I. Need for additional data structures. As mentioned in Sect. 3.2 SE schemes sometimes use indexes to speed up the search process. These indexes come with the cost of additional pre-processing steps (e.g. selecting keywords, set up the index data structure, etc.), require index maintenance and often an extra round of communication (first for querying the index and then for getting the actual results). Sometimes the underlying index data structure is hardly manageable for WCS databases without much effort (e.g. tree structures).

II. Support for Updates. When used in databases, a scheme for performing SE needs the ability to process updates[2], since in most practical cases datasets tend

---

[2] Mainly meaning adding items to the dataset. SE schemes capable of doing so are commonly referred to as being "dynamic".

to grow or change. As it turned out, only a surprisingly small number of SE schemes is capable of handling this.

III. Algorithm Requirements. Encryption and checking for searchword matches does not work with native database operations like in DET or OPE. Instead, more complex procedures are necessary, involving e.g. lookups in auxiliary data structures like bloom filters or traversing trees, using cryptographic primitives or concatenation of strings. For efficiency reasons this should not become too complex.

IV. Security vs. V. Search Efficiency. Security for index-based (hence, search efficient) schemes and not index-based schemes is hardly comparable due to the information leakage connected to querying the indexes. It consists of information about the index itself (e.g. number of words per document, number of documents, lengths of documents, document-IDs), search pattern information (what was searched for?) and access pattern information (how much answers do I get from executing a certain query compared to executing another one?). This leads to two competitive requirements: avoiding an index leads to more security but is generally slow for searches. Using an index may slow down encryption and leaks additional information, but can speed up querying significantly.

Table 2. Evaluation of the practical feasibility of popular SE schemes based on the criteria introduced in Sect. 4.3, ordered by date of publication

SE scheme	I	II	III[a]	IV[b]	V[c]
[45] (SWP)	++	++	− (XOR, PRF)	+ (IND-CPA)	$O(n)$
[39]	−−	−−	−− (XOR, SC, DED)	+ (IND-CKA2)	$O(m)$
[40]	−	+	− (XOR)	+ (IND-CKA2)	$O(log(u))$
[41]	−−	++	−− (XOR, PRF, HSH)	+ (IND-CKA2)	$O(log(u))$
[42]	−−	+	−− (XOR, PRF, HOM)	+ (IND-CKA2)	$O(m)$
[43] (SUISE)	−	++	− (SC, PRF)	+ (IND-CKA2)	$O(n/u)$
[44]	−−	−−	−− (XOR, SC, DED)	+ (IND-CKA2)	$O(m)$

[a] XOR = bitwise exclusive OR operations, PRF = pseudo-random functions, SC = string concatenation, DED = deterministic encryption/decryption, HOM = homomorphic encryption, HSH = hash functions.
[b] IND-CPA = indistinguishability under chosen-plaintext attacks, IND-CKA2 = adaptive indistinguishability under chosen keyword attacks (for both, see [52]).
[c] Searching on a dataset with size of n words (u of which are unique), resulting in m matches.

Table 2 gives an overview and brief evaluation of the SE schemes that we investigated in regard to the given criteria. Based on it, we selected to implement the schemes of [45] (*"SWP"*, an abbreviation of the three author's names Song, Wagner and Perrig) and [43] (securely updating index-based searchable encryption = *"SUISE"*) for further experiments (see Sect. 6.2).

Like we previously did for OPE, we point out reasons for why we do not consider other schemes from Table 2 being practical. [37] proposes to use bloom filters as indexes per document. Bit-wise bloomfilter operations are hard to handle for a WCS and a bloomfilter limits the number of words per document. [38] presented a similar approach, but using pre-built dictionaries, which they propose to store either on clientside (which defeats the purpose of remote searchable encryption) or on server side (which basically has the same shortcomings as in [37]). The approach of [39] achieves sub-linear search time, using a complex and for a database barely manageable index structure. Furthermore, [37–39] do not support updates. Hence, [42] proposed an extension for [39] with two additional laborious serverside data structures. [40] proposes a scheme in which not only searches require multiple rounds of communication between client and server, but also storage. Realizing this problem, the authors of [41] got rid of the interactivity by introducing even more client side computation.

5   Data Management on Server Side

We now have selected feasible PPE schemes, but just storing the PPE-encrypted values would of course leak information instantly (like equality and relative order between values) that is not supposed to leak when not querying. Therefore the authors of [53] proposed a so-called onion layer model (OLM) for their SQL-based architecture "CryptDB". The idea is to encrypt every value with a PPE scheme of each category (DET, OPE and SE) separately that leaks just enough information to still be able to perform certain operations over the encrypted data (as described in Sect. 3.2) and wrap it into another layer of random encryption (in the following: "RND") for not leaking any information. Then, the RND layer is only removed, if a query requires it. In this way the database is still able to process queries, but it learns only the minimal necessary amount of information.

5.1   An Onion Layer Model for WCSs

We adapt the basic idea of CryptDB's OLM, but the data model of WCSs requires some changes. As explained in Sect. 3.1, a fundamental working principle of WCSs is keeping all rows of a table sorted by the content of the row identifier column. Thus, the database must be able to compare the row identifier of a new row to be inserted with already existing ones in the table. Therefore OPE columns of row identifiers are not allowed to have a RND layer. Otherwise the WCS data model would be broken. That means, row identifier columns must be treated differently from all other columns regarding the onion layer design. They must leak the order of values to allow for row sorting independent of querying.

All data types supported by the databases can be grouped in three categories: strings (e.g. text, characters), numerical values (e.g. integers, timestamps) or byte blobs (e.g. byte arrays, raw files). Figure 1 presents this work's OLM design for string columns, before (left) and after (right) queries involving equality checks

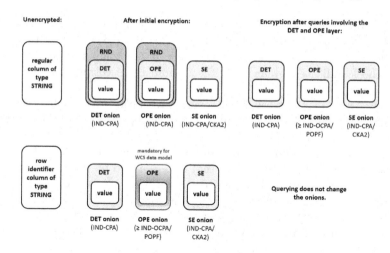

Fig. 1. Transformation of a plain string column into onion-layered ciphertext columns

and order comparisons were executed. The upper row shows the onions for regular string columns. Note the missing RND layers after a query like Q3–Q5 (see Sect. 5.2) was processed and the SE onion not requiring a RND layer at all, because schemes for SE usually provide the same security guarantees as RND layer encryption. The row below shows the onion in case of the string column is a row identifier column. Note that in this case the DET onion comes without a RND layer, because the mandatory OPE column already leaks equality[3], rendering a RND layer wrapped around the DET onion useless. However, keeping a DET onion still makes sense, because having implemented AES in hardware on the majority of modern processors, its decryption process works much faster than the decryption algorithms of other PPE schemes.

While Fig. 1 illustrates the situation for string columns as most complex cases, the OLM for the other two type categories is different, since not all operations make sense for all types of data. For instance, numerical values do not need the SE onion, because searching for words in numbers is neither somehow defined nor possible, even in unencrypted databases. Thus, numerical values can be encrypted faster than strings in the OLM. We investigate the difference in encryption performance in Sect. 6.2.

5.2 Querying the Encrypted Data

In our framework, the work flow for executing queries against PPE encrypted data organized in an OLM as described in Sect. 5.1 is as follows.

Step 1: A query usually contains one or more conditions that have to be met by a row to be included in the result set. Such a condition is always of

---

[3] Apart from rare exceptions (e.g. [51]) OPE schemes are deterministic. Hence they leak not only the relative order between plaintexts, but also equality.

the form *[columnname, compare operation, comparator]*. For every condition the client checks which onion is involved and whether the RND layer of this onion still exists and has to be removed or not (e.g. if the compare operation is an equality check indicated by an "=", the column representing the DET onion gets checked).

Step 2: After all necessary RND layers have been removed by the client (step 1) the set of all columns is identified, that have to read from the database. This set consists of two subsets, that might overlap. The first subset consists of all onion columns that are involved in query conditions as described above. The second subset consists of all columns, that were selected by the user's query.

Step 3: Having collected that information, the client constructs the query against the encrypted database by doing the following. All plaintext column names are replaced with their ciphertext counterparts according to the information that was collected previously (step 2). Furthermore all plaintext literals in conditions are replaced by their PPE encrypted counterparts. The query now contains no plaintext information anymore and can be carried out by the server.

Step 4: The client receives and decrypts the results. Note that the removed RND layers stay off, because the database saw the underlying values anyway.

6   Evaluation

6.1   Implementation

All experiments in this section were run on an Intel Core i7-4600U CPU @ 2.10 GHz, 8 GB RAM, a Samsung PM851 256 GB SSD using Ubuntu 16.04. The PPE schemes were implemented in Java 8, using cryptographic primitives of the Java Cryptography Extension and The Legion of the Bouncy Castle Java Cryptography API[4]. In order to avoid measuring network effects local installations of the databases were used, as only the computation time of the schemes in combination with the speed of the databases was to be measured. Cassandra was used in version 3.9, Apache was used in version 1.3.

6.2   Performance

While AES is a performant option for encryption and decryption in the RND and DET layer, we evaluated the performance as well as strengths and weaknesses in previous work for the OPE [54] and SE [55] layer schemes. Depending on the application the user might want to exploit these strengths and minimize the impact of these weaknesses. That is why we group the schemes that we have selected in Sect. 4 in three profiles, each of which determines what PPE schemes are actually used in the OPE and SE layer during data insertion:

- OPTIMIZED READING: This profile prioritizes schemes that have advantages for read queries (e.g. like selections). Thus, it is the best choice for "write-once" databases. The OPE schemes best suited for fast reading are RSS

---

[4] Available at https://www.bouncycastle.org/.

and OACIS. They have the same type of index, which results in equal reading performance. However, RSS is the preferred choice for this profile, because it has some minor advantages in the encryption process. For the SE layer the SUISE scheme is used. It is faster than SWP in the process of searching, in particular for repeated queries.

- OPTIMIZED WRITING: This profile prioritizes schemes with fast encryption algorithms for scenarios, in which data insertion occurs more often than reading. The OPE scheme best suited for fast writing is OACIS. For the SE layer the SWP scheme is used. Since it does not have to maintain indexes, it can insert data faster than SUISE.
- STORAGE-EFFICIENT: This profile prioritizes storage needs over computation time and hence PPE schemes, that require the least amount of storage for data and indexes on client and server side. Hence, it uses mOPE for the OPE layer and SWP for the SE layer, both working without indexes.

Not affected by these profiles is only the OPE layer of row identifier columns, that must always be encrypted using RSS. The other schemes are not suited for row identifiers for the following reasons. OACIS cannot be used, because it produces mutable ciphertexts, but row identifiers are supposed to be final (see Sect. 3.1). mOPE cannot be used, because it adds a secret modular offset to the ciphertexts, that the database must not know about, but that has to be taken into account, when inserting values.

PPE scheme indexes (if existing) are maintained per column. This allows involving only the index data of columns actually required for answering a query.

For our tests we used a subset of the Enron email dataset[5], which reflects the practical scenario of using PPE for protecting sensitive mailbox data. We assume an average mailbox to have a size from 1,000 up to 10,000 emails. Hence, the measurements are started with 1,000 randomly chosen emails of the corpus, which we increase up to 10,000 emails (that contain $1.03 \cdot 10^7$ words in 180.000 fields of data) in order to estimate how the schemes and databases scale.

Encryption. We measure the time for encrypting and inserting up to 10,000 mails ($1.03 \cdot 10^7$ words) using the table profiles and OLM as introduced above and presented the results in Table 3. Since the complexity of encrypting different types of data in the OLM is different, we distinguish between the encryption of text and numerical data. Hence, in a first series of measurements we investigate the performance of text data, using the data fields of the Enron email dataset "as they are", meaning as strings, that are encrypted as shown in Fig. 1. Note that this is the worst case scenario for the proposed architecture, because OPE for strings and SE layer computations are the most expensive operations in the OLM. These measurements are indicated in Table 3 as "text data". In a second series of measurements we extract an equal amount of numerical values from the Enron emails, like their size, date, priority etc. Their OLM encryption is less

---

[5] Available at https://www.cs.cmu.edu/~./enron/.

Table 3. Time needed for onion layer encryption of 180.000 fields of text data and 180.000 fields of numerical data in seconds (OR = Optimized Reading, OW = Optimized Writing, S = Storage-efficient, PL = performance loss).

Database		Text data			Numerical data		
Cassandra	Unencrypted	9.71			8.91		
	Profile	OR	OW	S	OR	OW	S
	Encrypted	253.2	57.48	393.0	22.2	12.1	93.1
	Factor of PL	26.1	5.92	40.4	2.49	1.35	10.4
HBase	Unencrypted	10.5			10.1		
	Profile	OR	OW	S	OR	OW	S
	Encrypted	241.0	59.01	362.2	20.6	11.1	83.8
	Factor of PL	22.9	5.62	34.5	2.04	1.1	8.3

complex, because OPE-encryption can be computed faster (since the necessary numerical format is already given) and the SE layer is not involved at all.[6]

As could be expected, the profile for OPTIMIZED WRITING performs best, increasing the average insertion time for text data by a factor of 5.92 using Cassandra and 5.62 using HBase. When inserting numerical data, the performance is much less affected by encryption: 35% with Cassandra and even only 10% with HBase. The profile for OPTIMIZED READING performs well for numerical data, roughly doubling the insertion times, whereas the performance loss factors are 26.1 and 22.9 for text data. The STORAGE EFFICIENT-profile suffers from the slow mOPE scheme in the OPE layer, which results in overall performance loss factors of 40.4/34.5 and 10.4/8.3. In real-world scenarios the performance will be somewhere in between the extremes for text and numerical data, that are shown in Table 3, depending on the composition of the dataset. In summary, it can be said that the OPTIMIZED WRITING-profile is a justifiable option in relation to the security that can be gained. Since in most scenarios query performance is much more crucial than writing performance, even the profile for OPTIMIZED READING should be sufficient for the majority of use cases. However, the STORAGE EFFICIENT-profile might not be feasible in practice due to its computationally expensive PPE schemes.

Concerning the databases it can be observed, that there are only non-significant differences. HBase is always a little faster than Cassandra. The smallest difference (5.3%) between both occurs when using the profile for OPTIMIZED WRITING for text data, the biggest difference (25.3%) when using STORAGE EFFICIENT-profile for numerical data.

Querying. Querying is more complicated than encrypting. The runtime of a query depends on many aspects, for example the query type, the state of the

---

[6] We do not perform test for byte blob data, since in the proposed OLM this would only result in performing in AES encryption.

onion layers and the PPE schemes used. These aspects are considered during the benchmarks in the following ways. First of all, five queries (Q1–Q5) are tested, that involve dealing with different combinations of PPE schemes and are based on real world use cases. Q1 asks for all emails of a certain sender. Thus, it requires one check for equality and involves the DET layer once (see Sect. 3.2). Q2 asks for all emails larger than a certain size. Hence it requires OPE once, because the order relation between two values has to be determined (also, see Sect. 3.2). Q3 asks for all emails with a certain word in their body. That means a word has to be searched for in encrypted text. That involves the SE layer once (again, see Sect. 3.2). Q4 combines all filter criteria from Q1–Q3 and Q5 is a more complex query, that asks for all mails from a certain sender (DET) in a certain period of time (2x OPE for the start and end point of this period) with certain words in the subject and body fields (2x SE). When performing one of them, it makes a significant difference in terms of runtime whether the same or a similar query was performed before or not, which the following two reasons are responsible for. Firstly, columns being involved in an equality check or order comparison before already lost their RND layer. The effort of removing it is not necessary again. Secondly, when the SUISE scheme is involved in SE more than once, it might already have the results in a second index it maintains, which also improves the query runtime significantly. For these two reasons every query is executed twice with Qx.1 indicating the first execution of the query Qx after encryption and Qx.x indicating the runtime, that can be expected from all future executions of Qx.

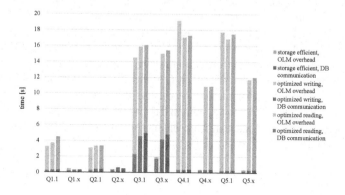

Fig. 2. Query runtime with Cassandra

The results are presented in Fig. 2 for Cassandra and Fig. 3 for HBase. In these figures "DB communication" denotes the pure runtime of the databases' communication mechanisms (which is the execution time of driver calls in case of Cassandra and the execution time of HBase's native API calls) and "OLM overhead" denotes everything that is a direct or indirect consequence of the onion layer model, for example query rewriting, RND layer removal or the SE

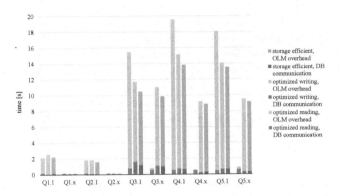

Fig. 3. Query runtime with HBase

processing as described in Sect. 5. Decryption time of the resultset (as obtained in step 4 of Sect. 5.2) is not explicitly shown, since it is insignificant (under 5ms for all queries). All queries have been conducted with the largest dataset that has been used in the previous encryption benchmark, having a volume of $1.03 \cdot 10^7$ words in 10,000 emails.

The following observations can be made. If encryption was done using the profile for OPTIMIZED READING, all Qx.x queries perform well under one second, except for Q3.x in combination with Cassandra. This can be considered practically feasible performance [15]. Qx.x queries are always faster than Qx.1 queries. That means the performance always improves, if similar queries are executed. Performing SE is very expensive compared to requiring DET or OPE functionality only. However, performing SE on small fields of data has barely a performance impact (compare Q4 and Q5; Q4 searches in bodies, Q5 adds SE only in the small subject field), but can slow down querying significantly, if done on a large subset of the data (compare Q1 and Q2 against Q3). HBase seems to have a slight overall performance advantage. A possible explanation for this might be the RND layer removal, which HBase is doing almost twice as fast compared to Cassandra. This can be seen in Q1 and Q2, in which most of the time is need for the RND layer removal.

7   Security

A formal security analysis of the PPE schemes that we used can be found in their originating publications. In this section we briefly discuss security aspects arising from putting the schemes into practice.

For AES, the only scheme used and needed in the DET layer, there are neither known attacks of practical relevance, nor any leakage besides the intentional determinism.

In SE schemes mainly three kinds of information can leak unintentionally: index information (e.g. number of words per document, number of documents, lengths of documents, document-IDs), search pattern information (what word was searched for?) and access pattern information (how much answers do I get from executing a certain query compared to executing another one?). The leakage of the used SE schemes can be described as follows. SWP [45] does not need an index and thus does not leak any index information and also hides the true length of plaintext words, because it uses fixed length trapdoors. However it leaks search patterns, since it passes pre-encrypted search words to the database, that are deterministically encrypted, which allows linking them to actual plain words. SWP also leaks access patterns, which is unavoidable in the client server scenario, in which a request (pre-encrypted searchword) can always be linked to the corresponding answer (the result set). The SUISE scheme [43] leaks the number of unique words per document. Since search tokens are generated deterministically, the search pattern leaks like in SWP. Trapdoors have a constant length, which also hides the length of the plaintext words.

Concerning the OPE layer neither of the schemes can leak any index information, because the index (if exists) resides on clientside. Search patterns can leak, because the used OPE schemes produce deterministic ciphertexts that can be tracked. Access patterns leak for the same reasons explained earlier for SE. Another security risk for OPE is to encrypt either very few or very much values of a domain: on the one hand, if only two values of a domain p_1 and p_2 are encrypted, they can easily be mapped to their corresponding ciphertexts c_1 and c_2. Obviously the smaller p value is encrypted in the smaller c value and the larger p value is encrypted in the larger c value. On the other hand, it is equally severe if all values of a specific domain have been encrypted. The ordered ciphertexts can simply be mapped to the ordered plaintexts (note that both problems also occur in non-deterministic OPE schemes). That means e.g. it makes sense to store a date in form of a unix timestamp, not split in individual characteristics like day (domain size only 31), month (domain size 12), etc.

8 Extensions and Future Work

The topic of this work leaves a lot of room for additional work. An evaluation in a real world cloud environment is needed. Fragmentation over independent databases (which we already implemented, but not presented for reasons of complexity) can be of further use to impede statistical attacks based on PPE leakage. We also plan to support other databases and data models. WCS tables can easily be transformed to fit e.g. in key value or document stores. An additional onion for homomorphic encryption can be used for data aggregations like sums and averages. Schemes for fuzzy/similarity search are of interest for the SE layer.

9 Conclusion

We analysed the requirements that PPE schemes have to meet in order to be feasible for NoSQL WCSs and evaluated various available schemes for OPE and SE regarding these requirements. Based on our findings we identified feasible

PPE schemes, proposed an OLM to handle PPE encrypted data on serverside and implemented both for a practical evaluation using the two popular WCSs Cassandra and HBase. We quantified the performance impact of PPE encryption and characterized practical security issues. We showed that choosing PPE schemes corresponding to the read/write needs of the scenario leads to still practically feasible performance.

References

1. O'Reilly, T.: What is web 2.0: design patterns and business models for the next generation of software. Commun. Strat. **65**(1), 17–37 (2007)
2. Han, J., Haihong, E., Le, G., Du, J.: Survey on NoSQL database. In: 2011 6th International Conference on Pervasive Computing and Applications (ICPCA), pp. 363–366. IEEE (2011)
3. Tudorica, B.G., Bucur, C.: A comparison between several NoSQL databases with comments and notes. In: 2011 10th Roedunet International Conference (RoEduNet), pp. 1–5. IEEE (2011)
4. Chang, F., Dean, J., Ghemawat, S., Hsieh, W.C., Wallach, D.A., Burrows, M., Chandra, T., Fikes, A., Gruber, R.E.: Bigtable: a distributed storage system for structured data. ACM Trans. Comput. Syst. **26**(2), 4 (2008)
5. Lakshman, A., Malik, P.: Cassandra: a decentralized structured storage system. ACM SIGOPS Operating Syst. Rev. **44**(2), 35–40 (2010)
6. Alex, H., Schoen, S., Heninger, N., Clarkson, W., Paul, W., Calandrino, J., Feldman, A., Appelbaum, J., Felten, E.: Lest we forget - cold boot attacks on encryption keys (2008)
7. Corkery, M.: Once Again, Thieves Enter Swift Financial Network and Steal, New York Times (2016). http://www.nytimes.com/2016/05/13/business/dealbook/swift-global-bank-network-attack.html. Accessed 20 Mar 2017
8. Lennon, M.: Hackers Used Sophisticated SMB Worm Tool to Attack Sony, Security Week (2016). http://www.securityweek.com/hackers-used-sophisticated-smb-worm-tool-attack-sony. Accessed 20 Mar 2017
9. Quinn, B., Arthur, C.: Playstation network hackers access data of 77 million users, The Guardian (2011). https://www.theguardian.com/technology/2011/apr/26/playstation-network-hackers-data. Accessed 20 Mar 2017
10. Crawford, D., Fuhrmans, V., Ball, D.: Germany Tackles Tax Evasion, Wall Street Journal (2010). http://www.wsj.com/articles/SB10001424052748704197104575051480386248538. Accessed 20 Mar 2017
11. Solid-IT: DB-engines ranking (2017). https://db-engines.com/de/ranking. Accessed 20 Mar 2017
12. Borthakur, D., Gray, J., Sarma, J.S., Muthukkaruppan, K., Spiegelberg, N., Kuang, H., Ranganathan, K., Molkov, D., Menon, A., Rash, S., et al.: Apache hadoop goes realtime at Facebook. In: Proceedings of the 2011 ACM SIGMOD International Conference on Management of Data, pp. 1071–1080. ACM (2011)
13. Hacigümüş, H., Iyer, B., Li, C., Mehrotra, S.: Executing SQL over encrypted data in the database-service-provider model. In: Proceedings of the 2002 ACM SIGMOD International Conference on Management of data, pp. 216–227. ACM (2002)
14. Popa, R.A., Redfield, C., Zeldovich, N., Balakrishnan, H.: CryptDB: processing queries on an encrypted database. Commun. ACM **55**(9), 103–111 (2012)

15. Shahzad, F., Iqbal, W., Bokhari, F.S.: On the use of CryptDB for securing electronic health data in the cloud: a performance study. In: 2015 17th International Conference on E-health Networking, Application and Services (HealthCom), pp. 120–125. IEEE (2015)
16. Tetali, S.D., Lesani, M., Majumdar, R., Millstein, T.: MrCrypt: static analysis for secure cloud computations. ACM SIGPLAN Not. 48(10), 271–286 (2013)
17. Akin, I.H., Sunar, B.: On the difficulty of securing web applications using CryptDB. In: 2014 IEEE Fourth International Conference on Big Data and Cloud Computing (BdCloud), pp. 745–752. IEEE (2014)
18. Tu, S., Kaashoek, M.F., Madden, S., Zeldovich, N.: Processing analytical queries over encrypted data. Proc. VLDB Endowment 6, 289–300 (2013)
19. Pappas, V., Krell, F., Vo, B., Kolesnikov, V., Malkin, T., Choi, S.G., George, W., Keromytis, A., Bellovin, S.: Blind seer: a scalable private DBMS. In: IEEE Symposium on Security and Privacy, vol. 2014, pp. 359–374. IEEE (2014)
20. Sarfraz, M.I., Nabeel, M., Cao, J., Bertino, E.: DBMask: fine-grained access control on encrypted relational databases. In: Proceedings of the 5th ACM Conference on Data and Application Security and Privacy, pp. 1–11. ACM (2015)
21. Li, J., Liu, Z., Chen, X., Xhafa, F., Tan, X., Wong, D.S.: L-EncDB: a lightweight framework for privacy-preserving data queries in cloud computing. Knowl.-Based Syst. 79, 18–26 (2015)
22. Kasten, A., Scherp, A., Armknecht, F., Krause, M.: Towards search on encrypted graph data. In: Proceedings of PrivOn (2013)
23. Yuan, X., Wang, X., Wang, C., Qian, C., Lin, J.: Building an encrypted, distributed, and searchable key-value store. In: Proceedings of the 11th ACM on Asia Conference on Computer and Communications Security, pp. 547–558. ACM (2016)
24. Poddar, R., Boelter, T., Popa, R.A.: Arx: a strongly encrypted database system. IACR Cryptology ePrint Arch. 2016, 591 (2016)
25. Arasu, A., Blanas, S., Eguro, K., Kaushik, R., Kossmann, D., Ramamurthy, R., Venkatesan, R.: Orthogonal security with cipherbase. In: CIDR (2013)
26. Bajaj, S., Sion, R.: TrustedDB: a trusted hardware-based database with privacy and data confidentiality. IEEE Trans. Knowl. Data Eng. 26(3), 752–765 (2014)
27. Khetrapal, A., Ganesh, V.: HBase and hypertable for large scale distributed storage systems. Department of Computer Science, Purdue University, pp. 22–28 (2006)
28. Sawyer, S.M., O'Gwynn, B.D., Tran, A., Yu, T.: Understanding query performance in Accumulo. In: 2013 IEEE High Performance Extreme Computing Conference (HPEC), pp. 1–6. IEEE (2013)
29. Cooney, M.: IBM touts encryption innovation, Network World (2009). http://www.networkworld.com/article/2259168/data-center/ibm-touts-encryption-innovation.html. Accessed 20 Mar 2017
30. Kadhem, H., Amagasa, T., Kitagawa, H.: A secure and efficient order preserving encryption scheme for relational databases. In: KMIS, pp. 25–35 (2010)
31. Boldyreva, A., Chenette, N., O'Neill, A.: Order-preserving encryption revisited: improved security analysis and alternative solutions. In: Rogaway, P. (ed.) CRYPTO 2011. LNCS, vol. 6841, pp. 578–595. Springer, Heidelberg (2011). doi:10.1007/978-3-642-22792-9_33
32. Liu, D., Wang, S.: Programmable order-preserving secure index for encrypted database query. In: 2012 IEEE 5th International Conference on Cloud Computing (CLOUD), pp. 502–509. IEEE (2012)
33. Popa, R.A., Li, F.H., Zeldovich, N.: An ideal-security protocol for order-preserving encoding. In: 2013 IEEE Symposium on Security and Privacy (SP), pp. 463–477. IEEE (2013)

34. Wozniak, S., Rossberg, M., Grau, S., Alshawish, A., Schaefer, G.: Beyond the ideal object: towards disclosure-resilient order-preserving encryption schemes. In: Proceedings of the 2013 ACM Workshop on Cloud Computing Security Workshop, pp. 89–100. ACM (2013)
35. Kerschbaum, F., Schröpfer, A.: Optimal average-complexity ideal-security order-preserving encryption. In: Proceedings of the 2014 ACM SIGSAC Conference on Computer and Communications Security, pp. 275–286. ACM (2014)
36. Chenette, N., Lewi, K., Weis, S.A., Wu, D.J.: Practical order-revealing encryption with limited leakage. In: Peyrin, T. (ed.) FSE 2016. LNCS, vol. 9783, pp. 474–493. Springer, Heidelberg (2016). doi:10.1007/978-3-662-52993-5_24
37. Goh, E.J., et al.: Secure indexes. IACR Cryptology ePrint Arch. **2003**, 216 (2003)
38. Chang, Y.-C., Mitzenmacher, M.: Privacy preserving keyword searches on remote encrypted data. In: Ioannidis, J., Keromytis, A., Yung, M. (eds.) ACNS 2005. LNCS, vol. 3531, pp. 442–455. Springer, Heidelberg (2005). doi:10.1007/11496137_30
39. Curtmola, R., Garay, J., Kamara, S., Ostrovsky, R.: Searchable symmetric encryption: improved definitions and efficient constructions. In: Proceedings of the 13th ACM Conference on Computer and Communications Security, pp. 79–88. ACM (2006)
40. Sedghi, S., Van Liesdonk, P., Doumen, J.M., Hartel, P.H., Jonker, W.: Adaptively secure computationally efficient searchable symmetric encryption (2009)
41. van Liesdonk, P., Sedghi, S., Doumen, J., Hartel, P., Jonker, W.: Computationally efficient searchable symmetric encryption. In: Jonker, W., Petković, M. (eds.) SDM 2010. LNCS, vol. 6358, pp. 87–100. Springer, Heidelberg (2010). doi:10.1007/978-3-642-15546-8_7
42. Kamara, S., Papamanthou, C., Roeder, T.: Dynamic searchable symmetric encryption. In: Proceedings of the 2012 ACM Conference on Computer and Communications Security, pp. 965–976. ACM (2012)
43. Hahn, F., Kerschbaum, F.: Searchable encryption with secure and efficient updates. In: Proceedings of the 2014 ACM SIGSAC Conference on Computer and Communications Security, pp. 310–320. ACM (2014)
44. Jho, N.S., Chang, K.Y., Hong, D., Seo, C.: Symmetric searchable encryption with efficient range query using multi-layered linked chains. J. Supercomputing **72**, 1–14 (2015)
45. Song, D.X., Wagner, D., Perrig, A.: Practical techniques for searches on encrypted data. In: 2000 IEEE Symposium on Security and Privacy, S&P 2000, Proceedings, pp. 44–55. IEEE (2000)
46. Anderson, R., Biham, E., Knudsen, L.: Serpent: a proposal for the advanced encryption standard. NIST AES Proposal **174**, 1–23 (1998)
47. Roche, D., Apon, D., Choi, S.G., Yerukhimov, A.: POPE: Partial order-preserving encoding. Technical report, Cryptology ePrint Arch. 2015/1106 (2015)
48. Liu, Z., Chen, X., Yang, J., Jia, C., You, I.: New order preserving encryption model for outsourced databases in cloud environments. J. Netw. Comput. Appl. **59**, 198–207 (2014)
49. Mavroforakis, C., Chenette, N., O'Neill, A., Kollios, G., Canetti, R.: Modular order-preserving encryption, revisited. In: Proceedings of the 2015 ACM SIGMOD International Conference on Management of Data, pp. 763–777. ACM (2015)
50. Boldyreva, A., Chenette, N., Lee, Y., O'Neill, A.: Order-preserving symmetric encryption. In: Joux, A. (ed.) EUROCRYPT 2009. LNCS, vol. 5479, pp. 224–241. Springer, Heidelberg (2009). doi:10.1007/978-3-642-01001-9_13

51. Kadhem, H., Amagasa, T., Kitagawa, H.: MV-OPES: multivalued-order preserving encryption scheme: a novel scheme for encrypting integer value to many different values. IEICE Trans. Inf. Syst. **93**(9), 2520–2533 (2010)

52. Bösch, C., Hartel, P., Jonker, W., Peter, A.: A survey of provably secure searchable encryption. ACM Comput. Surv. (CSUR) **47**(2), 18 (2015)

53. Popa, R.A., Redfield, C., Zeldovich, N., Balakrishnan, H.: CryptDB: protecting confidentiality with encrypted query processing. In: Proceedings of the Twenty-Third ACM Symposium on Operating Systems Principles, pp. 85–100. ACM (2011)

54. Waage, T., Homann, D., Wiese, L.: Practical application of order-preserving encryption in wide column stores. In: Proceedings of the 13th International Joint Conference on e-Business and Telecommunications - SECRYPT, pp. 352–359 (2016)

55. Waage, T., Jhajj, R.S., Wiese, L.: Searchable encryption in apache cassandra. In: Garcia-Alfaro, J., Kranakis, E., Bonfante, G. (eds.) FPS 2015. LNCS, vol. 9482, pp. 286–293. Springer, Cham (2016). doi:10.1007/978-3-319-30303-1_19

Towards a JSON-Based Fast Policy Evaluation Framework
(Short Paper)

Hao Jiang and Ahmed Bouabdallah[(✉)]

IMT Atlantique, Site of Rennes, 35510 Cesson-Sévigné, France
{hao.jiang,ahmed.bouabdallah}@imt-atlantique.fr

Abstract. In this paper we evaluate experimentally the performance of JACPoL, a previously introduced JSON-based access control policy language. The results show that JACPoL requires much less processing time and memory space than XACML by testing generic families of policies expressed in both languages.

Keywords: Low-latency access control · Efficient policy evaluation · JSON

1 Introduction

Policies represent sets of properties of information processing systems [1]. Their implementation mainly rests on the IETF architecture [2] initially introduced to manage QoS policies in networks. It consists in two main entities namely the PDP (Policy Decision Point) and the PEP (Policy Enforcement Point). The first one which is the smart part of the architecture acts as a controller the goal of which consists in handling and interpreting policy events, and deciding in accordance with the policy currently applicable, what action should be taken. The decision is transmitted to the PEP which has to concretely carry it out.

Access control policies are a specific kind of security policies aiming to control the actions that principals can perform on resources by permitting their access only to the authorized ones. Typically, the access requests are intercepted and analyzed by the PEP, which then transfers the request details to the PDP for evaluation and authorization decision [2]. In most implementations, the stateless nature of PEP enables its ease of scale. However, the PDP has to consult the right policy set and apply the rules therein to reach a decision for each request and thus is often the performance bottleneck of policy-based access control systems. Therefore, a policy language determining how policies are expressed and evaluated is important and has a direct influence on the performance of the PDP.

Especially, in nowadays, protecting private resources in real-time has evolved into a rigid demand in domains such as home automation, smart cities, health care services and intelligent transportation systems, etc., where the environments are characterized by heterogeneous, distributed computing systems exchanging

© Springer International Publishing AG 2017
H. Panetto et al. (Eds.): OTM 2017 Conferences, Part II, LNCS 10574, pp. 22–30, 2017.
https://doi.org/10.1007/978-3-319-69459-7_2

enormous volumes of time-critical data with varying levels of access control in a dynamic network. An access control policy language for these environments needs to be very well-structured, expressive but lightweight and easily extensible [3].

In the past decades, a lot of policy languages have been proposed for the specification of access control policies using XML, such as EPAL [4], X-GTRBAC [5] and the standardized XACML [6]. Nevertheless, it is generally acknowledged that XACML suffers from providing poorly defined and counterintuitive semantics [7], which makes it not good in simplicity and flexibility. On the other hand, XML performs well in expressiveness and adaptability but sacrifices its efficiency and scalability, compared to which JSON is considered to be more well-proportioned with respect to these requirements, and even simpler, easier, more efficient and thus favored by more and more nowadays' policy designers [8–11]. To address the aforementioned inefficiency issues of the XML format, the XACML Technical Committee recently designed the JSON profile [12] to be used in the exchange of XACML request and response messages between the PEP and PDP. However, the profile does not define the specification of XACML policies, which means, after the PDP parses the JSON-formatted XACML requests, it still needs to evaluate the parsed attributes with respect to the policies expressed in XML.

Leigh Griffin and his colleagues [10] have proposed JSONPL, a policy vocabulary encoded in JSON that semantically was identical to the original XML policy but stripped away the redundant meta data and cleaned up the array translation process. Their performance experiments showed that JSON could provide very similar expressiveness as XML but with much less verbosity. On the other hand, as much as we understand, JSONPL is merely aimed at implementing XACML policies in JSON and thus lacks its own formal schema and full specification as a policy specification language [13]. Major service providers such as Amazon Web Services (AWS) [14] have a tendency to implement their own security languages in JSON, but such kind of approaches are normally for proprietary usage thus provide only self-sufficient features and support limited use cases, which are not suitable to be a common policy language.

To the best of our knowledge, there are very few proposals that combine a rich set of language features with well-defined syntax and semantics, and such kind of access control policy language based on JSON has not even been attempted before. According to these observations, we proposed in a previous paper [15] a simple but expressive access control policy language (JACPoL) based on JSON. JACPoL by design provides a flexible and fine-grained ABAC (Attribute-based Access Control), and meanwhile it can be easily tailored to express a broad range of other access control models. We carefully positioned JACPoL in comparison with the traditional policy language XACML and we showed that JACPoL is more lightweight, scalable and flexible [15].

This paper presents a quantitative evaluation of JACPoL. We focus experimentally on the PDP performances. Using a common Python implementation, we assess the speed of writing, loading and processing generic families of policies expressed in JACPoL and XACML together with their relative memory consumption. The comparisons show that JACPoL systematically requires much less time and memory space.

The rest of the paper provides a quick overview of JACPoL in Sect. 2, a detailed performance evaluation in Sect. 3 and a conclusion in Sect. 4.

2 Fundamentals of JACPoL

This section recalls the foundations of JACPoL, and then introduces its structures with an overview of how an access request is evaluated with respect to JACPoL policies. A detailed introduction of JACPoL can be found in [15].

JACPoL is JSON-formatted [16]. It is also attribute-based by design but meanwhile supports RBAC [17]. When integrating RBAC, user roles are considered as an attribute (ARBAC) [18], or attributes are used to constrain user permissions (RABAC) [19], which obtains the advantages of RBAC while maintaining ABAC's flexibility and expressiveness. JACPoL adopts hierarchically nested structures similar to XACML. The layered architecture as shown in Fig. 1 not only enables scalable and fine-grained access control, but also eases the work of policy definition and management for policy designers. JACPoL supports *Implicit Logic Operators* which make use of JSON built-in data structures (*Object* and *Array*) to implicitly denote logic operations. This allows a policy designer to express complex operations without explicitly using logical operators, and makes JACPoL policies greatly reduced in size and easier to read and write by humans. Finally JACPoL supports *Obligations* to offer a rich set of security and network management features.

JACPoL uses hierarchical structures very similar to the XACML standard [20]. As shown in Fig. 1, JACPoL policies are structured as *Policy Sets* that consist of one or more child policy sets or policies, and a *Policy* is composed of a set of *Rules*.

Because not all *Rules*, *Policies*, or *Policy Sets* are relevant to a given request, JACPoL includes the notion of a *Target*. A *Target* determines whether a *Rule/Policy/Policy Set* is applicable to a request by setting constraints on attributes using simple Boolean expressions. A *Policy Set* is said to be *Applicable* if the access request satisfies the *Target*, and if so, then its child *Policies* are evaluated and the results returned by those child policies are combined using the policy-combining algorithm; otherwise, the *Policy Set* is skipped without further examining its child policies and returns a *Not Applicable* decision. Likewise, the *Target* of a *Policy* or a *Rule* has similar semantics.

The *Rule* is the fundamental unit that is evaluated eventually and can generate a conclusive decision (*Permit* or *Deny* specified in its *Effect* field). The *Condition* field in a rule is a simple or complex Boolean expression that refines the applicability of the rule beyond the predicates specified by its target, and is optional. If a request satisfies both the *Target* and *Condition* of a rule, then the rule is applicable to the request and its *Effect* is returned as its decision; otherwise, *Not Applicable* is returned.

For each *Rule, Policy,* or *Policy Set,* an *id* field is used to be uniquely identified, and an *Obligation* field is used to specify the operations which should be performed (typically by a PEP) before or after granting or denying an access request, while a *Priority* is specified for conflict resolution between different *Rules, Policies,* or *Policy Sets.*

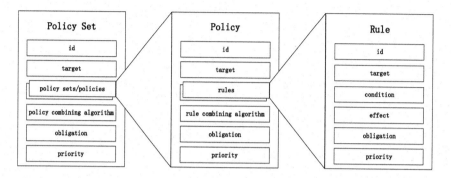

Fig. 1. JACPoL's hierarchical nested structure

3   Performance Evaluation

This section presents the results of performance tests on JACPoL and XACML. We assessed respectively how both languages' evaluation time and memory usage are affected with regards to the increase of nesting policies (policy depth) and the increase of sibling ones (policy scale). The values are ranged from 0 to 10000 with a step of 1000, as shown in Fig. 2.

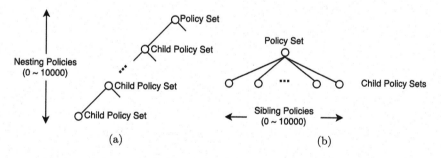

Fig. 2. Examples of nesting policies (a) and sibling policies (b)

We used the two policy languages to express the same policy task and compared their performances with different depth (number of nesting policies) and scale (number of sibling policies). Figure 3 provides XACML and JACPoL policy examples used for assessment. The two ellipses in each example indicate nesting (the upper and inner ellipsis) and sibling (the lower and outer ellipsis) child policies respectively. When processing a given request, the Boolean expression in the *Target* field of each policy is evaluated to verify the applicability of its child policies until the last child policy is evaluated. Note that the JSON profile of XACML defines XACML requests and responses but not policies, therefore we are unable to compare JACPoL with JSON-formatted XACML.

```
<PolicySet Id="0" PolicyCombiningAlgorithm="firstApplicable" Update="2017-03-14 17:18:31" Version="1">
  <Target>
    <Subjects>
      <Subject>
        <SubjectMatch MatchId="...#string-equal">
          <AttributeValue DataType="...#string">Sam</AttributeValue>
          <SubjectAttributeDesignator AttributeId="...:subject:subject-id" DataType="...#string" />
        </SubjectMatch>
      </Subject>
    </Subjects>
  </Target>
  <PolicySet Id="1" PolicyCombiningAlgorithm="firstApplicable" Update="2017-03-14 17:18:31" Version="1">
    ...
  </PolicySet>
  ...
</PolicySet>
```

(a)

```
{
  "Target": {"Subject": {"equals": "Sam"}},
  "Update": "2017-03-14 17:18:31",
  "PolicyCombiningAlgorithm": "firstApplicable",
  "Version": 1,
  "Id": 0,
  "Policies": [
    {
      "Target": {"Subject": {"equals": "Sam"}},
      "Update": "2017-03-14 17:18:31",
      "PolicyCombiningAlgorithm": "firstApplicable",
      "Version": 1,
      "Id": 1,
      "Policies": [...]
    },
    ...
  ]
}
```

(b)

Fig. 3. Examples of XACML policies (a) and JACPoL policies (b)

As shown in Fig. 4, the policy evaluation performance of the PDP is very related to the writing, loading and processing performance of a policy language. We agree that the schemes and technologies used in requesting/responding operations would also influence the performance, which is however beyond the scope of current JACPoL's specification.

Therefore, we have conducted 4 sets of tests in order to evaluate the writing, loading and processing speed and the memory consumption of the two policy

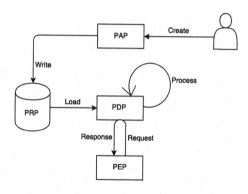

Fig. 4. Policy evaluation procedure

languages. Respectively, the writing test measures the average time (in seconds) to write policies from the memory into a single file; the loading test measures the average time (in seconds) to load policies from a single file into the memory; the processing test measures the average time (in seconds) to process a request against policies that are already loaded in the memory; and the last test measures the space consumption (in Megabytes) of policies in the memory. Each test evaluates the two languages with different policy nesting depth and sibling scale, and was repeated 10000 times conducted using Python on a Windows 10 ASUS N552VW laptop with 16G memory and a 2.6 GHz Intel core i7-6700HQ processor.

Figure 5a shows the writing time versus the number of policies. We observe that there are near linear correlations between the average writing time and number policies (both in depth and in scale) for both languages. Compared to XACML, JACPoL consumes only approximately half of the time taken by the former.

Figure 5b shows the average loading time versus the number of policies in depth and in scale. Similar to Fig. 5a, there is an almost linear correlation between the loading time and the number of policies. We can see that the loading time of JACPoL becomes much less than XACML as the number of policies increases.

Figure 5c shows the processing time versus the number of policies. Similar as the writing time and loading time, there is also an almost linear correlation between the processing time and the number of policies. On the other hand, unlike Figs. 5a and b, we can see that the policy structure would influence the processing time given the same policy size. We also observe that JACPoL is processed faster than XACML policies.

Figure 5d shows the memory consumption versus the number of policies. Similar as all figures above, the memory consumption increases also almost linearly with the growth of the number of policies. Given the same policy size, the memory space used by JACPoL is nearly half of that used by XACML.

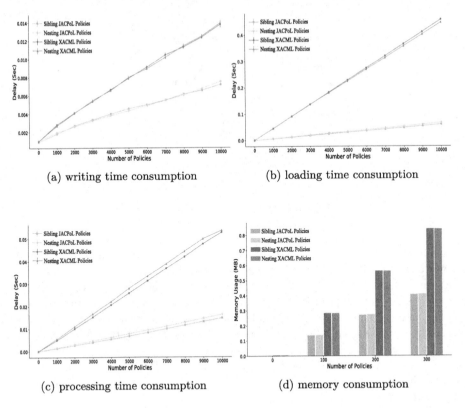

(a) writing time consumption   (b) loading time consumption

(c) processing time consumption   (d) memory consumption

Fig. 5. Comparative performance evaluation between the JACPoL and XACML

These figures demonstrate that JACPoL is highly scalable and efficient in comparison with XACML, with much faster writing, loading, processing speed and lower memory consumption.

4   Conclusion

Real-time requirements rarely intervene in the design of access control systems. Applications are emerging, however, that require policies to be evaluated with a very low latency and high throughput. We proposed in a previous paper [15] JACPoL, as a new JSON-based, attribute-centric and light-weight access control policy language which was showed through a comprehensive qualitative analysis, as expressive as XACML but more simple, scalable and flexible. We introduce in this paper a first level of quantitative evaluation focusing on the PDP performances by assessing the speed of writing, loading and processing generic families of policies expressed in JACPoL and XACML together with their relative memory consumption. The comparisons show that JACPoL systematically requires much less time and memory space.

Acknowledgement. We acknowledge the reviewers of C&TC'17 for their constructive comments.

References

1. Clarkson, M.R., Schneider, F.B.: Hyperproperties. J. Comput. Secur. **18**, 1157–1210 (2010)
2. Yavatkar, R., Pendarakis, D., Guerin, R.: A Framework for Policy-based Admission Control. IETF, RFC 2753, January 2000
3. Borders, K., Zhao, X., Prakash, A.: CPOL: high-performance policy evaluation. In: The 12th ACM Conference on Computer and Communications Security. ACM (2005)
4. Ashley, P., Hada, S., Karjoth, G., Powers, C., Schunter, M.: Enterprise privacy authorization language (EPAL). IBM Research, March 2003
5. Bhatti, R., Ghafoor, A., Bertino, E., Joshi, J.B.: X-GTRBAC: an XML-based policy specification framework and architecture for enterprise-wide access control. ACM Trans. Inf. Syst. Secur. (TISSEC) **8**(2), 187–227 (2005)
6. OASIS XACML Technical Committee: eXtensible access control markup language (XACML) Version 3.0. Oasis Standard, OASIS (2013). http://docs.oasis-open.org/xacml/3.0/xacml-3.0-core-spec-os-en.html. Last accessed 17 May 2017
7. Crampton, J., Morisset, C.: PTaCL: a language for attribute-based access control in open systems. In: Degano, P., Guttman, J.D. (eds.) POST 2012. LNCS, vol. 7215, pp. 390–409. Springer, Heidelberg (2012). doi:10.1007/978-3-642-28641-4_21
8. Crockford, D.: JSON — The fat-free alternative to XML (vol. 2006). http://www.json.org/fatfree.html. Last accessed 17 May 2017
9. El-Aziz, A.A., Kannan, A.: JSON encryption. In: 2014 International Conference on Computer Communication and Informatics (ICCCI). IEEE (2014)
10. Griffin, L., Butler, B., de Leastar, E., Jennings, B., Botvich, D.: On the performance of access control policy evaluation. In: 2012 IEEE International Symposium on Policies for Distributed Systems and Networks (POLICY), pp. 25–32. IEEE (2012)
11. W3schools: JSON vs. XML. www.w3schools.com/js/js_json_xml.asp. Last accessed 24 May 2017
12. Brossard, D.: JSON Profile of XACML 3.0 Version 1.0. XACML Committee Specification 01, 11 December 2014. http://docs.oasis-open.org/xacml/xacml-json-http/v1.0/cs01/xacml-json-http-v1.0-cs01.pdf. Last accessed 26 May 2017
13. Steven, D., Bernard, B., Leigh, G.: JSON-encoded ABAC (XACML) policies. FAME project of Waterford Institute of Technology. Presentation to OASIS XACML TC concerning JSON-encoded XACML policies, 30 May 2013
14. Amazon Web Services: AWS Identity and Access Management (IAM) User Guide. http://docs.aws.amazon.com/IAM/latest/UserGuide/introduction.html. Last accessed 27 May 2017
15. Jiang, H., Bouabdallah, A.: JACPoL: a simple but expressive JSON-based access control policy language. In: The 11th WISTP International Conference on Information Security Theory and Practice (WISTP 2017), 28–29 September 2017, Heraklion, Crete, Greece. Springer (2017, to appear)
16. ECMA International: ECMA-404 The JSON Data Interchange Standard. http://www.json.org/. Last accessed 27 May 2017
17. Ferraiolo, D.F., Kuhn, D.R.: Role-based Access Controls, 12 March 2009. arXiv preprint: arXiv:0903.2171

18. Obrsta, L., McCandlessb, D., Ferrella, D.: Fast semantic attribute-role-based access control (ARBAC) in a collaborative environment. In: 2012 8th International Conference on Collaborative Computing: Networking, Applications and Worksharing (CollaborateCom), Pittsburgh, PA, USA, 14–17 October 2012
19. Jin, X., Sandhu, R., Krishnan, R.: RABAC: role-centric attribute-based access control. In: Kotenko, I., Skormin, V. (eds.) MMM-ACNS 2012. LNCS, vol. 7531, pp. 84–96. Springer, Heidelberg (2012). doi:10.1007/978-3-642-33704-8_8
20. David, F., et al.: Extensible access control markup language (XACML) and next generation access control (NGAC). In: Proceedings of the 2016 ACM International Workshop on Attribute Based Access Control. ACM (2016)

Gibbon: An Availability Evaluation Framework for Distributed Databases

Daniel Seybold[(✉)], Christopher B. Hauser, Simon Volpert,
and Jörg Domaschka

Institute of Information Resource Management, Ulm University, Ulm, Germany
{daniel.seybold,christopher.hauser,simon.volpert,
joerg.domaschka}@uni-ulm.de

Abstract. Driven by new application domains, the database management systems (DBMSs) landscape has significantly evolved from single node DBMS to distributed database management systems (DDBMSs). In parallel, cloud computing became the preferred solution to run distributed applications. Hence, modern DDBMSs are designed to run in the cloud. Yet, in distributed systems the probability of failures is the higher the more entities are involved and by using cloud resources the probability of failures increases even more. Therefore, DDBMSs apply data replication across multiple nodes to provide high availability. Yet, high availability limits consistency or partition tolerance as stated by the CAP theorem. As the decision for two of the three attributes in not binary, the heterogeneous landscape of DDBMSs gets even more complex when it comes to their high availability mechanisms. Hence, the selection of a high available DDBMS to run in the cloud becomes a very challenging task, as supportive evaluation frameworks are not yet available. In order to ease the selection and increase the trust in running DDBMSs in the cloud, we present the Gibbon framework, a novel availability evaluation framework for DDBMSs. Gibbon defines quantifiable availability metrics, a customisable evaluation methodology and a novel evaluation framework architecture. Gibbon is discussed by an availability evaluation of MongoDB, analysing the take over and recovery time.

Keywords: Distributed database · Database evaluation · High availability · NoSQL · Cloud

1 Introduction

The landscape of database management systems (DBMSs) has evolved significantly over the last decade, especially when it comes to large-scale DBMSs installations. While relational database management systems (RDBMS) have been the common choice for persisting data over decades, the raise of the Web and new application domains such as *Big Data* and *Internet of Things (IoT)* drive the need for novel DBMS approaches. NoSQL and only recently NewSQL

© Springer International Publishing AG 2017
H. Panetto et al. (Eds.): OTM 2017 Conferences, Part II, LNCS 10574, pp. 31–49, 2017.
https://doi.org/10.1007/978-3-319-69459-7_3

DBMSs [15] have evolved. Such DBMSs are often designed as a distributed database management system (DDBMS) spread out over multiple *database nodes* and supposed to run on commodity or even virtualised hardware.

Due to their distributed architecture, DDBMSs support horizontal scalability and consequently the dynamic allocation and usage of compute, storage and network resources [1] based on the actual demand. This fact makes Infrastructure as a Service (IaaS) cloud platforms a suited choice for running DDBMSs, as it provides elastically, on-demand, self-service resource provisioning [19].

Even though distributed architectures can be used to improve the availability of the overall system, this is not automatically the case. In particular, in distributed systems the probability of failures is the higher the more entities are involved, *i.e.* in the case of DDBMSs the more data nodes are used. When cloud resources are used, the situation is worsened, as failures on lower level may affect multiple virtualised resources and cause mass failures [13,16].

With respect to DDBMSs, the common approach to availability is to replicate data items across multiple database nodes to ensure high availability in case of resource failures. However, the desire for availability is hindered by the CAP theorem stating that availability is achieved by scarifying consistency guarantees or partition tolerance [6]. However, the choice between two of these three attributes is not a binary decision and offers multiple trade-offs [5]. This has led to a very heterogeneous DDBMS landscape not only with respect to the sheer number of systems, but also the availability mechanisms they provide [11].

Hence, we find ourselves in the situation that more DDBMSs exist than ever, each of them promising availability features and often enough even high availability. Further, many of these DDBMS are operated on IaaS infrastructures with an increased risk of mass failures. Yet, for none of the DDBMSs it is known how it actually behaves under failure conditions and how the failure condition affects the availability of the DDBMSs. At the same time, no supportive frameworks for evaluating availability of DDBMSs exist [24] and in consequence, selecting a DDBMS becomes a gamble for users if availability is a major selection criterion.

In order to increase the trust in running DDBMSs on cloud resources and easing the selection of high available DDBMSs, we present Gibbon, a novel framework for evaluating the availability of DDBMSs. It explicitly supports cloud failure scenarios. Hence, our contribution is threefold: (1) we identify quantifiable metrics to evaluate availability; (2) we define an extensible evaluation methodology; (3) we present a novel availability evaluation framework architecture.

The remainder is structured as follows: Sect. 2 introduces the background on availability, DDBMS and cloud computing, while Sect. 3 analyses the impact of failures. Section 4 defines the availability metrics while Sect. 5 presents the evaluation methodology. Section 6 presents the framework architecture. Section 7 discusses the presented framework and Sect. 8 presents related work. Section 9 concludes.

2   Background

In order to consolidate the context of the Gibbon framework, we introduce in this section the background on DDBMSs, availability, and DDBMSs on IaaS.

2.1   Distributed Database Systems

Per definition, a DBMS manages data items grouped in collections or databases. For DDBMS these data items are spread out over multiple database nodes each of which runs management logic. Hence, the databases nodes communicate and cooperate in order to realise the expected functionality.

The use of distribution has two conceptionally unrelated benefits: (i) more data can be stored and processed, as the overall available capacity is the sum of the capacity of the individual database nodes. In this usage scenario the *sharding (partitioning) strategy* defines which data items are stored on which of the database nodes. (ii) data items can be stored redundantly on multiple database nodes protecting them against failures of individual database nodes. A *replication degree* of n denotes that a data item is stored n times in the system.

Replication. When sharding is used without replication, no tolerance against node failures exists. On the other hand, using replication without sharding means that all data is available on all database nodes. This is also referred to as full replication. When sharding and replication is used in parallel, each database node will contain only a subset of all data items [9]. Depending on the *replication strategy* [22] a user is allowed to interact with only one of the replicas (master-slave replication) or all of them (multi-master).

In the master-slave approach one of the n physical copies of a data item has the master role. Only this item can be changed by users. It is the task of the database node hosting this item to synchronize slave nodes. The latter only execute read requests. A failure of the master will require the re-election of a new master amongst the remaining slaves. In the multi-master approach, any replica can be updated and the hosting database nodes have to coordinate changes.

Geo-replication caters for mirroring the entire DDBMS cluster to a different location in order to protect the data against catastrophic events.

Replica Consistency. Having multiple copies of the same data item in the system requires to keep the copies in sync with each other. This is particularly true for multi-master approaches. Here, two approaches exist: synchronous propagation ensures consistency is coordinated amongst all database nodes hosting a replica before any change is confirmed to a client. Asynchronous propagation in turn delays this so that multiple diverging copies of the item can exist in the system and clients can perform conflicting updates unnoticed.

Node and Task Types. Besides storing data items, a DDBMS has several other tasks to do: *management tasks* keep track of the location of data items, routing queries to the right destination, and detecting node failures. *Query tasks* process queries issued by the client and interact with the database nodes according to the distribution information the management task stores. Depending on the actual DDBMS these tasks are executed by all database nodes in peer-to-peer manner, isolated in separate nodes, or even part of the database driver used by the client.

Storage Models. For DBMS three top-level categories are currently in use [15]: *relational, NoSQL* and *NewSQL* data stores. Relational data stores target transactional workloads, providing strong consistency guarantees based on the ACID paradigm [17]. Hence, relational data stores are originally designed as single server DBMS, which have been extended lately to support distribution (*e.g.* MySQL Cluster[1]). NoSQL storage models can be classified into key-value stores, document-oriented stores, column-oriented stores and graph-oriented stores. Compared to relational, their consistency guarantees are weaker and tend to towards BASE [21]. This weakening consistency makes those DDBMS better suited for distributed architectures and eases the realisation of features such as scalability and elasticity. NewSQL data stores are inspired by the relational data model and target strong consistency in conjunction with a distributed architecture.

2.2 Availability for DDBMSs

Generally, *availability* is defined as *the degree to which a system is operational and accessible when required for use* [14]. Besides *reliability* (the *"measure of the continuity of correct service"* [3]), availability is the main pillar of many fault-tolerant implementations [11].

In this paper, we solely focus on the availability aspect and assume that DDBMSs are reliable, *i.e.* work as specified. As our primary metric, we use what Zhong et al. call *expected service availability* and define availability of a DDBMS as the proportion of all successful requests [27] over all requests.

The availability of the DDBMS can be affected by two kinds of conditions [11]: (i) A high number of requests issued concurrently by clients, overloading the DDBMS such that the requests of clients cannot be handled at all or are handled with an unacceptable *latency* $> \Delta t$. (ii) Failures occur that impact network connectivity or availability of data items. The failure scenarios we consider are subject to Sect. 2.3.

2.3 Cloud Infrastructure

IaaS clouds have become a preferable way to run DDBMSs. Due to its cloud nature, IaaS offers more flexibility than bare metal resources. IaaS provides processing, storage, and network to run arbitrary software [19]. The processing and storage resources are typically encapsulated in a virtual machine (VM) entity that also includes the operating system (OS). VMs run on hypervisors on top of the physical infrastructure of the IaaS provider. As cloud providers typically operate multiple data centres, IaaS eases to span DDBMSs across different geographical locations. The location of cloud resources can be classified

---

[1] https://www.mysql.com/products/cluster/.

into geographical locations *(regions)*, data centres inside a region *(availability zones)*, *racks* inside a data centre and *physical hosts* inside a rack.

This set-up heavily influences availability as failures on different levels of that stack can have impact on individual database nodes (running in one VM), but also on larger sets of them. For instance, the failure of a hypervisor will lead to the failure of all VMs on that hypervisor.

An exemplary DDBMS deployment on IaaS is depicted in Fig. 1. Here, an 8-node DDBMS is deployed across two regions of one cloud provider. Each database node is placed on a VM and the VMs rely on different availability zones of the respective region. The example also illustrates the use of heterogeneous physical hardware: availability zones B and C are built upon physical servers without disks and dedicated storage servers. Availability zones A and D are built upon servers with built-in disks.

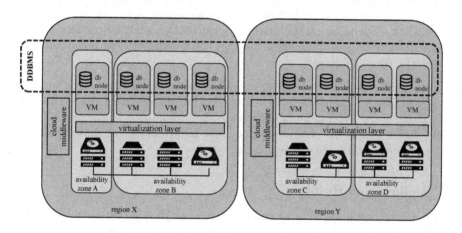

Fig. 1. DDBMS on IaaS

3   Failure Impact Analysis

Section 2.2 stated that two types of events impact the availability of DDBMSs: overload and failures. Dealing with overload has seen much attention in literature (*cf.* Sect. 8) so that this paper will focus on the latter that has barely received attention in the past [24]. In particular, our work addresses the capabilities of DDBMSs availability mechanisms to overcome failures in IaaS environments.

3.1   Replication for Availability

In Sect. 2.1, we introduced replication and partitioning as two major, but basically unrelated concepts used in DDBMSs. This section investigates the impact of their use under failure conditions.

No Replication. Apparently, when a database node fails and no replication is used, all data items stored on that database node become unavailable. The impact on the overall availability of the DDBMS depends on the access pattern of the failed shard, but for uniform distribution, the availability will drop to $\frac{N-1}{N}$ while this database node is unavailable.

Master-Slave Replication. When master-slave replication is used, the failure of a single database node has less impact, as copies of the data item are in the system. Yet, the process of detecting the failure and finding a new master for all data items from the failed database node needs time no matter if the new master is elected manually or automatically [11]. Hence, for uniform distribution of access, availability will still drop to $\frac{N-1}{N}$. Yet, hopefully for a shorter time.

Multi-master Replication. When using multi-master replication the failure of a single database node does not affect the overall availability, as any other database node hosting a replica of the requested data item can be contacted. Nevertheless, depending on the driver implementation and the routing, a small portion of requests may fail, when they are connected with the failed database node at the time of failure.

Functional Nodes. Some DDBMSs make use of additional hosts that function as entry points for clients or as a registry storing the mapping from data item to database node. The failure of these node also has impact on the overall availability. In particular, if configured wrong, the failure of any of those nodes can render the entire DDBMSs unavailable.

3.2 Failures and Recovery

This section investigates failures and recovery of DDBMSs hosted on Clouds. In particular, it derives how to emulate the failure of certain resources for the sake of evaluating availability metrics of different DDBMSs.

From Fig. 1 we can see that a cloud-operated DDBMS sits on top of several layers of hard- and software. Hence, even assuming that both DDBMS code and the operating system surrounding it are correct, the large stack leaves opportunities that can go wrong: On the infrastructure level, servers, network links, network device, power supplies, or cooling may fail. Similarly, on software level, management software and hypervisors can fail; algorithms, network, and devices can be buggy, misconfigured, or in the process of being restarted.

Any of these failures can affect virtual machines and virtual networks, but also physical servers, physical networks, entire racks, complete availability zones, or even entire data centers. Table 1 lists these failure levels with a way to emulate the failure and an action that helps to recovery from the failure.

The failure of a *database node* or a *virtual machine* can be represented as virtual machine unavailability and can be emulated by forcibly terminating the virtual machine. Here, it is important to ensure that no additional clean-up tasks, *e.g.* closing network connections get executed. The failure of a *physical server* leads to the unavailability of all virtual machines hosted on that

Table 1. DDBMS failures in a Cloud Infrastructure

Failure level	Emulate	Recovery
DDBMS node	Forcibly terminate VM	Replace VM
VM	Forcibly terminate VM	Replace VM
Physical server	Forcibly terminate all VMs on server	Replace VMs/move zone
Availability zone	Forcibly terminate all VMs in zone	Move zone or region
Region	Forcibly terminate all VMs in region	Move region/provider
Cloud provider	Forcibly terminate VMs hosted by provider	Move provider

server. The failure of a full *availability zone* or even an entire *region* leads to the unavailability of multiple physical servers and hence unavailability of all hosted virtual machines.

4   Availability Metrics

From the previous sections we derive input parameters and output metrics to evaluate the availability of DDBMSs running on cloud infrastructures. Input parameters describe the deployment and evaluation specifications of the DDBMS, while output metrics describe the experienced availability after the input parameters have been applied. Hence, a tuple of input parameters and output metrics provide the base for the availability evaluation (*cf.* Sect. 5).

4.1   Input Parameters

The input parameters as listed in Table 2 comprise the *deployment* and *evaluation* specification. The deployment specification combines the replication and partitioning characteristics (*cf.* Sect. 3.1), failure and recovery characteristics (*cf.* Sect. 3.2) and deployment information of the DDBMS nodes. The first two input parameters define the *replication* setting of the DDBMS, *i.e.* node replication level and cross data centre geo-replication level. None, one, or both replication levels might be configured for the DDBMS under observance. The replication first is defined by the strategy, defines the amount of replicas the replication will have in normal, healthy state, and the update laziness, how the write requests are synchronized between replicas.

Partitioning defines the setting for data partitioning, if present. If partitioning takes place, the amount of partitions are specified, and how the distribution strategy of data items to partitions is handled (group based, range based, with hashing functions). For accessing the distributed data items, the data access is of importance, namely if the client connects to the correct partition directly, via a proxy or requests are routed automatically.

Table 2. Input parameters

Input parameter		Description
Deployment	Node replication	Strategy (single-/multi-master, selection), replicas, laziness
	Geo-replication	Equals "node replication"
	Partitioning	Number of partitions, strategy (range, hash), data access (client, proxy, routing)
	Resources	Hierarchical infrastructure model of DDBMS (*cf.* Fig. 1)
Evaluation	Failure spec	Number of failing nodes per level (cf. Table 1), number of failing nodes per types
	Recovery Spec	Restart policies, number of database nodes to add (per node type if existent)
	Workload spec	Requests per second, read/write request ratio, number of data items

The *resources* parameter contains a full model of the allocated infrastructure resources for the DDBMS. This model includes all infrastructure entities involved from geographic location, to physical servers, virtual machines and DDBMS nodes (cf. Fig. 1). If the DDBMS differentiates node types, the type is reflected in the resource information as well.

Further, Table 2 also presents the *evaluation specification* parameters. The *failure specification* parameter defines a failure scenario which will be emulated by the Gibbon framework. For each level of potential failures described in Table 1, the amount of failing resource entities and (optionally) the DDBMS node type to fail is defined. The *recovery specification* parameter on the other hand describes the emulated recovery plan such as restarting virtual machines or adding new nodes to the DDBMS. The input parameters can skip the optional failure recovery parameter.

For simulating failure and recovery specification, the DDBMS is continuously under an artificial workload, defined by the *workload specification* parameter. This parameter defines the amount of read and write requests, as well as the total amount of data items stored in the DDBMS.

4.2 Output Metrics

The output metrics presented in Table 3 represent the experienced availability after the input parameters deployment and evaluation specification have been applied. The output metrics are results of continuously monitoring the metrics while input parameters are applied.

The *accessibility* α defines if the database is still reachable by clients (accepting incoming connections) and accepts read and write requests. While accessibility represents `boolean` values over time, the *performance impact* ϕ represents

Table 3. Output metrics

Output metric		Description
Statistics	Accessibility	DDBMS is accessible for client requests (read and write)
	Performance impact	Throughput (read/write requests per second), latency of requests
	Request error rate	Amount of failed requests due to data unavailability
Times	Take over time	Time until the failure spec is being masked by the DDBMS
	Recovery time	Time until the recovery spec is applied by the DDBMS

the throughput the DDBMS can handle during the evaluation scenario, including the amount of requests and the latency for request handling. For instance the performance may be decreased if replicas are down. In case of node failures, not all data partitions of a DDBMS may be available until the failure is handled. The *request error rate* ϵ describes as output metric the amount of failed requests due to data unavailability over the evaluation time.

The output metrics *take over time* and *recovery time* specify the measured time the DDBMS required to identify the applied failure specification, and the time it takes to apply the recovery specification. The accessibility, the performance impact and the data loss rate are time series values over the time the evaluation scenario is being applied. These values are measured periodically at runtime, are then aggregated and statistically represented in an percentile ranking over the time separately for the time it takes to apply (i) the failure specification and (ii) the recovery specification.

From the described output metrics, the overall availability metric of the evaluated DDBMS can be calculated. From the output metrics, only a configurable amount of percentiles (e.g. > 90) are considered. Each of the three metrics accessibility, performance impact and data loss rate gets its configurable weighting factor $W_\alpha, W_\epsilon, W_\phi$ resulting in the overall availability metric defined as $\Theta = \alpha * W_\alpha + \epsilon * W_\epsilon + \phi * W_\phi$.

5   Availability Evaluation

In this section we present an extensible methodology to evaluate the availability capabilities of DDBMSs based on the defined input parameters and output metrics (*cf.* Sect. 4). Therefore we define an adaptable evaluation process, which emulates the previous failure levels and enacts the respective recovery actions. This process enables the monitoring of the defined availability metrics in order to analyse the high availability efficiency of the evaluated DDBMS. First, we introduce the evaluation process, defining the required evaluation states and transitions. Second, we present an algorithm to inject cloud resource failures on different levels, based on a predefined failure specification.

5.1 Evaluation Process

Emulating cloud resource failures and monitoring the defined availability metrics, requires the definition of a thorough and adaptable evaluation process, from the DDBMS deployment in the cloud over the simulation of cloud resource failures to the recovering of the DDBMS. A fine-grained evaluation process is depicted in Fig. 2, where the monitoring periods of the availability metrics are presented in the white box, the evaluation process state in the yellow box and the framework components in the blue box. In the following we introduce the evaluation process states and the monitoring periods, while the framework components are described in Sect. 6.

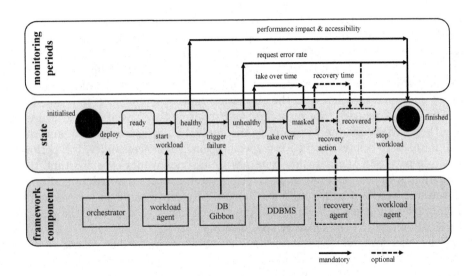

Fig. 2. Evaluation process

The depicted evaluation process in Fig. 2 illustrates an exemplary evaluation, which runs through all defined states exactly once by executing the respective transitions (cf. Tables 4 and 5). Yet, it is also possible to leave out dedicated transitions such as *recovery action*. The Gibbon framework executes each evaluation process and it also supports the combination of multiple evaluation processes into an *evaluation scenario*.

Throughout each evaluation process, the defined availability metrics (cf. Sect. 4) need to be monitored. Yet, the monitoring period for each availability metric depends on the evaluation process state as depicted in Fig. 2. The *performance impact* and *accessibility* is monitored from the *healthy* to the finished state to analyse the performance development over the intermediate states and compare it with the performance at the beginning. The *request error rate*

Table 4. Evaluation states

State	Description
Initialised	A new evaluation process is triggered
Ready	All nodes of the DDBMS deployed in the cloud and the configuration is finished, *i.e.* the DDBMS is operational
Healthy	All nodes of DDBMS are serving requests
Unhealthy	n nodes of the DDBMS are not operational due to cloud resource failures, the DDBMS has not yet started take over actions.
Masked	The DDBMS initiated automatically the take over of n replicas to reestablish the availability of all data records. The DDBMS is operational again but with $-n$ nodes
Recovered	The number of nodes complies again with the initial number of nodes, all nodes of DDBMS are serving requests
Finished	The evaluation process is finished

Table 5. Evaluation transitions

Transition	Description
Deploy	The DDBMS is being deployed and configured, *i.e.* dedicated VMs are allocated in specified locations
Start workload	A constant workload is started against the deployed DDBMS
Trigger failure	A specified cloud resource failure is emulated by the DB Gibbon component (*cf.* Sect. 5.2, provoking the failure of n nodes
Take over	DDBMS recognizes the failure of n nodes and initialises the take over of the remaining replicas by propagating the new locations for the currently unavailable data records
Recovery action	the DDBMS is restored to its actual number or nodes by adding n new nodes to the DDBMS. In this process the new nodes are integrated in the running DDBMS and the data is redistributed
Stop workload	The workload is stopped

is monitored during the *unhealthy* to the *recovered* state to analyse the development of the failed request rate. The *take over time* is monitored in the transition from the *unhealthy* to the *masked* state while the *recovery time* is monitored in the transition from the *masked* to the *recovered* state.

5.2   DB Gibbon Algorithm

In order to emulate failing cloud resources on the different levels, we build upon the concepts of Netflix's Simian Army [26] and adapt these concepts for DDBMS in the cloud. Therefore, we define an algorithm, which is able to enact the introduced cloud failure levels (cf. Sect. 3.2) for DDBMSs. Following the Simian Army [26] concepts, we name our algorithm the *DB Gibbon*. The algorithm is depicted in Listing 1.1 and requires as input parameters a list of `failures` `<List<failure >>` and the `dbDeployment`. The `failures` list contains n `failure` objects with the attributes `failureLevel`, indicating the cloud resource failure level (*cf.* Table 1), `failureQuantity` specifying the number of failures to be enacted by the DB Gibbon and `nodeType` specifying the failing nodes type. Possible `nodeTypes` are `<ANY, DATA, MANAGEMENT, QUERY>` The `dbDeployment` parameter contains the information of the deployed DDBMS topology, *i.e.* the mapping of each node to the cloud resource. Based on these parameters the DB Gibbon algorithm enacts the specified failures in the transition to the *unhealthy* state as depicted in Fig. 2. After enacting all failures, the algorithm returns the updated deployment, which is used to trigger the *recovery action* by calculating the difference to the initial deployment and deriving the required recovery actions.

Listing 1.1. DB Gibbon Algorithm

```
input: failures <List<failure >>, dbDeployment
output: dbDeployment
begin
 for each failure in failures
  if failure.failureLevel == failureLevel.VM
   def VMs List<VM> ← dbDeployment.getVMsOfNodeType(failure.nodeType)
   for (int i : failure.failureQuantity)
    def failedVM VM ← failRandomVM(VMs)
    dbDeployment ← updateDeployment(dbDeployment, failedVM)
   end

  else if failure.failureLevel == failureLevel.AVAILABILITY_ZONE
   for (int i : failure.failureQuantity)
    def VMs List<VM> ← dbDeployment
    .availabilityZone(i).getVMsOfNodeType(failure.nodeType)
    def failedVMs List<VM> ← failVMs(VMs)
    dbDeployment ← updateDeployment(dbDeployment, failedVMs)
   end

  else if failure.failureLevel == failureLevel.REGION
   for (int i : failure.failureQuantity)
    def VMs List<VM> ← dbDeployment.region(i)
    .getVMsOfNodeType(failure.nodeType)
    def failedVMs List<VM> ← failVMs(VMs, failureSeverity)
    dbDeployment ← updateDeployment(dbDeployment, failedVMs)
   end

  //only private cloud deployments
  else if failure.failureLevel == failureLevel.PHYSICAL_HOST
   for (int i : failure.failureQuantity)
    def VMs List<VM> ← dbDeployment.physicalHost(i)
    .getVMsOfNodeType(failure.nodeType)
    def failedVMs List<VM> ← failVMs(VMs, failureSeverity)
    dbDeployment ← updateDeployment(dbDeployment, failedVMs)
   end
  else
   return FAIL
 end
 return dbDeployment
end
```

6   Architecture

This section presents the architecture of the novel Gibbon Framework for executing the introduced evaluation methodology (*cf.* Fig. 2). A high-level view on the architecture is depicted in Fig. 3, introducing the technical framework components and their interactions between each other. The entry point to the Gibbon framework represents the evaluation API, expecting the evaluation scenario specification. This specification comprises four sub specifications for the respective framework components. In the following each framework component is explained with respect to the required specification and its technical details.

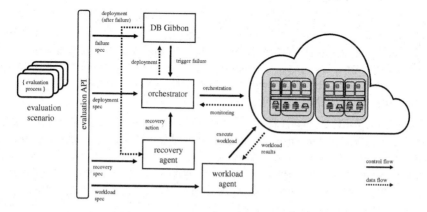

Fig. 3. Gibbon evaluation framework architecture

Orchestrator. Orchestrator receives the *deployment specification*, which comprises the description required cloud resources and the actual DDBMS with its configuration. Hence, the orchestrator interacts with the cloud provider APIs to provision the cloud resources and to orchestrate the DDBMS on these resources. As carved out in our previous work, the usage of advanced Cloud Orchestration Tools (COTs) over basic DevOps tools is preferable as COTs abstract cloud provider APIs and provide monitoring and adaptation capabilities during runtime [4]. The monitoring capabilities comprise general system metrics as well as customisable application specific metrics. Hence, the monitoring capabilities of COTs can be exploited to measure metrics such as the DDBMS nodes state or ongoing maintenance operations, which are required to express the availability metric (*cf.* Sect. 4). COTs offer run-time adaptations such as deleting or suspending cloud resources or DDBMS nodes, adding new cloud resources and DDBMS nodes or the execution of additional applications on the existing DDBMS nodes. These capabilities are used by the DB Gibbon and recovery agent components. The Gibbon framework builds upon the Cloudiator COT[2] [10], which supports

---

[2] http://cloudiator.org/.

the main public IaaS providers (Amazon EC2[3], Google Compute[4], Microsoft Azure[5]) as well as private clouds built upon OpenStack[6] and provides advanced cross-cloud monitoring and adaptation capabilities [12].

DB Gibbon. This component is responsible to emulate cloud resources failures based on the provided *failure specification*. It queries the COT for the current DDBMS deployment information, *i.e.* the mapping of nodes to cloud resources and their location. Based on the deployment information and the failure specification, it executes the DB Gibbon algorithm (*cf.* Sect. 5.2). In the execution phase of the algorithm the DB Gibbon interacts with the COT to enact the actual failures on cloud resource level, *e.g.* deleting all VMs of the DDBMS which rely in availability zone A.

Recovery Agent. The component receives the *recovery specification*, which comprises the defined recovery actions (*cf.* Sect. 3.2). It collects the DDBMS deployment state from the DB Gibbon after its execution and map the recovery actions to the failed resources. To enact the actual recovery actions, the recovery agent interacts with the COT, which executes the cloud provider API calls and orchestrates the DDBMS nodes.

Workload Agent. It keeps the DDBMS under constant workload during the evaluation process to measure the performance development during the different evaluation states by receiving a *workload specification*, describing the targeted operation types, the data set size and the number of operations. Further, the workload agent is responsible to measure the performance metrics and store them for further analysis in the context of the availability metrics. Our framework uses the Yahoo Cloud Serving Benchmark (YCSB) as workload agent [8] as the YCSB supports distributed execution, multiple DDBMSs and easy extensibility.

7   Discussion

In this section, we discuss the effectiveness of the Gibbon framework based on a concrete evaluation scenario for MongoDB[7]. First, we describe the applied deployment, failure, recovery, and workload specifications and second, we discuss early evaluation results of the accessibility, take over and recovery time.

7.1   Evaluation Scenario: MongoDB

We select MongoDB as the most prevalent document-oriented data store[8] as the preliminary DDBMS to evaluate. MongoDB is classified as CP in the context of

---

[3] https://aws.amazon.com/ec2/.
[4] https://cloud.google.com/compute/.
[5] https://azure.microsoft.com.
[6] https://www.openstack.org/.
[7] https://www.mongodb.com/.
[8] http://db-engines.com/en/ranking_trend.

the CAP theorem but still provides mechanisms to enable high availability [11], such as automatic take over in case of node failures. MongoDB's architecture is built upon three different services: (1) `mongos` act as query router, providing an interface between clients and a sharded cluster, (2) `configs` store the metadata of a sharded cluster, (3) `shards` persist the data. A group of shards build a replica set with one `primary` handling all requests and n `secondaries`, synchronizing the primary data and taking over in case the `primary` becomes unavailable.

The applied evaluation scenario builds upon a single evaluation process as depicted in Fig. 2. Yet, in this preliminary evaluation process we do not apply a constant workload during the evaluation but use three different data set sizes which are inserted in MongoDB during the deployment of MongoDB. Hence, we only measure the metrics *take over time* and *recovery time*. Accessibility is measured by periodic connection attempts by a MongoDB client.

The *deployment specification* comprises one `mongos` and three shards nodes, building a MongoDB replica set with one `primary` and two `secondaries` with full-replication and no sharding. For the sake of simplicity we did not deploy a production-ready setup with multiple `mongos` and `config` nodes. All nodes are provisioned on a private cloud based on OpenStack version Kilo with full and isolated access to all physical and virtual resources. All nodes are provisioned within the region *Ulm* and the availability zone *University*. Each node runs on a VM with 2 vCPUs, 4GB RAM, 40GB disk and Ubuntu 14.04.

As *failure specification* we applied one `failure` object with the attributes `failureLevel=VM`, `failureQunatity=1 nodeType=data` to the DB Gibbon.

The *recovery specification* defines to add of a new data node (*i.e.* secondary in MongoDB), as soon as MongoDB reaches the *masked* state after a node failure. MongoDB is configured to elect a new primary if the recent primary node failed.

As stated above, we do not apply a constant workload during the evaluation, the *workload specification* only defines the data set by number of `records=100K`, `400K`, `800K` and `record size=10KB`.

7.2 Evaluation Results

The preliminary evaluation results reveals two insights: the behaviour in case of node failures and in the case of adding a new replica to the system.

In case of a node failure, the behaviour depends on the failed node type. If a secondary fails, connected clients will loose their read-only connections and have to reconnect to another secondary or to the primary node. In this case we can assume, that at least the primary node is still *accessible*, so clients can reconnect immediately. If the primary fails, clients will loose their read/write connections, but may connect immediately to a secondary node for read requests. Remaining secondaries will recognize that the primary fails and will elect the new primary after a configurable timeout. During this timeout and election phase, no clients can issue write requests, the DDBMS is hence only *accessible* for read requests and not for write requests. Per default, the primary failover timeout is configured to ten seconds. In repeated experiments an average duration for election and primary take over of five seconds (± one second) is measured. Hence, the overall *take over time* is 15 s.

Whenever a node failure happens, the evaluation scenario adds a new replica after the remaining nodes elected a primary. The new secondary node has to synchronize the stored data from other nodes, for consistency reasons from the primary node. The replication time depends on the size of stored data. For 100 k, 400 k, and 800 k items the median for replication time with 30 runs takes 31 s, 300 s, and 553 s with a standard deviation of 7s, 15 s, and 25 s. The *recovery time* hence depends on the amount of data to replicate, plus a fixed amount of time it takes to allocate new resources on the Cloud. The DDBMS is *accessible* throughout the replication, yet with reduced resources due to the ongoing synchronisation.

8   Related Work

The view on availability in distributed systems evolved over the last two decades. The CAP theorem published in 2000, states that any networked shared-data system can only have two of the three properties consistency, availability and partition tolerance [6]. A revisited view on the CAP theorem is presented in 2012 [5], reflecting how emerging distributed systems such as DDBMSs adapted their consistency, availability and partition tolerance properties as the decision for two out of the three properties is not a binary decision [5].

The classification of DDBMSs according to their CAP properties in CA or CP DDBMSs is a widely discussed topic in database research. A first overview and analysis of emerging DDBMSs is provided by [7], analysing various DDBMSs with respect to their availability capabilities in the context of the CAP theorem.

DDBMSs mechanism to provide high availability are discussed by [15], breaking down the technical replication strategies from master-slave replication to masterless, asynchronous replication of 19 DDBMSs. Yet, the high availability mechanisms are only discussed on a theoretical level and no evaluation of their efficiency is proposed. Further, cloud resources are introduced as the preferable resources to run DDBMSs but the different failure levels and their implication to the availability of the DDBMS are not considered.

A similar approach is followed by [23], adding a dedicated classification of common DDBMSs with respect to their CAP properties, *i.e.* AP or CP. Further, the usage of cloud resources is discussed with respect to virtualisation and data replication across multiple regions. Yet, the focus relies on enabling consistency guarantees of wide-area DDBMSs while side-effects by using cloud resources that effect as well as ensure availability in DDBMS are not considered in detail.

An availability- and reliability-centric classification of DDBMSs is presented by [11]. Hereby, the challenges to provide non-functional requirements such as replication, consistency, conflict management, and partitioning are broken down in a fine grained classification schema and a set of 11 DDBMS are analysed. Two availability affecting issues and solutions are presented, overloading and node failures: *(i)* Is a DDBMS not available due to overloading, the DDBMS needs to be scaled out. *(ii)* Replicas need to be in place to overcome database node failures. Yet, the proposed solutions are on an architectural level and the actual capabilities of DDBMSs are not evaluated in real-world scenarios.

While theoretical classifications of DDBMSs provide a valuable starting point for a first selection of DDBMSs, the final selection still remains challenging due to the heterogeneous DDBMSs landscape. Hence, database evaluation frameworks provide additional insights in DDBMS capabilities by evaluation dedicated evaluation tiers based on different workload domains. While available evaluation frameworks such as YCSB [8] or YCSB++ [20], focus on the evaluation performance, scalability, elasticity and consistency, the availability tier is not yet considered by these frameworks [24]. First approaches in availability evaluation are based on the YCSB and focus the on decreased availability due to an overloaded database [18,25]. While an evaluation framework focusing on the availability of cloud-hosted DDBMSs is not yet available, an approach to enact synthetic failures on cloud resources is described by [26] and implemented at Netflix. Yet, this approach only describes the failure scenarios in the cloud and does not propose evaluation metrics or an evaluation methodology for cloud applications in general and DDBMS in particular. A first approach towards the availability metrics is presented by [2], yet the focus relies on the resilience of DBMSs, while DDBMSs and their availability mechanism are not considered. Existing failure-injection tools such as the DICE fault injection tool[9] or jepsen[10] either inject failures on node or DBMS level but do not support the injection of resource failures in different granularity.

9   Conclusion and Future Work

In the last decade the database management system (DBMS) landscape grew fast, resulting in a very heterogeneous DBMSs landscape, especially when it comes to distributed database management systems (DDBMSs). As cloud computing is the preferable way to run distributed applications, cloud computing seems to be the choice to run DDBMSs. Yet, the probability of failures increases with the number of distributed entities and cloud computing adds another layer of possible failures. Hence, DDBMSs apply data replication across multiple DDBMS nodes to provide high availability in case of node failures. Yet, providing high availability comes with limitations with respect to consistency or partition tolerance as stated by the CAP theorem. As these limitations are not binary, a vast number of high availability implementations in DDBMSs exist. This makes the selection of a DDBMS to run in the cloud a complex task, especially as supportive availability evaluation frameworks are missing.

Therefore, we present Gibbon, a novel availability evaluation framework for DDBMSs to increase the trust in running DDBMSs in the cloud. We describe levels of cloud resource failures, existing DDBMS concepts to provide high availability and distill a set of five quantifiable availability metrics. Further, we

---

[9] https://github.com/dice-project/DICE-Fault-Injection-Tool.
[10] https://github.com/jepsen-io/jepsen.

derive the DDBMSs specific technical details, affecting the availability evaluation processes. Building upon these findings, we introduce the concept of extensible evaluation scenarios, comprising n evaluation processes. Further, we present the DB Gibbon, which emulates cloud resource failures on different levels.

The Gibbon framework executes the defined evaluation scenarios for generic DDBMSs and cloud infrastructures. Its architecture comprises an orchestrator to deploy the DDBMS in the cloud, a workload agent, a recovery agent and the DB Gibbon to inject cloud resources failures. As preliminary evaluation, we evaluate the take over and recovery time of MonogDB in a private cloud, by injecting failures on the virtual machine level.

Future work will comprise an in-depth evaluation of multiple well-adopted DDBMSs (See footnote 8) based on the Gibbon framework. Further, the definition of a minimal evaluation scenario to derive a significant availability rating, is in progress. In this context, the statistical calculations of the overall availability rating index will be refined. Finally, the portability of Gibbon evaluate the availability of generic applications running in the cloud will be evaluated.

Acknowledgements. The research leading to these results has received funding from the EC's Framework Programme HORIZON 2020 under grant agreement numbers 644690 (CloudSocket) and 731664 (MELODIC). We also thank Daimler TSS for the encouraging and fruitful discussions on the topic.

References

1. Abadi, D., Agrawal, R., Ailamaki, A., Balazinska, M., Bernstein, P.A., Carey, M.J., Chaudhuri, S., Chaudhuri, S., Dean, J., Doan, A., et al.: The Beckman report on database research. Commun. ACM **59**(2), 92–99 (2016)
2. Almeida, R., Neto, A.A., Madeira, H.: Resilience benchmarking of transactional systems: experimental study of alternative metrics. In: PRDC (2017)
3. Avizienis, A., Laprie, J.C., Randell, B., Landwehr, C.: Basic concepts and taxonomy of dependable and secure computing. TDSC **1**(1), 11–33 (2004)
4. Baur, D., Seybold, D., Griesinger, F., Tsitsipas, A., Hauser, C.B., Domaschka, J.: Cloud orchestration features: are tools fit for purpose? In: UCC (2015)
5. Brewer, E.: Cap twelve years later: how the "rules" have changed. Computer **45**(2), 23–29 (2012)
6. Brewer, E.A.: Towards robust distributed systems. In: PODC (2000)
7. Cattell, R.: Scalable SQL and NoSQL data stores. ACM Sigmod Rec. **39**(4), 12–27 (2011)
8. Cooper, B.F., Silberstein, A., Tam, E., Ramakrishnan, R., Sears, R.: Benchmarking cloud serving systems with YCSB. In: SoCC (2010)
9. DeCandia, G., Hastorun, D., Jampani, M., Kakulapati, G., Lakshman, A., Pilchin, A., Sivasubramanian, S., Vosshall, P., Vogels, W.: Dynamo: Amazon's highly available key-value store. ACM SIGOPS Oper. Syst. Rev. **41**(6), 205–220 (2007)
10. Domaschka, J., Baur, D., Seybold, D., Griesinger, F.: Cloudiator: a cross-cloud, multi-tenant deployment and runtime engine. In: SummerSOC (2015)
11. Domaschka, J., Hauser, C.B., Erb, B.: Reliability and availability properties of distributed database systems. In: EDOC (2014)

12. Domaschka, J., Seybold, D., Griesinger, F., Baur, D.: Axe: a novel approach for generic, flexible, and comprehensive monitoring and adaptation of cross-cloud applications. In: ESOCC (2015)

13. Ford, D., Labelle, F., Popovici, F.I., Stokely, M., Truong, V.A., Barroso, L., Grimes, C., Quinlan, S.: Availability in globally distributed storage systems. In: OSDI (2010)

14. Geraci, A., Katki, F., McMonegal, L., Meyer, B., Lane, J., Wilson, P., Radatz, J., Yee, M., Porteous, H., Springsteel, F.: IEEE standard computer dictionary: Compilation of IEEE standard computer glossaries: 610. IEEE Press (1991)

15. Grolinger, K., Higashino, W.A., Tiwari, A., Capretz, M.A.: Data management in cloud environments: NoSQL and NewSQL data stores. In: JoCCASA (2013)

16. Gunawi, H.S., Hao, M., Suminto, R.O., Laksono, A., Satria, A.D., Adityatama, J., Eliazar, K.J.: Why does the cloud stop computing? Lessons from hundreds of service outages. In: SoCC (2016)

17. Haerder, T., Reuter, A.: Principles of transaction-oriented database recovery. In: CSUR (1983)

18. Konstantinou, I., Angelou, E., Boumpouka, C., Tsoumakos, D., Koziris, N.: On the elasticity of NoSQL databases over cloud management platforms. In: CIKM (2011)

19. Mell, P., Grance, T.: The NIST definition of cloud computing. Technical report, National Institute of Standards & Technology (2011)

20. Patil, S., Polte, M., Ren, K., Tantisiriroj, W., Xiao, L., López, J., Gibson, G., Fuchs, A., Rinaldi, B.: YCSB++: benchmarking and performance debugging advanced features in scalable table stores. In: SoCC (2011)

21. Pritchett, D.: Base: an acid alternative. Queue **6**, 48–55 (2008)

22. Sadalage, P.J., Fowler, M.: NoSQL Distilled: A Brief Guide to the Emerging World of Polyglot Persistence. Pearson Education, London (2012)

23. Sakr, S.: Cloud-hosted databases: technologies, challenges and opportunities. Cluster Comput. **17**(2), 487–502 (2014)

24. Seybold, D., Domaschka, J.: Is distributed database evaluation cloud-ready? In: ADBIS (2017)

25. Seybold, D., Wagner, N., Erb, B., Domaschka, J.: Is elasticity of scalable databases a myth? In: IEEE Big Data (2016)

26. Tseitlin, A.: The antifragile organization. Commun. ACM **56**(8), 40–44 (2013)

27. Zhong, M., Shen, K., Seiferas, J.: Replication degree customization for high availability. In: EuroSys (2008)

Locality-Aware GC Optimisations for Big Data Workloads

Duarte Patrício[1,2], Rodrigo Bruno[1,2], José Simão[1,3], Paulo Ferreira[1,2], and Luís Veiga[1,2(✉)]

[1] INESC-ID Lisboa, Lisbon, Portugal
{dpatricio,rbruno}@gsd.inesc-id.pt,
{paulo.ferreira,luis.veiga}@inesc-id.pt
[2] Instituto Superior Técnico, Universidade de Lisboa, Lisbon, Portugal
[3] Instituto Superior de Engenharia de Lisboa, Instituto Politécnico de Lisboa, Lisbon, Portugal
jsimao@cc.isel.ipl.pt

Abstract. Many Big Data analytics and IoT scenarios rely on fast and non-relational storage (NoSQL) to help processing massive amounts of data. In addition, managed runtimes (e.g. JVM) are now widely used to support the execution of these NoSQL storage solutions, particularly when dealing with Big Data key-value store-driven applications. The benefits of such runtimes can however be limited by automatic memory management, i.e., Garbage Collection (GC), which does not consider object locality, resulting in objects that point to each other being dispersed in memory. In the long run this may break the service-level of applications due to extra page faults and degradation of locality on system-level memory caches. We propose, LAG1 (short for Locality-Aware G1), an extension of modern heap layouts to promote locality between groups of related objects. This is done with no previous application profiling and in a way that is transparent to the programmer, without requiring changes to existing code. The heap layout and algorithmic extensions are implemented on top of the Garbage First (G1) garbage collector (the new by-default collector) of the HotSpot JVM. Using the YCSB benchmarking tool to benchmark HBase, a well-known and widely used Big Data application, we show negligible overhead in frequent operations such as the allocation of new objects, and significant improvements when accessing data, supported by higher hits in system-level memory structures.

Keywords: Cloud infrastructure · Java virtual machine · Garbage collection · Locality-aware · Big data

1 Introduction

Managed languages (such as Java) are gaining space as the choice to implement Big Data processing and storage frameworks [10,14,15,19], as they facilitate application development, This is mostly due to its automated memory management capabilities, flexible object-oriented design and quick development cycle.

© Springer International Publishing AG 2017
H. Panetto et al. (Eds.): OTM 2017 Conferences, Part II, LNCS 10574, pp. 50–67, 2017.
https://doi.org/10.1007/978-3-319-69459-7_4

These languages, and Java in particular, run on top of a runtime system (the Java Virtual Machine, JVM, is one such case) that manages code execution and memory management. Memory management is governed by the Garbage Collector (GC), a component that controls how objects are allocated and collected. Despite the considerable development benefits of automatic memory management, the GC can lead to serious performance problems in Big Data applications.

In particular, some of these performance problems are caused by the fact that the GC does not respect application's working set locality. In fact, while the application is running, the GC will move application objects throughout memory, possibly separating objects that belong to the same dataset and that, therefore, should be close to each other. This is a consequence of throughput oriented management mechanisms implemented by the GC that, however, hinder co-locality of related objects and the way objects are represented and placed in memory [5,6,9].

Space locality is known to have a relevant impact in performance [10,16,28]. For example, Wilson et. al. [28] exploited the hierarchical decomposition of data structure trees to reorganize the tracing algorithm, instead of strict depth-first or breadth-first tracing. Dynamic profiling was also studied by Chen [6], so that information on frequency of access is gathered and used in the placement of those objects. Huang [11], on the other hand, showed different strategies for online object reordering during GC, in order to improve program locality. Also, Ilham's work [12] shows increased locality in system-level memory structures, such as the L1D[1] cache and the dTLB[2], when ordering schemes for children object placement are accounted for, i. e., Depth-First (DF), Breath-First (BF) and Hot Depth-First (HDF). However, these works either apply a similar approach to all objects, which makes it difficult to tailor for storage-specify data-structures, or are hard to scale to very large heaps given the impact of per-object profiling in execution time. Furthermore they were not evaluated with modern parallel GC algorithms.

To reduce the impact of GC in the context of Big Data applications, others have made extensive modifications to the way certain objects are created and managed in special propose memory spaces, either requiring compiler and GC modifications, or application-specific data structures [5,17].

In this work, we are focused on large-scale key-value databases such as HBase [1], Cassandra [14], and Oracle KVS [3]. These databases tend to hold massive amounts of objects (key-value pairs) in memory, which end up being scattered in memory due to poor GC techniques, that completely disregard object locality. As the application graph grows, the number of misses and faults in memory increase with clear negative impact on applications performance.

We propose a novel approach by enhancing GC with locality awareness. In other words, we propose *LAG1*, a modified GC which goal is to keep highly related groups of objects (i.e., objects that have many references between each other, for example, a data structure) close to each other in memory, leading to improved locality. Thus, *LAG1* takes takes into account object references when

---

[1] The first level of the CPU data cache.
[2] The data Translation-Lookaside-Buffer.

moving objects in memory. *LAG1* is implemented by modifying a state-of-art GC algorithm, the Garbage First (G1), the new by-default GC for the OpenJDK JVM. *LAG1* does not require the use of new data structures, or even changes to existing code, thus making the solution easier to adopt in current and new systems.

In sum, the main contributions of this paper are:

i. A tracing algorithm, designed for automatically managed heaps, to identify highly related groups of objects.
ii. A garbage collector extension that copies highly related groups of objects to specific memory segments.

The rest of the paper is organized as follows. Section 2 presents the main building blocks of modern garbage collectors, and the factors that hinder locality in NoSQL Big Data storages. Section 3 presents the architecture of *LAG1*, a novel extension, for an existing GC, to co-locate object graphs in memory. Section 4 shows the modifications made to G1, a modern parallel GC, on top of which we built *LAG1*, and heap organisation to avoid the previous problems without modifications to the application. Section 5 presents the evaluation of *LAG1* showing its benefits, the small overheads of this solution and the benefits at application level. Section 7 draws final conclusions.

2   Background

Many of today's most used NoSQL databases are written in high-level languages [19], such as Java and C#. By doing so, developers rely on services supported by these managed runtimes, in particular, automatic memory management, GC. However, GC introduces several performance issues. As already mentioned, the lack of object locality compromises application performance; in this paper, this is exposed in the context of NoSQL databases, and a solution (*LAG1*) is proposed.

This section is used to further motivate for the problem and to provide sufficient background for the next sections, which describe the proposed solution.

2.1   NoSQL Databases and Object Locality

Currently, NoSQL databases are a popular tool to store massive amounts of data. Examples of these storage systems include HBase [1], Cassandra [14] and Oracle KVS [3]. These are distributed, column-oriented NoSQL databases, whose data model is a distributed key/value map where the key is an identifier for the row and the value is a highly structured object.

NoSQL databases use large caches to hold hot accessed data. However, it is a challenge for the GC to efficiently manage such large in-memory data structures. In fact, when running the YCSB benchmark framework [7] with a dataset of 12 GB, we noticed an excessive use of *page swapping* resulting from poor GC decisions that cause long application pauses (the result of this is shown in Sect. 5).

Previous works, such as Bu *et al.* with their *bloat-aware design* [5] have also alerted to this problem; we believe that going a step further to co-locate related objects achieves even better locality on system-level memory structures to benefit the overall execution time.

As time goes by and operations are performed on top of a NoSQL database, the GC needs to reclaim unreachable objects in order to free space for new application objects. By doing so, the GC copies objects in memory in order to free segments of memory. Since the GC does not account for object locality, it copies groups of highly connected objects (i.e., objects with many references between each other, for example, a data structure) into distant memory locations. This degrades application performance as cache locality does not hold in these scenarios. As datasets become larger, the amount of memory consumed by a NoSQL database grows, leading to an increased distance between highly connected objects.

2.2 Garbage Collection Algorithms and Heap Layouts

Garbage Collection (GC) is a well-known and widely used technique to automatically manage memory, i.e., programmers do not need to free objects after using them [13]. The GC operates over a large memory space called heap. All application objects reside in the heap, and is the job of the GC to provide memory for new application objects and to collect memory used by unreachable application objects.

Modern garbage collectors are generational, meaning that they follow the assumption that *most objects die young* [25]. Thus, most popular GC implementations divide the heap into two generations, one that holds newly allocated objects (the *young* generation) and one to hold objects that survived for at least a number of GC cycles (the *old* generation).

The young generation is further divided into three spaces: *eden, to* and *from*. The *eden* is used to fulfill object allocation requests while the *to* and *from* are used to hold objects that survived at least one GC cycle and that will be eventually copied to the *old* generation.

G1 [8], the baseline of our work, is one such generational GC with a region-based heap. A heap of this kind is split in small fixed-sized regions, instead of strictly dividing the heap as in Fig. 1(a). This is illustrated in Fig. 1(b), where several regions can be seen, each with its purpose. Thus, regions with an *O* are old regions (abstractly they belong to the *old generation*); regions with an *S* are survivor regions; and regions with a *Y* are young regions, both abstractly belonging to the *young generation*. G1 also keeps a per-region remembered-set. For each region, this set describes inter-region references between objects. Its use is important during garbage collection to know if there are objects from survivor or old regions that reference objects in young regions (thus allowing to collect only young regions).

Recent GC implementations, including G1, provide two main types of GC cycles: minor and full. A minor collection is designed to collect the *young* generation and to copy/promote survivor objects into the *old* generation. Since objects

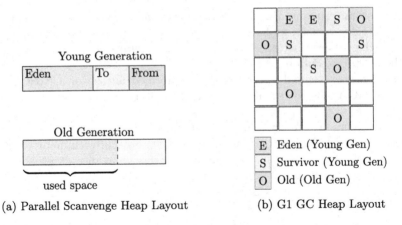

(a) Parallel Scanvenge Heap Layout   (b) G1 GC Heap Layout

Fig. 1. GC Heap Layouts

that survive only one minor collection might not be automatically promoted into the *old* generation (this is a GC configurable option), survivor spaces are used to hold these objects until they are old enough to be promoted. A full collection, as the name suggests, collects the whole heap (both *young* and *old* generations).

3   Gang Promotion in Heap Management

This section presents *LAG1*, a GC extension to co-locate dependent objects in order to improve the locality in system-level memory structures. With *LAG1* we introduce the concept of gang promotion to achieve this result. Also, we introduce a new memory region definition, which manages an unique segment of memory, called *container-region*.

The concept of gang promotion consists in copying live objects (in contrast with dead objects, i.e., unused objects in the heap) in a manner where the sub-graph of a family of objects does not intersect another sub-graph of a family of objects. We define a family as a group of objects belonging to the same sub-graph (for example, a data structure such as a linked list). The result is a single segment of memory for each family of objects. These segments of memory are *container-regions*.

For gang promotion to work, several challenges need to be addressed. These challenges are the following: (i) how to identify sub-graphs of highly-related objects, and (ii) how to efficiently promote sub-graphs of highly related objects, without causing overhead on the existing promotion mechanism, to a specific memory region. The following sections explain how we tackled these challenges in the two participating sub-systems for object management: the runtime (Sect. 3.1), the system that executes the application code and allocates new objects; and the garbage collector (Sect. 3.2), the system that collects objects no longer in use by the application.

3.1   Identifying Relevant Object-Graphs

One of the challenges in gang promotion is to identify sub-graphs of highly-related objects to co-locate, non-intrusively to the runtime sub-system. Naively, bookkeeping every allocation call and, at a *safepoint*[3], filter the sub-graph of allocated objects could suffice. But for a very large number of object sub-graphs, such as in a Big Data environment, that would largely consume system resources.

We have taken the approach of instrumenting only relevant allocation sites. Relevant allocation sites are those that allocate objects of a type deemed to be the head of a highly-related object sub-graph. This goes along with the fact that the head of an object sub-graph is usually the first object to be allocated in the system (another terminology is to refer the head of the sub-graph as the *root* of a structured tree). Therefore, identifying the root is enough for *LAG1* to move the whole tree to a special memory region in a later stage. In order for *LAG1* to identify the root, the user of the program must specify what is the type of the object that is the root. The set of types for the root objects is called *LAG1-RS*. The heuristics behind the identification of the root's type, and thus what can be included in the *LAG1-RS*, is left for the user to decide.

To avoid conflicts among mutator threads, *LAG1* saves identified roots on a thread-local array shown in Fig. 2. This figure shows what the instrumented allocation site does; it queries if the class of the object being instantiated is a *LAG1-RS*class and, if so, it inserts the object in the thread-local array. The term *LAG1-RS*(abbreviated form of *LAG1* Root Set) is the set of sub-graph roots across all mutator threads. This step is important for the GC stage, i.e., when the live object-graph is moved to a survivor space.

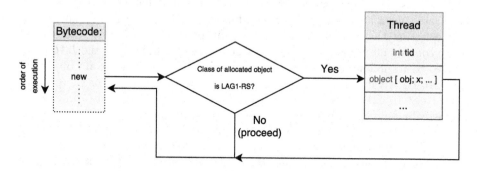

Fig. 2. Bookkeeping head of an object sub-graph at instrumented allocation site

3.2   Gang GC on a Large Heap

To correctly promote object sub-graphs identified through the mechanism presented in Sect. 3.1, two more challenges need to be addressed: (i) avoid unreachable sub-graphs referenced by *LAG1-RS*; and (ii) integrate sub-graph promotion

---

[3] The mechanism used in HotSpot to create Stop-the-World pauses. Garbage collection cycles run inside a safepoint, during which all application threads are stopped.

in *LAG1* with the existing promotion mechanism in G1. We address these two challenges in the paragraphs below.

Avoid Unreachable Sub-Graphs. Saving allocated heads of sub-graphs in thread-local arrays can have adverse effects, if these references are used as roots for tracing the remaining sub-graph at GC time. This is so because the references may belong to objects used as temporary storage (use cases for this are VM warm-ups, cloning data, etc.) and can become unreachable by the tracing algorithm. *LAG1* uses the *LAG1-RS* to create *container-regions*, instead of using it for tracing live objects. There is no memory region associated with newly created *container-regions*, therefore obliging that the sub-graph must be reachable by the root-set (threads' stacks, globals and statics) in order to have heap space effectively assigned.

Another important aspect of *LAG1* is associating the identifier of the created *container-region* with each object in the *LAG1-RS*. With this technique, *LAG1* does not lose the associative relationship between the head of a sub-graph and the *container-region* that will hold the objects. It also provides a way for *LAG1* to propagate this identifier to this root's followers or children (i.e., the objects belonging to the root's sub-graph).

Promotion of *LAG1* sub-graph. The first phase of a GC cycle is to trace from the root-set, i.e., the threads' stack frames, the global variables and the static variables in the system. The order of the operations is non-important since garbage collectors are designed to be throughput-oriented during collection phases. This policy does not conform with *LAG1* policy that every object should be located near its siblings. To avoid this, *LAG1* inserts a checkpoint phase before letting the rest of the tracing algorithm do its work. The checkpoint phase consists on propagating the container region identifier that the *LAG1-RS* created to its followers. This will associate any follower of a certain object in the *LAG1-RS* with the identifier for the same *container-region* of the parent, recursively. In *LAG1*, this phase is called *pre-marking*. *LAG1* also provides a work-stealing model for parallel garbage collectors during the *pre-marking* phase to speed-up computation. Although *pre-marking* may add additional overhead to the GC, experiments show that heads of object sub-graphs are allocated rarely during the application. Thus, the overhead is negligible for a long running application.

After the *pre-marking* phase, the second phase of the GC, which consists on the tracing and promotion of root-set followers, can proceed normally with no constraints. *LAG1* no longer intervenes on the GC phases until it sees a live object associated with a *container-region* and targeted for promotion to a *tenured* space. For these objects, it targets its destination space to the *container-region* instead of the default old region. Since *container-regions* are on the same space as the old regions, *LAG1* complies with the object age while still providing a target space next to its siblings of "tenured" age in a transparent manner.

Figure 3 shows a rundown of the operations executed as part of gang promotion. In Fig. 3(a), a GC Thread is shown accessing the thread-local array

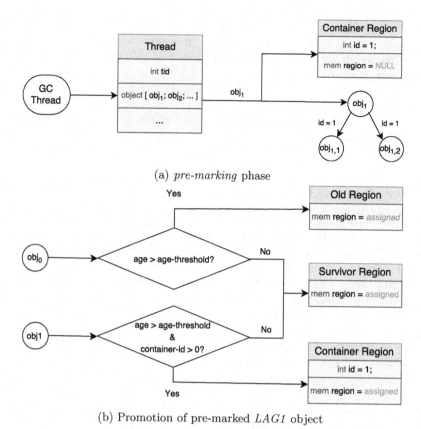

(a) *pre-marking* phase

(b) Promotion of pre-marked *LAG1* object

Fig. 3. A rundown of the operations executed for gang promotion

of a mutator thread, which previously saved obj_1 and obj_2. The thread then fetches the obj_1 reference, creates a new *container-region* with an unique identifier and propagates this identifier to the siblings of obj_1. Figure 3(b), on the other hand, shows the promotion step. It illustrates a regular object obj_0 being promoted, with no need to check if it has any *container-id* association (in the *pre-marking* phase, *LAG1-RS* objects and its siblings were associated with their *container-region*). But, since obj_1 has an association with a *container-region* it goes through a different condition. Therefore, obj_1 is copied to the memory region of its *container-region* if it is old enough.

4   Deployment of Gang Promotion on Hotspot JVM

LAG1 is implemented on OpenJDK 8 HotSpot JVM. The OpenJDK HotSpot JVM [2] is the state-of-the-art Java virtual machine used in most Java deployments. It is a highly portable and highly optimized virtual execution environment for Java-bytecode based languages (Java, Scala, Clojure, etc.).

The new by-default garbage collector is the Garbage First (G1) GC [8], a low-pause collector, with a soft real-time pause guarantee, while still achieving high throughput. G1 is the baseline garbage collector of this work's prototype, thus we take advantage of some of its features such as: the generational heap space and its region-based division of the heap, meaning that the heap space is split into small fixed-sized regions.

In this section we describe our modifications to the Java runtime sub-system of HotSpot JVM (Sect. 4.1) and how we modified G1 to implement *LAG1* (Sect. 4.2).

4.1   Java Runtime Instrumentation

In our prototype, we tackled the problem of identifying relevant object sub-graphs (Sect. 3.1) in the Java runtime. The Java runtime sub-system of HotSpot is divided into three components: (i) the assembly interpreter, (ii) a lightly optimizing bytecode compiler (C1), (iii) a highly optimizing bytecode compiler (C2). C1 is better suited for client-machine applications, thus we disregarded its application. We only considered the assembly interpreter and C2, the latter for methods with high invocation count.

Instrumentation for *LAG1* was tackled on the allocation site of the root of a relevant object sub-graph (e.g., a data structure). Relevant object types are left for the user to decide, using a command line option. For example, if the relevant object sub-graph is a LinkedList structure, the user should specify the full qualified class name (i.e., java.util.LinkedList). Since class loading is prior to object allocation for any given type, we first register the user-specified class to be *LAG1-RS* by placing a bit on the virtual machine class-representation. Thus, during allocation, all we do is a fast check for the bit on the class to be installed on the object. If it is present, then we add the object address to a thread-local indexed array. This requires only three operations at assembly-level, a compare (for the presence of the bit in the class), a load (to load the object address in the thread-local array) and an increment (to increment the thread-local array index).

4.2   LAG1 — Locality-Aware Extension of G1

To implement *container-regions*, *LAG1* takes advantage of the regionalized architecture of the heap that G1 already provides. Since old regions (G1 heap regions belonging to the old generation) are already present in G1, *container-regions* are specialized old regions. The reason for this decision is that *LAG1* handles large object sub-graphs, preferably long-lived, thus it would be impracticable to use the young generation regions.

During minor GC (a GC that collects only the young regions), before the tracing of the reachable live object graph is initiated, *LAG1* checks if there are saved references in the *LAG1-RS*. If there are references, the *pre-marking* phase is initiated. The *pre-marking* phase comprises both creating *container-regions*

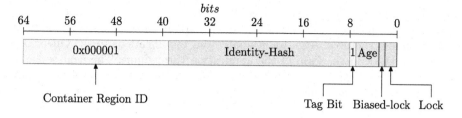

Fig. 4. The *pre-marking* phase tagging an object with a container region identifier

for *LAG1-RS* objects and propagating the *container-region* identifier to its followers. This identifier is simply an integer to index the *container-region* array. Additionally, *pre-marking* also includes tagging relevant objects. The identifier integer and tag bit are installed on unused bits in the header, such as illustrated in Fig. 4, where it shows the *pre-marking* of an object header. In Fig. 4 some lower-level details are shown, regarding the header of a Java object, with the important bits in the *tag-bit* (an unused bit in the HotSpot JVM, which we use to mark a *LAG1-RS* object and its sub-graph) and in the *container-region* identifier.

While implementing *LAG1*, we took into consideration that may exist *LAG1-RS* objects already promoted in a previous GC. This means that may exist new objects (allocated since the last GC), children of a *LAG1-RS* object already promoted in an earlier stage. To mark these newly allocated objects with the *container-region* identifier, we intercepted the remembered-set operations of G1 (also called *old-to-young* on other garbage collectors [9]) and added one more instruction to install the identifier of the referent (the parent in an older generation) on the follower in the young generation. Therefore, there is no possibility of losing the follower to another space by not being *pre-marked* in time, because the remembered-set operations are always executed before the promotion of the followers. Another favorable aspect of this approach is that it no longer requires a checkpoint barrier before regular tracing, such as explained in Sect. 3.2.

The last stage for *LAG1* is to promote (i.e., copy) objects according to the *container-region* identifier. Since G1 already checks the object's age to decide if it should promote to a survivor region or to an old region, *LAG1* only adds an additional check. The check consists on looking at the object header and see if it has a container region identifier installed. It is a fast bit mask operation, so no overhead is inflicted.

5   Evaluation

To evaluate *LAG1*, we considered the fact that hot object sizes in Java are as big as L1[4] and L2[5] cache line sizes, and thus very few of them fit in those caches.

---

[4] L1 is the 1st level of CPU cache: 32 KB in size and 64 B per line in modern models.
[5] L2 is the 2nd level of CPU cache: 256 KB in size and 64 B per line in modern models.

Therefore, our experiments consisted in observing the virtual memory performance, more specifically the dTLB (CPU-level) and the *page-table* (Kernel-level) system structures. Also, we evaluated our modifications to the OpenJDK 8 HotSpot JVM, in the form of the application throughput. The next sections present the setup we used (Sect. 5.1), the program locality achieved with our solution (Sect. 5.2), and the high-level behaviour of the application (Sect. 5.3).

5.1   Evaluation Setup

Experimental runs were executed on a 4-core machine with 8 logical cores, 3 levels of cache with a 8 MB L3 and 16 GB of memory, running a 64-bit Linux 4.4.0 kernel. To test the locality effects in system-level memory structures, such as the dTLB and page-table, we resorted to performance monitoring counters in the Linux tools package[6]. The target of our experiments was HBase [1], a widely used large-scale data store for Big Data processing, using YCSB [7] as a client application. YCSB is a highly configurable cloud benchmarking tool, widely used to benchmark large-scale data stores. The following paragraphs present the configurations we used on YCSB to benchmark HBase running our modified JVM.

YCSB can be configured with a large number of parameters, including: number of operations, number of records to load on the data store, the ratio of operations for each action (insert, update, read, scan) and the size of each record. The size of each record was fixed to 1 KB for all experiments. Also, the number of operations to perform on the data was also fixed to $1 * 10^5$. On the other hand, the number of records to load and the ratio of operations was varied. Since the JVM was configured to have a maximum size of 12 GB for the Java heap, the number of records (*load*) used was: (i) 6 GB, (ii) 8 GB, (iii) 10 GB and (iv) 12 GB. For this evaluation, the configuration for YCSB consisted of two workloads with memory loading characteristics: (i) a read-intensive (RI) workload and (ii) a scan-intensive (SI) workload. The detailed characteristics of each workload is described below.

Workload RI 70% of reads, 25% of scans and 5% of updates

Workload SI 25% of reads, 70% of scans and 5% of updates

5.2   Program Locality in System

In this section we show the improvements that our solution has in system-level memory structures. The focus is given to the dTLB and the page-table, because Big Data Java applications (which handle large sub-graphs of objects) will not see a big improvement in CPU caches given their size.

Key-value stores, such as HBase [1], usually use multi-level map where, given a *table* name, a *row* name and a *field* name, a value can be inserted, read or updated [19]:

---

[6] http://linux.die.net/man/1/perf.

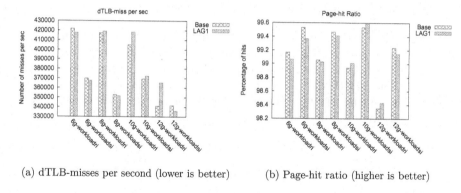

(a) dTLB-misses per second (lower is better)   (b) Page-hit ratio (higher is better)

Fig. 5. Locality on system-level memory structures

$$\texttt{map <table-name, map<row-key, sortedmap<field-key, data>>>} \qquad (1)$$

With *LAG1* we expect that the *field-keys* and the actual data, represented in Equation (1), be closer in memory. Figure 5 shows the results obtained, for each pair `<size of data-set>-<workload type>`, where the bars for *Base* refer to the baseline JVM and *LAG1* our modified JVM. We begin our analysis with the observation that, the first step in virtual-memory address translation will start with: a dTLB load, then a page-table query (if the dTLB load misses) and then, if the requested address is not in the page-table, a page-fault is issued. Figure 5(a) shows that the dTLB misses per second is stable for workloads that do not cause pressure in the heap, i.e., 6 GB and 8 GB of dataset size. Therefore, variations in the page-table, shown in Fig. 5(b), are mostly related to external factors (e.g., OS virtual-memory policies, GC, etc.).

However, as the size of the dataset — workloads of 10 GB and 12 GB of dataset size — gets closer to the Java heap size (12 GB), we begin to see the dTLB is stressing. That means the dTLB cache no longer has the capability of saving that many translated virtual-memory addresses to physical addresses, thus this mechanism no longer becomes important. The responsibility is passed to the page-table, where the OS will do a *page-walk*[7]. At this point, we see that with *LAG1* the page-table hit-ratio, shown in Fig. 5(b), is increased in comparison with the baseline JVM. This is more evident in read-intensive (RI) than scan-intensive (SI) workloads, because scan-intensive workloads read multiple values sequentially. And, as referred previously, Java objects may be large in size, when compared with system-level memory structures, thus spanning multiple page entries (and consequently, multiple dTLB entries). The test with 12 GB of dataset size and a scan-intensive workload (*12g-workloadsi*) is the only that does not follow the pattern, but that is because it already has low dTLB-misses as shown in Fig. 5(a).

---

[7] A page-walk consists on querying page-table entries, to see if the address the CPU is trying to load is present in physical memory.

5.3   Application Behaviour

In this section, we present the results for the application throughput when running HBase with *LAG1*. The results are from the timeseries output of YCSB, which ran 100 000 (100 k) operations on an HBase instance with a load of 6 GB, 8 GB, 10 GB and 12 GB records. We first ran a warm-up phase over the entries, therefore all results are the best obtained across 3 tests, in the percentile shown. The workloads were the same as in Sect. 5.1.

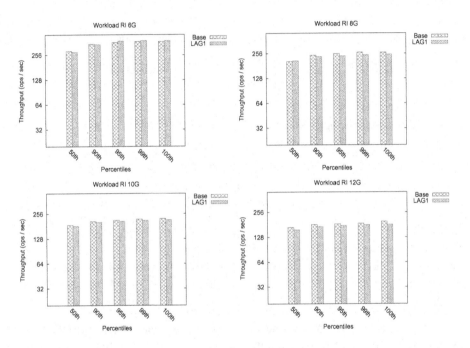

Fig. 6. Throughput on HBase with Workload RI for a variety of datasets

Figures 6 and 7 show the comparison of throughput between *LAG1* and the baseline JVM. It can be observed that, although *LAG1* added complexity to the baseline JVM, for all tests it did not influence throughput significantly (and in some cases, nothing at all). We believe that this is a positive result, because improvements in program locality outweigh the added complexity, and that future research could benefit from focusing on program locality aspects.

6   Related Work

Research in automatic memory management has proven that there is no unique solution that fits all classes of applications. The best choice of GC is, in many

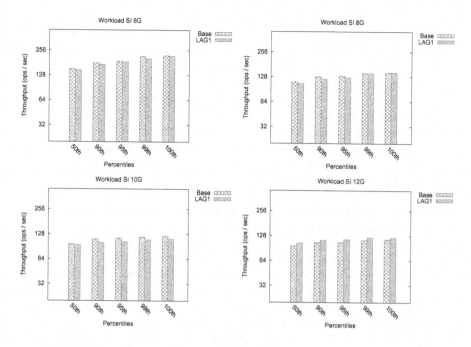

Fig. 7. Throughput on HBase with Workload SI for a variety of datasets

cases, application and input-dependent [23,24]. This has spanned a vast collection of algorithms, in many cases combinations of older ones, which can be stacked with application-specific profiles [22].

Parallel, stop-the-world algorithms have been making a successful entry in the field of big-data applications, since they can efficiently collect a whole heap within shorter pauses and do not require constant synchronization with the mutator, as it is the case with concurrent collection [9]. However, Java-supported Big Data applications in general, and storage in particular, stress the GC with lack of locality in large heaps and bloat of objects. This is mainly tackled using three kinds of approaches [5,15,17]: (i) avoiding per-object headers and imposing new memory organizations at the framework-level, (ii) speeding-up garbage collection by identifying objects that are created and destroyed together and, (iii) coordinating the stop-the-world moment in inter-dependent JVM instances. Because most works focus on reducing overheads by dramatically changing the layout of objects and out-of-heap specially crafted structures, these solutions need changes both to the compiler and the GC system or rely on complex static analysis which is hard to prove correct and complete.

Facade [17] is a compiler and augmented runtime that reduces the number of objects in the heap by separating data (fields) from control (methods) and putting data in an off-heap structure without the need to maintain the bloat-causing header. Hyracks [5] is a graph processing framework that also uses a scheme where small objects are collapsed into special-purpose data structures.

Because this is done at the framework-level, and not at the JVM-level, it is difficult to reuse the approach. Overhead can also be caused by GC operations running uncoordinated inter-dependent JVM instances [15]. When each of these instances needs to collect unreachable objects, if it does so regardless of each other; this can lead to significant pause times.

On the other hand, previous work about object ordering schemes [6,11,16] have shown that taking advantage of placement strategies, can increase locality in system-level memory structures and achieve better performance, especially when using guided techniques for optimal object placement. However, current approaches rely either on static analysis of fine-tuned dynamic profiling to avoid an excessive overhead. Instead, *LAG1* only relies on the user to specify the class of objects that hold the data, since it is already a low overhead solution.

NG2C [4] is a new GC algorithm that combines pretenuring with user-defined dynamic generations. It allocates objects with similar lifetime profiles in the same generation; by allocating such objects close to each other, *i.e.* in the same generation, it avoids object promotion (copying between generations) and heap fragmentation (which leads to heap compactions) both responsible for most of the duration of HotSpot GC pause times. Compared to *LAG1*, NG2C takes another approach to the issue of object locality, which may result in objects that point to each other being dispersed in memory. In the long run, contrary to *LAG1*, this may lead to extra page faults and degradation of locality on system-level memory caches.

7   Conclusion

Several Big Data frameworks and storages are executed on a managed runtimes, taking advantage of parallel garbage collection and Just-In-Time (JIT) compilation. However, modern parallel memory management and throughput-oriented techniques can hinder locality. Our approach was to promote objects' co-locality which minimizes the number of memory pages used, taking more advantage of system-level data and translation caches. This was done with an extension to the Garbage First (G1) GC promotion mechanism and algorithmic modifications to the runtime system, which we named *LAG1*.

The results provide positive conclusions on the use of *LAG1* on state-of-the-art JVM, the OpenJDK 8 HotSpot. First, we showed that the promotion efforts to co-locate highly-related object sub-graphs favourably increase page-table hits with real world executions. This is evident in large datasets with demanding workloads for the available memory, a common practice today. Second, we demonstrated that program locality outweighs added complexity on the runtime system with locality-aware policies. This was demonstrated with stable throughput across a variety of workloads and dataset sizes.

In the future, we would like to assess how the improvements provided by *LAG1* can also enhance performance transversally to other work on Java VM-based mechanisms and middleware, whose operation is also heavily dependent on object graph locality and on performing graph transversals, e.g., object replication [26,27], checkpoint and replay, [20,21], and dynamic software update [18].

Finally, although RAM memory is cheaper nowadays, the dataset sizes are growing faster than the available memory in cloud systems. Vendors cannot always comply with the agreed SLAs, because of the chaotic layout of objects in memory, when the latter is under pressure. It is our belief that given our results, future research could be more focused on program locality aspects. On the other hand, we are also focused on future work, including the evaluation with more specialized hardware, such as NUMA architectures, on larger datasets.

Acknowledgements. This work was supported by national funds through Fundação para a Ciência e a Tecnologia with reference PTDC/EEI-SCR/6945/2014, and by the ERDF through COMPETE 2020 Programme, within project POCI-01-0145-FEDER-016883. This work was partially supported by Instituto Superior de Engenharia de Lisboa and Instituto Politécnico de Lisboa. This work was supported by national funds through Fundação para a Ciência e a Tecnologia (FCT) with reference UID/CEC/50021/2013.

References

1. http://hbase.apache.org/. Visited 16 Feb 2017
2. http://openjdk.java.net/. Visited 16 Feb 2017
3. http://www.oracle.com/technetwork/database/database-technologies/nosqldb/overview/index.html. Visited 16 Feb 2017
4. Bruno, R., Oliveira, L.P., Ferreira, P.: NG2C: pretenuring garbage collection with dynamic generations for hotspot big data applications. In: Proceedings of the 2017 ACM SIGPLAN International Symposium on Memory Management, ISMM 2017, NY, USA, pp. 2–13 (2017), http://doi.acm.org/10.1145/3092255.3092272
5. Bu, Y., Borkar, V., Xu, G., Carey, M.J.: A bloat-aware design for big data applications. In: Proceedings of the 2013 International Symposium on Memory Management, ISMM 2013, pp. 119–130. ACM (2013)
6. Chen, W.K., Bhansali, S., Chilimbi, T., Gao, X., Chuang, W.: Profile-guided proactive garbage collection for locality optimization. In: Proceedings of the 27th ACM SIGPLAN Conference on Programming Language Design and Implementation, pp. 332–340. ACM (2006)
7. Cooper, B.F., Silberstein, A., Tam, E., Ramakrishnan, R., Sears, R.: Benchmarking cloud serving systems with YCSB. In: Proceedings of the 1st ACM Symposium on Cloud Computing, pp. 143–154. ACM (2010)
8. Detlefs, D., Flood, C., Heller, S., Printezis, T.: Garbage-first garbage collection. In: Proceedings of the 4th International Symposium on Memory Management, ISMM 2004, NY, USA, pp. 37–48 (2004), http://doi.acm.org/10.1145/1029873.1029879
9. Gidra, L., Thomas, G., Sopena, J., Shapiro, M.: A study of the scalability of stop-the-world garbage collectors on multicores. In: Proceedings of the Eighteenth International Conference on Architectural Support for Programming Languages and Operating Systems, ASPLOS 2013, pp. 229–240. ACM (2013)
10. Gidra, L., Thomas, G., Sopena, J., Shapiro, M., Nguyen, N.: Numagic: a garbage collector for big data on big NUMA machines. In: Proceedings of the Twentieth International Conference on Architectural Support for Programming Languages and Operating Systems, pp. 661–673. ACM (2015)

11. Huang, X., Blackburn, S.M., McKinley, K.S., Moss, J.E.B., Wang, Z., Cheng, P.: The garbage collection advantage. In: Proceedings of the 19th Annual ACM SIG-PLAN Conference on Object-Oriented Programming, Systems, Languages, and Applications - OOPSLA 2004, New York, USA, p. 69. ACM, New York (2004)
12. Ilham, A.A., Murakami, K.: Evaluation and optimization of java object ordering schemes. In: 2011 International Conference on Electrical Engineering and Informatics (ICEEI), pp. 1–6. IEEE (2011)
13. Jones, R., Hosking, A., Moss, J.E.B.: The Garbage Collection Handbook: The Art of Automatic Memory Management, 1st edn. Chapman & Hall/CRC (2011)
14. Lakshman, A., Malik, P.: Cassandra: a decentralized structured storage system. ACM SIGOPS Oper. Syst. Rev. **44**(2), 35–40 (2010)
15. Maas, M., Asanović, K., Harris, T., Kubiatowicz, J.: Taurus: a holistic language runtime system for coordinating distributed managed-language applications. In: Proceedings of the Twenty-First International Conference on Architectural Support for Programming Languages and Operating Systems, ASPLOS 2016, NY, USA, pp. 457–471. ACM, New York (2016)
16. Moon, D.A.: Garbage collection in a large lisp system. In: Proceedings of the 1984 ACM Symposium on LISP and Functional Programming, NY, USA, pp. 235–246. ACM, New York (1984)
17. Nguyen, K., Wang, K., Bu, Y., Fang, L., Hu, J., Xu, G.H.: FACADE: a compiler and runtime for (almost) object-bounded big data applications. In: ASPLOS, pp. 675–690. ACM (2015)
18. Pina, L., Veiga, L., Hicks, M.W.: Rubah: DSU for java on a stock JVM. In: Black, A.P., Millstein, T.D. (eds.) Proceedings of the 2014 ACM International Conference on Object Oriented Programming Systems Languages & Applications, OOPSLA 2014, Part of SPLASH 2014, Portland, OR, USA, 20–24 October, 2014, pp. 103–119. ACM (2014), http://doi.acm.org/10.1145/2660193.2660220
19. Redmond, E., Wilson, J.R.: Seven Databases in Seven Weeks: A Guide to Modern Databases and the NoSQL Movement. Pragmatic Bookshelf (2012)
20. Silva, J.M., Simão, J., Veiga, L.: Ditto – deterministic execution replayability-as-a-service for Java VM on multiprocessors. In: Eyers, D., Schwan, K. (eds.) Middleware 2013. LNCS, vol. 8275, pp. 405–424. Springer, Heidelberg (2013). doi:10.1007/978-3-642-45065-5_21
21. Simão, J., Garrochinho, T., Veiga, L.: A checkpointing-enabled and resource-aware java virtual machine for efficient and robust e-science applications in grid environments. Concurrency Comput. Pract. Exp. **24**(13), 1421–1442 (2012), https://doi.org/10.1002/cpe.1879
22. Singer, J., Brown, G., Watson, I., Cavazos, J.: Intelligent selection of application-specific garbage collectors. In: Proceedings of the 6th International Symposium on Memory Management, pp. 91–102. ACM (2007)
23. Soman, S., Krintz, C.: Application-specific garbage collection. J. Syst. Softw. **80**, 1037–1056 (2007), http://dx.doi.org/10.1016/j.jss.2006.12.566
24. Tay, Y.C., Zong, X., He, X.: An equation-based heap sizing rule. Perform. Eval. **70**(11), 948–964 (2013)
25. Ungar, D.: Generation scavenging: a non-disruptive high performance storage reclamation algorithm. ACM Sigplan Not. **19**(5), 157–167 (1984)
26. Veiga, L., Ferreira, P.: Incremental replication for mobility support in OBIWAN. In: ICDCS, pp. 249–256 (2002), https://doi.org/10.1109/ICDCS.2002.1022262

27. Veiga, L., Ferreira, P.: Poliper: policies for mobile and pervasive environments. In: Kon, F., Costa, F.M., Wang, N., Cerqueira, R. (eds.) Proceedings of the 3rd Workshop on Adaptive and Reflective Middleware, ARM 2003, Toronto, Ontario, Canada, 19 October 2004, pp. 238–243. ACM (2004), http://doi.acm.org/10.1145/1028613.1028623

28. Wilson, P.R., Lam, M.S., Moher, T.G.: Effective static-graph reorganization to improve locality in garbage-collected systems. SIGPLAN Not. **26**(6), 177–191 (1991)

FairCloud: Truthful Cloud Scheduling with Continuous and Combinatorial Auctions

Artur Fonseca[1,2], José Simão[1,3], and Luís Veiga[1,2(✉)]

[1] INESC-ID Lisboa, Lisbon, Portugal
[2] Instituto Superior Técnico, Universidade de Lisboa, Lisbon, Portugal
artur.fonseca@ist.utl.pt, luis.veiga@inesc-id.pt
[3] Instituto Superior de Engenharia de Lisboa, Instituto Politécnico de Lisboa,
Lisbon, Portugal
jsimao@cc.isel.ipl.pt

Abstract. With Cloud Computing, access to computational resources has become increasingly facilitated and applications could offer improved scalability and availability. The datacenters that support this model have a huge energy consumption and a limited pricing model. One way of improving energy efficiency is by reducing the idle time of resources - resources are active but serve a limited useful business purpose. This can be done by improving the scheduling across datacenters. We present FairCloud, a scalable Cloud-Auction system that facilitates the allocation by allowing the adaptation of VM requests (through conversion to other VM types and/or resource capping - degradation), depending on the User profile. Additionally, this system implements an internal reputation system, to detect providers with low Quality of Service (QoS). FairCloud was implemented using CloudSim and the extensions CloudAuctions. FairCloud was tested with the Google Cluster Data. We observed that we achieved more quality in the requests while maintaining the CPU Utilization. Our reputation mechanism proved to be effective by lowering the Order on the Providers with lower quality.

Keywords: Cloud computing · Energy · Pricing · Auctions · Scheduling · Reputation · User profiles

1 Introduction

The concept of Cloud Computing is related with shared computer resources on demand. It facilitates the management of resources and increased the scalability and availability of applications. However, those properties are associated with low energy efficiency and a pricing model that does not consider heterogeneous participants.

Energy. Up to 30% of the servers are on idle - less than three percent average daily utilization [7]. They still consume energy but serve a limited useful business

© Springer International Publishing AG 2017
H. Panetto et al. (Eds.): OTM 2017 Conferences, Part II, LNCS 10574, pp. 68–85, 2017.
https://doi.org/10.1007/978-3-319-69459-7_5

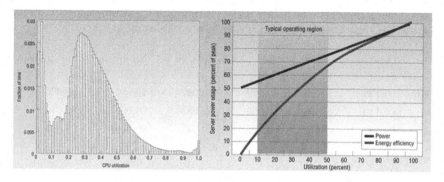

Fig. 1. (a) CPU utilization observed in 5.000 servers. (b) Server power usage per utilization. Source: [3]

purpose. The typical average daily server utilization is 6%, creating capital and energy loss. In Fig. 1, we observe that the majority of servers use 30% CPU and the energy efficiency increases with the utilization.

Previous studies [3,4] analyse the reasons. Even during periods of low service demand, servers cannot be terminated because they still run background tasks: distributed databases, small network tasks.

On the other hand, applications cannot run on fully utilized servers, as even non-significant workload fluctuation or any internal disruptions, such as hardware or software faults, will lead to performance degradation and failing to provide the expected QoS.

As detailed in the work [9], energy efficiency can be achieved by reducing energy loss (e.g. hardware, overhead supporting systems) and by improving energy efficiency (e.g. increase utilization).

Cloud Pricing. Cloud providers can apply different pricing models, as described in [2]. Table 1 compares the different prices in Amazon EC2 instances. On the **Pay-as-you-go** model the user is charged hourly without any long-term commitments or upfront payments. On **Subscription** model, the total price is reduced but an upfront payment and a long time commitment (e.g. 1 year) are required. On **Pay-for-resources**, the price is based on resource consumption. Even though it is a fair system, it is hard to monitor the consumption.

On fixed price approach, it is not possible to reflect demand and supply trends (i.e. not market-driven). It does not leverage users that are willing to pay more. Furthermore, overpricing can lead to resource waste.

Objectives. Our contribution will focus on giving load to the idle servers, increasing their utilization. This will be achieved by creating a Cloud-Auction mechanism that works in a similar way of a scheduler, but considering allocation in multiple datacenters. A dynamic pricing algorithm must be used to accommodate heterogeneous users and providers, and market conditions.

Table 1. Comparison between hourly prices in Amazon EC2 instances for 3 VM types for Linux. The region is EU(London). Prices were recorded on 21/12/2016.

Type	On-demand	Spot	Spot (1 h.)	Spot (6 h.)	No-upfront (1 y.)	Full-upfront(1 y.)
m4.large	$0.125	$0.014	$0.069	$0.087	$0.095	$0.079
c4.large	$0.119	$0.0152	$0.065	$0.083	$0.091	$0.076
d2.xl	$0.54	$0.077	$0.425	$0.54	$0.469	$0.393

In summary, the contributions of this work are the following:

– Analysis of previous research works in Cloud-Auctions and related topics (e.g. auction theory, scheduling, energy efficiency);
– Formulation of a taxonomy;
– Architecture of a prototype enforcing the algorithm;
– Evaluation with real User requests.

Document Structure. This work is organized as follows: In Sect. 2, we present and analyse the related work. Section 3 describes the Application Programming Interface (API) and the architecture of the solution, focusing on the main components and their interactions. Section 4 describes our implementation. In Sect. 5, we explain the evaluation methodology and describe the results. Section 6 presents some concluding remarks.

2 Related Work

In this section, we describe a Cloud-Auction taxonomy and study relevant related research and commercial systems. The main participants in a Cloud-Auction are the user and the provider. The key elements of analysis are: Negotiation, Time Dimension, Bid Representation, and Goals.

Negotiation. On Single Auctions only one participant (either user or provider) submits bids. It is mostly used in non-automated contexts (e.g. real state). Particularly, on sealed first-price auctions, the bids are private and users cannot adjust their bids. The highest bidder pays the price they submitted. The commercial system Amazon EC2 Spot Instances[1] implements a Single sealed auction - only users bids.

On Double Auctions both participants bid until an agreement is reached. The main difficulty is to find a competitive equilibrium. This type of auction is conducted by an independent participant - the auctioneer.

---

[1] http://docs.aws.amazon.com/AWSEC2/latest/UserGuide/using-spot-instances.html.

Bid improvement. If an agreement is not possible, the auction may terminate or allow **bid improvement**. In the system in [16], there is a price and time-slot negotiation: the user and provider send concurrent pairs (price, time). The downside of bid improvement is the participant still needs to be active after submitting the initial bid.

Time Dimension

Reservation. The Reservation can be on Spot or Forward. On Spot auctions the allocation starts immediately, unlike Forward auctions. In the work in [6,16], the user can bid for a future spot. On other systems like Compatible Online Cloud Auction mechanism (COCA) [19], the provider is free to decide when the allocation will take place, but considering the user's time restrictions.

Market. The Market is called Call if it waits for a set of bids and processes them as a batch (e.g. every hour). Particularly, in [6] the Forward auctions are done once a day. Amazon EC2 Spot Instances is also a Call Market but with shorter periods (e.g. 5 min).

On Continuous auctions, the bid is processed as soon as it arrives. One example is COCA [19].

Bid Representation Language. Users and providers must be able to specify the domain and interpret a Bid Representation Language.

Valuation. Valuation is how much it is worth for an entity for having a service performed. Most systems represent only the current bid and others a price range. COCA [19] allows the specification of a function: price per instance. The approach overcomes the Bid Improvement limitations. The bid information is sufficient to make a decision - the participant can be offline after bidding.

Desired resources. Finally, the participants can choose a custom type (e.g. VM with 1 GB RAM and 1 vCPU) or be forced to choose a fixed type (e.g. Small-VM). The quantity of each type can be fixed or an interval. In a Combinatorial auction the participants can bid on combinations of items (i.e. bundles).

Goals - Mutual

Valuation. The majority of the algorithms try to maximize the participant Valuation, following the scheduling restrictions. Utility is the level of the participant satisfaction regarding a dimension (e.g. time, price) or the total system (i.e. attributing weights to each dimension).

Truthfulness. Truthfulness is the property of algorithms that incentivizes participants to reveal their true valuation.

Privacy Preservation. The work in [5] formulates a threat model and presents a system that ensures Privacy Preservation.

Goals - SLA

Social Welfare. Regarding the **user**, the **social welfare** is related with the QoS. In Ginseng [1], there is only a guaranteed base and in the case of similar bids, the current users are preferred to avoid the transfer costs. On [20], the providers are rated and ordered (i.e. more paying users have better quality). Finally, Amazon EC2 Spot instances does not assure Social Welfare. The VM is lost if the spot price increases above the bid. Similarly, Google Preemptible VMs[2] may terminate immediately.

Goals - Provider

Energy Efficiency. A high Energy Efficiency is directly related with the provider's profit. The auction algorithm can perform the allocations in a way that the overall energy efficiency is maximized.

Time-slot. Simultaneously, the **provider** is interested in the **time-slot** allocation: when the service is low, as soon as possible; or best fit, in order to optimize resource utilization.

3   Architecture

In this Section, we present FairCloud, a Cloud-Auction system. After analysing the state-of-the-art, we extracted the desired properties to be able to formulate the requirements. Then, we present the FairCloud bid representation language, the entities and events, and detail the auctioneer core algorithms.

3.1   Architecture Design Requirements

FairCloud complies with the following requirements. It is Continuous, Truthful and Combinatorial. The algorithm should also consider the allocation QoS, in particular:

- VM conversion - receive a lower capacity for a price discount (i.e. requests a Medium, but receives a Small);
- VM degradation - partial service where only a minimum % of resources is guaranteed;
- Reputation mechanism - only based on internal metrics, used to sort the providers and allow the User to filter those under a minimum reputation score.

   The specification of those features is done using a textual bid representation language.

---

[2] https://cloud.google.com/compute/docs/instances/preemptible.

3.2   Bid Representation Language

VM Types. The VM types were normalized to 5 general-purpose types: **nano**, **micro**, **small**, **medium** and **large**. Each one specifies: weight (based on the computational power relations), CPU speed (measured in Millions of Instructions Per Second (MIPS)), RAM size, bandwidth and storage capacity.

User Bid. A User Bid contains the following parameters:

– **ID** Unique Identifier;
– **Degradation Profile** One of the following: `Demanding` (100% resources), `Restricted` (80%), `Relaxed` (60%). It represents the minimum partial service required;
– **Timestamp** When the bid should be visible to the auctioneer;
– **Conversion** Indicates if the user accepts conversions;
– **Reputation** The minimum desired of the provider - number from 0 to 100.

For each VM type requested:

– **VMType** As described above;
– **Amount**;
– **Price per hour** The maximum price accepted.

Provider Bid. A Provider Bid follows a similar structure:

– **ID** Unique Identifier;
– **MIPS** Maximum available quantity;
– **RAM** Maximum available quantity;
– **Degradation profile** One of the following: `Demanding` (100% resources), `Restricted` (80%), `Relaxed` (60%). It represents the minimum partial services guaranteed.

For each VM type available:

– **VMType** On of the five types;
– **Price per hour** The minimum price requested.

3.3   Entities and Events

FairCloud has 3 types of entities: BidderDatacenterBroker (representing the user), Auctioneer and BidderDatacenter. They communicate through asynchronous events. An event specifies the target (can be self), the delay (can be 0), the operation ID code and the parameters. Figure 2 summarizes, through a sequence diagram, the protocol of interactions involved among different participants.

The participants start by registering in the auction using their Unique Identifier. Later, the Auctioneer sends an AuctionStart event, waits for the Bids and sends Bid Acknowledges. Next, the allocation algorithm executes, as detailed in the next section, and the participants are notified with the Allocation Publication event. According to the results, each DatacenterBroker creates the requested VM in the corresponding Datacenter and waits for the execution to return. Additionally, the BidderDatacenter notifies the Auctioneer that the execution ended. This allows the Auctioneer to release the resources from the bid, allowing a new allocation.

The Auctioneer is cyclic, therefore it sends self repeat events with a delay of 20 min. This entity measures all datacenters utilization of CPU and Random Access Memory (RAM). The Monitoring occurs every 5 min. These metrics are transparent to the Datacenter and do not affect the algorithm.

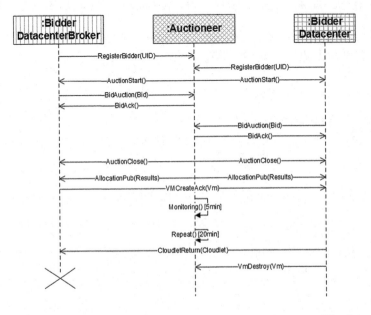

Fig. 2. Sequence diagram describing FairCloud entities and events. The delay is represented inside the [].

3.4 Auctioneer Algorithms

FairCloud implements multiple algorithms. Algorithm 1 receives and matches the user and the provider bids. For each valid pair (user, provider), the Normal, Conversion and Degradation allocation algorithms execute. Algorithm 2 updates the reputation data structure. Finally, Algorithm 3 is used to calculate the reputation value for each provider.

Bid Sorting and Matching. The bid sorting and matching (Algorithm 1) is triggered in regular periods of 20 min.

Initially, the UserBidList is ordered descending by bid density. The bid density is a measure of how much a user bids per unit of allocation and it is given by Eq. 1. The limit K represents the number of different VM types in the Bid. In the UserBid, the amount parameter is in the bid. However, on the ProviderBid, the amount is obtained by dividing the total capacity for the VM type capacity. Assuming a datacenter with 1500 MIPS of capacity and a nano VM with 75 MIPS, the amount is 20.

$$BidDensity = \frac{Bid.TotalPrice}{\sqrt{\sum_{k=1}^{K} Bid.VMType[k].Amount * weight(Bid.VMType[k])}} \tag{1}$$

Then, a Join operation is made between ProviderBidList and Reputation-Map. The ProviderBidList is ordered ascending by $\frac{BidDensity}{Reputation}$. By using this ratio, if two Providers have the same price, the one with more QoS is preferred.

For each UserBid, the algorithm filters ProviderBidList by the UserBid minimum reputation desired, and iterates the ProviderBidList. For each pair (User-Bid, ProviderBid), if the user average price is greater than the provider average price, and the UserBid timestamp is greater than the CurrentTime, then the allocation algorithms are executed. The three allocation algorithms: Normal, Conversion and Degradation are executed sequentially.

Algorithm 1. Bid Sorting and Matching. Simultaneously, handles user and provider bids.

Input Bid, UserBidList, ProviderBidList, ReputationMap
Output UserBidList, ProviderBidList, ReputationMap
 1: **if** Bid **is** $UserBid$ **then**
 2:     Add Bid to UserBidList
 3: **else**
 4:     Add Bid to ProviderBidList
 5: **end if**
 6: UserBidList.OrderDescendingBy($BidDensity$)
 7: ProviderBidList.Join (ReputationMap)
 8: ProviderBidList.OrderAscendingBy($\frac{BidDensity}{Reputation}$)
 9: **for** UserBid u in UserBidList **do**
10:     $ProviderBidList.Filter(Reputation > u.Reputation)$
11:     **for** ProviderBid p in ProviderBidList **do**
12:         **if** $u.AveragePrice$ $>=$ $p.AveragePrice$ **and** $u.Timestamp$ $>=$ $CurrentTime$ **then**
13:             NormalAllocation()
14:             ConversionAllocation()
15:             DegradationAllocation()
16:         **end if**
17:     **end for**
18: **end for**

Normal Allocation. First, the algorithm assures that the provider degradation profile is `Demanding`. Then, it iterates all requested VM. The available amount is obtained by analysing how many units can still fit in the Provider. Finally, the assigned amount is the minimum between requested and available amounts. The algorithms give the maximum reputation to the provider, for each allocation. The final unit price is the average between requested and offered price.

Conversion Allocation. The condition for a conversion is the User bid accepting conversions. The algorithm iterates the user requested VM types and the provider offered types. If a conversion is possible (i.e. the offered type is one and only one size above the requested type), the available and assigned amount are calculated. The Reputation given is 0.98. In this type of allocation, the user offered price is reduced to half. The final price is $Average(OfferedPrice * 0.5, RequestedPrice)$.

Degradation Allocation. First, the algorithm assures that the degradation is possible (i.e. the provider offered quality is higher than the requested quality). Then, it iterates all requested VM. The available and assigned amount are obtained the same way as the previous allocations. The reputation score depends on the quality offered: `Restricted`, 0.98 and `Relaxed`, 0.95.

The price depends on the user profile and the offered degradation profile. We defined a partial utility matrix described in [14,15]. The combinations marked with X are not possible.

$$factor = \begin{array}{l|ccc} User\backslash Provider & Demanding & Restricted & Relaxed \\ \hline Demanding & 1 & 1 & 1 \\ Restricted & X & 0.8 & 0.9 \\ Relaxed & X & X & 0.8 \end{array}$$

The final price is $Average(OfferedPrice * Factor, RequestedPrice)$.

Reputation Update and Calculation. The allocation algorithms call the reputation update algorithm, and the reputation calculation is needed to sort the bids. The reputation updating algorithm removes the oldest value of the queue, to ensure it always has the 20 most recent scores. Then, it adds the new score. For convenience, it is possible to add multiple scores in one algorithm call, using the parameter Number.

Algorithm 3 describes how to calculate the final reputation value by multiplying each partial value.

4   Implementation

FairCloud has 3 layers: CloudSim, CloudAuctions (extension) and Bid Representation Language.

Algorithm 2. Reputation Updating. The parameter Number allows the attribution of multiple scores in one call.

Input ReputationMap, Provider, Value, Number
Output ReputationMap
1: **for** i = 0 **to** Number **do**
2:    $ReputationMap.Get(Provider).RemoveFirst()$
3:    $ReputationMap.Get(Provider).Add(Value)$
4: **end for**

Algorithm 3. Calculate Reputation.

Input ReputationMap, Provider
Output Value
1: $Value \leftarrow 100$
2: **for** Value v **in** ReputationMap.Get(Provider) **do**
3:    $Value \leftarrow Value * v$
4: **end for**

CloudSim[3] is a simulator that has been widely used in works related with energy efficiency, work-flows, scalability and pricing policies [18]. This layer offers the API for VM management and for running the User Cloudlets (i.e. user code). It also provides metrics (e.g. utilization, energy consumption, profit, latency). There are many extensions to CloudSim developed by third parties, and it can be made parallelized and distributed [8].

We used the extension CloudAuctions, proposed and used on [11]. This extension was updated, fine-tuned and new functionalities were implemented. Finally, a Bid Representation Language was created, in order to support the user and provider bid API described in Sect. 3.2. Figure 3 summarizes these layers.

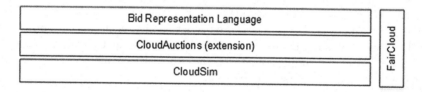

Fig. 3. FairCloud layers.

The FairCloud classes can be divided into Bid related and SimEntities. The class diagrams in Figs. 4 and 5 represent the two groups, respectively.

The Parser is responsible to create and configure the bids according to the API. A BidCalculation contains a Bid and additional information, such as the bid density and the assigned amount. The Auctioneer operates with the BidCalculations.

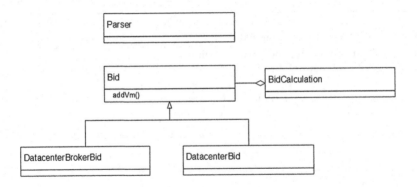

Fig. 4. Class diagram with the core Bid related classes.

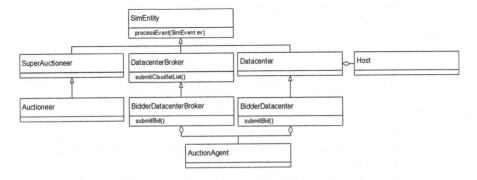

Fig. 5. Class diagram with the core SimEntities related classes.

The Parser also creates the BidderDatacenterBrokers (and submits the Cloudlets) and the BidderDatacenters (and the Hosts). These two entities delegate the Bidding communication to their AuctionAgent.

The SuperAuctioneer contains shared functionality, such as the monitoring algorithm and constants definition. All SimEntities implement the processEvent method. After the simulation starts, the processEvent method receives and processes the events described in Sect. 3.3.

5   Evaluation

To evaluate FairCloud, we used a sub-set of the Google Cluster Data (2011 2).[4] This data provides information regarding Machines, Jobs and Tasks, and Resource Usage. It represents 29 day's worth of cell information from May 2011. Table 2 shows the tables we used and their schema.

The first transformation was to select the entries with timestamp between 0 and 1 hour. The provider data was obtained from the table **Machine events** and

---

[4] https://github.com/google/cluster-data.

Table 2. Google cluster data tables used and their schema.

	Machine events	Task events
1	Timestamp	Timestamp
2	Machine ID	Job ID
3	Event type	Event type
4	Capacity: CPU	Scheduling class
5	Capacity: Memory	Priority
6		Requested CPU
7		Requested RAM

we selected the `Add` events, that represent the machines added to the cluster. The user data was gathered from the `Task events` table, focusing on the entries in the `Submit` transition. The bids were obtained by grouping the tasks by `Job ID`. Each job represents a Userbid and its tasks the VMs requested in that UserBid.

After the filtering and grouping, each Dataset contains 9 ProviderBids and 10.452 UserBids (requesting 178.692 VMs).

Datasets. We created 6 Datasets with different strategies and variables. Table 3 summarizes their configurations.

Table 3. Datasets configuration summary.

	Datacenters	Degradation	Price	Conversions	Reputation
1	Homogeneous	`Restricted`	Equal	No	0
2	Round robin	`Restricted`	Equal	No	0
3	Round robin	Variable	Equal	No	0
4	Round robin	Variable	Random	No	0
4b	Round robin	Variable	Random	No	Disabled
5	Round robin	Variable	Random	Yes	Variable

Datacenters. All datasets have 9 datacenters but their capacity is different. Dataset 1 has an homogeneous configuration while the remaining datasets have a different capacity. In the last configuration, we used a round robin strategy (e.g. entries 1,10,19,28... belong to Datacenter 1).

Degradation Profile. On datasets 1 and 2, the user and provider degradation profile is always `Restricted`. On the remaining datasets, the datacenter profile is the following: [1–3] - `Relaxed`; [4–7] - `Restricted`; [8–9] - `Demanding`. The user profile is based in the priority field (integer from 0 to 11). Each job contains a priority which was normalized according to [10] with the following sets: production, middle and gratis. The degradation profile assignment is the following: production priority - `Demanding`; middle priority - `Restricted`; gratis priority - `Relaxed`.

Price. We define the base price as the Amazon EC2 Spot Instances price, observed on 25/06/2017, for the Predefined 1 h duration. Depending on the dataset, the price is multiplied by a factor. The factor allows us to simulate participants that are willing to pay more than others.

On datasets 1, 2 and 3, the provider factor is 1 and the user factor is 1.5. This represents that the users pay 50% more than the providers' request.

On the remaining datasets, the provider's factor is obtained from a Random Normal Distribution (Avg = 1; STD = 0.1) and the user's from a Random Normal Distribution (Avg = 1.5; STD = 0.3).

Conversions. Only on dataset 5 conversions can be allowed. The UserBids resulting from jobs with a priority between 0 and 9 allow conversions.

Reputation. On the datasets 1 to 4 the minimum reputation required is 0. On dataset 5, users require a minimum of reputation. We used the field *scheduling class* that represents how latency-sensitive the job is. We did the following scheduling class - minimum reputation assignment: [0, 1] - 35; [2] - 65; [3] - 100.

Dataset 4b configuration is similar to 4, however the reputation mechanism is disabled.

Metrics. The evaluation metrics are divided in two categories. Global Quality assesses the auction algorithm and Efficiency is focused on the system.

Global Quality

- **Average price** per VM type determines if the auction approach is beneficial compared to the traditional/static approach;
- **Allocation success rate** is the rate of bids that are successfully allocated;
- The **execution time** is the period from the bid being visible to the system and the allocation return.

Efficiency

- **CPU Utilization** allows us to assess if the auction is distributing the allocations. A low value on this metric will lead to a low Resource Utilization, and consequently, to low **Energy Efficiency** as more machines will be on idle state [12,13]. The CPU utilization is measured every 5 min;
- **Strict Allocation Quality** Represents the ratio of requests that follow the user required degradation profile;
- **Relaxed Allocation Quality** Represents the ratio of requests that do not follow the user required degradation profile;
- **Providers rating order** This metric allows us to measure the benefits of the reputation system. The order should change dynamically in each round when the provider quality changes.

Benchmarks. We compare our system with one relevant research work and with one widely used commercial system.

CloudAuctions. CloudAuctions is an auction system that does not consider the QoS. To provide a fair comparison, the system was improved to repeat the algorithm and the available amount is calculated dynamically (based on capacity). CloudAuctions is described in [17] and the source code is available.

AWS EC2 On-demand. We implemented a system similar to Amazon EC2 On-demand. The User requests are sorted by their timestamp and the Provider offers are sorted by degradation profile (better quality first). The matching is possible if the User price is higher than the Provider price. This system is not an auction: the final price is always the Provider price. The prices are available online[5].

Results - Allocation vs Quality. With this test we aim to assess the allocation rates and their quality when the conditions change. Figure 6 compares, for each dataset, the different types of allocations (Strict, Relaxed or Not Allocated) and the 3 algorithms. In the first graphic (Datasets 1 and 2 - D1/D2), all allocations were successful because the requested and offered quality were the same, and user price was higher than the provider price.

On Dataset 3, only FairCloud offers total Strict quality. The other two algorithms do not consider the quality, therefore nearly 40% of the allocations are Relaxed. The trade-off for the FairCloud Strict quality is execution time. Regarding dataset 3, FairCloud average execution time is 4745 min. AWS and CloudAuctions process the same requests in 3958 min (17% less).

Dataset 4 simulates participants with heterogeneous prices. In this scenario, the Not Allocated ratio increased, representing the user bids with the price too low. Particularly, the Not Allocated % in AWS is due to their bid matching criteria: the User price must be higher than the Provider predefined price, that is extremely high.

On the last graphic, Dataset 5, we see that FairCloud allocated 2% less UserBids than in Dataset 4. This represents the UserBids that could not be allocated due to the minimum required reputation by them. Naturally, we do not observe any difference between AWS and CloudAuctions on the last two graphics. The minimum required Reputation is not considered by them.

Results - CPU Utilization vs Execution Time. The previous section provided us the total allocations. Now, we will detail that information by analysing the cumulative CPU utilization. In Fig. 7, the first graphic represents the full allocation - all requests are accepted.

On Dataset 3, FairCloud provides a lower utilization but, comparing with the other two algorithms, the overall utilization is the same. This behaviour happens because FairCloud delays requests to ensure quality.

---

[5] https://aws.amazon.com/ec2/pricing/on-demand/.

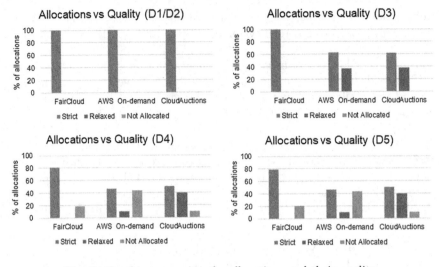

Fig. 6. Graphic comparing the allocations and their quality.

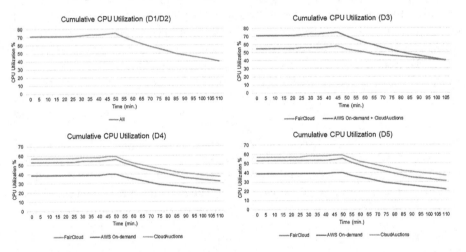

Fig. 7. Graphic showing the Cumulative CPU Utilization. The information was gathered every 5 min.

On the last two graphics, CloudAuctions provides 5% more allocations by ignoring the Users requested quality.

Results - Reputation Order. Our algorithm implements a reputation system. With this test, we can see the benefits of using a hybrid ordering approach (price and reputation). Figure 8 shows the providers' order (ranking) over time. In the first round (Time = 0), the providers were ordered by price only (i.e. all had the maximum reputation). In the following round, the order changed: the providers with lower quality are now in the last places. As the time passes, we still see some minor adjustments.

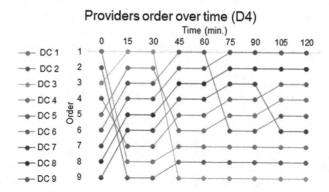

Fig. 8. Graphic showing the Providers order in each algorithm execution. As the horizontal axis suggests the algorithm is executed every 15 min.

We demonstrated that our reputation mechanism is able to detect Providers with unsatisfactory quality and benefit those with slightly higher price but better QoS.

6 Conclusion

In this work, we have presented FairCloud, a novel approach to Cloud-Auctions. We started by describing the problems in Cloud Computing, Energy Efficiency and Cloud Pricing that serve as motivation. Next, we presented the proposed objectives. After analysing the current literature, we formulated a taxonomy and presented the main research and commercial systems.

FairCloud key insights to improve allocation include allowing the adaptation of VM requests (through conversion of VM types, resource capping) based on User profile, and maintaining an internal reputation system, in order to detect providers with low QoS. FairCloud is built on top of CloudSim and the CloudAuctions extensions.

In testing with Google Cluster Data, FairCloud achieves more QoS in the requests while maintaining CPU Utilization. The reputation mechanism is effective in avoiding Providers with lower quality.

In future work, we intend to address supporting multiple geo-distributed auctioneers and a history of prices. Lastly, we could implement the AWS Spot Instances (auction algorithm) and compare the providers' revenue.

Acknowledgements. This work was supported by national funds through Fundação para a Ciência e a Tecnologia with reference PTDC/EEI-SCR/6945/2014, and by the

ERDF through COMPETE 2020 Programme, within project POCI-01-0145-FEDER-016883. This work was partially supported by Instituto Superior de Engenharia de Lisboa and Instituto Politécnico de Lisboa. This work was supported by national funds through Fundação para a Ciência e a Tecnologia (FCT) with reference UID/CEC/50021/2013.

References

1. Agmon Ben-Yehuda, O., Posener, E., Ben-Yehuda, M., Schuster, A., Mu'alem, A.: Ginseng: market-driven memory allocation. ACM SIGPLAN Not. **49**(7), 41–52 (2014)
2. Al-Roomi, M., Al-Ebrahim, S., Buqrais, S., Ahmad, I.: Cloud computing pricing models: a survey. Int. J. Grid Distrib. Comput. **6**(5), 93–106 (2013)
3. Barroso, L.A., Hölzle, U.: The case for energy-proportional computing. Computer **40**(12), 33–37 (2007)
4. Beloglazov, A., Buyya, R., Lee, Y.C., Zomaya, A.: A taxonomy and survey of energy-efficient data centers and cloud computing systems. Adv. Comput. **82**, 47–111 (2010)
5. Chen, Z., Chen, L., Huang, L., Zhong, H.: On privacy-preserving cloud auction. In: 2016 IEEE 35th Symposium on Reliable Distributed Systems (SRDS), pp. 279–288. IEEE, September 2016
6. Fujiwara, I., Aida, K., Ono, I.: Applying double-sided combinational auctions to resource allocation in cloud computing. In: Proceedings - 2010 10th Annual International Symposium on Applications and the Internet, SAINT 2010, pp. 7–14 (2010)
7. Kaplan, J., Forrest, W., Kindler, N.: Revolutionizing data center energy efficiency. McKinsey & Company, Technical report, July 2008
8. Kathiravelu, P., Veiga, L.: Concurrent and distributed CloudSim simulations. In: IEEE 22nd International Symposium on Modelling, Analysis & Simulation of Computer and Telecommunication Systems, MASCOTS 2014, Paris, France, 9–11 September 2014, pp. 490–493. IEEE Computer Society (2014). https://doi.org/10.1109/MASCOTS.2014.70
9. Mastelic, T., Oleksiak, A., Claussen, H., Brandic, I., Pierson, J.M., Vasilakos, A.V.: Cloud computing. ACM Comput. Surv. **47**(2), 1–36 (2014)
10. Reiss, C., Tumanov, A., Ganger, G.R., Katz, R.H., Kozuch, M.: Heterogeneity and dynamicity of clouds at scale: Google trace analysis. In: Proceedings of the Third ACM Symposium on Cloud Computing, SoCC 2012, pp. 1–13 (2012)
11. Samimi, P., Teimouri, Y., Mukhtar, M.: A combinatorial double auction resource allocation model in cloud computing. Inf. Sci. **357**, 201–216 (2016)
12. Sharifi, L., Cerdà-Alabern, L., Freitag, F., Veiga, L.: Energy efficient cloud service provisioning: keeping data center granularity in perspective. J. Grid Comput. **14**, 299–325 (2016)
13. Sharifi, L., Rameshan, N., Freitag, F., Veiga, L.: Energy efficiency dilemma: P2P-cloud vs. datacenter. In: 2014 IEEE 6th International Conference on Cloud Computing Technology and Science, pp. 611–619. IEEE, December 2014
14. Simão, J., Veiga, L.: Flexible SLAs in the cloud with a partial utility-driven scheduling architecture. In: IEEE 5th International Conference on Cloud Computing Technology and Science, CloudCom 2013, Bristol, United Kingdom, 2–5 December 2013, vol. 1, pp. 274–281. IEEE Computer Society (2013). https://doi.org/10.1109/CloudCom.2013.43

15. Simão, J., Veiga, L.: Partial utility-driven scheduling for flexible SLA and pricing arbitration in clouds. IEEE Trans. Cloud Comput. **4**(4), 467–480 (2016). https://doi.org/10.1109/TCC.2014.2372753
16. Son, S., Sim, K.M.: A price- and-time-slot-negotiation mechanism for cloud service reservations. IEEE Trans. Syst. Man Cybern. B Cybern. **42**(3), 713–728 (2012)
17. Wang, H., Tianfield, H., Mair, Q.: Auction based resource allocation in cloud computing. Multiagent Grid Syst. **10**(1), 51–66 (2014)
18. Wang, Y.H., Wu, I.C.: CloudSim: a toolkit formodeling and simulation of cloud computing environments and evaluation of resource provisioning algorithms. Softw. Pract. Exp. **39**(7), 701–736 (2009)
19. Zhang, H., Jiang, H., Li, B., Liu, F., Vasilakos, A.V., Liu, J.: A framework for truthful online auctions in cloud computing with heterogeneous user demands. IEEE Trans. Comput. **65**(3), 805–818 (2016)
20. Zhao, Y., Huang, Z., Liu, W., Peng, J., Zhang, Q.: A combinatorial double auction based resource allocation mechanism with multiple rounds for geo-distributed data centers. In: 2016 IEEE International Conference on Communications, ICC 2016 (2016)

A Novel WebGIS-Based Situational Awareness Platform for Trustworthy Big Data Integration and Analytics in Mobility Context

Susanna Bonura[✉], Giuseppe Cammarata, Rosolino Finazzo,
Giuseppe Francaviglia, and Vito Morreale

R&D Laboratory - Engineering Ingegneria Informatica S.p.A., Viale Regione
Siciliana, 7275 Palermo, Italy
{susanna.bonura,giuseppe.cammarata,rino.finazzo,giuseppe.francaviglia,
vito.morreale}@eng.it
http://www.eng.it

Abstract. The availability of big amounts of dynamic data from several sources in mobility context and their real time integration can deliver a picture for emergency management in urban and extra-urban areas. A WebGIS portal is able to support the perception of all elements in current situation. However, in general, during the observation decision maker's attention capacity is not sufficient to address concerns due to information overload. A situational picture is necessary to go beyond the simple perception of the elements in the environment, supporting the overall comprehension of the current situation and providing predictions and decision support. In this paper we present *MAGNIFIER*, a WebGIS-based intelligent system for emergency management, to entirely support the real-time situational awareness. Starting from the current situation and by using the practical reasoning model by Bratman, *MAGNIFIER* is able to suggest the appropriate course of actions to be executed to meet decision maker's goals.

Keywords: Big data real-time analytics · Big data integration · Decision support system · Emergency management · Intelligent system · Practical reasoning · Situational awareness · WebGIS

1 Introduction

The huge amount of dynamic data from several sources as well as geospatial and temporal information and their integration in real time can serve to deliver a picture of main events to be monitored for emergency management in urban and extra-urban areas. A crucial concept in the field of emergency management is that of *Situational Awareness* (SA) which, in general, with regards to being aware of what is happening around one in terms of where one is, where one is supposed to be, and whether anyone or anything around is a threat to one's health and safety.

© Springer International Publishing AG 2017
H. Panetto et al. (Eds.): OTM 2017 Conferences, Part II, LNCS 10574, pp. 86–98, 2017.
https://doi.org/10.1007/978-3-319-69459-7_6

Several models exist in literature on SA. Dominguez defines individual SA as the continuous extraction of environmental information to directing and anticipating future events [1].

Bedny and Meister consider a continuous loop on which SA directs the interaction with the world and such an interaction modifies SA. This interaction is motivated by the disparity between the decision maker's goals and the current perceived situation [2].

Smith and Hancock proposed a model stating that SA is neither resident on individuals nor in the world but rather on the interactions that are motivated by decision maker's schemata; the outcome of that interaction will modify existing schemata, which in turn directs further exploration [3].

But the model that has received most attention is the Ensdley's three-levels model: in the first level, training and experience directs attention to critical elements in the environment; the second level integrates elements that aid understanding the meaning of critical elements; and the last level considers understanding the possible future scenarios [4].

Today, the most advanced WebGIS tools (Geospatial Information Systems that use web technologies to communicate between a server and a client), are able to support the situational awareness in what regards the perception of all elements in current situation (according to the first level of the Ensdley's model). Thus, WebGIS can help decision makers in data integration and visualization, its localization to real-time analysis, and mapping of potential disasters to show vulnerable areas, critical situations and potential harm.

However, in general, in dynamic environments, decision maker's attention capacity is not sufficient to address attention demands resulting from information overload. Therefore, a situational picture is necessary which is able to go beyond the simple perception of the elements in the environment, supporting the overall comprehension of the current situation and the user's decision making process (by providing alerts, predictions and recommendations).

Currently, the existing open source WebGIS platforms just refer to real-time geospatial analysis and visualization about a certain phenomena (e.g. Ushahidi[1] and Sahana Eden[2] for disaster events). Other advanced commercial solutions (e.g. archGIS Platform by Ensri[3]) are able to integrate information from different sources and to manage real-time data analysis, but they do not offer decision support with regards to the delivered situational picture. Furthermore, some GIS-based decision support systems exploit multicriteria decision analysis (MCDA) techniques to evaluate possible alternatives, but they are lacking a proper scientific foundation and some methods involve stringent assumptions which are difficult to substantiate in real-world situations [5,6].

In this paper we present *MAGNIFIER*, an open source WebGIS-based intelligent system for emergency management, designed and implemented to entirely support the real-time situational awareness, which starts from an appropriate

---

[1] https://www.ushahidi.com.

[2] http://eden.sahanafoundation.org/.

[3] http://www.esri.com/software/arcgis.

perception of all elements in current situation, to come to an understanding of the meaning of those elements in an integrated form, and to the ability to project future states of the environment that are valuable for decision making.

In addition, *MAGNIFIER* provides decision maker with a decision support: it is able to suggest the appropriate course of actions to be executed to meet decision maker's goals, by reasoning on the current situation according to the practical reasoning model, namely the reasoning process directed towards actions [7]. Through such a model, *MAGNIFIER* is endowed with the ability to reason on the states of the environment where it is situated, its goals and its plans, so to meet its objectives while reasoning about failures (e.g. when some unwanted states occur), in order to learn new desired behaviors.

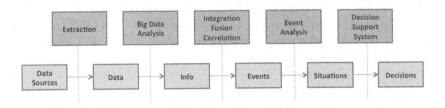

Fig. 1. Intelligent information processing chain

MAGNIFIER is compliant with our data processing model we named *Intelligent Information Processing Chain* (Fig. 1). According to such a model, data collected from different sources (sensor networks, news, rss feeds, comments, tweets etc.) are processed to extract useful information (by means of advanced analytics). Some pieces of such information, properly integrated, fused and correlated, will result in events. Events are evaluated in order to create the situational picture that is at the basis of decision maker activities (possibly supported by automatic reasoning services).

In order to better explain how *MAGNIFIER* works, the following scenario is provided (Fig. 2): in Palermo city two urban areas are highly dangerous due to flooding. Such areas include critical infrastructures (schools, roads, etc.) with different vulnerability levels which need to be continuously monitored in case of adverse weather conditions. Usually, WebGIS platforms allow decision makers to choose a certain number of layers which, overlapped, form a situational picture. This picture helps decision maker to *perceive* the current situation, in terms of dangerousness, infrastructure vulnerabilities and potential risk.

MAGNIFIER proactively builds and provides the layer to help decision maker to better *understand* the situation picture by showing information regarding the danger and the vulnerability of infrastructures and suggesting the actions to be performed in order to restore situations in their normality.

Such a picture is called *MAGNIFIER Situational Picture* and shows a recommendation, characterized by a level of danger (high, medium, low), vulnerable infrastructures, type of risk, the action to be taken to prevent the risk and finally by the risk status (warning, alert, emergency).

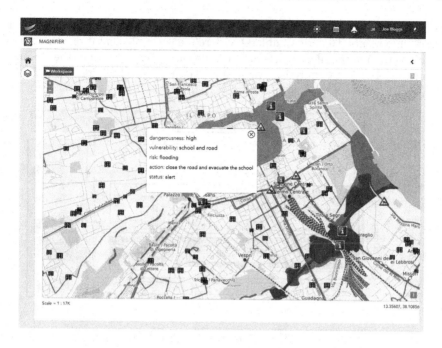

Fig. 2. An example of *MAGNIFIER* situational picture

This paper is organized as follows: in Sect. 2 we give a brief overview of the MAGNIFIER WebGIS-based Situational Awareness Platform and its functional components; then in Sects. 3 and 4 we provide a description of the main technological choices focusing on the PRACTIONIST DSS component; finally in Sect. 5 we describe the DSS execution logic through the scenario above mentioned.

2   The MAGNIFIER WebGIS-Based Situational Awareness Platform

MAGNIFIER has been designed and implemented with the objective to store and manage data from several sources, relating to urban and extra-urban areas of interest and, therefore, provide support in the real-time investigation of situations of potential danger and in the decision making process.

For this aim, the TOREADOR environment[4] was adopted to support a trustworthy big data integration and analytics.

More in detail, such environment (Fig. 3) comprehends an integration layer, where some components from the *FIWARE* Catalogue were exploited[5].

---

[4] The TOREADOR environment is part of a bigger platform under development through the TOREADOR initiative (http://www.toreador-project.eu/).

[5] *FIWARE* is a PPP European initiative (supported by 25 ICT players; Engineering Ingegneria Informatica S.p.A. is one of them) aiming to build a software platform

The *FIWARE* Catalogue contains a rich library of public, royalty-free and open source components (Generic Enablers) with reference implementations that allow developers to put into effect functionalities such as the connection to the Internet of Things or Big Data analysis, making programming much easier.

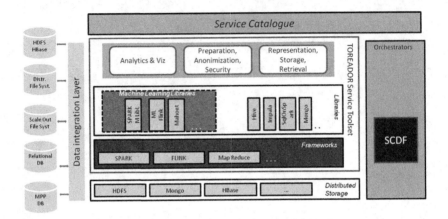

Fig. 3. The TOREADOR environment

In particular, in order to support the integration of data coming from several sources, we have properly installed in the integration layer, the *ORION* Context Broker, which is an implementation of the Context Broker GE, providing the NGSI9 and NGSI10 interfaces. Using these interfaces, clients can do several operations: to register context producer applications, to update context information, being notified when changes on context information take place or with a given frequency, to query context information.

Together with *ORION*, we used *Cygnus*, which implements a connector for context data coming from the *ORION* Context Broker and aimed to be stored in a specific persistent storage (in our case it is Mongo DB). Furthermore, we integrated *PROTON*, the CEP GE which analyzes event data in real-time, generates immediate insight and enables instant response to changing conditions. It provides means to expressively and flexibly define and maintain the event processing logic of the application, and at runtime it is designed to meet all the functional and non-functional requirements without taking a toll on the application performance, so reducing application developers and system managers concerns.

Besides, the TOREADOR environment is able to make available a huge set of services for batch and stream data analytics, as well as for anonymization,

---

dedicated to the creation of Future Internet applications, by leveraging on advanced technologies such as Cloud Computing, Internet of Things, Engineering Services, Data & Content Management, Advanced User Interfaces and Future Networks security (https://www.fiware.org/).

preparation and security and for representation, storage and retrieval. Such applications, using the most cutting-edge Big Data libraries and frameworks (e.g. Spark MLlib, ML Flink, Mahout as libraries, and Spark, Flink, MapReduce as frameworks), are integrated into the environment for the deployment of applications in several domains and scenarios.

The Service Catalogue presents the services available in the TOREADOR environment with REST API.

Finally, the TOREADOR environment is endowed with Spring Cloud Data Flow (SCDF)[6], a cloud-native programming and operating model for composable data microservices, aiming at creating and orchestrating data pipelines for common use cases such as data ingestion, real-time analytics, and data import/export.

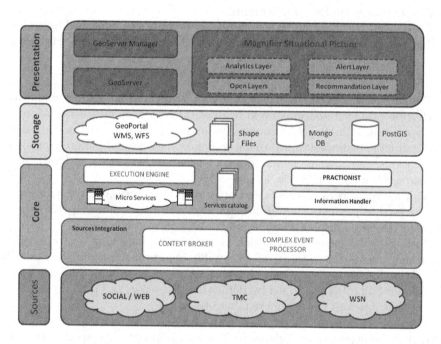

Fig. 4. Conceptual multi-layer architecture of *MAGNIFIER*

Several data sources are managed by the *MAGNIFIER* (Fig. 4). First of all, it is noteworthy that a Wireless Sensor Network prototype was designed and implemented so as to be installed in vehicles. The WSN consists of four sensor nodes, a gateway node and a controller for the network management. The gateway is responsible for forwarding the measures by each individual sensor to the controller. The sensor nodes are respectively composed by the following sensors: (i) noise, light, rain; (ii) speed and vibration; (iii) air quality, pressure, temperature, dust, humidity, carbon monoxide; (iv) infra-red light, soil and air infra-red

---

[6] https://cloud.spring.io/spring-cloud-dataflow/.

temperature, ultra sonic distance. The controller is in charge of generating other virtual sensors starting from the measurements of real sensors, through sensor fusion techniques. Furthermore, it is responsible for decoding messages from the gateway and for the WSN management.

Through a Traffic Message Channel (TMC) receiver[7], real-time information which are transmitted on FM frequencies about traffic events and state of the roads/infrastructure, are received, properly decoded (according to the TMC protocol) and standardized (according to NGSI9 and NGSI10 standards).

Other kinds of data managed by *MAGNIFIER* are:

- news, RSS feeds, comments from web sources. This data is extracted from public web sources (i.e. institutions and organizations which are in charge for the control of territory) that provide real-time data on the state of roads and infrastructures, and on traffic events.
- Tweets from the TWITTER source. We access to tweets published by profiles that release information about the status of road infrastructure; the access to tweets is via public APIs provided by the TWITTER platform.
- Traffic open dataset, provided by different public bodies.

The core of *MAGNIFIER* is represented by several technological components which were extended and integrated to meet the system requirements. More in detail, drawing from the data sources, such layer is in charge of making data more and more elaborated until to get to define a situational picture composed by situations, alerts, recommendations (according to the *Intelligent Information Processing Chain* shown in Fig. 1).

Information, generated events and situational awareness acquired by the system are used to populate the various stores: postgis tables and shape files, used by geoserver to publish thematic layers.

PostGIS is the main free relational database with a geographical extension, which implements the support for geographic objects in PostgreSQL. PostGIS follows the directives of the Open Geospatial Consortium Simple Features Specification for SQL; it is developed by Refractions Research, and is an open source project that develops the spatial database technology.

The shapefile format is a popular geospatial vector data format for GIS software. It is developed and regulated by Esri as a (mostly) open specification for data interoperability among Esri and other GIS software products. The shapefile format can spatially describe vector features: points, lines, and polygons, representing, for example, water wells, rivers, and lakes. Each item usually has attributes that describe it, such as name or temperature.

---

[7] Traffic Message Channel is a technology for delivering traffic and travel information to motor vehicle drivers. It is digitally coded using the ALERT C protocol into RDS Type 8A groups carried via conventional FM radio broadcasts. It can also be transmitted on Digital Audio Broadcasting or satellite radio. TMC allows silent delivery of dynamic information suitable for reproduction or display in the user's language without interrupting audio broadcast services. Both public and commercial services are operational in many countries. When data is integrated directly into a navigation system, traffic information can be used in the system's route calculation.

The store MongoDB is used primarily to store all data acquired from the sources and any information or derived events, for further processing by the core components to make predictions and learning.

The layer on top of the Fig. 4 includes all the elements which characterize the user interface, that is the situational picture composed by situations, alerts, recommendations as interpreted by *MAGNIFIER*.

In the following section, the components of each layer will be detailed and the relation among them explained.

3   The MAGNIFIER Technological Architecture

MAGNIFIER is the result of the integration (and in certain cases of the extension) of a number of innovative open source technological solutions (Fig. 5).

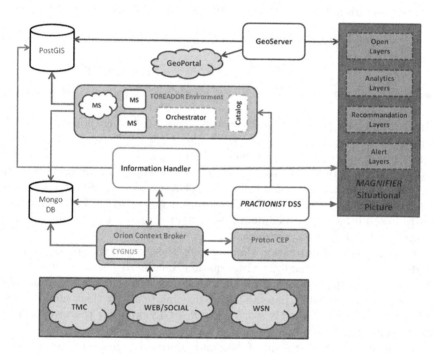

Fig. 5. Technological architecture of *MAGNIFIER*

The data acquired from the sources previously explained, is referenced with respect to time and space, normalized according to the NGSI9 and NGSI10 standards, and sent to *ORION*.

Besides these sources, other producers for *ORION* are:

– *PROTON*, which sends to it the complex events generated;
– the Information Handler, which send to it information properly elaborated.

The components subscribed to *ORION* as consumers, are:

- the *Cygnus* connector, which stores all elements passing through *ORION* (raw data, information, simple events, complex events) on MongoDB;
- *PROTON*, which is subscribed for certain events to generate complex events;
- the Information Handler, which is in charge of the transformation of information coming from sources and *PROTON* in more complex information by invoking the advanced analytics engine (e.g. R engine (https://www.r-project.org/)).
- *PRACTIONIST*, which manages high level events (coming from *PROTON* and the Information Handler), to generate alerts and recommendations so to recognize and analyze situations needed to be monitored (please refer to the Sect. 4 for a brief understanding of PRACTIONIST).

The following pseudo-code shows three examples of rules to generate complex events starting from events notified by *ORION* to the *PROTON* CEP:

$$fog \Leftarrow dust\_poll \geq 0.30 \, mg/m^3 \, and \, air\_qual \equiv fresh. \tag{1}$$

$$snow \Leftarrow soil\_temp \leq 4 \, and \, son\_dist \leq max\_son\_dist. \tag{2}$$

$$flooding \Leftarrow soil\_temp \geq 4 \, and \, son\_dist \leq max\_son\_dist \, and \, son\_dist \leq ir\_dist. \tag{3}$$

In the first case, the *fog* complex event is generated when pollution level is greater than $0.30 \, mg/m^3$ and air is classified as *fresh*. In the second case, if the temperature of soil is less or equal than 4 °C, and the sonar distance from soil is less than maximum pre-configured sonar distance, then *PROTON* generates the *snow* complex event. In the last case, *PROTON* creates the *flooding* complex event if it comes to know the soil temperature is greater than 4 °C, sonar distance from soil is less than maximum pre-configured sonar distance and infrared distance greater than sonar distance.

As an example of Information Handler task, the tilt and the speed measures provided by sensors are used by the Information Handler to generate a *hole* event. The hole event includes information about the status of the hole (Unavoidable, easly and hardly avoidable, potentially missing) and entities (low, medium and hight shock). The *hole* event is stored in a PostGIS table for the visualization by layer, other than stored on MongoDB for the at-rest analysis.

Periodically the huge amount of data and information generated at any level are processed and made available on MongoDB and PostGIS, to prepare data for the visualization and decision maker's query by and *MAGNIFIER* DSS, in order to become more aware of the situation.

The information stored in PostGIS is to be used by GeoServer so to be displayed in the form of thematic layers. The set of layers generated at run time, together with other thematic layers gathered from geo-portals as geo-services, will allow decision maker to have an understanding of the situation. *PRACTIONIST* DSS through the management and aggregation of high-level events from *ORION*, is able to define the danger and vulnerability and possibly identify risk situations for people and/or things present within a certain area. Furthermore, *PRACTIONIST* DSS, through the huge amount of data processed can run simulations to predict the evolution of the situation.

Situational awareness gained by the *MAGNIFIER* system is made available by means of layers, alerts and recommandations to end users, who can benefit from this support to manage risks and emergencies.

4   PRACTIONIST DSS

We exploited the *PRACTIONIST* framework [8–12] to design and implement a goal-oriented decision support system, based on practical reasoning model [13–15] and able to reinforce the decision maker's awareness in regards with the current situational picture; the system reasons about the status of the environment, its internal status, and the high-level goals defined by domain experts (the decision maker or everybody who is able, starting from the analysis of the environment and its dynamics, to highlight system needs in terms of desired states and goals) in order to point out desired states and to achieve and maintain them, and to support the decision maker by suggesting him alerts and recommendations.

The *PRACTIONIST* framework aims at supporting the programmer in developing agents endowed with the following elements: *(i)* a set of perceptors able to listen to some relevant perceptions; *(ii)* a set of beliefs, which represents the information the agent has got about both its internal state and the external world; *(iii)* a set of goals, which are some objectives related to some states of affairs to bring about or actions to perform; *(iv)* a set of plans that are the means to achieve its intentions; *(v)* a set of actions the agent can perform to act over its environment; *(vi)* and a set of effectors that actually support the agent in performing its actions.

In short, we modeled situation awareness levels according to the Endsley's model [4] by exploiting some specific components of our framework:

– Perception of the elements in the environment: in *PRACTIONIST*, *perceptors* listen to some relevant external stimuli, while *beliefs* represent the information about these stimuli; we de-fined a customizable perception logic which is able to adapt to the decision maker's needs and priorities, in order to focus the perceptors' attention on specific elements of the environment, which are consequently represented by beliefs and stored in a knowledge base;
– Comprehension of the current situation: in *PRACTIONIST*, the belief logic is built upon a prolog-like language able to infer and deduct new beliefs by means of *formulas*; that means the system is able to comprehend patterns of elements and to integrate them, creating new information and beliefs;
– Projection of future status: in *PRACTIONIST* we use some *formulas* and *belief revision rules* to identify situations that could become very dangerous in the future, in order to investigate about their projection by means of some prediction algorithms developed in R; moreover, feedbacks generated by these algorithms will be managed by *plans* designed to *(i)* understand the feedbacks,

(ii) review the perception and attention logic, *(iii)* notify the decision maker regarding dangerous situations, *(iv)* suggest decisions to the decision maker.

We chose to represent ideal states by means of policies that the system has to apply and maintain over time; in case a policy is no longer satisfied, probably a dangerous situation is occurring, so the system will suggest decisions to the decision maker, and will require the prediction component to inquire into the projection of the situation.

Furthermore, we took advantage of the goal-orientation programming to specify ideal states of the system by means of state of affairs to be either achieved and maintained, or ceased and avoided. In *PRACTIONIST* we exploit the *Maintain* and the *Avoid* goals to define policies: in details, the first type of goal is used to define a situation to be maintained, while the last is used to model a situation to be avoided. These goals are always active and in execution, so every time they are no longer successful, some plans could be executed to restore their successful conditions. In the following figure, a general view of the elements involved to design the decision support system is shown (Fig. 6):

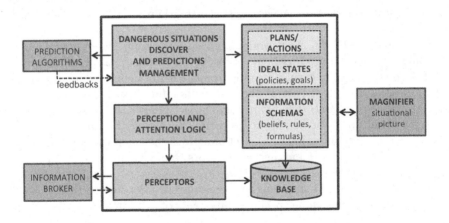

Fig. 6. MAGNIFIER DSS designed by using PRACTIONIST

Blocks with green background represent entities that are external to the decision support system boundaries, but which are included in the MAGNIFIER solution: that are the information broker (it represents the central node in charge of information routing), the prediction algorithms, and the MAGNIFIER situational picture, which includes GUIs used to show both alerts and suggested decisions.

In the case of the scenario described above, the PRACTIONIST DSS associates different vulnerability levels to infrastructures depending on several factors: their proximity to highly dangerous areas, their position, who attends the infrastructure (children, adult etc.). Furthermore, the DSS monitors communication routes, whose vulnerability level due to flooding, also depends on low quality of the itself infrastructure and on the presence of potholes.

Because of the presence of dangerous elements and vulnerable infrastructure, the DSS associates to the area and in particular to infrastructure monitored a level of risk that depends on several factors. Depending on the risk level and the presence of other negative factors which may occur (accidents, narrowing the carriageway and/or work in progress), the DSS begins assessing measures to be taken in case of danger. The monitoring of an area already begins with the presence of negative elements.

In this case the system by exploiting predictive analysis techniques, evaluates the conditions of the area in case of adverse weather conditions with different degrees; according to the prediction results, it starts showing alerts and recommendations to restore the initial conditions of the area. In case of imminent weather alert, the system begins showing an alert.

The result of the DSS reasoning process results in recommendations, which are located near critical infrastructure. As already mentioned, any recommendation is characterized by a level of danger (high, medium, low), vulnerable infrastructures, type of risk, the action to be taken to prevent the risk and finally by the risk status (warning, alert, emergency).

5   Conclusion and Future Works

In this paper we presented *MAGNIFIER*, a novel WebGIS-based platform for emergency management. Through practical reasoning mechanisms, it is able to provide decision maker with a complete situational picture, enriched with predictions, alerts, recommendations. In the mobility context, *MAGNIFIER* aims at being an innovative Decision Support System. It results from the integration of several innovative open source technologies enabling the *Intelligent Information Processing Chain* model which is showed in this paper.

In particular, *PRACTIONIST* was adopted to implement the intelligent goal-oriented decision support module, based on practical reasoning model, in order to reinforce the decision maker's awareness with regard to the current situational picture while suggesting alerts and recommendations to prevent or mitigate risks. TOREADOR environment provides support in data management for both *at-rest* and *in-motion* analysis. Other advanced technologies in Cloud Computing and Data & Content Management contexts which were adopted, are from the *FIWARE* Catalogue. *FIWARE* Catalogue contains a rich library of public, royalty-free and open source components (Generic Enablers) with reference implementations that facilitate developers' service innovation activities.

However, some further work should be done with respect to several *MAGNIFIER* open issues. Among them, our intention is to improve the DSS learning capability: when an unexpected alert arrives, the log analysis and the past situations investigation (by zooming through several time intervals) could help the MAGNIFIER system to identify cause-effect relations, thus improving the DSS performance. Finally, we are working to provide decision maker with a tool set to interact in real-time with the system so to customize the operating logic.

Acknowledgments. This work is partially supported by the EU's Horizon 2020 Research and Innovation Programme through the TOREADOR project.

References

1. Salmon, P.M., Staton, N.A., et al.: What really is going on? a review of situation awareness models for individuals and teams. Theor. Issues Ergon. Sci. **9**(4), 297–323 (2008)
2. Bedny, G., Meister, D.: Theory of activity and situation awareness. Int. J. Cogn. Ergon. **3**(1), 63–72 (1999)
3. Smith, K., Hancock, P.A.: Situation awareness is adaptive, externelly directed consciousness. Hum. Factors **37**, 137–148 (1995)
4. Endsley, M.R., Robertson, M.M.: Situation awareness in aircraft teams. Int. J. Ind. Ergon. **26**, 310–325 (2000)
5. Geneletti, D.: A GIS-based decision support system to identify nature conservation priorities in an Alpine valley. Land Use Policy **21**(2), 149–160 (2004)
6. Malczewski, J.: GIS-based multicriteria decision analysis: a survey of the literature. Int. J. Geogr. Inf. Sci. **20**(7), 703–726 (2006)
7. Wooldridge, M.: Introduction to Multiagent Systems. Wiley (2002)
8. Morreale, V., Bonura, S., Francaviglia, G., Cossentino, M., Gaglio, S.: PRACTIONIST: a new framework for BDI agents. In: Proceedings of the 3rd European Workshop on Multi-agent Systems (EUMAS 2005), Brussels, Belgium (2005)
9. Morreale, V., Bonura, S., Francaviglia, G., Centineo, F., Cossentino, M., Gaglio, S.: Reasoning about goals in BDI agents: the PRACTIONIST framework. In: Proceedings of the Workshop on Objects and Agents (WOA 2006), Catania, Italy (2006)
10. Morreale, V., Bonura, S., Francaviglia, G., Centineo, F., Cossentino, M., Gaglio, S.: Goal-oriented development of BDI agents: the PRACTIONIST approach. In: Proceedings of Intelligent Agent Technology, Hong Kong, China. IEEE Computer Society Press (2006)
11. Morreale, V., Francaviglia, G., Centineo, F., Puccio, M., Cossentino, M.: Goal-oriented agent patterns with the PRACTIONIST framework. In: Proceedings of the 4th European Workshop on Multi-agent Systems (EUMAS 2006), Lisbon, Portugal (2006)
12. Morreale, V., Bonura, S., Francaviglia, G., Centineo, F., Puccio, M., Cossentino, M.: Developing intentional systems with the PRACTIONIST framework. In: Proceedings of the 5th IEEE International Conference on Industrial Informatics (INDIN 2007), Vienna (2007)
13. Rao, A.S., Georgeff, M.P.: BDI agents: from theory to practice. In: Proceedings of the First International Conference on Multi-Agent Systems, pp. 312–319. MIT Press, San Francisco (1995). http://www.uni-koblenz.de/fruit/LITERATURE/rg95.ps.gz
14. Bratman, M.E.: Intention, Plans, and Practical Reason. Harvard University Press, Cambridge (1987)
15. Winikoff, M., Padgham, L., Harland, J., Thangarajah, J.: Declarative and procedural goals in intelligent agent systems. In: KR, pp. 470–481 (2002)

On the Verification of Software Vulnerabilities During Static Code Analysis Using Data Mining Techniques

(Short Paper)

Foteini Cheirdari$^{(\boxtimes)}$ and George Karabatis

University of Maryland, Baltimore County, MD 21250, USA
{fcheirdl,georgek}@umbc.edu

Abstract. Software assurance analysts deal with thousands of potential vulnerabilities many of which could be false positives during the process of static code analysis. Manual review of all such potential vulnerabilities is tedious, time consuming, and frequently impractical. Several experiments were conducted using a production code base with the aid of a variety of static code analysis tools. A data mining process was created, which employed different classifiers for comparison. Furthermore, a selection process identified the most important features that led to significant improvements in accuracy, precision, and recall, as evidenced by the experimental data. This paper proposes machine learning algorithms to minimize false positives with a high degree of accuracy.

Keywords: Software assurance · Vulnerability discovery · Data mining · Feature selection

1 Introduction

Static analysis tools for the discovery of vulnerabilities have a high number of false positive results. Developers and software assurance analysts need to manually examine the code to identify the true positive findings, a task that is often not feasible due to limited resources and time constraints.

In the past, software assurance was not a mandatory requirement; therefore, developers were not trained to consider it during system development. Software assurance analysts have noticed that a significant number of software systems contain vulnerabilities. Work on cybersecurity was mostly geared towards operating system and network hardening. The overwhelming number of potential vulnerabilities in conjunction with the high number of false positives may discourage developers from fixing vulnerabilities in their code. In addition, software assurance analysts cannot manually review all vulnerabilities. Therefore, it would be beneficial if machine learning could label vulnerabilities with accuracy. Assurance analysts view each finding on the file where the vulnerability exists without knowledge of the dependencies and structure of the system. To overcome this limitation, this study suggests the analyst to work with the system developer in order to correctly label the training set used to run the classifier.

H. Panetto et al. (Eds.): OTM 2017 Conferences, Part II, LNCS 10574, pp. 99–106, 2017.
https://doi.org/10.1007/978-3-319-69459-7_7

The most important impact of the suggested approach, is the higher accuracy in predicting the true vulnerabilities. Higher accuracy increases the value added from the static analysis tools and the work performed by the software assurance analysis teams. This study proposes an approach that labels with high degree of accuracy the static code analysis findings by identifying specific attributes, algorithms, and processes to label the training set. The benefits of the proposed approaches include (1) improving the overall security posture of the system, (2) increasing the accuracy of predictions (3) minimizing human input (4) saving time and resources, and (5) increasing the value generated from static analysis tools.

In this research, we compare 13 classifiers and we use multiple static analysis tools for the experiments. We also concluded that just three fields are adequate to provide the best results for the classifier. In addition, based on our experience in the field of static analysis we determine that human input is absolutely necessary when labeling the training set.

2 Related Work

Machine learning for software assurance has been studied in the past and it has proven to offer benefits in the area of true and false positive prediction. Below are listed some of the relevant studies that are dealing with static code analysis and false positive prediction:

In [1] the authors use a combination of techniques for code analysis (slicing, Iterative Context Extension, loop abstraction for Bounded Model Checking), while [2] proposes thread specialization for pruning false positives of static data-race detection. In [4] we see a combined abstract interpretation and source code bounded model checking. The study in [5] detects and corrects vulnerabilities and provides classifier suggestions. The research in [6] used sensitivity analysis for feature selection and compared Artificial Neural Networks (ANN) to Support Vector Machines SVMs. The study determined that SVM had better performance. In [7] the authors used SPAR-ROW as a static analysis tool and java open source projects. A feature vector was extracted using Abstract Syntax Tree (AST) and as classifier the SVM was used. The study in [8] uses logistic regression models that predict if warnings constitute action-able faults. In [10] the authors compare different classifiers and use Airac, a bug finding C analyzer. For the study, the classifiers that offer the best results are random forest and boosting. The work in [11] is focused on alert patterns and ranking actionable alerts. The alerts are generated using Findbugs. The study in [12] compares 15 machine learning algorithms and identifies 14 alert characteristics, the accuracy range was 88–97%. The research in [13] ranks the warnings using the Automated Warning Application for Reliability Engineering (AWARE) that uses historical data from the filtering of alerts that were deemed false positive from the developers. The authors in [17] prioritize warnings based on software change history. If warnings are resolved quickly then the warnings on this category are important. A continuation of [17] with more detailed explanation can be seen in [18]. In [19] the authors explore the z-ranking approach that uses a statistical model to rank warnings. Partitioning techniques are used in [23] to identify redundant warnings in order to reduce the manual review effort of the warnings. The research in [21] is a continuation of the previous study and groups the redundant warnings into clusters. A comparison in [23] of 34 classifiers is performed and 10 features are identified that provide the highest accuracy up to 90%. In [28] the authors implement a clustering algorithm on a realistic buffer-overflow analyzer and achieved on reducing 54% the alarm reports. Airac (Array Index Range Analyzer for C) is introduced in [30] that collects all true buffer-overrun points in ANSI C programs. It ranks the alarms based on probabilities.

3 Approach

Static code analysis tools find potential vulnerabilities only by accessing the source code and without compiling it. Our effort is geared towards helping the software assurance analysts to label the vulnerabilities with higher precision. The static analysis tools do not always see the whole picture, e.g. when the vulnerability has been mitigated somewhere else in the code. Also, it is not uncommon for the static tools to make mistakes when identifying vulnerabilities. Therefore, human analysis and developer input are necessary when trying to predict true from false positives. However, human analysis on all the results is not always possible due to the high number of vulnerabilities (usually they range in the thousands). A solution that this research proposes is to combine human analysis with machine learning to predict vulnerabilities with high degree of accuracy.

The approach has the following phases. First, the code base was scanned using multiple existing tools, the results were aggregated into a CSV file of vulnerabilities. A subset was then selected to manually examine the results as true and false positives. Following that, a training and a test set were created and different classifiers were compared. The classifier with the highest accuracy was used in order to determine the best feature selection. Finally, all the classifiers were tested again, the relevant features identified and the results were recorded. Below is a chart of the process used (Fig. 1):

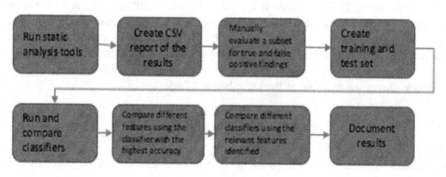

Fig. 1. Approach overview

The dataset used for this study is a subset of a system which have undergone manual assessment to identify potential security related defects. The subject software was scanned by using Code DX, a framework that employs multiple static analysis tools. The analysts went through the most significant findings based on the severity of the potential vulnerabilities. For this study only a limited portion of the reviewed findings were used. Specifically, 540 findings of the assessment findings were used for this research. The training set contained 54 instances out of the 540. This study used only 10% as a training test instead of the common 70% because the goal is to provide a feasible solution on real life situations that software assurance teams can implement. The dataset used for this study had 355 findings marked as true positive and 185 marked as false positive. Code DX aggregated the results of the following tools: Checkstyle, Dependency-Check, FindBugs, Fortify, JSHint, PHP_CodeSniffer. Since only a subset of the results was used, the training and test datasets included the results of the following tools: Fortify, PMD and Findbugs. The majority of the findings of the dataset used for the experiments were located in Java files.

The WEKA software was used to run the classifiers and to select the sample for the training set, via stratified sampling. Different algorithms were tested and it was determined after many tests that SGD algorithm provided the best results [10].

The original dataset had the following attributes per vulnerability:

- ID, Severity, CWE, Rule, description, Tool, Location, Path.

CWE refers to the Common Weakness Enumeration industry standards that list software weakness types [1]. After performing different experiments with subsets of the original attributes shown above it was determined that three of the attributes when used to run the classifiers had higher recall, accuracy, and precision:

- CWE, Path (without the filename), Description.

After multiple experiments, it was determined that the Description field can be substituted for only a part of the values it originally contained, further improving the results. The reason is that the noise was eliminated from the Description field and only the data that uniquely identifies the vulnerability remains. This dataset had the following characteristics that made the elimination of extra details of the findings irrelevant and noisy: It is a small dataset and the vulnerabilities are repeated in different places in the code. The same code is copied and paste in different places so the same mistakes are repeated. By removing the filenames and any other detailed relevant to the location (for example class name or line members) the algorithm successfully identifies vulnerability patterns. Instead of using the Description as it generated from the tools set, the Description column was replaced with the method, variable or type of vulnerability please see example at the table below (Table 1):

Table 1. Field replacement examples

Original "Description" field	Replaced
The method 1 in filename1.java can crash the program by dereferencing a null pointer on line 1	Method 1
The function function1 in filename 1 java sometimes fails to release a system resource allocated by connection 1 on line 2	Connection 1

Accuracy is the percentage of the vulnerabilities that were identified correctly as either true positive or true negative. Recall is the percentage of true positive vulnerabilities that were identified. Precision is the percentage of the vulnerabilities identified as true positive that were correct. The F-measure of the system is defined as the weighted harmonic mean of its precision p and recall R, that is:

$$F = \frac{1}{a\frac{1}{p} + (1-a)\frac{1}{R}}, \text{ where the weight } \alpha \in [0,1].$$

The balanced F-measure, commonly denoted as F1 or just F, equally weighs precision and recall, which means $\alpha = 1/2$. The F1 measure can be written as $F1 = \frac{2PR}{P+R}$. The F-measure can be viewed as a compromise between recall and precision [15]. It is high only when both recall and precision are high. It is equivalent to recall when $\alpha = 0$ and precision when $\alpha = 1$. The F-measure assumes values in the interval [0, 1]. It is 0 when no relevant documents have been retrieved, and is 1 if all retrieved documents are relevant and all relevant documents have been retrieved. Table 2 shows the accuracy, recall and precision obtained using the original 8 features, using the three (location, CWE and Description) and using the 3 but replacing the Description with the subtype field which is a part of the original fields as shown in Table 3 examples. The algorithm used to compare the features was the SGD algorithm. Stochastic Gradient Descent (SGD), also known as incremental gradient descent, is a stochastic approximation of the gradient descent optimization method for minimizing an objective function that is written as a sum of differentiable functions [27]. In other words, SGD tries to find minima or maxima by iteration. It implements stochastic gradient descent for learning various linear models (binary class SVM, binary class logistic regression, squared loss, Huber loss and epsilon-insensitive loss linear regression) [26]. Globally replaces all missing values and transforms nominal attributes into binary ones. It also normalizes all attributes, so the coefficients in the output are based on the normalized data. It is observed that the new field with the location and CWE have the highest accuracy and the highest F measure.

Table 2. Feature comparison

Features	Accuracy	Recall	Precision	F measure
Original features (8)	0.972	0.966	0.991	0.978
Location, CWE, Description	0.981	0.972	1	0.986
Location, CWE, Subtype	0.994	1	0.992	0.996

Table 3 displays the precision, recall, F measure, and accuracy results using the selected fields of the study. Specifically, SGD and ADABoost are the higher performing classifiers for Recall. Both classifiers correctly identified all positive vulnerabilities from the dataset, meaning that the errors for both classifiers were on the false positives domain. In total 10 out of 13 algorithms offer precision over 99%. Some of the classifiers have 100% precision.

Table 3. Classifier comparison results

Algorithm	Recall	Precision	F measure	Accuracy
ADABoost	1.000	0.915	0.956	0.954
SGD	1.000	0.992	0.996	0.994
Random tree	0.997	0.912	0.953	0.935
Random committee	0.997	0.913	0.953	0.935
Random forest	0.975	0.997	0.986	0.981
SMO	0.972	1.000	0.986	0.981
Bagging	0.958	0.997	0.977	0.970
Simple logistics	0.949	1.000	0.974	0.967
J48	0.949	1.000	0.974	0.967
Logistics	0.944	1.000	0.971	0.963
Naïve Bayes	0.918	1.000	0.957	0.946
Voted perceptron	0.918	1.000	0.957	0.946
One R	0.527	0.995	0.689	0.689

In general, the precision is higher than the recall for the majority of the classifiers, as seen on Table 3. The higher precision signifies that the majority of the classifiers correctly found the true positives. However, many of them missed true vulnerabilities and identified them as false. The SGD identified 3 false vulnerabilities as true, and it correctly identified all false positives. All of the other classifiers missed true findings and identified them as false. From a cyber security perspective, the SGD algorithm provides more accurate results with lower risk, since it did not miss any true vulnerabilities.

One observation of the study is high accuracy. In order to make sure that the results were correct, the experiments were run multiple times validating the results, and possibly it is attributed to the dataset.

The SGD algorithm is the one that offers the highest F Measure. We do not want to miss any true positive findings while making sure we capture as many false positive findings as possible. However, it is better to identify a false finding as true, rather than the other way around, to avoid severe consequences on the software security.

4   Conclusion, Discussion, and Future Work

Static code analysis, given the limitations of existing technology, leads to a lot of false positive findings. Software assurance analysis teams that are not familiar with the system and its dependencies may not be able to recognize that a vulnerability is mitigated in another part of the code. This is a major disadvantage of static code analysis. This study proposes the use of machine learning during the static analysis process in order to eliminate the high percentage of false positives. The approach proposed in this study increases the value of the reports generated to depict system security posture, and saves critical time and resources. This study used a production code base and scanned it through multiple static code analysis tools. A small sample of the findings from the toolset was then manually reviewed and labeled. The sample was then divided into training and test sets and run through different potential classifier algorithms. The Stochastic Gradient Descent (SGD) was determined as the algorithm that provides the highest accuracy, precision, recall, and F measure. This research proposes: (a) manual review of a training set of raw findings by both the developer and the software analyst, (b) using three features (CWE, Location, and Subtype) for the training and test sets that are necessary to achieve the highest prediction accuracy (c) Using the SGD algorithm for predicting true and false positive findings.

In the future, we plan to add a percentage column next to the prediction so the developers can rank the results based on the possibility to be true or false vulnerabilities. In addition, we will perform experiments with more datasets that include a variety of languages.

References

1. Chimdyalwar, B., Darke, P., Chavda, A., Vaghani, S., Chauhan, A.: Eliminating static analysis false positives using loop abstraction and bounded model checking. In: Bjørner, N., de Boer, F. (eds.) FM 2015. LNCS, vol. 9109, pp. 573–576. Springer, Cham (2015). doi:10.1007/978-3-319-19249-9_35
2. Chen, C., Lu, K., Wang, X., Zhou, X., Fang, L.: Pruning false positives of static data-race detection via thread specialization. In: Wu, C., Cohen, A. (eds.) APPT 2013. LNCS, vol. 8299, pp. 77–90. Springer, Heidelberg (2013). doi:10.1007/978-3-642-45293-2_6
3. Common Weakness Enumeration. https://cwe.mitre.org/
4. Post, H., Sinz, C., Kaiser, A., Gorges, T.: Reducing false positives by combining abstract interpretation and bounded model checking. In: IEEE (2008)
5. Medeiros, I., Neves, N., Correia, M.: Detecting and removing web application vulnerabilities with static analysis and data mining. In: IEEE (2016)
6. Gondra, I.: Applying machine learning to software fault-proneness prediction. In: JSS (2007)
7. Yoon, J., Jin, M., Jung, Y.: Reducing false alarms from an industrial strength static analyzer by SVM. In: APSEC (2014)
8. Ruthruff, J.R., Penix, J., Morgenthaler, J.D., Elbaum, S., Rothermel, G.: Predicting accurate and actionable static analysis warnings: an experimental approach. In: ACM (2008)
9. KD nuggets, Precision and Recall Calculation. http://www.kdnuggets.com/faq/precision-recall.html
10. Yi, K., Choi, H., Kim, J., Kim, Y.: An empirical study on classification methods for alarms from a bug-finding static C analyzer. Inf. Process. Lett. 102(2–3), 118–123 (2007)
11. Hanam, Q., Tan, L., Holmes, R., Lam, P.: Finding patterns in static analysis alerts: improving actionable alert ranking. In: MSR (2014)
12. Heckman, S., Williams, L.: A model building process for identifying static analysis alerts. In: IEEE (2009)
13. Heckman, S., Williams, L.: Automated ranking and filtering of static analysis alerts. In: IEEE (2006)
14. SCIKIT, Stochastic Gradient Descent. http://scikit-learn.org/stable/modules/sgd.html
15. Springer Link, F Measure. https://link.springer.com/referenceworkentry/10.1007/978-0-387-39940-9_483
16. Stat Trek, Statistics and Probability Dictionary. http://stattrek.com/statistics/dictionary.aspx?definition=Stratified_sampling

17. Kim, S., Ernst, M.D.: Prioritizing warning categories by analyzing software history. In: IEEE Xplore (2007)
18. Kim, S., Ernst, M.D.: Which warnings should I fix first? In: ACM SIGSOFT (2007)
19. Kremenek, T., Engler, D.: Z-ranking: using statistical analysis to counter the impact of static analysis approximations. In: Cousot, R. (ed.) SAS 2003. LNCS, vol. 2694, pp. 295–315. Springer, Heidelberg (2003). doi:10.1007/3-540-44898-5_16
20. Muske, T.B., Baid, A., Sanas, T.: Review efforts reduction by partitioning of static analysis warnings. In: 2013 IEEE 13th International Working Conference on Source Code Analysis and Manipulation (SCAM) (2013)
21. Muske, T.: Improving review of clustered-code analysis warnings. In: 2014 IEEE International Conference on Software Maintenance and Evolution (2014)
22. Muske, T., Serebrenik, A.: Survey of approaches for handling static analysis alarms. In: SCAM (2016)
23. Yuksel, U., Sozer, H.: Automated classification of static code analysis alerts: a case study. In: ICSM (2013)
24. UFLDL Tutorial, Optimization: Stochastic Gradient Descent. http://ufldl.stanford.edu/tutorial/supervised/OptimizationStochasticGradientDescent/
25. Weka 3: Data Mining Software in Java. http://www.cs.waikato.ac.nz/ml/weka/
26. WEKA. weka.sourceforge.net/doc.dev/weka/classifiers/functions/SGD.html
27. WIKIPEDIA, Stochastic Gradient Descent. https://en.wikipedia.org/wiki/Stochastic_gradient_descent
28. Lee, W., Lee, W., Yi, K.: Sound non-statistical clustering od static analysis alarms. In: VMCAI (2012)
29. Kim, Y., Lee, J., Han, H., Choe, K.-M.: Filtering false alarms of buffer overflow analysis using SMT solvers. In: Infosof (2009)
30. Jung, Y., Kim, J., Shin, J., Yi, K.: Taming false alarms from a domain-unaware C analyzer by a bayesian statistical post analysis. In: Hankin, C., Siveroni, I. (eds.) SAS 2005. LNCS, vol. 3672, pp. 203–217. Springer, Heidelberg (2005). doi:10.1007/11547662_15

International Conference on Ontologies, DataBases, and Applications of Semantics (ODBASE) 2017

ODBASE 2017 PC Co-chairs' Message

Adrian Paschke, Nick Bassiliades, and Hans Weigand

We are delighted to present the proceedings of the 16th International Conference on Ontologies, DataBases, and Applications of Semantics (ODBASE) which was held in Rhodes (Greece) 24–25 October 2017. The ODBASE Conference series provides a forum for research and practitioners on the use of ontologies and data semantics in novel applications, and continues to draw a highly diverse body of researchers and practitioners. ODBASE is part of the OnTheMove (OTM 2017) federated event composed of three interrelated yet complementary scientific conferences that together attempt to span a relevant range of the advanced research on, and cutting-edge development and application of, information handling and systems in the wider current context of ubiquitous distributed computing. The other two co-located conferences are CoopIS'17 (Cooperative Information Systems) and C&TC'17 (Cloud and Trusted Computing). Of particular relevance to ODBASE 2017 are papers that bridge traditional boundaries between disciplines such as artificial intelligence and Semantic Web, databases, data analytics and machine learning, social networks, distributed and mobile systems, information retrieval, knowledge discovery, and computational linguistics.

This year, we received 46 paper submissions and had a program committee of 48 dedicated colleagues, including researchers and practitioners from diverse research areas. Special arrangements were made during the review process to ensure that each paper was reviewed by 3-4 members of different research areas. The result of this effort is the selection of high quality papers: twenty regular papers, six short papers, and four posters. Their themes included studies and solutions to a number of modern challenges such as querying, cleaning, publishing, benchmarking and visualizing linked data, RDF documents and graph databases, ontology engineering, semantic mapping, social network analysis, and semantics-based applications to various domains, such as health, tourism, smart cities, law, etc. The scientific program is complemented with a very interesting keynote speech by Markus Lanthaler on Pragmatic Semantics at Web Scale.

We would like to thank all the members of the Program Committee for their hard work in selecting the papers and for helping to make this conference a success. We would also like to thank all the researchers who submitted their work. Last but not least, special thanks go to the members of the OTM team for their support and guidance.

We hope that you enjoy ODBASE 2017 and have a wonderful time in Rhodes!

Linked Data and Ontology Reference Model for Infectious Disease Reporting Systems

Olga Streibel, Felix Kybranz, and Göran Kirchner[✉]

Data Management Unit, Department of Infectious Disease Epidemiology,
Robert Koch Institute, Seestr. 10, Berlin, Germany
{streibelo,kybranzf,kirchnerg}@rki.de
http://www.rki.de

Abstract. Linked data and ontologies are already in wide use in many fields. Especially systems based on medical data can be valuably improved by enhancing their contents and meta-models semantically, using ontologies in their backbone. This semantic enhancement brings an add-on value to standard systems, enabling an overall better data management and allowing a more intelligent data processing. In our work we focus on such a standard system, which we enhance semantically, transferring its classic relational models together with its data into a semantic model. This information system processes and analyzes data related to infectious disease reports in Germany. Data from reports on infectious diseases not only contains specific parts of microbiological and medical information, but also a combination of various aspects of contextual knowledge, that is needed in order to take measures preventing a wider spread and reducing further transmissions. In this paper we describe our practical approach for transferring the relational data models into ontologies, establishing an improved data standard for the current system in use. Moreover, we propose a semantic reference model based on different contexts, covering the requirements of semantified data from infectious disease reports.

Keywords: Semantic model · Ontologies · Linked data for epidemiology

1 Introduction

Public health data management increases in importance, in particular, in our globalized society. In case of an outbreak, the relevant data and information have to be reported in shortest time to the public health authorities in order to diminish the further spread by taking appropriate interventions and control measures. With regard to prevention and early detection, sophisticated approaches for forecasting and estimation of critical epidemiological situations are needed. When it already has come to an infection, that—if not handled on time—may result in an outbreak, fast reaction to the situation is necessary. This reaction is usually based on a rapid and reliable data collection, as well as a trustworthy

© Springer International Publishing AG 2017
H. Panetto et al. (Eds.): OTM 2017 Conferences, Part II, LNCS 10574, pp. 109–124, 2017.
https://doi.org/10.1007/978-3-319-69459-7_8

information processing. The notification and reporting process usually follows the regulations of the national public health laws.

In Germany, the relevant *Law for Protection against Infection Act* (IfSG) formulates rules and defines the frame for necessary actions to be taken, describes the duties for the obliged notifiers and specifies the notifiable evidence for pathogens and diagnoses that have to be reported in case of their occurrence. The Robert Koch Institute[1] receives and analyzes data of various communicable diseases in Germany[2]. These data are collected by the local health authorities from various sources, like microbiological labs, treating physicians and community facilities. The existing reporting system is maintained with help of a surveillance software, SurvNet@RKI [4], developed at Robert Koch Institute for the purpose of infectious and epidemiological case data collection, transmission and analysis. However, more effort is required to meet the needs of an extended system in which local health departments are capable of receiving electronic reports from all the notifiers and their various information systems at hand. This demands an extension of the system which guarantees interoperability and enables automatic data processing. All involved formal rule systems and exhaustive semantic data services have to be compliant with the national law regulations.

With regard to the requirement on establishing a profound and sophisticated semantic model, in this stage of the project we focus on a *semantification* of the relational data from current systems' meta and master data, transferring it into a consistent semantic model. The most important outcome of this semantification is the fact that semantified data, thus our concepts defined in the context of infectious disease and epidemiology, is explicitly and publicly available in a *machine-readable* and *human-understandable* format and *de-reference-able* on the web (via Unified Resource Identifier). On the other hand, we are incorporating the semantic description of main concepts into the system in a way that allows for further extension of those descriptions by existing medical vocabularies hence it allows for further enhancement of systems interoperability.

In this paper we report on a part of our work done on semantification of infectious diseases systems' data during the initial stage of our project and introduce the general semantic reference model for data from infectious disease reports. In Sect. 2 we start with the description of the general architecture of our semantic component and discuss the main goal of data semantification. The description of our semantic reference model follows in Sect. 3. We continue the discussion of details of the model in Sect. 4 mainly by the example of our geo-political ontology. Our paper closes with the short review of work relevant to ours as well as the description of evaluation frame for our semantic reference model by the example of geo-political ontology evaluation.

---

[1] http://www.rki.de.
[2] http://goo.gl/eZ50HH.

2   Goal of Semantics Usage

Our aimed linked data and semantics standard for linkable and de-referenceable data on infectious disease - control and reporting, has one main following goal: Making tacit knowledge about infectious disease data control and reporting explicit, machine readable, re-usable and available for services and users.[3]

This includes in general following steps:

1. Transforming explicit (literal) concepts included in the current surveillance system (SurvNet@RKI) into semantic concepts
2. Transforming tacit knowledge included between current system's database table relations into explicit semantic concepts and relations
3. Linking the transferred knowledge model with relevant external knowledge (e.g. DBpedia, relevant ontologies from OBO foundry[4], or e-health standards, e.g. ICD10[5], SNOMED-CT[6])
4. Enriching the knowledge model by creating formal (Description Logic-based) set of rules, describing e.g. case reports, distributions and privacy enforcing rules etc.
5. Publishing the resulting knowledge as linked data graph in order to make it available for other system components and services, or provide access through an external terminology server (we use the CTS2[7] based terminology server[8]) (Fig. 1).

Fig. 1. Five steps that we follow by bringing semantics into our relational data.

Hereby we follow these steps while modeling the relevant background knowledge and embedding it through RDFS and OWL2 based ontologies into our semantic component. The semantic component is currently implemented basically on the Virtuoso-based service.

---

[3] Here we speak in terms of the Robert Koch Institute's system, however we furthermore hope to motivate also a general, canonical and re-usable linked data standard that may be successfully applied to any information system handling infectious disease data.
[4] http://www.obofoundry.org/.
[5] https://goo.gl/gDGHVA.
[6] http://www.snomed.org/snomed-ct.
[7] http://www.omg.org/spec/CTS2/.
[8] https://publicwiki-01.fraunhofer.de/CTS2-LE/index.php/Hauptseite.

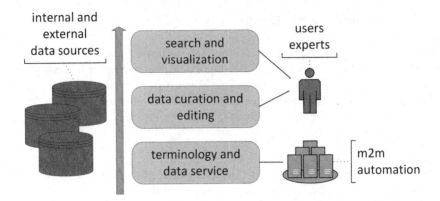

Fig. 2. Semantic component–three levels for different user groups

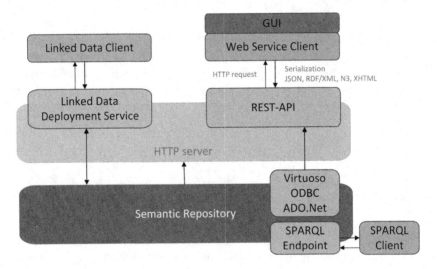

Fig. 3. Semantic component–general architecture

Figure 2 shows three main levels of our semantic component which include three different user interactions with the overall service that we plan to offer. The upper level, search and visualization, is meant to provide non-technical users, e.g. epidemiology experts or health officials, with the interface through which they can browse the knowledge and visualize respectively relevant parts of it. The middle level, data curation and editing, is meant for the ontology and master data experts who take care internally about the updating of ontology models with regard to the requirements and team agreements on data changes. The bottom level, which we preliminary implemented in the initial stage of our project, is meant for direct service use as well as for use by the technical experts and system developers. This part of our component should feed other system components with the meaningful data.

The general technical architecture of the semantic component is illustrated in Fig. 3. In the initial stage of our project, we used Virtuoso server (OpenVirtuoso Vers. 6.01.3127) and implemented the belonging functions as C# applications communicating with both, our current internal SQL servers databases and the Virtuoso-based semantic data service.

3   Semantic Reference Model

The linked data and semantic reference model is implemented into the semantic component which we described in the previous section. In the current project stage the bottom level of our component, the terminology and data service, has been filled with semantified data. In the following we provide a simple example of relational data and what we mean by semantifying it.

3.1   Semantifying Relational Data

An example of an explicit literal concept is shown in Listing 3.1. A String literal is used as a disease name in SurvNet database 'Meta.Disease' table in the column named 'DiseaseName'.

Listing 3.1: Malaria as a literal concept.

```
Malaria.
```

An example of an explicit semantic concept, thus semantified data is shown in Listing 3.2.

Listing 3.2: Malaria semantified.

```
@prefix demis: <http://rki.de/demis/example-disease-ontology#>.
@prefix owl: <http://www.w3.org/2002/07/owl#>.
@prefix rdf: <http://www.w3.org/1999/02/22-rdf-syntax-ns#>.

demis:Malaria rdf:type owl:Class.
```

An example of explicit knowledge as derived from the database is shown in Listing 3.3.

Listing 3.3: Explicit knowledge about Malaria.

```
Malaria is a disease.
```

An example of explicit knowledge expressed in formal semantics is shown in Listing 3.4.[9]

Listing 3.4: Explicit knowledge formally expressed.

```
demis:Disease rdf:type owl:Class.
demis:Malaria rdf:type owl:Class;
    rdfs:subClassOf demis:Disease.
```

---

[9] From here on we will drop all prefix definitions and assume them to be found on https://prefix.cc/.

An example of tacit knowledge is shown in Listing 3.5. The tacit knowledge: 'Malaria' is a String literal in SurvNet database table in the column 'Meta.Disease'. 'Plasmodium spp.' is a String literal in SurvNet database table in the column 'SpecimenName'. A relation 'caused-by' is not written explicitly anywhere, thus it is tacit. A relation 'has-name' is inferred from 'SpecimenName' as well. This example of tacit knowledge is expressed in turtle in Listing 3.6.

Listing 3.5: Tacit knowledge.

```
Malaria is caused by a pathogene named plasmodium spp.
```

Listing 3.6: Tacit knowledge, semantified.[a]

```
:caused-by rdf:type owl:ObjectProperty;
  rdfs:domain :Specimen;
  rdfs:range :Disease.

:Pathogen rdf:type owl:Class.
:plasmodium_spp rdf:type owl:NamedIndividual, :Pathogene.

:Malaria :caused-by :plasmodium_spp.
```

[a] Malaria, which was defined above as an OWL class, is used here as a participant of an object property relation. Under an OWL interpretation this would imply that Malaria is an individual, which is possible if we use OWL2 punning.

An example of linking to external medical codes and external knowledge is shown in Listing 3.7[10].

Listing 3.7: Linking to external knowledge.

```
:has_ICD10code rdf:type owl:DatatypeProperty;
  rdfs:domain :Disease;
  rdfs:range xsd:string.

:Malaria rdf:type owl:Class;
  rdfs:subClassOf :Disease;
  rdfs:comment "https://goo.gl/eYx5bZ"^^xsd:anyURI;
  rdfs:seeAlso "https://de.wikipedia.org/wiki/Malaria"^^xsd:anyURI.

:Malaria_quartana rdf:type owl:NamedIndividual, :Malaria;
  :has_ICD10code "B52"^^xsd:string.

:Malaria_tertiana rdf:type owl:NamedIndividual, :Malaria;
  :has_ICD10code "B51"^^xsd:string.

:Malaria_tropica rdf:type owl:NamedIndividual, :Malaria;
  :has_ICD10code "B50"^^xsd:string.
```

[10] We have used here *rdfs:seeAlso* to follow the weaker SKOS approach. We could have used a stronger relation instead, e.g. *owl:sameAs*.

The examples above show only a very tiny excerpt of our work done on fostering the data and estimating its belonging to different contexts. For the purpose of transferring the relational data into a linked data graph, we implemented automatic concept and relation extraction in C# via dotnetRDF[11]. As far as it was feasible, we retrieved the data automatically from the system's master data service by translating the table names into OWL classes, table columns into relations and table rows into instances. An example is given in Listings 3.8 and 3.9 respectively.

Listing 3.8: Table 'States' with columns in master data service.

```
- [StateNameShortDEU]
- [StateNameShortENG]
- [StateNameFullGER]
- [StateNameFullENG]
- [ISO3]
- [ISO2]
- [CitizenshipGER]
- [CitizenshipENG]
- [ExistentFrom]
- [ExistentTo]
```

Listing 3.9: Semantic description in turtle for one of table's data row with its belonging relations data extracted from tables columns.

```
demisgeo:Country a owl:Class.
demisgeo:Togo a demisgeo:Country;
  demisgeo:StateNameShort 'Togo'@de ;
  demisgeo:StateNameShort 'Togo'@en ;
  demisgeo:StateNameFull 'die Republik Togo'@de ;
  demisgeo:StateNameFull 'Togolese Republic'@en ;
  demisgeo:hasISO3 'TGO';
  demisgeo:hasISO2 'TG';
  demisgeo:hasCitizenship 'togoisch'@de ;
  demisgeo:hasCitizenship 'Togolese national'@en ;
  demisgeo:ExistendFrom '1960-04-27'^^xsd:date .
```

The diagram below summarizes the steps taken while automatically retrieving the ontology concepts, relations and instances from the SQL server (Fig. 4):

We modeled these relational data into several different RDFS graph snippets and figured out which of the existent and publicly available ontologies we may use at hand in order to support creation of our own semantic model. As a result of preliminary review of existing relevant ontologies and vocabularies, we decided to use (either full or partially) mainly the following:

- XSD - for the purpose of using the simple values
- RDF and RDFS - it offers the 'light' representation for our semantics in cases, where we mainly use *subClassof* relations

---

[11] http://www.dotnetrdf.org/.

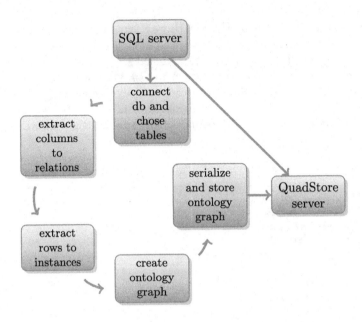

Fig. 4. Transforming relational data to RDF.

- OWL - it is currently our main language in which we code our ontologies
- FOAF - it offers simple and handy definition of such concepts as *Person*, *Address*, etc.
- SKOS - it offers structured definition of abstract concepts and offers useful relations, e.g. *skos:broader*, *skos:narrower* that let us combine abstract concepts in logical groups in a semiformal manner
- Geo related ontologies, i.e. FAO[12], W3C - WGS84[13]

Following, we decided to use at least the semantics as spanned by the existing relations (and in these relations included concepts) which we list here:

- XSD xsd:boolean, xsd:date, xsd:decimal, xsd:integer,xsd:string, xsd:time
- RDF rdf:about, rdf:datatype, rdf:resource, rdf:type
- RDFS rdfs:Class, rdfs:comment, rdfs:domain, rdfs:isDefinedBy, rdfs:label, rdfs:seeAlso, rdfs:subClassOf, rdfs:range
- OWL owl:AssymetricProperty (optional–high complexity while reasoning), owl:Class, owl:equivalentClass, owl:import, owl:inverseOf (optional due to high complexity while reasoning), owl:NamedIndividual, owl:ObjectProperty, owl:onProperty, owl:Ontology, owl:Restriction (optional–high complexity while reasoning), owl:sameAs, owl:versionIRI, owl:TransitiveProperty

---

[12] http://www.fao.org/countryprofiles/geoinfo/en/.
[13] https://www.w3.org/2003/01/geo/wgs84_pos.

3.2 Consistent Model

While aiming at consistent knowledge model, we tackled many different contexts for our knowledge base resulting from the system data. They include in general:

1 Data that contextually fits under the general and specific knowledge of infectious diseases
2 Data and rules that mainly originate in the Law for Protection against Infection Act (IfSG)
3 Data which results from epidemiological definitions of case reporting as specified by Robert Koch Institute
4 Geo-political data

Therefore, we propose to structure the reference model that we use for our knowledge base in a way that puts centrally the semantified data from the existing relational databases. These data, which mainly delivers the instances for our ontology, is semantically described based on our own semantic concepts and relations as well as the concepts and relations resulting from the vocabularies and schemas as described in the previous section. In the Fig. 5 the outer border visualizes the meta level of our model. It is based so far mainly on our own meta concept which we call *demis:Epiconcept*. The core ontology, illustrated in the second outer border unites all the knowledge needed for maintaining the system (the current SurvNet data), thus it includes concepts (for complexity reasons, we list here only exemplary few concepts and omit the belonging relations) *Person, Patient, Pathogen, Symptom* and similar as well as the disease relevant knowledge. Core ontology is being extracted from current SurvNet's meta model. Directly bound to the core ontology are IfSG/law, case definition, and geo-political ontologies. They can be either included as aspects (bound through meta relations, e.g. *aspect_of, context_of* to the meta concept, *demis:Epiconcept*) or represented by different ontologies that are connected to core ontology via imports. Only most relevant concepts from each particular model (IfSG/law, case definition, geo-political) need to be included in these ontologies.

Mapping is an ontology that contains mappings to extern knowledge bases. These are either knowledge bases linked through CTS2 server or DBpedia. Instances/raw data in the middle of the picture visualize the center of our model that are the particular instances which are connected to our ontologies and to the extern models.

We believe that this reference model built on several ontologies as presented here, which still is a subject for further enhancements over the project duration, might be successfully applied to similar systems–as in our case–that aim to handle infectious disease data in semantic manner.

The concepts which are belonging to the core ontology are inalienable regarding the general infectious disease semantic model whereas the different contextual ontologies, i.e. geo-political ontology allow for meaningful combination of the concepts and putting them in different (contextual) relations. In the following Section, we focus on detailed description of geo ontology which we created for our reference model and which is our current show case for the semantic reference model in infectious disease reporting system.

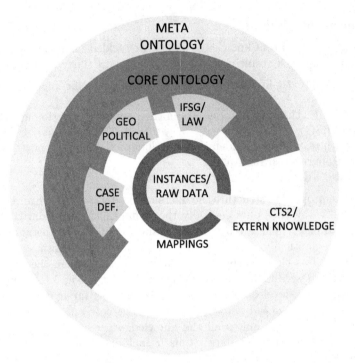

Fig. 5. The overall semantic reference model with different contexts for desired ontologies covering relevant knowledge parts necessary for the infectious diseases reporting.

4   Geo Political Ontology

Geo-political knowledge is of a high relevance when it comes to the infectious disease reporting. When diseases are being reported, the respective patients and disease cases are always connected to a place, which is located not only in the context of this patient's current habitation but also geographically, e.g. *Europe*, or administratively, e.g. *City, State, Country* or politically, e.g. *West Europe* etc. However, the structuring of geo relevant knowledge can be fast covered by using many existent concepts or vocabularies. In creating the first version of our geo ontology we considered: internal concepts used in the epidemiological context with regard to infectious disease reporting, existing data from current system's Master Data Service databases as well as standard concepts used widely in other available vocabularies. Foremost we consider the geographical data which relates to Germany. Furthermore, our geo ontology offers the geo related knowledge with regard to global data. We reviewed several vocabularies and standards relevant and estimated their relevance based on concepts and relations as well as potential instance's included. In Table 1, we list the relevant ontologies.

The resulting ontology is based on most important concepts and relations used routinely in the epidemiological context at Robert Koch Institute and

Table 1. The relevant ontologies and vocabularies.

Ontology/Vocabulary	Relevance for geo ontology
FAO	High
Geographical entity ontology	Relevant (US data)
GEO names	High (data service avail.)
Linked geo data	Relevant (data service avail.)
NeoGeo	Relevant (rel. W3C Geo-Spat.)
SKOS	High
WGS84 Geo Pos.	Medium
W3C Geo-Spat.	Relevant (rel. WGS84)

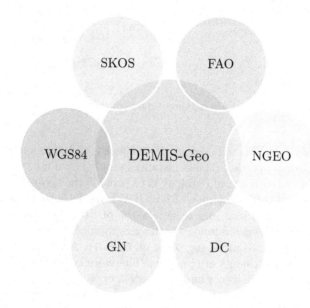

Fig. 6. Ontologies used in the geo ontology.

extracted from current system's master data service databases as well as on concepts and relations from standard ontologies and vocabularies which we estimated as high relevant. The geo-political ontology graph of our *DEMIS-Geo* ontology is illustrated in the diagram below (Fig. 6).

The importance of the concepts is always determined by discussions in team, in consultation with technical epidemiology experts. As far as it is possible, these concepts and relations are matched with existing relevant concepts and relations from standard models of W3C (e.g. SKOS, Geo-spatial vocabulary). In order to ensure the high quality of the geo ontology, we created a basis version of it and test it against a set of SPARQL queries that describe our pre-defined competency questions.

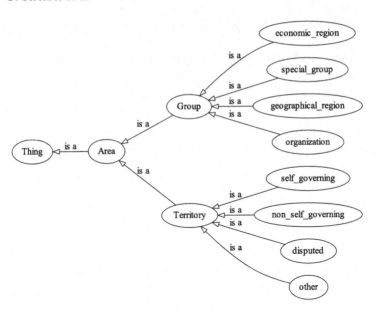

Fig. 7. FAO ontology–main concepts.

The geo ontology we created consists of 48 classes, 42.100 axioms, 50 object properties, 146 data properties, 33.998 instances and is based on 383.100 triples in total. The goal of geo-political ontology is to provide the geo-political knowledge in the context of epidemiology as RDFS/OWL model/s that can be stored in our Quadstore on Virtuoso and answer a set of SPARQL queries. One of important aspects for the geo ontology is the geographical and political context for respective concepts/instances, e.g. *Area* or *Country*. According to it, different 'divisions' of sub continents are possible. Therefore countries like *Spain* are regarded as *South Europe* in a geographical context or as *West Europe* in political context. It is important to offer to users and services of geo ontology the possibility of distinguishing this context for the respective concepts. A preliminary list of concepts that are affected: *Sub-continent, Country, City, Area, Population*. In order to satisfy this requirement, we defined two specific relations that are being used often with regard to geo-political classification of world regions (Fig. 7).

Listing 4.1: Defining necessary relations for geographical and political context.

```
demisgeo:geographicalBroader rdf:instanceOf skos:broader.
demisgeo:politicalBroader rdf:instanceOf skos:broader.
demisgeo:geographicalNarrower rdf:instanceOf skos:narrower.
demisgeo:politicalNarrower rdf:instanceOf skos:narrower.
```

The definition as illustrated in Listing 4.1 implies the fact that both, *demisgeo:geographicalBroader, demisgeo:geographicalNarrower, demisgeo:politicalBroader, demisgeo:politicalNarrower* are transitive properties since

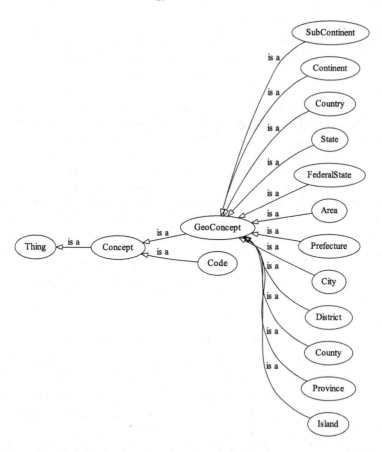

Fig. 8. Geo ontology–main concepts.

they are derived from *skos:broader, skos:narrower* that per se are defined as *owl:instanceOf owl:transitiveProperty*. In detail defined as shown in Listing 4.2 (Fig. 8).

Listing 4.2: Defining necessary relations for geographical and political context in detail.

```
demisgeo:geographicalBroader rdfs:subPropertyOf skos:broader.
demisgeo:politicalBroader rdfs:subPropertyOf skos:broader.
skos:broader rdfs:subPropertyOf skos:broaderTransitive.
skos:broaderTransitive rdf:instancesOf owl:TransitiveProperty.

demisgeo:geographicalNarrower rdfs:subPropertyOf skos:narrower.
demisgeo:politicalNarrower rdfs:subPropertyOf skos:narrower.
skos:narrower rdfs:subPropertyOf skos:narrowerTransitive.
skos:narrowerTransitive  rdf:instancesOf owl:TransitiveProperty.
skos:narrower owl:inverseOf skos:broader.
skos:narrowerTransitive owl:inverseOf skos:broaderTransitive.
```

5 Relevant Work

There are many works related to the general technical approach that we apply in our work, the transformation of the relational data into semantic knowledge base. We are highly aware of the classic approach that allows for mappings from a RDB to RDF as described in W3C[14]. We also find the work about *Ontop* [2] highly interesting to our technical problem since it offers a system architecture with an added ontology layer that still allows for preserving the relational databases if needed. We are still testing our semantic component and improving its overall use (in this project stage–internally) so that we do not completely exclude the relational database layer. In [1], we found very useful summary of challenges in data integration with which we have to deal either. Most related to our domain is the work included in [3], since it tackles the problems of epidemiology in Semantic Web terms. To best of our knowledge, we have not found any general semantic reference model for the infectious disease control and reporting as proposed here in our work. Therefore we believe that we are developing an useful and re-usable reference model that may be applied to any other countries' epidemiological surveillance.

6 Evaluation Through Competency Questions

6.1 Use Case

While the main use case for the geo-political ontology is to define geo-political knowledge (in the domain of epidemiology) in a machine-readable format and to consume geo-political knowledge (in the domain of epidemiology) by the respective system components and services of our system, the preliminary application of geo-political ontology allows for improvement of fostering the current system's master data (of geo-political context). The very particular case of use for geo-political ontology is to provide to the team of developers a novel possibility in applying versioned geo knowledge per one-click when creating relational databases needed on-the-fly during system's maintenance. Users–developers while accessing the geo ontology should be able to get updated, complete, correct and consistent geo related data (concepts and relations) out of which they can easily create relational databases for their (daily) use (e.g. further data analysis applications).

6.2 Competency Questions

For the evaluation purposes we developed an initial set of questions, thus SPARQL queries, which we applied onto our ontology model. Hereby we preliminary lied our focus on the recall, not the precision of query answering. Hence our main ontology purpose is to support and partially replace the current system's master data (meta information), we focused mainly on questions of that we

---

[14] http://www.w3.org/TR/2012/REC-rdb-direct-mapping-20120927/.

think are helpful in automatic retrieval of relatively complex knowledge pieces. Among others, these are exemplary:

Listing 6.1: What is the sub-continent to which geographically does Kazakhstan belong?

```
SELECT ?y
  WHERE {
    ?y rdf:type demisgeo:SubContinent .
    ?y demisgeo:geographicNarrower demisgeo:Kazakhstan . }
```

Listing 6.2: What is the name/ISO3-code of a country that has Suva as capital and is located in the Pacific Ocean?

```
SELECT ?y ?x
  WHERE {
    ?y rdf:type demisgeo:Country .
    ?y skos:broader demisgeo:Pacific_Ocean .
    ?y demisgeo:capital demisgeo:Suva .
    ?y demisgeo:isoCountryCode3 ?x . }
```

The preliminary set of competency questions that we used for our geo ontology includes (we list here only exemplary few):

1 What is the federal state to which the city of Rüdesheim belongs?
2 On which sub-continent are the Marshall-Islands and were did they historically belong to?
3 Which are the Asian countries that were politically belonging to the former Soviet Union?
4 What is the name of former country named Zaire?
5 What is the capital of Togo?
6 What are the official languages spoken in Taiwan?
7 Which countries are associated with the Pacific Ocean?
8 Which cities are belonging to Germany?
9 Which islands are located in the Pacific Ocean?
10 To which state does Tybee-Island belong?

Our competency questions in form of SPARQL queries can be answered successfully. In the current stage of development, we are combining the preliminary extracted disease related data with the geo ontology in order to answer questions like:

In which countries does Ebola virus typically spread and what are the main symptoms of this disease?

Step by step, we are combining more context (as described in our reference model) into our queries while constantly improving our knowledge base.

Acknowledgments. This project is an ongoing work under the funding of the German Federal Ministry of Health. The work described above is part of DEMIS (Deutsches Elektronisches Melde- und Informationssystem für den Infektionsschutz, German Electronic Reporting and Information System for Infectious Disease Control) project work. We would like to thank all our project colleagues involved in it, especially our technical DEMIS project leader, Hermann Claus, for making our contribution possible.

References

1. Scott Marshall, M., Boyce, R., Deus, H.F., Zhao, J., Willighagen, E.L., Samwald, M., Pichler, E., Hajagos, J., Prud'hommeaux, E., Stephens, S.: Emerging practices for mapping and linking life sciences data using RDF: a case series, vol. 14, pp. 2–13 (2012)
2. Calvanese, D., Cogrel, B., Komla-Ebri, S., Kontchakov, R., Lanti, D., Rezk, M., Rodriguez-Muro, M., Xiao, G.: Ontop: answering SPARQL queries over relational databases. J. Semant. Web **8**(3), 471–487 (2017). doi:10.3233/SW-160217
3. Ferreira, J.D., Pesquita, C., Couto, F.M., Silva, M.J.: Bringing epidemiology into the semantic web. In: Proceedings of ICBO: International Conference on Biomedical Ontology, vol. 897 (2012). ISSN: 1613-0073
4. Krause, G., Altmann, D., Faensen, D., Porten, K., Benzler, J., Pfoch, T., Ammon, A., Kramer, M.H., Claus, H.: SurvNet electronic surveillance system for infectious disease outbreaks, Germany. Emerg. Infect. Dis. **13**(10), 1548–1555 (2007)

PFSgeo: Preference-Enriched Faceted Search for Geographical Data

Panagiotis Lionakis[1,2($\boxtimes$)] and Yannis Tzitzikas[1,2]

[1] Institute of Computer Science, FORTH-ICS, Heraklion, Greece
{lionakis,tzitzik}@ics.forth.gr
[2] Computer Science Department, University of Crete, Heraklion, Greece

Abstract. In this paper we show how an *exploratory search process*, specifically the *Preference-enriched Faceted Search* (PFS) process, can be enriched for exploring datasets that also contain *geographic* information. In the introduced extension, that we call *PFSgeo*, the objects can have geographical coordinates, the interaction model is extended, and the web-interface is enriched with a map which the user can use for inspecting and restricting his focus, as well as for expressing *preferences*. Preference inheritance is supported as well as an automatic scope-based resolution of conflicts. We detail the implementation of the interaction model, elaborate on performance and report the positive results of a task-based evaluation with users. The value of *PFSgeo* is that it provides a generic and interactive method for aiding users to select the desired option(s) among a set of options that are described by several attributes including geographical ones, and it is the first model that supports map-based preferences.

Keywords: RDF and geospatial data · Faceted search · Exploratory search · Preferences · Map visualization

1 Introduction

A plethora of datasets contain geographic information and there are several approaches that combine Linked Data[1] and spatial data, create virtual geospatial RDF graphs on top of geographical databases [4] or use well-established existing controlled vocabularies (thesauri) and ontologies to enhance metadata documents with synonyms and translated terms, as well as location names for an improved discovery and linked-data eligibility using bounding box or text-based search [21]. For example, LinkedGeoData[2] is an effort to add a spatial dimension to the Web of Data/Semantic Web. LinkedGeoData uses the information collected by the OpenStreetMap[3] project and makes it available as an RDF

---

[1] http://lod-cloud.net/.
[2] http://linkedgeodata.org/.
[3] http://www.openstreetmap.org/.

© Springer International Publishing AG 2017
H. Panetto et al. (Eds.): OTM 2017 Conferences, Part II, LNCS 10574, pp. 125–143, 2017.
https://doi.org/10.1007/978-3-319-69459-7_9

dataset according to the Linked Data principles. Similar to this, GeoLinked-Data [9] is an open initiative of the Ontology Engineering Group (OEG) whose aim is to enrich the Web of Data with Spanish geospatial data. Moreover, there are works such as GeoNames that provides a geographical database available and accessible under a Creative Commons Attribution 3.0 License, containing over 10 million geographical names corresponding to over 9 million unique features [19]. Great Britain's national mapping agency, Ordnance Survey[4], has been the first national mapping agency committed to make publicly available various kinds of geospatial data from Great Britain as open Linked Data. Similar to this, data.geohive.ie[5] aims to provide an authoritative platform for serving Ireland's national geospatial data including Linked Data. Consequently there is a need for generic, flexible and interactive methods that allow users to explore such datasets and get an overview of the results on the map which is the most prominent technique in various domain-specific commercial platforms (e.g. Booking.com, Airbnb etc.).

Faceted search [12] is the more prominent technique in e-commerce and tourism services, and has been generalized also for RDF datasets (see [17] for a survey). The enrichment of faceted search with *preferences*, hereafter *Preference-enriched Faceted Search* [18], for short PFS, has been proven useful for recall-oriented information needs, because such needs involve decision making that can benefit from the gradual interaction and expression of preferences. PFS supports user clicks that correspond to either *hard* or *soft* constraints. The first kind corresponds to the actions of the classical faceted search where the user can *restrict* (i.e. filter) his focus gradually while getting an overview of the information space (just like in Booking.com). The second kind of actions corresponds to actions that specify *preferences* that *rank* accordingly the information space, e.g. a user can say that 2-star is the most preferred category and this statement will *rank first* the 2-star hotels but will not vanish the rest. The distinctive features of PFS is that it allows expressing preferences over attributes whose values can be hierarchically organized (and/or multi-valued), it supports preference inheritance, and it offers scope-based rules for resolving automatically the conflicts that may arise. As a result the user is able to restrict his current focus through the faceted interaction scheme (hard constraints) that lead to non-empty results, and rank the objects of his focus according to preference (soft constraints). Recently, PFS has been used in various domains, e.g. for offering a flexible process for the identification of fish species [16], as well as a Voting Advice Application [15].

However the plain PFS does not exploit the geographical aspect(s) of the data. To tackle this requirement, in this paper we introduce PFSgeo, an extension of PFS appropriate for exploring datasets with objects that have *spatial coordinates*. Consider a small information base comprising information about hotel rooms where each room is described by various attributes (e.g. location, stars, room type, price, accessories, address, etc.) including two spatial attributes, i.e. Longitude and Latitude. The Longitude-Latitude pair defines a point on

---

[4] http://data.ordnancesurvey.co.uk/.

[5] http://data.geohive.ie/.

the map. We adopt the coordinate system that is used by Google Maps (i.e. the WGS84 standard). For example, the way these two attributes are represented, as well as the coordinates of one hotel, are shown in the corresponding RDF in the TURTLE serialization format:

```
@prefix hippalus: <http://ics.forth.gr/isl/hippalus/#>.
@prefix xsd: <http://www.w3.org/2001/XMLSchema#>.
@prefix rdfs: <http://www.w3.org/2000/01/rdf-schema#>.
@@prefix owl: <http://www.w3.org/2002/07/owl#>.
@@prefix rdf: <http://www.w3.org/1999/02/22-rdf-syntax-ns#>.
...
hippalus:Longitude a rdf:Property;
rdfs:domain hippalus:Hippalus_Id;
rdfs:range xsd:float.

hippalus:Latitude a rdf:Property;
rdfs:domain hippalus:Hippalus_Id;
rdfs:range xsd:float.
...
hippalus:Lato_Boutique_Hotel
a hippalus:hippalusID;
hippalus:Latitude ''35.341751'' ^^xsd:float;
hippalus:Longitude ''25.136610'' ^^xsd:float.
```

The extension of PFS for geographical data raised several issues and challenges including (a) how to come up with an intuitive user interface that can be used easily by casual users, (b) how to avoid cluttering the map with too many objects, and how to make evident the rank of these objects on the map, and (c) how to distribute the required computational tasks to the server-side and external sources (map APIs) for achieving good performance.

We introduce PFSgeo an extension of PFS that offers the ability (a) to show the corresponding objects on a map, and (b) to use the map for restricting the information space and/or expressing preferences (by selecting an area of the map and then issuing a statement). To grasp the idea, consider the simple scenario shown in Fig. 1 where the user has zoomed over the area of interest and has already expressed (through the facet exploration panel in the left) as most preferred the 2-star hotels amongst other actions. The updated hotels are shown not only as a list of textual entries in the upper half of the central part of the screen (where we can see three buckets of hotels based on the actions of this scenario), but also as markers on the map. The labels on the markers indicate the preference order, e.g. the label "Rank1" on a marker indicates the existence of a hotel in the first bucket, while the number on a cluster of markers (markerclusterer) indicates how many markers this cluster contains. PFS supports *preference inheritance in facets with hierarchically organized values and a scope-based method for* automatically resolving conflicts. The motivation is that it is very common (and practical) also in natural language to state something general and then to provide the exceptions. Indeed, this makes sense also in the geographic domain as it can offer flexibility which is evident also from the

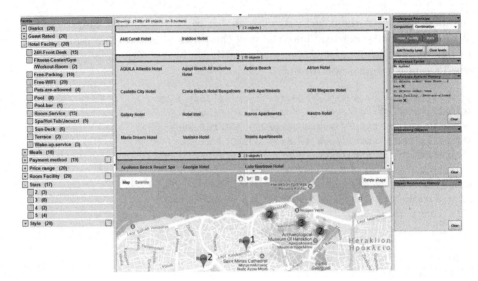

Fig. 1. A simple case scenario of exploration using both the facet exploration panel in the left and shapes over the map.

following example: Consider a user that prefers hotels in the broader area of Heraklion city but not in the area of the city center because it might be noisy there. To this end he could issue the following actions b1, b2:

```
b1: BEST  x1,y1,x2,y2      // broader area of Heraklion city
b2: WORST x1',y1',x2',y2'  // Heraklion city center
```

assuming that the first four coordinates capture the entire city of Heraklion, while those of $b2$ capture only a subarea (city center). Notice that the area of $b2$ is included in the area of $b1$, as shown in Fig. 2, therefore the produced ranking should have first all hotels of Heraklion except those in the subarea (rank 1), followed by those not in Heraklion (rank 2), and finally those in the city centre (rank 3). In general the user through left and right clicks can gradually formulate complex restrictions and rankings based on statements expressed over the facet values and the map (in any order). For example, the user can gradually (and without having to type anything) express information needs like: *"hotels in Heraklion with price less than 90 Euros per night and free wifi while preferring (a) 2-star hotels, (b) the areas that I have marked on the map, (c) tennis than basketball facilities"*. In a nutshell the key contributions of this paper are: (a) it is the first work about exploratory search with preferences over geographical data that supports preference inheritance and automatic resolution of preference conflicts, (b) it details an implementation of the model in the context of a publicly accessible prototype that supports PFS and uses Google Maps, (c) it reports experimental results regarding performance and scalability, and (d) it reports the results of a task-based evaluation with users.

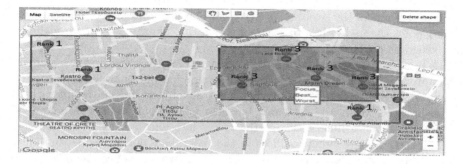

Fig. 2. Auto conflict resolution of preferences over intersected areas.

The introduced approach can be applied over RDF data, over the results of SPARQL queries, or over flat-structured files after a transformation to RDF. The proposed interaction can be very useful in e-commerce applications and geo-related services (e.g. in tourism). The rest of the paper is organized as follows. At first we describe background and related work, then we describe the approach and its implementation and finally we report experimental results about the performance and a task-based evaluation with users. Finally, we conclude and identify issues that are worth further research. A video demonstrating the PFS interaction model as implemented by the web system Hippalus (without maps) is available at https://www.youtube.com/watch?v=Cah-z7KmlXc while a video of PFSgeo (demonstrating PFS with maps) is available at https://youtu.be/h7Ov5QXajYM.

2   Related Work and Novelty

There are several works in the database world that provide a preference based ranking of spatial data, e.g. [1,11,22] study a number of algorithms and indexes for top-k spatial preference queries that rank spatial objects based on qualities of features in their spatial neighbourhood. Other works like GEORank [3] provide a location-aware and temporal ranking of news according to user's current or a fixed location set in the user profile. In GeoRank, this location is static, hence they do not provide exploration services which is the focus of our work. Efforts such as 3cixty [14], deal with the process of building comprehensive knowledge bases that contain spatial data about events and activities of cities, collected from numerous local and global data providers.

As regards systems that support exploratory search, the system Facete [13] provides faceted search over data that contain geographic information. Mappify [8] is a web application to create map views displaying concept based points of interest. Providing faceted exploration capabilities, Mappify allows one to define custom concepts on a SPARQL endpoint. In comparison to Facete or Mappify, we do not use maps just for displaying objects; the user can use the map also for restricting the objects (by selecting the desired area) and also for defining his/her preferences (by selecting preferred or non preferred areas). Furthermore at any point the user is free to issue an action through the map or through the non geographical facets.

In brief, and to the best of our knowledge, PFSgeo is the first work about exploratory search with preferences and geographic information. It uses geographical maps not only for displaying the focus objects during the interaction (as in previous works and systems), but also as an input method. Moreover it supports preference inheritance in the hierarchies and a scope-based method for automatically resolving conflicts in the geographic domain based on inference. Table 1, categorizes the aforementioned frameworks and commercial booking/real estate platforms (Booking.com, Airbnb, rightmove.co.uk) and PFSgeo, according to various aspects like Faceted restriction, Preferences, display of focus on the map, shape-based (map) restriction and preferences.

Table 1. Related approaches and comparison with PFSgeo.

Feature	Research prototypes		Commercial platforms			PFSgeo
	Facete	GEORank	Booking.com	Airbnb	Rightmove	PFSgeo
Faceted restrictions	✓	-	✓	✓	✓	✓
Preferences	-	-	-	-	-	✓
Display of focus on map	✓	✓	✓	✓	✓	✓
Shape-based (map) restrictions	-	-	-	-	✓	✓
Shape-based (map) preferences	-	-	-	-	-	✓

3   The PFSgeo Approach

3.1   Extensions of the Language of PFS

Note that PFS offers actions that allow the user to order facets, values, and objects using *best, worst, preferTo* actions (i.e. relative preferences), *aroundTo* actions (over a specific value), or actions that order them lexicographically, or based on their values or count values. Furthermore, the user is able to *compose* object related preference actions, using *Priority, Pareto, Pareto Optimal* (i.e. skyline) and other. To realize PFSgeo we extended the language described in [18] with actions that have *spatial scope*. Based on the current implementation, the user's actions when interacting with the map is BEST and WORST, in contrast to the typical exploration paradigm of PFS that also supports other actions such as *preferTo*. The latter kind of actions would be difficult to be expressed over the map. We extended the PFS grammar with geo-anchored actions and we adapted accordingly the algorithms so that geo-anchored actions can be combined with actions over the rest facets. In the syntax "geo" denotes something with a spatial representation, e.g. rectangle, polygon, circle or point (our prototype focuses on rectangles for the time being). For example the extension of the *syntax* for expressing an action "BEST lat1, lon1, lat2, lon2" for a rectangle specified by the coordinates of its bottom-left and upper-right corners, expressed according to WGS84[6] standard, has the following form:

```
object order: geo rectangle [(lat1,lon1),(lat2,lon2)] best
```

3.2 Extensions of the System Hippalus

Then we extended Hippalus[7] which is a publicly accessible web system that implements the PFS interaction model. The information base that feeds Hippalus is represented in RDFS[8] (using a schema adequate for representing objects described according to dimensions with hierarchically organized values). We have implemented a tool called HDT (Hippalus Data Transformer) that transforms CSV files as well as the results of SELECT SPARQL results to an RDF file that is directly loadable by Hippalus. The values of an attribute can be hierarchically organized, and set-valued attributes are supported. The entire process is sketched in Fig. 3. For loading and querying such files, Hippalus uses Jena[9], a framework for building Semantic Web applications. Hippalus offers a web interface for Faceted Search enriched with preference actions offered through HTML 5 context menus[10]. The performed actions are internally translated to statements of the preference language described in [18], and are then sent to the server through HTTP requests. The server analyzes them, using the language's parser, and checks their validity. If valid, they are passed to the appropriate preference algorithm. Finally, the respective preference bucket order is computed and the ranked list of objects according to preference, is sent to the user's browser.

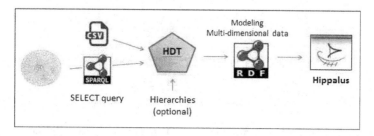

Fig. 3. The process for selecting and transforming data for being loadable by Hippalus.

---

[6] https://en.wikipedia.org/wiki/World_Geodetic_System.

[7] http://www.ics.forth.gr/isl/Hippalus/.

[8] http://www.w3.org/TR/rdf-schema/.

[9] http://jena.apache.org/.

[10] Available only to firefox 8 and up.

For supporting PFSgeo, apart from extending the language, we extended Hippalus with Google Maps. We used libraries of Google Maps JavaScript API, specifically the *drawingManager Library*[11] which provides a graphical interface for drawing shapes such as rectangles, and the *Geometry Library* (See footnote 11) that offers utility functions for computations involving polygons and polylines e.g. for getting the markers within a shape. Figure 4 shows the new architecture that shows how Google Maps are involved as well as the extensions that were made at the server-side. For tackling cases where there are too many objects which are very close, the system implements the *MarkerClusterer* $v3$[12], which allows the display of multiple GEO locations as a labeled cluster. This cluster is basically a united marker indicating the number of the containing single markers within this area. The combined use of Google Map and Marker-Clusterer allows the system to update the displayed results both on demand or on the fly. Both *restrictions* and *preferences* over geographical data are possible through a pop-up menu, using the *right click* after a *shape* selection. As regards preferences, the user can issue either *BEST* or *WORST* preference actions. Each action is translated into a HTTP POST request method to Hippalus server using the JSON data-interchange format. This object contains the long-lat pair of the markers within this area, given by Google Maps JavaScript API V3.

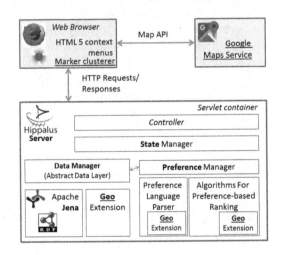

Fig. 4. Client - Server architecture of Hippalus with geo-extension.

3.3   Algorithmic Perspective

The *algorithmic perspective* concerns how restriction and preference actions are managed efficiently in case these actions have been issued through the map. We have identified two methods that we shall describe through an example.

---

[11] https://developers.google.com/maps/documentation/javascript/reference.
[12] https://developers.google.com/maps/articles/toomanymarkers.

Without loss of generality, let's assume that the user selects an area in the map through a rectangular shape, defined by its coordinates as described previously and through right click he selects one option, say *BEST*.

Method 1. In this method at first (and at client-side, i.e. browser-side) we find the hotel markers that fall in selected area of the map (by issuing a call to Google Maps). Suppose that these markers correspond to a set of k hotels denoted by $Hshape = \{h_1, \ldots, h_k\}$. Then the browser sends to the (Hippalus) server a set of preference actions, specifically $|Hshape|$ in number $BEST$ actions, i.e. the following set of actions: $\{BEST\ h \mid h \in Hshape\}$. According to Method 1 Longitude and Latitude are treated as ordinary facets from the server-side. As we shall see in the section with the experiments, Method 1 proved inefficient for preference actions issued through the map. The first reason is that a lot of data are sent to the server. The second reason is that the complexity of the algorithm for preference-based ranking (as proposed in [18]) that runs on the server side of Hippalus, has time complexity $O(|A||B|^2)$ where A is the set of objects in the focus, and B is the number of preference actions. The algorithm is also given below in Algorithm 1, adapted to our context. It follows that by increasing the number of actions (e.g. in our example although the user issued 1 *BEST* action, the server received $|Hshape|$ in number $BEST$ actions) we actually affect negatively the performance of the algorithm. This is the motivation for Method 2.

Method 2. This method follows a different approach that is better aligned with the PFS and its algorithms. The key point is that instead of sending to the server the set of actions $\{BEST\ h \mid h \in Hshape\}$ we sent just one action: $BEST\ x1, y1, x2, y2$ and this is the motivation for extending the language of [18] with actions that have geographical anchors (as mentioned in Subsect. 3.1). We can already see the benefits: less data have to be transferred and smaller $|B|$, thus faster production of the ranked list. It is also worth noting that the extension of the language with such actions does not require changing the core algorithms of [18] since that framework already captures actions with *scope* (for supporting preference inheritance in facets with hierarchically organized values). In our case the scope is spatial. Moreover, the scope-based method for automatically resolving conflicts makes sense also in the geographic domain as explained in the introductory section. Note that in the geographical context an action b is narrower than a b', denoted by $b \sqsubseteq b'$, if and only if the area of b defined by $(x1, y1, x2, y2)$ is included in the area of b' defined by $(x1', y1', x2', y2')$. Since we use the rectangle shape as a proof of concept, it is not hard to see that this holds if $(x1 \geq x1') \wedge (y1 \geq y1') \wedge (x2 \leq x2') \wedge (y2 \leq y2')$ and this is actually how $CheckSubScopeOf$ (of Algorithm 1) has to be implemented for working over preference actions anchored to geographical areas. It follows that the geographical areas do not add any cost in the computation of $(B, \sqsubseteq)$.

The second part of the algorithm computes the *active scope* of each $b \in B$. In our case the active scope of an action b, is defined by excluding from the area of b all areas that are narrower than b. To implement this part of the algorithm we need to adapt $IsInScope(e, b)$ for our case. This function should return True if e belongs to the area of b. Obviously this can be decided very fast, since a point

(x, y) falls in the area defined by $(x1, y1, x2, y2)$ iff $(x1 \le x \le x2) \wedge (y1 \le y \le y2)$. Subsequently we use the active scopes, that we have just computed, for extending B to a set B', specifically for an action $b = BEST\ (x1, y1, x2, y2) \in B$, we add to B' the following set of actions: $\{BEST\ h \mid h \in ActiveScope[b]\}$.

The third part the algorithm just parses the set B' in order to get the sets Be, Wo, i.e. collecting those hotels with $BEST$ and those with $WORST$ and then produces the ranked hotels by topological sorting over the graph that contains the following preference relationships $\{e \succeq e' \mid e \in Be, e' \in Wo\}$. This is all that is required for producing the preference-based ranking of hotels.

As noted, conflict resolution based on scope is supported. However, it is possible to have two actions, e.g. one BEST and one WORST, over two *overlapping* areas. In that case none of these actions is narrower than another. If there is not any object (hotel in our case) that falls in the intersection of these areas, then this conflict does not have any impact on ranking of objects. If however, one or more objects belong to the intersection of the overlapping areas, then we have

---

Algorithm 1. AlgClearOpt(E, B)
Input: the set of elements E, the set of actions B
Output: a bucket order over E

1: /** *Part (1): Computation of $(B, \sqsubseteq)$ where $\sqsubseteq$ orders actions according to their scope*/
2: $Visited \leftarrow \emptyset$
3: $R_\sqsubseteq \leftarrow \emptyset$ ▷ $R_\sqsubseteq$ corresponds to $\sqsubseteq$
4: **for** each $b \in B$ **do**
5:     **for** each $b' \in B \setminus Visited$ **do**
6:         **if** $CheckSubScopeOf(b, b')$ **then**   ▷ if the scope of b is included in the scope of b'
7:             $R_\sqsubseteq \leftarrow R_\sqsubseteq \cup \{(b \sqsubseteq b')\}$
8:         **else if** $CheckSubScopeOf(b', b)$ **then**
9:             $R_\sqsubseteq \leftarrow R_\sqsubseteq \cup \{(b' \sqsubseteq b)\}$
10:        **end if**
11:        $Visited \leftarrow Visited \cup \{b\}$
12:     **end for**
13: **end for**
14: /** *Part (2): Efficient Computation of Active Scopes* */
15: **for** each $b \in B$ **do**
16:     $C(b) \leftarrow$ direct children of b wrt $R_\sqsubseteq$
17:     $ActiveScope[b] \leftarrow \{e \in E \mid IsInScope(e, b) \wedge$
18:     $(\forall c \in C(b)$ it holds $IsInScope(e, c) =$ False$)\}$
19: **end for**
20: /** *Part (3): Derivation of the final bucket order* */
21: Use the active scopes to expand the set of actions B to a set of actions B'
22: $(Be, Wo) \leftarrow$ Parse(B')   ▷ Be the best, Wo the worst
23: Sort the elements of E by applying topological sorting over the Be-Wo graph.
24: **return** the produced bucket order of E

---

a non-resolvable conflict. In that case, the algorithm prevents cyclic preferences and produces the bucket order up to the objects that have the conflict.

Another important (for scalability) characteristic of the algorithm is that it never computes the entire scope of any action, i.e. its scope in the entire information base. Instead, it checks whether the elements of the current focus E belong to the scopes of issued actions. This means that its computational complexity does not depend on $|Obj|$, but on E (where $E \subseteq Obj$) which we can assume that is not big because the user applies preferences after he has focused on a smaller set of objects. The same is true for restrictions performed through the map. Furthermore, it is not useful (for the user) to show on the map more than a few hundreds objects (since the markers would clutter the space). For this reason, only if the focus is less than a configurable threshold (say 100) it is useful to show the objects on the map and rank them according to the issued preference actions. The algorithms can handle this load since for producing the preference-based ranking for foci of even 1,000 objects takes less than 3 s.

4   Performance Evaluation

4.1   Testing Scalability

Information Base. We used a synthetically produced dataset with information about hotels in the region of the Crete island where each hotel is represented by various attributes (as described in the introductory section). Two main scenarios were designed each having the objective to measure the performance of the system over geo data. Each scenario is essentially a sequence of requests that simulates a user that interacts with the system using the map. To conduct the experiments, we used a laptop PC (dual-core 1.8 GHz CPU, 6 GB of main memory) that deployed PFSgeo locally. The **First Scenario** concerns *restrictions* using the map, i.e. it simulates a user that issues multi restriction actions using a shape. The **Second Scenario** concerns *preferences* using the map, i.e. it simulates a user that issues preference actions using a shape. We used the following datasets: (a) Dataset 1 containing 1,000 hotels, (b) Dataset 2 containing 5,000 hotels and (c) Dataset 3 containing 20,000 hotels.

Simulation. The simulation platform considers the area and the information base described previously. In order to evaluate the system's performance we apply the scenarios on predefined shapes on the map that include a set of markers. The number of markers differs depending on the size of the dataset. In each of these two scenarios we issue 10 requests from 9 different (simulated) users. For this reason we have created 9 different non-intersecting shapes within the boundaries of the simulation area, as shown in Fig. 5. The division of the area ensures that any possible subarea will contain markers, while the sequence simulates a user that issues actions with gradually bigger shapes, meaning that the corresponding load becomes heavier since more instances have to be passed as parameters. Note that in shape 9 the user issues an action over all instances of the dataset. Each shape is represented by the action of the respective user

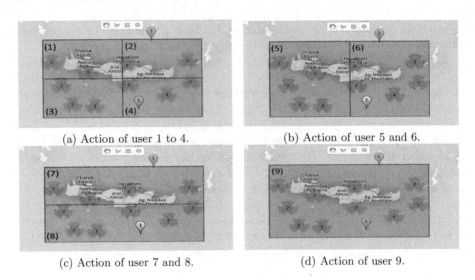

(a) Action of user 1 to 4.   (b) Action of user 5 and 6.

(c) Action of user 7 and 8.   (d) Action of user 9.

Fig. 5. The simulated different user's actions for preferences and restrictions.

accordingly. We should mention that a restriction action based on intersecting shapes, say a shape A and a shape B, eventually has the same outcome as the restriction action using as shape the intersection $A \cap B$.

4.2 Analysis of the Performance

Standalone Experiment Results. The analysis aims to evaluate the performance by measuring the delay that a user experiences (note that the delay while exploring a geo dataset should not overcome the user's tolerance [7]). To this end, we introduce the following metrics:

- **initTime:** this is the time the user experiences for loading the instances on startup and for displaying them on the map.
- **restriction time-*1st scenario:*** this is the total response delay for restricting the focus using a multi *restriction query*.
- **preference time-*2nd scenario:*** this is the total response delay when the user issues a *preference query* using a shape.

Based on the current implementation these metrics do not concern only the management of the geographic information, in the sense that during the exploration of a spatial dataset, *initTime, restriction time* and *preference time* are affected not only by the geo aspect but also by the various delays of Hippalus. Specifically, both **restriction time** and **preference time** include the time required for *"updateHierarchy"* and *"updateHistory"* which is the total time required to update the facet-values and history displays, as well as, *"getGeoLocations"* and *"updateInstances"* which is the total required time to get the locations from Hippalus server and the time to create-update the markers on the map and update the instances in buckets. The reported times include the time required for rendering the map using third party APIs calls, i.e. Google Maps.

As regards the results, as we can see in Fig. 6a the average **init time** for datasets with 20,000 instances is 1.8 s. Figures 6b, c and d show the results of the **restrictions** for Dataset 1, Dataset 2 and Dataset 3.

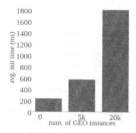

(a) Average **init time** over the *3 datasets*.

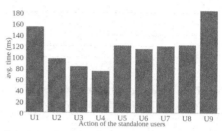

(b) Average **restriction times** of *Dataset 1* for each standalone user.

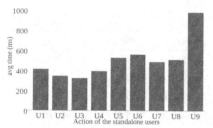

(c) Average **restriction times** of *Dataset 2* for each standalone user.

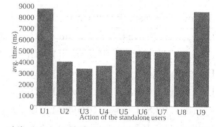

(d) Average **restriction times** of *Dataset 3* for each standalone user.

Fig. 6. **Init times** and **restriction times** using Method 1.

4.3 Comparative Experimental Results

Here we report experimental results for comparing Method 1 with Method 2. Notice that for more than 1,000 instances the system with Method 2 has an average preference time of 2.8 s in contrast to the 98.4 s of the system with Method 1, i.e. this method is one order of magnitude faster than the previous one (specifically 35 times faster). To this respect, Fig. 7a presents a comparison of those 2 methods regarding the preference action of the users, for the Dataset 1. Finally, Fig. 7b shows the average query time for 10 requests that get served by the system for each of the three metrics for all datasets, where Table 2 shows the results of the preference action of Method 2 compared to Method 1. Regarding the Method 1 note that for datasets 2 and 3 the system could not respond and terminated with a timeout error.

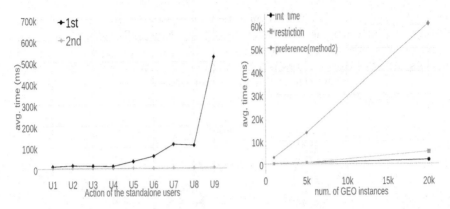

Fig. 7. (a) Average **preference** times of *Dataset 1* for each standalone user *(Method 1 compared to Method 2)* (b) Average times for the *3 metrics* of *3 datasets* for all users.

Table 2. Average **preference** times of the simulated users over the *3 datasets* (*Method 1 comparing to Method 2*).

	Method 1			Method 2		
Actions	Dataset 1	Dataset 2	Dataset 3	Dataset 1	Dataset 2	Dataset 3
User1	9,794 ms	(timeout)	(timeout)	1,880 ms	7,868 ms	14,792 ms
User2	14,265 ms	(timeout)	(timeout)	1,925 ms	8,343 ms	51,608 ms
User3	12,469 ms	(timeout)	(timeout)	1,637 ms	8,932 ms	51,403 ms
User4	11,018 ms	(timeout)	(timeout)	1,484 ms	8,569 ms	53,232 ms
User5	33,871 ms	(timeout)	(timeout)	3,007 ms	14,424 ms	2,396 ms
User6	58,785 ms	(timeout)	(timeout)	2,774 ms	15,211 ms	82,511 ms
User7	114,007 ms	(timeout)	(timeout)	3,709 ms	15,484 ms	47,586 ms
User8	108,983 ms	(timeout)	(timeout)	3,240 ms	15,684 ms	103,110 ms
User9	522,978 ms	(timeout)	(timeout)	5,682 ms	27,535 ms	137,783 ms

5   Evaluation with Users

Although Hippalus has been evaluated positively by users in various contexts (the interested reader can refer to [15,16]), since the extension with maps is novel, we conducted a comparative task-based evaluation in which users would have to explore and express their preferences and/or restrictions using both UIs (with and without maps). We prepared some tasks and asked the participants to carry them out. The tasks are based on the following general scenario: *You are planning to stay in a hotel either for vacations or for a business trip in a specific area. You want to discover hotels in that area, however you do not have any a-priori knowledge about the location names in that area. You have at your disposal a set of restrictions (e.g. price, stars, facilities) but you cannot apply a restriction based on location name because you do not have such knowledge.*

5.1 Preparation

Purpose. Our objective was to investigate whether even in a small dataset, with only 20 hotels, the map extension would make search more effective and users more satisfied with the interaction. We distinguish two different UIs:

(a) Classical model of PFS interaction model UI1
(b) PFS exploration using the map both as visualization and input for the set of the user's actions UI2

The different UIs can be selected by the user using the display options menu as shown in the video tutorial. The default view is "Text and Map" which displays the information space both as textual and map-based representation. The evaluation performed for both UIs in order to extract the users' satisfaction level and exploit any possible problems and difficulties.

Training. We created a Google form and we invited all the candidate participants to fill it in. Initially, the users were asked to watch two video tutorials: one demonstrating Hippalus without maps (the video is available at https://www.youtube.com/watch?v=Cah-z7KmlXc) and one demonstrating Hippalus with maps (the video is available at https://youtu.be/h7Ov5QXajYM).

Participants. According to [6,10,20] Jakob Nielsen proposed that for a single usability test, 10 participants are enough for revealing severe usability problems. However since we are not interested only in usability problems we decided to involve more users. In total 20 persons participated, (75% male and 25% female), with various age groups (20% were between the ages of 20 and 25, 65% were between the ages of 25 and 30, 5% were between the ages of 30 and 35, while 10% were 35 and over) and various professions (65% CS-related employees, 20% researchers and the rest 15% were other). The evaluation started on March 10 and ended on April 6 of 2017.

Tasks. We asked the users to explore a dataset of Hotels (containing various information as in Booking.com) first without map and then with map, and carry out a sequence of basic tasks, including restrictions and expressions of preferences, as shown in Table 3. Note that we have 4 different tasks for each of the two different UIs *(IDs a,b)* which eventually gives a total number of 8. In each task the user had to provide the name of a single hotel (for tasks 2 and 3), two hotels (for tasks 1 and 4) or "none" (if no answer can be given) as possible answers. For example, for the task *find the hotels that are closer to the sea* it makes sense to assume that this could be a subjective issue since the statement "closer" in terms of distance may differ according to user's perspective. To this respect, we count every hotel name as correct whereas we only deemed the "none" answer

Overall, which one do you prefer?

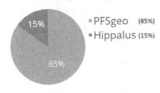

- PFSgeo (85%)
- Hippalus (15%)

Fig. 8. PFSgeo vs Hippalus.

Table 4. Number of wrong answers in the evaluation with users.

Task Id	PFSgeo	PFS
1	0/20	5/20
2	1/20	4/20
3	0/20	2/20
4	0/20	6/20

as mistaken, since the objective of this evaluation was to evaluate the usability of the system, using the *System Usability Score* **(SUS)** [2]. For this reason, we asked the users at the end to rate each system using the following scale ⟨Useless, Not Useful, Neutral, Useful, Very Useful⟩, and state which of these they prefer.

Results. As shown in Table 4, 98.75% of the participants managed to find a hotel using the PFSgeo, in contrast to 78.75% of plain PFS. In the overall rating, shown in Fig. 9, we can see that 50% of the participants rated the system Very Useful and 40% Useful, hence in total 90% of the participants were positive. Note that only 15% of the participants rated the PFS Hippalus Very Useful. It is also interesting to note that that 85% of the participants stated that they prefer the PFSgeo over the PFS Hippalus (Fig. 8). In terms of statistical significance, the exact (Clopper-Pearson) 95% Confidence Interval is (62%, 97%) indicating that the PFSgeo is strongly preferred by the users over the Hippalus.

Table 3. Tasks description for user evaluation over the two UIs.

ID	Task description
a1, b1	You are interested in "3-star" Hotels in "Heraklion", and you prefer these which are close to the sea. Browse the system and give the 2 hotels that you prefer most
a2, b2	You would like to visit the "Archaeological Museum of Heraklion" (located in the center of the city). Find the hotel that is closer to this museum and prefer the most economical
a3, b3	You would like to find a hotel in Heraklion with Price range "49–99" as BEST but you do not prefer those close to the sea. Use the system and provide one such hotel
a4, b4	You are interested in "3-star" Hotels in Crete but you do not like hotels located in cities (Heraklion) unless they are close to a sightseeing (Archaeological Museum). Use the system and return the two more preferred hotels

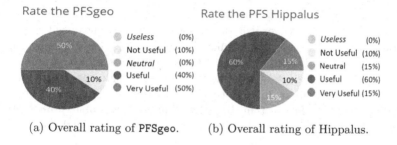

(a) Overall rating of PFSgeo.   (b) Overall rating of Hippalus.

Fig. 9. Overall ratings of PFSgeo and Hippalus.

6   Concluding Remarks

In this paper we showed how a model for exploratory search, the Preference-enriched Faceted Search (PFS), can be enriched for being appropriate for exploring datasets that contain geographical information. The extended model, that we call PFSgeo, uses geographical maps not only for displaying the objects of the focus during the interaction (as in previous works and systems), but also as an input means for restricting the focus and for defining preferences. We should also stress that the proposed framework supports preference inheritance in the hierarchies and a scope-based method for automatically resolving conflicts. Subsequently we presented an implementation of the proposed model as an extension of the system Hippalus. Since there is not any system that supports PFS over datasets with geographic information the challenging task is how to support this interaction for big datasets. For foci that contain large amounts of objects (which is expected in the context of exploratory search), we support marker clustering to avoid cluttering the map. Finally we elaborated on performance issues and we provided measurements over synthetically produced datasets about hotels. Based on this analysis, we identified those tasks that affect the scalability of the approach, and proposed methods that can be used for improving the efficiency of the approach. The user evaluation highlighted that when exploring datasets with geo aspect the functionality of the map is a very promising direction regarding the usability of the system. Overall 90% of the participants rated PFSgeo very positive (50% as Very Useful and 40% as Useful) and 98.75% of the total tasks were completed successfully.

Since this is the first work on extending Preference-enriched Faceted Search for geographical data, there are several directions that are worth further research. For instance, the current implementation supports only one geographical facet, however more than one are required for various kinds of objects (e.g. flights have two geographical facets "from" and "to"). Moreover, since a geographical area is continuous it is worth investigating also radius-based preferences, for instance if the user selects an point/area and issues a BEST action, then all objects could be ranked according to their distance to the point/area (e.g. "show me all the restaurants within 20 min driving time around me"). For achieving performance over very big data sets, one could investigate enriching the server-side with a spatial index (e.g. R-Tree, Quadtree, etc.), or adopting a triplestore that supports GeoSPARQL. Moreover it is worth investigating extensions of Triple Pattern Fragments for the geospatial Linked Data (like [5]). For enhancing applicability it is worth supporting more standards for representing coordinates and geometries. Finally, the interaction model assumes "single entity type objects" (according to the taxonomy in [17]), therefore it could be extended for objects of multiple entity-types. The system Hippalus is publicly accessible through http:// www.ics.forth.gr/isl/Hippalus/ (requires Firefox version 8 or higher).

References

1. de Almeida, J.P.D., Rocha-Junior, J.B.: Top-k spatial keyword preference query. J. Inf. Data Manage. **6**(3), 162 (2016)
2. Bangor, A., Kortum, P., Miller, J.: Determining what individual sus scores mean: adding an adjective rating scale. J. Usability Stud. **4**(3), 114–123 (2009)
3. Bao, J., Mokbel, M.F.: Georank: an efficient location-aware news feed ranking system. In: Proceedings of the 21st ACM SIGSPATIAL International Conference on Advances in Geographic Information Systems, pp. 184–193. ACM (2013)
4. Bereta, K., Koubarakis, M.: Ontop of geospatial databases. In: Groth, P., Simperl, E., Gray, A., Sabou, M., Krötzsch, M., Lecue, F., Flöck, F., Gil, Y. (eds.) ISWC 2016. LNCS, vol. 9981, pp. 37–52. Springer, Cham (2016). doi:10.1007/978-3-319-46523-4_3
5. Debruyne, C., Clinton, É., O'Sullivan, D.: Client-side processing of GeoSPARQL functions with triple pattern fragments. In: Proceedings of the Workshop on Linked Data on the Web, LDOW (2017)
6. Faulkner, L.: Beyond the five-user assumption: benefits of increased sample sizes in usability testing. Behav. Res. Meth. **35**(3), 379–383 (2003)
7. Galletta, D.F., Henry, R., McCoy, S., Polak, P.: Web site delays: how tolerant are users? J. Assoc. Inf. Syst. **5**(1), 1–28 (2004)
8. Lehmann, J., Athanasiou, S., Both, A., Garcia Rojas, A., Giannopoulos, G., Hladky, D., Le Grange, J.J., Ngonga Ngomo, A.C., Sherif, M., Stadler, C., Wauer, M., Westphal, P., Zaslawski, V.: Managing geospatial linked data in the GeoKnow project, January 2015
9. Lopez-Pellicer, F.J., Silva, M.J., Chaves, M., Javier Zarazaga-Soria, F., Muro-Medrano, P.R.: Geo linked data. In: Bringas, P.G., Hameurlain, A., Quirchmayr, G. (eds.) DEXA 2010. LNCS, vol. 6261, pp. 495–502. Springer, Heidelberg (2010). doi:10.1007/978-3-642-15364-8_42
10. Nielsen, J., Molich, R.: Heuristic evaluation of user interfaces. In: Proceedings of the SIGCHI Conference on Human Factors in Computing Systems, pp. 249–256. ACM (1990)
11. Rocha-Junior, J.B., Vlachou, A., Doulkeridis, C., Nørvåg, K.: Efficient processing of top-k spatial preference queries. Proc. VLDB Endow. **4**(2), 93–104 (2010)
12. Sacco, G.M., Tzitzikas, Y.: Dynamic Taxonomies and Faceted Search: Theory, Practice, and Experience, 1st edn. Springer, Heidelberg (2009)
13. Stadler, C., Martin, M., Auer, S.: Exploring the web of spatial data with facete. In: Proceedings of the Companion Publication of the 23rd International Conference on World Wide Web Companion, pp. 175–178. International World Wide Web Conferences Steering Committee (2014)

14. Troncy, R., Rizzo, G., Jameson, A., Corcho, O., Plu, J., Palumbo, E., Hermida, J.C.B., Spirescu, A., Kuhn, K.D., Barbu, C., Rossi, M., Celino, I., Agarwal, R., Scanu, C., Valla, M., Haaker, T.: 3cixty: Building comprehensive knowledge bases for city exploration. Web Semantics: Science, Services and Agents on the World Wide Web (2017). http://www.sciencedirect.com/science/article/pii/S1570826817300318

15. Tzitzikas, Y., Dimitrakis, E.: Preference-enriched faceted search for voting aid applications. IEEE Trans. Emerg. Topics Comput. PP(99), 1 (2016)

16. Tzitzikas, Y., Bailly, N., Papadakos, P., Minadakis, N., Nikitakis, G.: Using preference-enriched faceted search for species identification. Int. J. Metadata Semant. Ontol. 11(3), 165–179 (2016)

17. Tzitzikas, Y., Manolis, N., Papadakos, P.: Faceted exploration of RDF/S datasets: a survey. J. Intell. Inf. Syst. 48(2), 329–364 (2016)

18. Tzitzikas, Y., Papadakos, P.: Interactive exploration of multidimensional and hierarchical information spaces with real-time preference elicitation. Fundamenta Informaticae 20, 1–42 (2012)

19. Vatant, B., Wick, M.: Geonames ontology (2006). http://www.geonames.org/ontology

20. Virzi, R.A.: Refining the test phase of usability evaluation: how many subjects is enough? Hum. Factors J. Hum. Factors Ergon. Soc. 34(4), 457–468 (1992)

21. Vockner, B., Mittlböck, M.: Geo-enrichment and semantic enhancement of metadata sets to augment discovery in geoportals. ISPRS Int. J. Geo-inf. 3(1), 345–367 (2014)

22. Yiu, M.L., Lu, H., Mamoulis, N., Vaitis, M.: Ranking spatial data by quality preferences. IEEE Trans. Knowl. Data Eng. 23(3), 433–446 (2011)

Speeding up Publication of Linked Data Using Data Chunking in LinkedPipes ETL

Jakub Klímek[(⊠)] and Petr Škoda

Faculty of Mathematics and Physics, Charles University, Malostranské nám. 25,
118 00 Praha 1, Czech Republic
klimek@ksi.mff.cuni.cz

Abstract. There is a multitude of tools for preparation of Linked Data from data sources such as CSV and XML files. These tools usually perform as expected when processing examples, or smaller real world data. However, a majority of these tools become hard to use when faced with a larger dataset such as hundreds of megabytes large CSV file. Tools which load the entire resulting RDF dataset into memory usually have memory requirements unsatisfiable by commodity hardware. This is the case of RDF-based ETL tools. Their limits can be avoided by running them on powerful and expensive hardware, which is, however, not an option for majority of data publishers. Tools which process the data in a streamed way tend to have limited transformation options. This is the case of text-based transformations, such as XSLT, or per-item SPARQL transformations such as the streamed version of TARQL. In this paper, we show how the power and transformation options of RDF-based ETL tools can be combined with the possibility to transform large datasets on common consumer hardware for so called chunkable data - data which can be split in a certain way. We demonstrate our approach in our RDF-based ETL tool, LinkedPipes ETL. We include experiments on selected real world datasets and a comparison of performance and memory consumption of available tools.

Keywords: Linked data · RDF · ETL · Transformation · Data chunking

1 Introduction

After almost a decade, Linked Open Data (LOD) gains interest of more and more organizations as a way of publishing data, as evidenced by the growth rate of the LOD Cloud Diagram[1]. As part of the OpenData.cz initiative[2] we help organizations in the government, such as the Czech Trade Inspection Authority or the Czech Social Security Administration, with the process of LOD publishing.

This work was supported in part by the Czech Science Foundation (GAČR), grant number 16-09713S and in part by the project SVV 260451.

[1] http://lod-cloud.net/

[2] https://opendata.cz

H. Panetto et al. (Eds.): OTM 2017 Conferences, Part II, LNCS 10574, pp. 144–160, 2017.
https://doi.org/10.1007/978-3-319-69459-7_10

Naturally, one of the key factors in deciding whether to invest in publishing LOD in these organizations is the availability of tooling, their hardware requirements and the effort needed to publish LOD in addition to publishing CSV or XML.

There is a multitude of tools for LOD publishing. Some work well on commodity hardware - a typical piece of equipment of employees in government organizations - only with smaller datasets. Some work well even with larger datasets, but using powerful servers, which, however, are typically not available to government organizations and when they are, they are not available for publishing of open data. A considerably smaller number of tools works well on a consumer-grade computer even with with larger datasets, such as hundreds of megabytes large CSV files, and these tend to be rather specialized in the task, e.g. TARQL[3] for streamed conversion of a CSV file to an RDF file using a SPARQL CONSTRUCT query. Such tools then need to be integrated into a larger platform, which facilitates the whole LOD production process. These platforms are ETL (Extract Transform Load) platforms. They allow their users to specify repeatable transformation pipelines, which gather (extract) input data, transform it into its desired form and load it into a triplestore or a file system and make the whole process automatically repeatable.

From our experience with publishing of LOD using ETL tools in government organizations and with development of the ETL tools themselves, we have identified a subset of larger datasets, which are being transformed inefficiently. It is the type of datasets where with a large number of independent entities such as records in individual files or rows in a table, which can be transformed independently of each other. They can, therefore, be processed in smaller quantities, more efficiently than dealing with all records at once. Today, typically, each step of a transformation pipeline takes all input data, does a transformation, and produces output data, which is then used as a whole as input for the next step.

Contributions. In this paper we describe a method for data chunking, its benefits and limitations for publishing LOD, and its implementation in our ETL platform, LinkedPipes ETL [5]. We demonstrate that data chunking considerably speeds up the LOD publishing processes and saves hardware resources. We describe the effect on real world dataset transformation pipelines and we experiment with data chunk sizes and memory consumption to measure the speedup and, therefore, to determine the hardware requirements for specific types of datasets.

Outline. The rest of the paper is structured as follows. In Sect. 2 we describe the notion of data chunks and discuss its properties and limitations. In Sect. 3 we introduce LinkedPipes ETL, our RDF-based ETL platform, some of its features and the additional data chunk components. In Sect. 4 we show the real-world datasets and corresponding transformation pipelines which were improved by data chunks. In Sect. 5 we experiment with the new approach in terms of determining the correspondence among data chunk sizes, memory consumption

---

[3] https://tarql.github.io/

and time elapsed before the transformation is finished, and we discuss the results. In Sect. 6 we survey related work. Finally, in Sect. 7 we indicate our future plans and we conclude.

2 Data Chunking

The motivation for data chunking comes from our experience with ETL pipelines for publication of datasets as LOD in various government organizations. There is a distinct subset of datasets which consist of a list of records, which are later transformed independently of each other throughout the pipeline right until they are finally loaded into a triplestore and/or saved as an RDF dump file. An example of such a dataset can be a CSV table of inspections performed by the Czech Trade Inspection Authority where each row is a record of one inspection, containing its location, date and the ID of the inspected subject. When such dataset is published as LOD, it can be done by an ETL pipeline consisting of several transformation steps (see e.g. Figure 1) such as a direct transformation of the CSV table to RDF according to the CSV on the Web W3C Recommendation[4], transformation of the directly mapped RDF to a form properly modeled using LOD vocabularies, linking the subject IDs to business registry records for that company, etc. Each of those steps is typically represented by one or more components in the transformation pipeline where the data is passed from one component to another and each component performs a partial transformation of the data.

Due to the generic and reusable nature of the ETL components, the data is passed along the pipeline in a way where all input data of one component is loaded into a triplestore, the transformation, e.g. a SPARQL CONSTRUCT query, is performed on the data and the results are stored back in the triplestore, from where it is loaded as an input for the next component in the pipeline. This is to ensure that all intermediary results are preserved for debugging and errors in components do not affect data passed in other parts of the pipeline. However, while this approach is perfectly fine for smaller datasets, when it comes to larger ones, two main problems arise.

The first problem is memory consumption, because loading and querying larger datasets in triplestores can be very memory intensive. This becomes a problem for institutions willing to publish data as LOD, but having only regular workstations which can be dedicated to the process, not servers with large amounts of RAM. The second problem is the query execution time, as queries over larger datasets tend to take long time to execute, even for queries which should work with a single entity at a time, as they are processed independently of each other.

Our solution to both these problems is to help the processing of data about independent entities by splitting the processed data into *data chunks*. A *data chunk* is a group of one or more records, which can be processed individually or all at once, yielding the same result. Depending on the specific data format,

[4] https://www.w3.org/TR/csv2rdf/

one data chunk may be one or more rows of a CSV table, one or more CSV files, one or more XML files or one or more RDF files. These chunks can then be processed throughout the transformation pipeline really independently, i.e. one chunk at a time, not loading the whole dataset at once, saving memory consumption and speeding up the processing of queries, at the cost of additional I/O overhead, which can be seen later in Sect. 5. This process is in general similar to *streaming*, which is successful in processing of, e.g., large CSV files where rows are treated as independent records and are also processed independently. In the RDF world, streaming is also known (there is a W3C RDF Stream Processing Community Group[5]) and used especially for processing real-time data and events [1]. However, these solutions are currently in an experimental stage, often tailored to a specific task and not integrated with other functionality needed for a use case such as data publishing for government organization. Ease of use is crucial as typically, the open data publishing process in the Czech Republic is done by an office clerk with minimal IT support and a crash course in RDF and SPARQL. Therefore, integrating smaller scripts together, e.g. in shell, is out of the question. A true streaming solution would also be an overkill for, e.g., a simple case of publishing a larger Excel sheet as LOD. This is why we use a lighter approach with data chunks and integrate it into our existing, ETL-based solution. The chunked approach can be used whenever the data is a list of independent items.

One side-effect of the chunked approach is possible duplication of certain RDF triples in each data chunk. Let us, for instance, have a table of records about inspections of the Czech Trade Inspection Authority, containing an ID of the inspected company. When multiple records about inspections in the same company with the same ID are in different data chunks, they get to be processed separately. It may then happen that each of the resulting data chunks in RDF representation contains a triple such as `ex:company a org:Organization`, which leads to duplication, which is, however, resolved automatically when the data is loaded into a triplestore.

The idea of data chunking is quite straightforward, but to our best knowledge, there is no satisfactory, well documented and easily deployable implementation tackling it in the context of RDF.

3   Data Chunking in LinkedPipes ETL

Based on the feedback we got for UnifiedViews [6], our previous Linked Data ETL tool, we developed LinkedPipes ETL[6] (LP-ETL) [5] from scratch, addressing the identified issues. With LP-ETL, we focus on open APIs, where all supported operations are accessible by anyone using the API. Next we focus on easy deployment, all that is needed is Java, Node.js, a Git client and Apache Maven. We maintain a thorough documentation and the workflow is intuitive. Configuration representation is now based on the LOD principles. The aim of LP-ETL is to support the process of data publication and consumption, and

---

[5] https://www.w3.org/community/rsp/
[6] https://etl.linkedpipes.com

especially publication of internal data in relational databases or Excel, CSV, XML or JSON files as LOD and their catalogization.

From the user point of view, the main entity in LP-ETL is a *data transformation pipeline*, a repeatable process consisting of individual steps represented by reusable and configurable *components*. Each component in the pipeline can produce and/or consume either files, or RDF data. The components are interlinked, representing the data flow among them. There is a library of reusable components ready to be placed in the pipeline and it is also easy to develop a new component. Some of the features of LP-ETL include support for debugging of pipelines[7], intelligent suggestions of components which can follow a selected component in a pipeline and full API coverage of necessary actions. Moreover, the pipelines themselves and all their configuration are RDF data accessible through dereference of their IRIs, which makes their sharing easier. This can be seen in the demo instance[8]. All the RDF operations such as data serialization, deserialization and querying is done using the Eclipse RDF4J library[9].

In this section we introduce a set of new LP-ETL components, which are variants of the existing ones[10], and allow users to exploit data chunks in their data processing pipelines. The components work with a so called *chunked data unit* - a representation of the data chunks passed between the components. The data chunks in LP-ETL are passed throughout the pipeline as RDF data. This means that in each pipeline processing data chunks, there has to be a component which creates the data chunks and controls their size and number. So far, we have components for creating data chunks out of CSV files, RDF files and SPARQL endpoints. Nevertheless, many input data formats can be transformed to these, e.g. Excel files to CSV files, XML files to RDF files, etc. Next, the components actually exploiting the data chunks can be used. Right now, we have three SPARQL based components capable of exploiting data chunks. Finally, the data chunks can be either loaded into a triplestore, merged into a single RDF dump file or merged into a regular data unit, so that the pipeline can continue in a non-chunked way.

3.1 SPARQL Endpoint Extractor

The SPARQL endpoint component facilitates querying of a remote SPARQL endpoint. It requires the endpoint URL and a SPARQL CONSTRUCT query in its configuration. The chunked version of the component is focused on getting descriptions of a larger number of entities. Let us have, e.g., the RÚIAN (Sect. 4.4) dataset containing all addresses (`ruian:AdresníMísto`) in the Czech Republic (approx. 2 million entities) accessible via a SPARQL endpoint. Now if we want to work with all of them, e.g. we want to transform geocoordinates from one coordinate system to another, we need to query for data about each

---

[7] https://etl.linkedpipes.com/documentation/#debug

[8] https://demo.etl.linkedpipes.com

[9] http://rdf4j.org/

[10] https://etl.linkedpipes.com/components/

of them. This is, however, problematic, as getting this amount of data using a generic SPARQL query, e.g. `CONSTRUCT WHERE {?a a ruian:AdresníMísto; ?p ?o.}` can fail for various reasons, including timeouts, out of memory errors or in the case of OpenLink Virtuoso the error `Virtuoso 22023 Error SR...: The result vector is too large`[11].

Let us now assume we already have a `ruian:AdresníMísto` instance IRIs list. We could, for each known instance, ask `CONSTRUCT WHERE {address:21693951 a ruian:AdresníMísto; ?p ?o.}`, which is returned fast and the overall HTTP request and response is small. This is what the chunked version of the component does. In addition to its non-chunked version, it accepts a CSV file with one column containing a list of entity IRIs and the name of a SPARQL variable in the header. The SPARQL query itself then has to contain a special placeholder where a VALUES clause for the specified variable is inserted by LP-ETL. Another parameter is the chunk size, which determines how many entity IRIs will be placed into the VALUES clause of one query.

One potential problem remains, and that is getting the actual list of entity instance IRIs. Typically, the list will be obtained using a SPARQL query such as `SELECT DISTINCT ?a WHERE {?a a ruian:AdresníMísto}`, which works fine for smaller numbers of entities. However, when the number of entities itself gets big, e.g. greater than 1 million, we once again run into issues with getting the list intact. Either the SPARQL endpoint can have a maximum number of returned results set lower than that, or there are bugs[12] preventing us to get the full list this way. To overcome this, we added an implementation of a SPARQL SELECT with Scrollable Cursor[13], which can get the job done using multiple requests.

3.2   Tabular Extractor and Files to RDF

The *tabular* component maps input CSV files directly to their RDF representation according to the CSV on the Web W3C Recommendation unless specified otherwise. Normally, all output RDF data is passed to the next component as a whole. The chunked version of the component takes one more parameter, which says how many rows transformed to RDF should be contained in one output data chunk. The processing speed is unaffected, as the output is simply split into multiple chunks instead of one.

It may happen that we already have our RDF data split into files according to entity descriptions, e.g., by XSLT transformation in ARES (Sect. 4.2), or by downloading it from a data source published this way. In that case, we can use the chunked version of the *Files to RDF* component to build the chunked data unit. It can be configured with the number of input files which should form a single data chunk.

---

[11] https://github.com/openlink/virtuoso-opensource/issues/119
[12] https://github.com/openlink/virtuoso-opensource/issues/207
[13] http://vos.openlinksw.com/owiki/wiki/VOS/VirtTipsAndTricksHowToHandle
BandwidthLimitExceed

3.3   SPARQL Transformers

The chunked version of the SPARQL Construct component performs the config-ured query on each input data chunk, producing equal number of transformed output data chunks containing the transformed data. The chunked version of SPARQL Update component adds or removes data from each data chunk as configured by the query.

SPARQL Linker is a rather special data chunk processing component takes data chunks on one input and typically smaller, so called reference data on another input. Then, for each data chunk, it merges it with the reference data in an in-memory store and performs a SPARQL CONSTRUCT query, again producing a data chunk. This can be used e.g. to link entities to code lists (SKOS[14] Concept Schemes), as in MONITOR (Sect. 4.3).

3.4   Merger Component and Loaders

The chunked data can be merged into non-chunked RDF data to be processed further as a whole by the *Chunked merger* component. This, however, may not be possible for larger datasets, or may take a long time and consume large amounts of memory, as this component does the merge by loading the input data into a single repository. A merge of RDF files can also be done by a simple concatenation of files in, e.g. N-Triples serialization. This, however, does not solve duplicate triples, and therefore can result in unnecessarily larger files. A typical task where the chunked data is not enough is for instance generating VoID statistics such as the number of triples, number of entities and number of distinct subjects, predicates and objects present in a dataset. For such task we may need to have the whole dataset available in a single triplestore. Nevertheless, there are approaches such as RDFpro [2] which can perform this particular task on a data dump, which can be obtained much easier. Therefore, it might be sufficient to wrap RDFpro as a component in LP-ETL and use it instead, depending on the particular task, which is part of our future work (Sect. 7).

The chunked data can be directly loaded into a triplestore using a SPARQL Update INSERT DATA command, implemented in the chunked version of the SPARQL Update component. Often enough, it is not required to have the whole dataset loaded into a triplestore at all and producing a single RDF dump file is sufficient, as it is e.g. in RÚIAN (Sect. 4.4). In that case, a chunked version of the *RDF to file* component can be used to produce a single RDF dump file directly from the data chunks with low time and memory consumption.

4   Datasets and Pipelines for Experiments

The need for tackling the problem of excessive time and memory consumption during relatively simple data transformation and publishing process comes from

---

[14] https://www.w3.org/TR/skos-reference/

many real world use cases. As part of the COMSODE EU FP7 project[15], we transformed hundreds of datasets using our previous ETL tool, UnifiedViews. In this section, we will introduce some of the datasets and the transformation pipelines used to publish them as LOD in LinkedPipes ETL, with support for data chunks. We will later use the core parts of these pipelines to perform experiments showing the relationship between data chunk size, memory consumption and time spent on the transformation. Note that a typical ETL pipeline, as the acronym suggests, always consists of three main parts. The first part is where data gets extracted, usually from the web, however, not always in a straightforward way in a form of a simple download. Then there is the core part, where data gets transformed from the source format (CSV, XML, ...) to the target format (RDF) and additional transformations may be performed. This part is where we employ the data chunking approach and this part is marked by yellow components. Finally, there is the part where the data gets loaded, i.e. a dump file stored on a web server, triples loaded into a triplestore and datasets cataloged in a data catalog such as CKAN.

4.1   LISTID - List of Czech Business Entity IDs

This dataset is a list of all Czech company IDs and their names[16] including the date of creation and the date of invalidation of the ID. It is a zipped, pipe-separated, ISO 8859-2 encoded CSV file containing approximately 4.3 million rows in 200 MB. A company ID is a number of up to 8 digits. It is useful to have this dataset as Linked Open Data, as many other datasets contain information about companies and can link to this one, creating a company IRI can easily using the company ID. In the pipeline in Fig. 1 we first download the web page, in which the link to the latest file is present. The HTML CSS selector component takes the HTML file on the input, and a set of CSS selectors, used to parse out the download link, as configuration. We pass the link to the *HTTP Get list* component, which downloads the file, which then gets unzipped. The tabular component forms data chunks of directly mapped RDF data, which are transformed to the final RDF vocabulary using the SPARQL Construct component. Its results are merged into an RDF dump file and loaded into a repository on

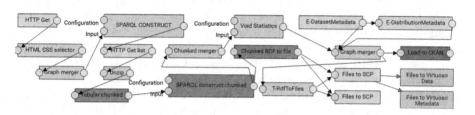

Fig. 1. LISTID dataset transformation pipeline

---
[15] http://comsode.eu
[16] http://www.statnipokladna.cz/cs/csuis/sprava-ciselniku

top of which the VoID statistics are computed. Finally, DCAT-AP metadata is added and both the data and metadata get loaded on a web server and into a triplestore and the dataset is cataloged in CKAN.

4.2 ARES - Czech Business Registry

This dataset is an unofficial mirror of the Czech Business Registry containing detailed information about Czech companies, their type and ownership structure. The dataset is accessible through a limited web API[17], through which a user can request information about a company based on its company ID. The API yields 1000 XML responses during the day and 5000 XML responses during the night. We are systematically downloading information about companies referenced in our other datasets. As of today, we have approximately 220 000 XML records cached, each a few kilobytes large, and we maintain the cache outside of LinkedPipes ETL. This is why the pipeline (see Fig. 2) starts with getting a zipped file containing the cache. In order to publish this data as LOD, we transform each XML file to RDF using XSLT, producing 220 000 RDF files. Because the data contains the ownership structure of each company including names of people, which may be considered sensitive information, to be on the safe side, we need to anonymize each record using two SPARQL queries, one for roles of people in companies and one for addresses of residence. Then we publish the data as an RDF dump, as triples in a triplestore and we catalog it in CKAN.

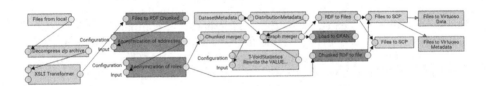

Fig. 2. ARES dataset transformation pipeline

4.3 MONITOR - Budget Information of All Levels of Czech State Administration

The MONITOR dataset is an official dataset containing aggregated budget and accounting information passed from all levels of the Czech state administration as open data[18]. It consists of two main parts, various codelists for classifications as a series of XML files called Master data and the transaction data, which are zipped, semi-colon separated CSV files. The dataset we use in this paper contains balance sheets of all units of government from 2010 to 2014, total of 12.5 million rows and 14 columns, spanning 1.08 GB of CSV data. The files are exported from an SAP[19] database, which causes negative numbers to be

---

[17] http://wwwinfo.mfcr.cz/ares/ares_xml.html.en
[18] http://monitor.statnipokladna.cz/en/2016/zdrojova-data/
[19] https://www.sap.com

represented with a minus sign at the end, i.e. "42.16-" and positive numbers to be represented with a space at the end, i.e. "42.16". First, the data needs to be downloaded, unzipped, relevant files selected and mapped from CSV to tabular RDF. Then, the data is converted to the RDF Data Cube Vocabulary[20] by a SPARQL CONSTRUCT query, which also handles the SAP peculiarity and generates data structure definitions. Next in the pipeline (Fig. 3) is the SPARQL Linker component (see Sect. 3.3), which links each individual balance sheet line to the used codelist items valid in the given time period. The codelists contain a dataset with yet another set of information about business entities, which are parts of the state administration. Theses are again linked to the dataset of all company IDs (Sect. 4.1). Finally, as usual, the VoID statistics are computed over the whole dataset, which is currently the most expensive part. DCAT-AP metadata is added, files loaded to a web server, triples loaded to a triplestore and the dataset gets cataloged in CKAN.

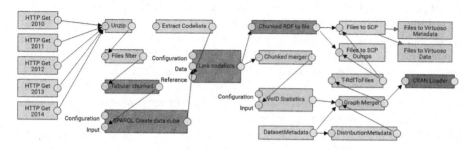

Fig. 3. MONITOR dataset transformation pipeline

4.4 RÚIAN - Registry of Territorial Identification, Addresses and Real Estate

This dataset is another key part of the future LOD infrastructure in the Czech Republic. It contains descriptions of all Czech regions, cities, villages, buildings, property lots, addresses and more, which makes it perfect for being a popular link target for other datasets, including ARES (Sect. 4.2). The data is published as open data[21].

First in the pipeline in Fig. 4 is a list of files for each town in the Czech republic (top branch) plus one extra for the higher level regions (bottom branch) has to be downloaded, totaling approx. 6300 file links. Second, each of those Gzipped XML files has to be downloaded and transformed using XSLT. Finally, the resulting RDF files have to be merged into a single RDF dump, which is later loaded into a triplestore. The files vary in size, which corresponds to the number of objects in the given town, ranging from a few kilobytes up to 1 GB for Prague.

---

[20] https://www.w3.org/TR/vocab-data-cube/

[21] http://www.cuzk.cz/Uvod/Produkty-a-sluzby/RUIAN/RUIAN.aspx (in Czech).

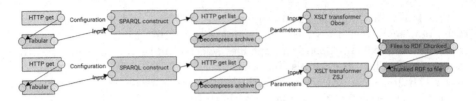

Fig. 4. RÚIAN dataset transformation pipeline

Once the dataset is transformed, it can be enriched, e.g. by transformation of the geocoordinates, which are represented in the S-JTSK (EPSG 5514) system to the commonly used WGS84 (EPSG 4937) system using the GeoTools library[22] wrapped as a LP-ETL component. There is a total of 28 million points in the RÚIAN dataset, so the challenge is to get them from the SPARQL endpoint, process them and load them back efficiently. Since this pipeline depends on too many variables including the implementation of the SPARQL endpoint (OpenLink Virtuoso in our case), the machine it runs on (8-core 16 GB RAM virtual machine) etc., we do not measure this pipeline in Sect. 5 and use it only to demonstrate the usage of our SPARQL extraction components. Since extracting the points and their geocoordinates directly using a SPARQL CONSTRUCT query does not work for various reasons (see Sect. 3.1), we need to get them in chunks. It is also not possible to get the data using a simple paged query, because that involves ordering, for which the number of points is also too big. Therefore, we need to get the list of IRIs of the points using a SELECT query and then ask for individual points in multiple queries. However, in Virtuoso, there is a bug preventing us from getting more than approx. 1 million rows as a result of a SELECT query. This brings us to the geocoding pipeline. It starts with a SPARQL select component, which allows us to get the whole list of 28 million point IRIs in a paged way. Next, using a SPARQL Construct query we query for the data about the points using specific IRIs from the input and the VALUES clause, resulting in data chunks. Then, the points in the data chunks are transformed using the GeoTools library and finally, the chunks containing the transformed geocoordinates are loaded back to the SPARQL endpoint.

5   Experiments

In this section, we will experiment on the real world datasets described in Sect. 4. In particular, we will show how the data size and chunk size affects memory consumption and time spent on the transformation of these datasets. The described experiments can be reproduced by following instructions on our web page https://etl.linkedpipes.com/conferences/odbase2017. The results of the experiments show that the chunked data approach is viable and allows us to better express minimum hardware requirements for the publishers willing to publish their data as Linked Open Data using LinkedPipes ETL.

---

[22] http://www.geotools.org/

The data transformation pipelines usually do more than just the transformation itself, they also download the source data, upload the target data, register it in a catalog, etc. The chunking approach has no effect on these parts, therefore, we omit those and use only the core parts of the transformation pipelines, shown as yellow components. The important measurement is the relative speedup in contrast to non-chunked processing and the differences among pipeline runs with various data chunk sizes and memory available to the triplestore. For the experiments, LP-ETL used OpenRDF Sesame version 4.0.1, in which the SPARQL transformations are executed.

As we are interested only in larger runtime differences, we can afford to run the experiments on a virtual machine, which, of course, can introduce some inaccuracy due to CPU and hard drive sharing. Nevertheless, it is also quite typical for government organizations to dedicate a virtual machine to the task of publishing open data rather than dedicating a physical computer. The important parameters of the experiments are the amount of memory available to the Java process of LP-ETL, which includes the Sesame triplestore, and the data chunk size. We measure the runtime of the core pipeline part, split into total time necessary for repository initialization, data load, the queries themselves and data write. Note that the data chunk size in the experiments is the number of entities described in the chunk, not the number of triples.

5.1 LISTID

The LISTID dataset (Sect. 4.1) represents a medium sized table. It had 4 333 047 rows (entities), resulting in 14 507 650 RDF triples after transformation. What can be seen in Fig. 5 is that without chunking, the transformation would require 16 GB of RAM, which is the first successful measurement for the 100M chunk size. The data chunk size of 10 entities results in 433 305 chunks and introduces a considerable overhead. On average, 1 input row becomes a chunk containing 3.3 RDF triples, which become 9.3 triples after transformation using the SPARQL query. In UnifiedViews with 96 GB RAM available, this pipeline

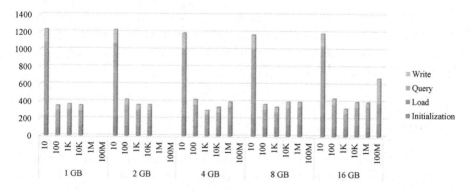

Fig. 5. LISTID memory consumption vs. chunk size and time elapsed in seconds

part ran for 5 h and 30 min (19 800 s). Using the latest, streamed version of TARQL, the transformation finished after 48 min (2 880 s) with at least 2 GB of memory available and in 58 min with 1 GB of memory and less, down to 64 MB.

5.2 ARES

The ARES dataset (Sect. 4.2) shows processing of data obtained by transformation of a larger number (approx. 220 000) of XML files to RDF. The XSLT transformation itself is not part of the measured part of the pipeline, we start with the resulting set of RDF files. In ARES, each entity represents a description of a company including its owner structure, which can have variable complexity. This means that each entity description will have different number of triples and therefore the chunks are variable in size. The maximum number of triples describing and entity is 856.4 and minimum is 29.1. The average number of triples is 119.8 before anonymization and 55.7 triples after anonymization.

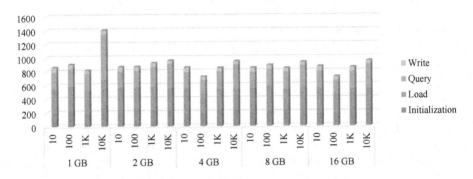

Fig. 6. ARES transformation memory consumption vs. chunk size and time elapsed in seconds

As can be seen in Fig. 6, for chunk sizes from 10 to 10 K entities, the total execution time of the core part of the pipeline in LP-ETL was more or less the same, about 800 s, with the only exception being 1 GB of memory, where the pipeline finished, but ran into excessive garbage collection for the 10 K chunk size. For the 1M chunk size and up, the pipeline would require more than 16 GB of memory. In UnifiedViews with 96 GB RAM available, this pipeline part did not finish even after 200 h.

5.3 MONITOR

In the MONITOR dataset (see Sect. 4.3) we are working with a uniform set of CSV tables having a total of 12.5 million rows. The average number of triples describing one entity is 12.8 in directly mapped tabular format and 15.9 triples after transformation, resulting in 200 million triples in total. Since this transformation is quite time consuming, we only ran the most interesting test configurations. The transformation consists of two steps, one transforms the directly

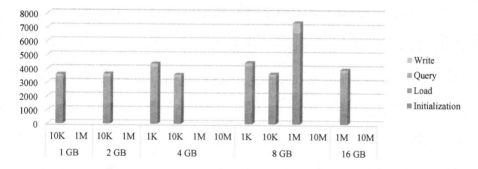

Fig. 7. MONITOR transformation memory consumption vs. chunk size and time elapsed in seconds

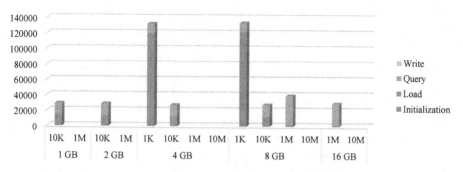

Fig. 8. MONITOR linking memory consumption vs. chunk size and time elapsed in seconds

mapped tabular data to the RDF Data Cube vocabulary (see Fig. 7) and the second one links the data to codelists (see Fig. 8) using the SPARQL linker (Sect. 3.3). With chunk size of 1 thousand entities, there is a bit of overhead during transformation, but a large amount of overhead in the linking part, where for each chunk, the codelists and the chunk need to be merged into a repository in which the SPARQL query is executed. Chunk size 1 million is possible from 8 GB of memory and up, resulting in some garbage collection with 8 GB of memory, which disappears with 16 GB of memory. Nevertheless, the chunk size of 10 000 seems to be the ideal choice for this dataset. In UnifiedViews with 96 GB of RAM available, the core part of the pipeline ran for 67 020 s (approx. 18 h) with usage of the so called "per graph" execution, which is similar to chunking, where 1 chunk corresponds to 1 input CSV file, which is roughly a chunk size of 2 million triples.

5.4 RÚIAN

In RÚIAN the chunks are split according to towns described. Therefore, while the chunk describing Prague is big, a chunk describing a small village is small. Average number of triples describing a town is 102 894, resulting in 669 million triples, which makes RÚIAN our largest dataset. Due to the size of the largest chunk representing Prague, the minimum amount of memory required is 16 GB, with chunk size of 1 introducing unnecessary overhead, leaving only two viable options, which are chunk size of 10 and 100, resulting in total execution time of 6400 s. In UnifiedViews with 96 GB of memory and no chunking support, this pipeline ran for approximately 2 days.

6 Related Work

LinkedPipes ETL is based on our experience gained from *UnifiedViews* (UV) [6]. Compared to UV, LP-ETL has a more efficient implementation of passing of files among components, stores its pipelines in RDF, offers open APIs for all actions and does not require a relational database or a servlet container to run. In this paper, however, we focus solely on the improvements achieved by the chunking approach and the other advantages of LP-ETL are out of scope of this paper. Other recent RDF aware ETL solutions are *LinDa Workbench* [12] and *DataGraft*[23], which are focused on transformation of tabular data only. For their comparison see [5]. Another ETL solution is DataLift [11], which works in an interactive fashion, which is not suitable for batch processing. TARQL is a simple, well-known tool for transformation of CSV files to RDF in a streamed way, but it does not offer anything else and can be only integrated in larger platforms. Karma [7] is a mapping assistant for mapping relational (CSV, Excel) and XML and JSON files and OWL ontologies to an internal semantic model and publishing the files on the web as RDF or JSON. The user does not have to know SPARQL, as the transformations are specified in a graphical user interface, and the specified transformations can be run offline, once specified. However, Karma lacks tools for handling the files before and after the transformation itself. The notion of pipeline based processing was first seen in DERI pipes [8], which is no longer developed and had extensibility limitations and no debugging support. Another approach to transformation of data to LOD is using a declarative language such as RML separately from the transformation engine [10]. Finally, there is Fusepool P3 [4] which is still under development, but claims to be based on the recent W3C Linked Data Platform specification[24].

In the area of working with larger RDF datasets on regular computer equipment more effectively than building a triplestore and using SPARQL, there is RDFSlice [9] focused on extracting subsets of RDF datasets using SliceSPARQL - a restricted subset of SPARQL. RDFpro [2] is a library or RDF processing stream-oriented tools which perform specific tasks like, e.g. computing VoID statistical metadata or RDF deduplication, efficiently, without building a triplestore. It is an ideal candidate to be wrapped as a set of LP-ETL components to provide support for fast, non-SPARQL operations. In addition, multiple MapReduce based solutions for querying large RDF datasets are present in the literature [3]. All of them are in experimental stage, with large overheads and strong focus

---

[23] https://datagraft.net/
[24] https://www.w3.org/TR/ldp/

on execution on a multi-node cluster, where they achieve best results. Therefore, they do not seem suitable for deployment in our use cases. Nevertheless, experimenting with these approaches is part of our future work.

7 Conclusions and Future Work

In this paper we presented a current problem with preparation of larger datasets for Linked Data publication and a solution targeting so called chunkable data. Next we briefly introduced LinkedPipes ETL, our lightweight ETL tool for Linked Data preparation, and its extensions adding support for faster and memory saving processing of chunkable data. We introduced real world datasets and their transformation pipelines and we experimented with data chunk size and how it affects the memory consumption of the used triplestore.

In every case presented, the chunked data approach performed better than the original version and considerably better than in UnifiedViews, and, where applicable, better than TARQL. This shows that the approach should be used whenever possible and has a potential for further optimization of RDF processing in ETL. LinkedPipes ETL is currently in use by the OpenBudgets.eu project[25] and for preparation of datasets published on the Czech Linked Open Data portal[26].

Part of our future work includes further investigation of RDF streaming approaches and whether and how data chunks can be processed in a streamed way. This means that all the affected components of a pipeline would run in parallel, processing the data chunks as they come, not waiting for all of them to be processed by the previous component. This should lead to a further speedup of parts of the transformation pipelines, not affecting parts of the pipeline processing data which is not chunkable, or is processed in a way requiring access to the whole dataset. Also, the streamed processing should be made more effortless to the user. A second line of experiments will focus on MapReduce based methods [3] for RDF storage and querying, which may be useful for large scale and distributed RDF processing. In addition, we plan to integrate the RDFpro library to complement our existing components for data chunks.

References

1. Calbimonte, J.-P., Aberer, K.: Reactive processing of RDF streams of events. In: Gandon, F., Guéret, C., Villata, S., Breslin, J., Faron-Zucker, C., Zimmermann, A. (eds.) ESWC 2015. LNCS, vol. 9341, pp. 457–468. Springer, Cham (2015). doi:10.1007/978-3-319-25639-9_56
2. Corcoglioniti, F., Aprosio, A.P., Rospocher, M.: Demonstrating the power of streaming and sorting for non-distributed RDF processing: RDFpro. In: Proceedings of the ISWC 2015 Posters & Demonstrations Track Co-located with the 14th International Semantic Web Conference (ISWC 2015), vol. 1486. CEUR Workshop Proceedings, Bethlehem, PA, USA, 11 October 2015. CEUR-WS.org (2015)

---

[25] http://openbudgets.eu
[26] https://linked.opendata.cz

3. Giménez-Garcia, J.M., Fernández, J.D., Martínez-Prieto, M.A.: MapReduce-based solutions for scalable SPARQL querying. Open J. Semant. Web (OJSW) 1(1), 1–18 (2014)
4. Gschwend, A., Neuroni, A.C., Gehrig, T., Combettoo, M.: Publication and reuse of linked data: the fusepool publish-process-perform platform for linked data. Innov. Public Sect. 22, 116–123 (2015)
5. Klímek, J., Škoda, P., Nečaský, M.: LinkedPipes ETL: evolved linked data preparation. In: Sack, H., Rizzo, G., Steinmetz, N., Mladenić, D., Auer, S., Lange, C. (eds.) ESWC 2016. LNCS, vol. 9989, pp. 95–100. Springer, Cham (2016). doi:10.1007/978-3-319-47602-5_20
6. Knap, T., Hanečák, P., Klímek, J., Mader, C., Nečaský, M., Nuffelen, B.V., Škoda, P.: UnifiedViews: an ETL tool for RDF data management. Semantic Web (Accepted 2017). http://semantic-web-journal.net/content/unifiedviews-etl-tool-rdf-data-management-0
7. Knoblock, C.A., Szekely, P., Ambite, J.L., Goel, A., Gupta, S., Lerman, K., Muslea, M., Taheriyan, M., Mallick, P.: Semi-automatically mapping structured sources into the semantic web. In: Simperl, E., Cimiano, P., Polleres, A., Corcho, O., Presutti, V. (eds.) ESWC 2012. LNCS, vol. 7295, pp. 375–390. Springer, Heidelberg (2012). doi:10.1007/978-3-642-30284-8_32
8. Le-Phuoc, D., Polleres, A., Hauswirth, M., Tummarello, G., Morbidoni, C.: Rapid prototyping of semantic mash-ups through semantic web pipes. In: Proceedings of the 18th International Conference on World Wide Web, WWW 2009, pp. 581–590. ACM, New York (2009)
9. Marx, E., Shekarpour, S., Auer, S., Ngomo, A.-C.N.: Large-scale RDF dataset slicing. In: Proceedings of the 2013 IEEE Seventh International Conference on Semantic Computing, ICSC 2013, pp. 228–235. IEEE Computer Society, Washington, DC (2013)
10. De Meester, B., Maroy, W., Dimou, A., Verborgh, R., Mannens, E.: Declarative data transformations for linked data generation: the case of DBpedia. In: Blomqvist, E., Maynard, D., Gangemi, A., Hoekstra, R., Hitzler, P., Hartig, O. (eds.) ESWC 2017. LNCS, vol. 10250, pp. 33–48. Springer, Cham (2017). doi:10.1007/978-3-319-58451-5_3
11. Scharffe, F., Atemezing, G., Troncy, R., Gandon, F., Villata, S., Bucher, B., Hamdi, F., Bihanic, L., Képéklian, G., Cotton, F., Euzenat, J., Fan, Z., Vandenbussche, P.-Y., Vatant, B.: Enabling linked data publication with the Datalift platform. In: Proceedings of AAAI Workshop on Semantic Cities, Toronto, Canada, July 2012
12. Thellmann, K., Orlandi, F., Auer, S.: LinDA - visualising and exploring linked data. In: Proceedings of the Posters and Demos Track of 10th International Conference on Semantic Systems - SEMANTiCS 2014, Leipzig, Germany, September 2014

A Particle Swarm-Based Approach for Semantic Similarity Computation

Samira Babalou[(✉)], Alsayed Algergawy, and Birgitta König-Ries

Heinz-Nixdorf Chair for Distributed Information Systems,
Friedrich Schiller University of Jena, Jena, Germany
{samira.babalou,alsayed.algergawy,birgitta.koenig-ries}@uni-jena.de

Abstract. Semantic similarity plays a vital role within a myriad of shared data applications, such as data and information integration. A first step towards building such applications is to determine concepts, which are semantically similar to each other. One way to compute this similarity of two concepts is to assess their word similarity by exploiting different knowledge sources, e.g., ontologies, thesauri, domain corpora, etc. Over the last few years, several approaches to similarity assessment based on quantifying information content of concepts have been proposed and have shown encouraging performance. For all these approaches, the Least Common Subsumer (LCS) of two concepts plays an important role in determining their similarity. In this paper, we investigate the influence the choice of this node (or a set of nodes) on the quality of the similarity assessment. In particular, we develop a particle swarm optimization approach that optimally discovers LCSs. An empirical evaluation, based on well-established biomedical benchmarks and ontologies, illustrates the accuracy of the proposed approach, and demonstrates that similarity estimations provided by our approach are significantly more correlated with human ratings of similarity than those obtained via related works.

Keywords: Semantic web · Biology ontology · Semantic similarity · Particle swarm optimization

1 Introduction

The computation of semantic similarity between pairs of concepts across different data sources and knowledge bases plays a crucial role in several applications such as data integration [19], information retrieval [27], information extraction [28], data clustering [12] and word sense disambiguation [14]. Generally speaking, similarity measures assess the degree of proximity between concepts, where a numerical score that quantifies this proximity as a function of the semantic evidence is computed. To achieve this goal, 'a large set of similarity measures have been proposed and developed. Among them are *corpora-based* and *ontology-based* [2,21] measures.

The ontology-based measures can be generally classified into three main categories: *feature-based*, *edge-based*, and *information content-based (IC)* measures.

© Springer International Publishing AG 2017
H. Panetto et al. (Eds.): OTM 2017 Conferences, Part II, LNCS 10574, pp. 161–179, 2017.
https://doi.org/10.1007/978-3-319-69459-7_11

Low accuracy of the edge-based methods [17,29], and unavailable properties in the feature-based methods [16] cause the IC-based methods to gain more attention. IC-based methods can further be distinguished into those that work on a single ontology and those that use multiple ontologies [2,21]. The latter were introduced to overcome the problem that a single ontology may not contain sufficient detail on all relevant aspects to effectively compute similarity. In all these approaches, the critical issue - significantly affecting the result of the similarity computation - is how to identify the Least Common Equivalent Subsumer (LCS) of compared concepts. This is even more challenging when using multiple ontologies, as it may happen that the first concept exists in one ontology and the second concept exists in another ontology. Selecting LCS with maximum IC [21] or more than terminologically matching of subsumer [2] have been proposed, nevertheless the similarity of proposed methods is far with human rating.

In this paper, we address the problem of improving ontology-based semantic similarity measure computation by focusing on this crucial step: how can the optimal LCSs be found that enhances similarity computation? To this end, we formalize this problem as an optimization problem: given two concepts find the optimal LCSs based on a discrete Particle Swarm Optimization-based algorithm (PSO). Given the two concepts c_1 and c_2 belonging to an ontology O_1 (or a set of ontologies), we start by collecting the set of ancestors for each concept. To construct a meaningful search space within which the optimal LCSs should be selected, the two candidate ancestor sets are combined. For this purpose, we develop an PSO-based approach that iteratively selects the optimal set (or a least one concept) of concepts that could be used during the similarity computation. Our experimental results show the effectiveness of the proposed algorithm. We have conducted several tests in the biomedical domain using established benchmarks. This domain is of particular interests, since the field boasts the availability of huge volumes of data resources. Therefore, identifying and measuring the similarity between biomedical terms in these resources will help greatly in utilizing and integrating information sources.

The paper is structured as follows: Sect. 2 provides some background knowledge in this context. Our proposed method will be introduced in Sect. 3. Experimental evaluation and results are discussed in the Sect. 4, and the paper is concluded in Sect. 5.

2   Background

The problem of semantically computing the relatedness between pairs of concepts has attracted a lot of attention. Therefore, a large number of approaches have been proposed [2,16,17,21]. Almost all of these approaches exploit different knowledge resources, such as thesauri, domain corpora, and ontologies to enhance the similarity computation. *Corpora-based* and *ontology-based* are common similarity measures. In this section, we first present different categories of *ontology-based* measures and then we report on different scenarios where ontologies can be used during similarity computation.

2.1 Ontology-Based Similarity Measures

In the ontology context, similarity approaches, as have been discussed in [2,22] can be classified as below:

- **Edge-based measures** depend on the graph representation of ontologies in which concepts are interrelated by means of semantic links [17,29]. These approaches are simple approaches since they only consider the minimum number of taxonomic links between concept pairs. This results in a limited accuracy where a large amount of taxonomical knowledge modeled in the ontologies is not being considered.
- **Feature-based measures** estimate similarity as a function of the amount of overlapping and non-overlapping knowledge features [16,22]. They potentially improve edge-counting approaches by considering additional semantic evidence. However, these measures can only be applied to the subset of the available ontologies, in which these information are available. In practice this seriously limits applicability of these approaches.
- **Information Content-based (IC) measures** quantify the amount of information provided by a given term based on its probability of appearance in a corpus [18]. Formally, the IC of a concept c is the inverse of its probability of occurrence, $p(c)$, as $IC(c) = -logp(c)$. The corpora-based IC measures are hampered by small size and high data sparseness of background data due to the need of manually tagging the sense for each word in the corpus.

To overcome the limitations of corpora-based IC calculus, ontology-based IC measures have been proposed [9,11,18]. The empirical evaluations have underlined that ontology-based IC methods are usually more accurate than corpora-based methods for similarity assessments [20,25].

For IC-based measures, the commonality of two concepts c_1 and c_2 belonging to an ontology/taxonomy is assessed according to the informativeness of their *Least Common Equivalent Subsumer* $(LCS(c_1, c_2))$ which is an estimator of the information shared between both concepts. The approaches differ in how they compute this informativeness (see Table 1 for a summary):

Table 1. A set of IC-based similarity measures.

No.	Method	Formula
1	Resnik [18]	$sim_{res}(c_1, c_2) = IC(LCS(c_1, c_2))$
2	Lin [11]	$sim_{lin}(c_1, c_2) = \frac{2 \times IC(LCS(c_1, c_2))}{(IC(c_1) + IC(c_2))}$
3	Jiang and Conrath [9]	$dis_{j\&c}(c_1, c_2) =$ $(IC(c_1) + IC(c_2)) - 2 \times IC(LCS(c_1, c_2))$

In Resnik's metric, the similarity of two concepts completely depends on their LCS, therefore any pair of concepts with the same LCS get the same similarity value. As a result, both Lin [11] and Jiang and Conrath (J&C) [9] also consider the IC of the compared concepts themselves. Lin measures the similarity as the ratio between the common information between concepts (i.e., $IC(LCS)$) and the information needed to fully describe them (i.e., the IC of each concept alone). In the J&C method [9], a distance measure (the inverse of similarity) has been proposed based on the difference between the sum of the individual ICs of the two concepts and the IC of their LCS.

In the ontology context, calculating the $IC(c)$ has been proposed by Seco et al. [25] as stated in Formula 1 by approximating its probability by the ratio between the size of $leaves(c)$, as a measure of the generality of c, and the number of subsumers of c including itself ($subs(c)$), as a measure of the concreteness of c. This ratio is divided by max_leaves, which represent the number of leaves of the ontology, which acts as a normalisation factor to obtain values in the 0..1 range. Finally, 1 is added to the numerator and denominator to avoid $log(0)$.

$$IC(c) = -log(\frac{\frac{|leaves(c)|}{|subs(c)|} + 1}{max\_leaves + 1}) \qquad (1)$$

Quite obviously, in the IC-based similarity methods, the result depends on how LCS in the ontologies is computed. As outlined above, this is particularly challenging when dealing with multiple ontologies, in particular when concepts occur in one ontology only. As will be shown in the evaluation section, the existing approaches leave room for improvement. To this end, we will discuss finding LCS methods in the next section.

2.2 Ontology-Based Similarity Computation Scenarios

To assess the similarity of two concepts in an ontology (or belonging to a set of ontologies), we face several scenarios, as also discussed by [1,3]. We distinguish between the two words: *term* and *concept*, such that the word *term* refers to a word the similarity of which user is interested to know, while *concept* refers to the corresponded entity of the user's term in the ontology. In the following, we discuss different scenarios related to computing the similarity between a pair of *terms* based on *ontology-based measures*.

- **Scenario 1: Both terms exist in one ontology:** suppose we would like to know $Sim(Pulmonary\ embolus, Myocardial\ infarction) = ?$ with both concepts being in one ontology (e.g. Fig. 1 SNOMED-CT). In this case, the similarity is computed, where their shared ancestor can be extracted directly from the existing path between these two concepts of the ontology.
- **Scenario 2: Both terms exist in several ontologies:** This scenario deals with overlapping knowledge. With the dissemination of information, a large number of available ontologies modelled the knowledge in different levels for each domain. For example, SNOMED-CT (Systemized Nomenclature of Medical Clinical Terms) [26] and MeSH (Medical Subject Headings) [13] are knowledge bases with different scopes and purposes, but both model biomedical concepts. For instance, as Fig. 1 shows, the two concepts *Delusion*

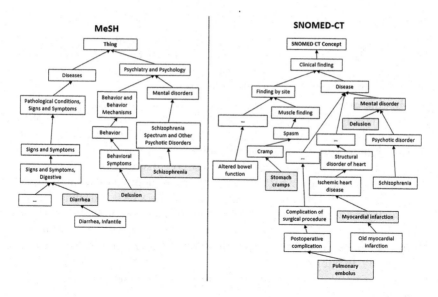

Fig. 1. Excerpts from two sample ontologies

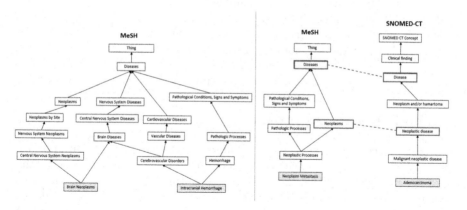

Fig. 2. SP in one and multiple ontologies (red boxes), dotted lines are bridges. (Color figure online)

and *Schizophrenia* exist in both MeSH and SNOMED-CT. In this case, some approaches [3, 21] calculate similarity in each ontology individually, then select the maximum value among them. Another proposed method by Al-Mubaid and Nguyen [1] considers the similarity of the primary ontology, which is selected by the user. In our experimental test (see Sect. 4.3, Third set) we compare other possibilities, e.g. minimum similarity or average aggregated value.

- **Scenario 3: Each term exists in different ontologies:** The most difficult scenario occurs when one of the concepts can be found in one ontology only, while the other can be found in another ontology, only. For example, suppose we want to compute $sim(Diarrhea, Stomach\ cramps)$ while $Diarrhea \in$MeSH and $Stomach\ cramps \in$SNOMED-CT as shown in Fig. 1. In this case, there is no physical path between the compared concepts; thus we need to find a virtual LCS. Possible approaches are as below:

- Merging ontologies: Two or more existing ontologies are first merged by creating a new node (called anything) which is a direct ancestor of their roots [1,16,19]. In [1] ontologies are connected by joining all nodes with the same textual label as a bridge node. This approach need a lot of effort to merge the ontologies, which cause several heterogeneities problem.
- Selecting a subset for each concept of an ontology, then comparing these sets [21]: Both sets are compared to find equivalent subsumers via terminologically approaches [3,21] or semantically approaches [2,23]. Perciesaly, to find IC value for this virtual LCS, this group of approaches instead of the compeletely merging the ontologies, try to connect them partially. In our experimental result, we follow of this approaches. We connect the ontologies by terminological label matching of similar concepts, as bridge (see Fig. 2).

To sum up, due to its importance, semantic similarity estimation has obtained a lot of attention. *Ontology-based* measures outperform *corpora-based*, however, and to a large extent, these measures depend on determining the optimal LCS. As will be shown in our evaluation, existing techniques are often far away from the human (reference) rating. Therefore, we introduce an optimization-based approach to determine the optimal LCS to enhance the similarity computation.

3   The Proposed Approach

To accurately compute the semantic similarity between pairs of concepts, the optimal LCS of these pairs should be identified. To achieve this goal, in this paper, we formulate the determination of the optimal LCS as an optimization problem making use of discrete Particle Swarm Optimization (PSO) to address this issue. The objective is to find an *optimal LCS* in the IC formula which increase the accuracy of similarity computation. A complex fitness function based on similarity measures of ontological entities, as well as a tailored particle update procedure are presented. The approach can easily process very large inputs, since an iterative traversal of large ontologies is avoided. Furthermore, due to the independent computation of each particle, the algorithm is highly parallelized, which is a crucial feature in processing large-scale ontologies.

PSO is an evolutionary algorithm (EA), more precisely a generic population-based metaheuristic optimization algorithm. It has been successfully applied to various problems in science and engineering. It is generally applicable to problems, where the global optimum of an objective function is to be found. The presented solution is inspired by the work presented in [4–6] where a discrete PSO was successfully applied in ontology alignment [4] as well as the selection of an optimal set of attributes for a classifier presented in [5,6].

3.1 Finding LCS as a Particle Swarm Optimization Problem

In general, PSO uses a number N of *particles*, called *population*, which evolves in a number I of *iterations*. During each iteration, every particle evolves using a *velocity vector*, which determines its new position in the parameter space. This evolution happens via a guided, randomised re-initialisation of each particle.

Let C be the set of concepts of an ontology O. We define *concept subsumption* ($\sqsubseteq^\top$) as a binary relation $\sqsubseteq^\top \colon C \times C$, where $\top$ is the interlink distance between a concept and its ancestors. Hence, having two concepts c_1 and c_2, $c_1 \sqsubseteq^\top c_2$ implies that c_2 is an nth taxonomical ancestor of c_1 and, inversely, c_1 is an nth specialisation of c_2. The *closure* of this relation ($\sqsubseteq^*$) is the union of the result of applying $\sqsubseteq^\top$ for $\top = 0, ..., d$, with d being he maximum number of taxonomical links between a concept in O and the root node. Consequently, $c_1 \sqsubseteq^* c_2$ is fulfilled if c_2 is an ancestor of c_1 at any level in the taxonomical tree or if $c_1 = c_2$.

The optimal LCS that accurately quantifies the similarity between pairs of concepts c_1 and c_2 should be one of the concepts' subsumers. Therefore, we define an LCS candidate ($clcs$) of a concept c as follows:

$$clcs(c) = \{c_i \in C | c \sqsubseteq^* c_i\} \tag{2}$$

To determine the candidate set of LCS for two concepts c_1 and c_2, we use the information in Formula 3, which can be interpreted based on each scenario. If c_1 and c_2 exist in one ontology, the shared ancestors for both concepts are considered one time only. This set does not have any redundancy, same concepts (common concepts in the two candidate sets) are only considered one time. In the case that c_1 exists in O_1 and c_2 exists in O_2, at first O_1 and O_2 are connected by shared ancestors of c_1 and c_2, (which we call bridges (see Sect. 3.2). We ensure non-redundancy by subtracting the bridge ($\backslash Bridges$).

$$cLCS(c_1, c_2) = \{c_i \in C | c_i \in (clcs(c_1) \cup clcs(c_2))\} \backslash \{c_r | (c_r, c_y) \in bridges(O_1, O_2)\} \tag{3}$$

In the traditional PSO, the length of a particle is a constant set by the user. For $i \in 1, ..., N$, a particle is defined as a vector:

$$\overrightarrow{p}_i = \{c_{(i,1)}, c_{(i,2)}, ..., c_{(i,n)}\} \tag{4}$$

where n is the number of concepts in each particle $\overrightarrow{p}_i$. Each particle is a subset of $cLCS$, which represents a candidate solution to the problem. $c_{(i,j)}$ represents the jth concept represented by the ith particle. A fitness function for each $c_{(i,j)}$ is formulated as follows:

$$f(c_{(i,j)}) = \Gamma(\overrightarrow{h}(c_{(i,j)}), \overrightarrow{w}) \tag{5}$$

This function evaluates $c_{(i,j)}$, according to an *evaluation strategy* Γ, a vector of rating functions $\overrightarrow{h}$ and a vector of weights $\overrightarrow{w}$, where the weights determine the influence of each rating function. The fitness of a particle $\overrightarrow{p}_i$ is defined as:

$$F(\overrightarrow{p}_i) = \frac{\sum_{i=1}^{|\overrightarrow{p}_i|} f(c_{(i,j)})}{|\overrightarrow{p}_i|}, c_{(i,j)} \in \overrightarrow{p}_i \tag{6}$$

where $F(\overrightarrow{p}_i)$ is the average fitness of the particle $\overrightarrow{p}_i$, where $|\overrightarrow{p}_i|$ denotes the size of the particle. Each particle maintains the configuration of the best position (personal best). The swarm also maintains the best performance of all particles (global best). As stated before PSO is a set of particles evolving based on a set of criteria. A particle moves from its initial position through a multidimensional search space to reach an optimal position. The movement of the particle is mainly affected by three factors: (i) the particle's own fitness, (ii) the particle's best position (personal (local) best, $pBest$ $\overrightarrow{B}_i$), and (iii) best position found by the particles in the surrounding of the particle (global best, $gbest$ $\overrightarrow{G}$).

At the beginning, the previous personal best position of the particle $\overrightarrow{p}_i$ is empty, i.e., $\overrightarrow{B}_i = \phi$. Therefore, once an initial particle is generated $\overrightarrow{B}_i$ is set to $\overrightarrow{p}_i$. After that at every iteration, if $\overrightarrow{p}_i$ is updated, $\overrightarrow{B}_i$ is also updated based on the values of the fitness vector. Also, at the beginning, $\overrightarrow{G}$ is empty. Therefore, once all the $\overrightarrow{B}_i$ have been determined, $\overrightarrow{G}$ is set to the fittest $\overrightarrow{B}_i$ previously computed. After that, $\overrightarrow{G}$ is updated if the fittest $f(\overrightarrow{B}_i)$ in the swarm is better than $f(\overrightarrow{B}_i)$. And, in that case, $f(\overrightarrow{B}_i)$ is set to $f(\overrightarrow{G}) = fittestf(\overrightarrow{B}_i)$. Otherwise, $\overrightarrow{B}_i$ remains as it is.

To determine all these updating strategies, we need to determine the fitness values inside the fitness vector. The $fitness\ vector$ of the particle $\overrightarrow{p}_i$ is denoted by an $2 - by - m$ array

$$\overrightarrow{F}_i = \begin{pmatrix} f_{(i,1)}\ f_{(i,2)} \cdots f_{(i,n)} \\ c_{(i,1)}\ c_{(i,2)} \cdots c_{(i,n)} \end{pmatrix} \tag{7}$$

Note that the vector is ordered by its fitness values.

In the standard PSO, each particle i is associated with a unique velocity vector $\overrightarrow{V}_i$ where the element $v_{(i,j)}$ determines the rate of change of each respective coordinate $p(i,j)$ in $\overrightarrow{p}_i$. Similar to the approach proposed in [5], we make use of the proportional likelihood instead of the vector of velocities. In that case, we use an $2 - by - m$ array of proportional likelihoods, where 2 is the number of rows in this array and m is the number of particle elements, represented as follows:

$$\overrightarrow{V}_i = \begin{pmatrix} v_{(i,1)}\ v_{(i,2)} \cdots v_{(i,n)} \\ c_{(i,1)}\ c_{(i,2)} \cdots c_{(i,n)} \end{pmatrix} \tag{8}$$

Initially, for each $v_{(i,j)}$, $c_{(i,j)}$ is set to 1. This initialisation is also done for new subsumers joining the particle during its evolution. The update of the proportional likelihoods is then done using the following formula:

$$v_{(i,j)}^{t+1} = v_{(i,j)}^t + \gamma + \beta \tag{9}$$

where β and γ are two real values $\in \mathbb{R}^+$ such that $\gamma = 0$ if the concept $c_{(i,j)} \notin \overrightarrow{B}_i$ and $\beta = 0$ if the concept $c_{(i,j)} \notin \overrightarrow{G}_i$ and t is the number of iteration.

If it is present in $\overrightarrow{G}$, add γ to $v_{(i,m)}$. These two parameters control the influence of the fact that a subsumer is also present in the personal best (β) or the global best (γ) LCS, respectively.

To build the final result, we need to keep best elements in each iteration for both local $pBest$ $\overrightarrow{B}_i$ and global $gbest$ $\overrightarrow{G}$. To this end, we make use of a $keep\text{-}set$ to save the best one for $\overrightarrow{G}$, while the $safe\text{-}set$ to save the best one for $\overrightarrow{B}_i$. The $keep\text{-}set$ is defined as follows:

$$K_{(p_i,\kappa)} = F_{(p_i,\kappa)} \cap V_{(p_i,\kappa)} \tag{10}$$

where the sets $F_{(p_i,\kappa)}$ and $V_{(p_i,\kappa)}$ contain those subsumers of a particle, which are the κ best evaluated, and highest ranked according to their proportional likelihood respectively. Similarly, the $safe\text{-}set$ is defined as follows:

$$S_{(p_i,\sigma)} = \{c_{(i,j_\mu)} | f_{(i,\mu)} > \sigma\} \tag{11}$$

a set of subsumers, which will never be replaced in this particle. Here $\sigma \in (0,1)$ is the fitness threshold for subsumers to be included in the safe-set. In each iteration, the particle keeps the set $S_{(p_i,\sigma)} \cup K_{(p_i,\kappa)}$ and replaces the remaining with new random ones. This behavior ensures a convergence of each particle towards an optimum according to 6, since the keep-set will steadily increase, and the fluctuation due to random re-initialisation will become less drastic as the swarm evolves.

Algorithm. The above mentioned steps and procedure can be summarized as shown in Algorithm 1. It starts by defining the set of inputs and the required result. It accepts a pair of concepts c_1 and c_2, a set of ontologies $\mathcal{O}$, and the required set of parameters. It returns the optimal set of LCS that enhance the similarity computation between the two concepts, *lines* 1–3. The algorithm extracts the candidate set for each concept given $\mathcal{O}$, *lines* 4 and 5 and it then constructs the search space for the optimization problem using the Eq. 3, *line* 6. A first step in the *PSO-based* algorithm is to initialize both local and global best sets $\overrightarrow{B}$ & $\overrightarrow{G}$, *line* 7.

The outer **for** loop, *lines* 8–32 represents the iterative process of the algorithm, which it has N iterations. In each iteration, a particle is defined, *lines* 9–14. It randomly selects a concept from the search space, *line* 10 and the fitness function is estimated for each new selected concept using Eq. 7, *line* 11. The velocity of each newly selected concept is initialized to 1, *line* 13. After that the fitness vector for the defined particle is defined and estimated, *lines* 15 and 16. The update process for the velocities of each concept in this particle is conducted according to the steps in *lines* 18–26. To update the set of concepts in this particle, both the *keep-set* ($K_{(p_i,\kappa)}$) and *safe-set* ($S_{(p_i,\sigma)}$) are defined based on the ordered values of the velocity and fitness vectors, *lines* 27–31. After the pre-defined number of iterations, N, the optimal set of concepts will be the elements of the global best set, *line* 33.

Algorithm 1. PSO algorithm to find LCS

Data: Two concepts c_1 & c_2

1 One or more ontologies, $\mathcal{O} = \{\mathcal{O}_1, \mathcal{O}_2, ..., \mathcal{O}_z\}$;

2 $\beta, \gamma, \kappa, \sigma$ parameters ;

Result: $\overrightarrow{G}$, *optimal LCS*

3 $clcs(c_1) \Leftarrow getCandSet(c_1, \mathcal{O})$;

4 $clcs(c_2) \Leftarrow getCandSet(c_2, \mathcal{O})$;

5 $cLCS(c_1, c_2) \Leftarrow getCandSet(clcs(c_1), clcs(c_2))$;

6 $\overrightarrow{B} \Leftarrow \emptyset; \overrightarrow{G} \Leftarrow \emptyset$;

7 **for** *(i = 1 to N)* **do**

8   **for** *(j = 1 to n)* **do**

9          $c_{(i,j)} \Leftarrow selectNext(cLCS(c_1, c_2))$;

10          $f(c_{(i,j)}) \Leftarrow \Gamma(\overrightarrow{h}(c_{(i,j)}), \overrightarrow{w})$;

11          $\overrightarrow{p}_i \Leftarrow \overrightarrow{p}_i \cup \{c_j\}$;

12          $\overrightarrow{V_{p_i}}(c_j) \Leftarrow 1$

13   **end**

14   Build $\overrightarrow{F}_i$ according to 7 ;

15   Compute $F(\overrightarrow{p}_i)$ according to 6 ;

16      $\overrightarrow{B}_i \Leftarrow \overrightarrow{p}_i$;

17   **for** *(m = 1 to |p|)* **do**

18   **if** $c_{(i,l_m)} \in \overrightarrow{B}_i$ **then**

19              $v_{(i,m)} \Leftarrow v_{(i,m)} + \beta$

20   **end**

21   **if** $c_{(i,l_m)} \in \overrightarrow{G}$ **then**

22              $v_{(i,m)} \Leftarrow v_{(i,m)} + \gamma$

23   **end**

24          $v_{(i,m)} \Leftarrow v_{(i,m)} \times rand()$;

25   **end**

26   Sort $\overrightarrow{V}_{p_i}$ by v_i in descending order ;

27   Sort $\overrightarrow{F}_{p_i}$ by f_i in descending order ;

28      $K_{(p_i, \kappa)} \Leftarrow F_{(p_i, \kappa)} \cap V_{(p_i, \kappa)}$;

29      $S_{(p_i, \sigma)} \Leftarrow \{c_{(i,j_\mu)} | f_{(i,\mu)} > \sigma\}$;

30      $\overrightarrow{p}_i \Leftarrow \{S_{(i,\sigma)} \cup K_{(i,\kappa)}\}$;

31 **end**

32 **return** $\overrightarrow{G}$

3.2   Fitness Scores of the LCS Evaluation

Algorithm 1 introduces main steps of the proposed approach. In order to estimate the fitness of a concept $c_{(i,j)}$ to the compared concepts, c_1 and c_2, we use a fitness function, Algorithm 1, *line* 10. In this section, we present a detailed description of the used fitness function. In general, a fitness function consists of three components: *rating function*, *evaluation strategy*, and *weighting scheme*.

Rating Function. To quantify the relatedness between the concept $c_{(i,j)}$ inside the search space and a pair of concepts c_1 and c_2, we make use of the following rating functions. These functions are selected to cover different aspects of the relatedness between the concept and the pair of concepts.

1. *Hierarchy-based rating.* This rating function is used to measure the distance between concept $c_{(i,j)}$ and the pair of concepts. To this end, we propose to use the following formula:

$$Hier\_rat(c_{(i,j)}) = \frac{1}{dist(c_{(i,j)}, c_1) + dist(c_{(i,j)}, c_2)} \tag{12}$$

   where $dist$ returns the shortest path length between two concepts.
2. *IC-based rating.* This function quantifies the amount of information the concepts have in common [9,11,18].

$$IC\_rat(c_{(i,j)}) = \frac{1}{IC(c_{(i,j)})} \tag{13}$$

3. *Hyponyms-based rating.* We also use a hyponyms-based rating function to consider the taxonomy in which concepts are modelled. Thus, we propose an intrinsic approximation of the term co-occurrence probability that solely evaluates ontological evidence of semantic equivalence. To this end, in the hyponyms-based measure as stated by Formula 14 each candidate solution gets a value by the number of its hyponyms. This value will be normalized by the total number of classes.

$$Hyp\_rat(c_{(i,j)}) = \frac{|hyponymes(c_{(i,j)})|}{|C|} \tag{14}$$

Evaluation Strategies. After defining rating functions, they need to be aggregated using an evaluation strategy Γ, defined as: $\Gamma : \overrightarrow{h}(c) \times \overrightarrow{w} \rightarrow (0,1)$. To implement this evaluation strategy, there are several variants, such as *minimum*, *maximum*, *average*, or *weighted-average*. In the current implementation, we make use of the *weighted average* strategy, defined as follows:

$$\Gamma(\overrightarrow{h}(c), \overrightarrow{w}) = \sum_{i=1}^{z} w_i h_i(c) \tag{15}$$

where $\sum_{i=1}^{z} w_i = 1$ and z is the number of rating functions. In our case $z = 3$, this formula becomes:

$$\overrightarrow{h}(c_{(i,j)}) = w_{hy} \times Hyp\_rat(c_{(i,j)}) + w_{in} \times Hier\_rat(c_{(i,j)}) + w_{ic} \times IC\_rat(c_{(i,j)}) \tag{16}$$

Weighting Scheme. Once we select the weighted average evaluation strategy, the arising question is how to set these weights. In the current implementation, we make use of the following rules:

- w_{ic} is set by the user.
- w_{in} is set to 1 if c_1 and c_2 are ancestors of each other, i.e. $\{c_1 \sqsubseteq^* c_2 \,\|\, c_2 \sqsubseteq^* c_1\}$, else set to 0.
- w_{hy} is set to 1 if the candidate ancestor belong to shared ancestors's set, SP, else set to 0.

  Hence, the shared hypernyms between two concepts give us a good clue about their semantic equivalence, i.e. as the overlap between their hypernym sets increases, their meanings become more equivalent. We define the shared ancestors (SA) in one and multiple ontologies respectively in Formulas 17 and 18.

  1- Discovering SA in one ontology: SA in one ontology is the set of elements which exists in $clcs(c_1)$ and $clcs(c_2)$ simultaneously as shown in Formula 17.

  2- Discovering SA among different ontologies: SA in multiple ontologies is the set of 2-tuples of elements from the two ontologies, where the elements within one tuple possess terminologically matching labels as shown in 18. We refer to these 2-tuples as *bridges*. These bridges also serve to connect several ontologies to be able to calculate the IC of LCS.

$$SA = \{sa | sa \in clcs(c_1) \cap clcs(c_2)\} \tag{17}$$

$$bridge(c_1, c_2) = \{(c_r, c_y) | c_r \equiv c_y, \ c_r \in O_i, c_y \in O_j, i \neq j\} \tag{18}$$

Here $\equiv$ means terminological equivalence of labels. Figure 2 shows the definition of SA in one ontology (left side) and definition of SA as *bridge* in multiple ontologies (right side).

4   Experimental Evaluation

To evaluate and validate the quality of the proposed approach, we conducted a set of experiments utilizing different datasets and ontologies. In the following, we first present used datasets and ontologies, then we describe experimental environments, and finally we report about experimental results.

4.1   Benchmarks and Ontologies

The most widely used way to objectively evaluate the accuracy of similarity assessments is to compare them with human judgements of similarity. To this end, we selected two widely used benchmarks in the biomedical domain, because several detailed benchmarks and partially overlapping ontologies are available in this domain.

– **Pedersen et al.** [15]: Dataset 1 consists of 30 medical term pairs whose similarity was assessed by a group of medical experts from the Mayo Clinic: 9 medical coders who were trained with the notion of semantic similarity and 3 physicians who rated terms without any special training. The similarity values provided by these experts were obtained on a scale from 1 (non-similar) to 4 (identical).
– **Hliaoutakis** [8]: Dataset 2 has originally been designed based on MeSH. It consists of 36 pairs of terms. This dataset was assessed by 8 medical experts from 0 (non-similar) to 1 (identical).

In our tests, we use the most comprehensive and precise clinical health terminologies namely:

– **SNOMED-CT:** Systematised Nomenclature of Medicine, Clinical Terms (SNOMED-CT) [26] as a domain-specific biomedical ontology which covers more than 360,000 medical concepts classified in 18 partially overlapping hierarchies.
– **MeSH:** Medical Subject Headings (MeSH) [13] as a domain-specific biomedical ontology provides around 25,000 medical and biological concepts hierarchically classified in 16 categories.

4.2 Implementation

The proposed *PSO-based* approach has been implemented using JAVA within the open source Semantic Measures Library[1] [7]. For the evaluation we make use of *SNOMED-CT* release of January 2017 and *MeSH* 2017 release. To run MeSH and SNOMED together, we follow the approach of [3,21] to connect two ontologies by bridges (see Formula 18). Furthermore, other choices of parameter values were $\kappa = 0.7$ and $\sigma = 0.9$ to determine the size of the *keep-set* and *safe-set*. β set by 0.4 and γ set by 0.5 to control the influence of personal and global best particle, respectively. These values were empirically determined in our experimental tests; but we make no claim that these are optimal values. Parameter optimization is a topic for future research.

4.3 Experimental Results

This section is devoted to present experimental results comparing them to state-of-the-art-approaches. To this end, we consider different experimental sets and scenarios.

First Set. In this set of experiments, we include the *PSO-based* approach into three well-known IC-based similarity measures and compare results to the results introduced in the approach [2]. Sánchez and Batet [21] compare the labels of c_1

---

[1] http://www.semantic-measures-library.org/sml.

Table 2. Correlation of concept similarity determined by different approaches to human ratings results

IC-based measure	Ontologies	LCS discovery	Correlation to physicians	Correlation to coder	Correlation to both	Evaluated in
Resnik	SNOMED-CT	None	0.553	0.598	0.602	[2]
	SNOMED-CT	This work	**0.617**	**0.825**	**0.745**	This work
	MeSH	None	0.608	0.668	0.670	[2]
	MeSH	This work	**0.700**	**0.856**	**0.808**	This work
	SNOMED-CT + MeSH	Sánchez and Batet	0.489	0.544	0.542	[2]
	SNOMED-CT + MeSH	Saruladha et al.	0.474	0.546	0.535	[2]
	SNOMED-CT + MeSH	Batet et al.	**0.617**	0.624	0.649	[2]
	SNOMED-CT + MeSH	This work	0.614	**0.705**	**0.685**	This work
	MeSH+ SNOMED-CT	Sánchez and Batet	0.444	0.534	0.512	[2]
	MeSH+ SNOMED-CT	Saruladha et al.	0.432	0.536	0.508	[2]
	MeSH+ SNOMED-CT	Batet et al.	**0.562**	**0.639**	**0.632**	[2]
	MeSH+ SNOMED-CT	This work	0.485	0.617	0.573	This work
Lin	SNOMED-CT	None	0.566	0.628	0.625	[2]
	SNOMED-CT	This work	**0.621**	**0.817**	**0.747**	This work
	MeSH	None	0.614	0.674	0.676	[2]
	MeSH	This work	**0.708**	**0.858**	**0.813**	This work
	SNOMED-CT + MeSH	Sánchez and Batet	0.512	0.561	0.561	[2]
	SNOMED-CT + MeSH	Saruladha et al.	0.501	0.569	0.560	[2]
	SNOMED-CT + MeSH	Batet et al.	**0.637**	**0.654**	**0.674**	[2]
	SNOMED-CT + MeSH	This work	0.612	0.639	0.650	This work
	MeSH+ SNOMED-CT	Sánchez and Batet	0.446	0.542	0.517	[2]
	MeSH+ SNOMED-CT	Saruladha et al.	0.432	0.543	0.511	[2]
	MeSH+ SNOMED-CT	Batet et al.	0.561	0.648	0.637	[2]
	MeSH+ SNOMED-CT	This work	**0.653**	**0.734**	**0.720**	This work
Jiang and Conrath	SNOMED-CT	None	0.538	0.612	0.602	[2]
	SNOMED-CT	This work	**0.567**	**0.731**	**0.674**	This work
	MeSH	None	0.618	0.670	0.676	[2]
	MeSH	This work	**0.710**	**0.826**	**0.798**	This work
	SNOMED-CT + MeSH	Sánchez and Batet	0.514	0.573	0.569	[2]
	SNOMED-CT + MeSH	Saruladha et al.	0.505	0.580	0.569	[2]
	SNOMED-CT + MeSH	Batet et al.	**0.637**	**0.651**	**0.673**	[2]
	SNOMED-CT + MeSH	This work	0.592	0.598	0.619	This work
	MeSH+ SNOMED-CT	Sánchez and Batet	0.423	0.527	0.498	[2]
	MeSH+ SNOMED-CT	Saruladha et al.	0.404	0.524	0.487	[2]
	MeSH+ SNOMED-CT	Batet et al.	0.542	0.638	0.622	[2]
	MeSH+ SNOMED-CT	This work	**0.653**	**0.725**	**0.716**	This work

and c_2 and select, as their LCS, the most specific and terminologically matching concept with maximum IC. Saruladha et al. [24] consider LCS with minimum IC. Batet et al. [2] look for a pair of subsumers with a higher degree of semantic equivalence than the terminologically matched one.

In this set of experiments, we consider different scenarios for using ontologies: single ontology ($MeSH \mid SNOWMED\text{-}CT$) or multiple ontologies ($MeSH +SNOWMED\text{-}CT \mid SNOWMED\text{-}CT+MeSH$). We applied these similarity measures to the Pedersen dataset, where 25 out of 30 word pairs can be found in both ontologies (to be similar with tests of [2]). The Pedersen dataset was evaluated by two different groups of people (physician, coder). So, we represent our result based on comparing with the coder rating, with the physician rating, and with both (average value for the both groups). Rows in boldface show the best results in each category.

Table 2 represents the Pearson correlation of our similarity value and the similarity value of physicians, coders and both in the Pedersen benchmark. In the single ontology scenario setting, our obtained result is significantly better than that of the others (SNOMED-CT 0.825, 0.817, 0.731 and MeSH 0.856, 0.858 and 0.826 in the IC of Resnik, Lin and J&C based on the coder rating respectively.) Results of multiple ontologies are not better than in the single ontology scenario setting, but in general, they have competitive performance to the state-of-the-art (e.g. in IC-Resnik for running SNOMED-CT+MeSH 0.705 v.s. 0.624 of Batet et al., or in IC-J&C for running MeSH+SNOMED-CT 0.725 v.s. 0.638 of Batet et al. based on the coder rating). In all cases, our results have better correlation than those of the coder rating, given that the coder rating has higher inter-agreement than physicians (0.78 v.s. 0.68). Also, coders, because of their training and skills, were more familiar than physicians to hierarchical classifications and semantic similarity which led to a better correlation with the design principles of our similarity approach.

Second Set. In this set of experiments, we evaluated the proposed approach using the Hliaoutakis et al. benchmark where 35 out of 36 terms pairs are found in the used ontologies. We also compare the quality of the approach to a set of variant similarity measures. To this end, we consider a set of *edge-based* measures [1,19,23]: Rada, W&P and L&C. Rada [17] quantifies the semantic distance of two concepts c_1 and c_2 as the sum of the minimum taxonomical path their LCS. W&P [29] considers the Rada strategy by normalizing with the maximum depth of the taxonomy, and L&C [10] is one more step following Rada method by using the number of *is-a* relations from the LCS to the root of the ontology. In these similarity measures [10,17,29], LCS discovery methods have evaluated by approaches of [1,19,23]. Al-Mubaid et al. [1] consider LCS as the deepest terminologically-equivalent subsumers. Approach of Sánchez et al. [23] is based on semantic overlapping and structural similarity. While the approach of Rodriguez et al. Petrakis et al. [16,19] joined the root nodes of the two ontologies and considered as the LCS. Moreover, the last part of Table 3 illustrated

Table 3. Comparing result with edge-based methods in Hliaoutakis's dataset.

Similarity measur	Ontologies	LCS discovery	Correlation to human rating	Evaluated in
Rada	MeSH	None	0.68	[23]
	WordNet	None	0.53	[23]
	MeSH+WordNet	Rodriguez et al., Petrakis et al.	−0.21	[23]
	MeSH+WordNet	Al-Mubaid	0.63	[23]
	MeSH+WordNet	Sánchez et al.	0.65	[23]
	MeSH+WordNet	Sánchez et al.	0.67	[23]
W& P	MeSH	None	0.69	[23]
	WordNet	None	0.53	[23]
	MeSH+WordNet	Rodriguez et al., Petrakis et al.	−0.16	[23]
	MeSH+WordNet	Al-Mubaid	0.67	[23]
	MeSH+WordNet	Sánchez et al.	0.67	[23]
	MeSH+WordNet	Sánchez et al.	0.67	[23]
L& C	MeSH	None	0.74	[23]
	WordNet	None	0.66	[23]
	MeSH+WordNet	Rodriguez et al., Petrakis et al.	−0.18	[23]
	MeSH+WordNet	Al-Mubaid	0.68	[23]
	MeSH+WordNet	Sánchez et al.	0.71	[23]
	MeSH+WordNet	Sánchez et al.	0.72	[23]
IC-Resnik	MeSH	This work	0.676	This work
	SNOMED-CT	This work	0.648	This work
	MeSH+SNOMED-CT	This work	0.499	This work
	SNOMED-CT+MeSH	This work	0.480	This work
IC-Lin	MeSH	This work	0.692	This work
	SNOMED-CT	This work	0.677	This work
	MeSH+SNOMED-CT	This work	0.564	This work
	SNOMED-CT+MeSH	This work	0.583	This work
IC-Resnik	MeSH	This work	0.686	This work
	SNOMED-CT	This work	0.616	This work
	MeSH+SNOMED-CT	This work	0.559	This work
	SNOMED-CT+MeSH	This work	0.539	This work
Batet et al.	SNOMED-CT	Connect direct ancestors	0.558	[3]
	MeSH	Connect direct ancestors	0.750	[3]
	WordNet	Connect direct ancestors	0.610	[3]
	SNOMED-CT+MeSH	Connect direct ancestors	0.727	[3]
	MeSH+WordNet	Connect direct ancestors	0.772	[3]
	SNOMED-CT+MeSH	Connect direct ancestors	0.740	[3]
	SNOMED-CT+MeSH+WordNet	Connect direct ancestors	0.786	[3]

the result of Batet et al. [3] as an edge based method with connecting direct ancestors as an LCS discovery method.

Approaches in Table 3 are obtained using the Hliaoutakis's dataset. Our results in the most but not all cases are better than the others. This can be explained as the created links (bridges), in the case when multiple ontologies are used, need to be improved. Also, it should be mentioned, they use different ontologies, which also is our future work.

Third Set. In this set of experiments, we have investigated the effect of apair of concepts appearing in different ontologies. As we discussed in Sect. 2.2, in Scenario 2 both terms appear in several ontologies. The arising question here is which combining strategy should be used in such a scenario. To this end, we have conducted this set of experiments to determine the final similarity using different combining strategies: *average value, maximum, minimum* or *weighted aggregation*. Results are shown in Fig. 3. These figures show weighted average has the potential to outperform others, if weights are set optimally. Without prior knowledge, average seems to be a reasonable choice outperforming maximum or minimum in most cases. Our test for weighted aggregation was done with 0.6 and 0.4 weights for MeSH and SNOMED respectively in the Pedersen dataset and 0.7/0.3 in Hliaoutakis dataset.

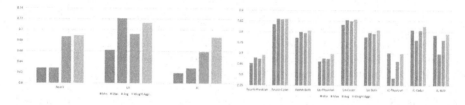

Fig. 3. Effect of different combination strategies on the correlation of the computed similarity and human ratings in Hliaoutakis's dataset (left), Pedersen's dataset (right).

5   Conclusion

In this paper, we addressed the problem of enhancing the computation of semantic similarity between pairs of concepts. To this end, we have introduced a *PSO-based* approach to effectively support the discovery of an optimal set of concepts that improve similarity computing. The proposed approach is an information content (*IC*)-based similarity assessment making use of ontologies. In that context, we demonstrated how either a single ontology or multiple ontologies can support the similarity computation. The proposed method has been implemented and tested using two well-known benchmarks. The experimental results show that the proposed method outperforms state-of-the-art approaches.

This work can be extended in several directions: a first step is to conduct more tests and evaluation of the proposed approach utilizing different datasets and ontologies from other domains. The second direction is the improvement of building bridges in the case of using multiple ontologies. Furthermore, we plan to apply this semantic similarity in several applications, such as data and knowledge integration.

Acknowledgments. The work has been (partly) funded by the *Deutsche Forschungsgemeinschaft* (DFG) as part of CRC 1076 AQUADIVA. S. Babalou is also supported by a scholarship from German Academic Exchange Service (DAAD).

References

1. Al-Mubaid, H., Nguyen, H.A.: Measuring semantic similarity between biomedical concepts within multiple ontologies. IEEE Trans. Syst. Man Cybern. Part C Appl. Rev. **39**(4), 389–398 (2009)
2. Batet, M., Harispe, S., Ranwez, S., Sánchez, D., Ranwez, V.: An information theoretic approach to improve semantic similarity assessments across multiple ontologies. Info. Sci. **283**, 197–210 (2014)
3. Batet, M., Sánchez, D., Valls, A., Gibert, K.: Semantic similarity estimation from multiple ontologies. Appl. Intell. **38**(1), 29–44 (2013)
4. Bock, J., Hettenhausen, J.: Discrete particle swarm optimisation for ontology alignment. Inf. Sci. **192**, 152–173 (2012)
5. Correa, E.S., Freitas, A.A., Johnson, C.G.: A new discrete particle swarm algorithm applied to attribute selection in a bioinformatics data set. In: Proceedings of the 8th Annual Conference on Genetic and Evolutionary Computation, pp. 35–42. ACM (2006)
6. Correa, E.S., Freitas, A.A., Johnson, C.G.: Particle swarm and Bayesian networks applied to attribute selection for protein functional classification. In: Proceedings of the 9th Annual Conference on Companion on Genetic and Evolutionary Computation, pp. 2651–2658. ACM (2007)
7. Harispe, S., Ranwez, S., Janaqi, S., Montmain, J.: The semantic measures library and toolkit: fast computation of semantic similarity and relatedness using biomedical ontologies. Bioinformatics **30**(5), 740–742 (2013)
8. Hliaoutakis, A.: Semantic similarity measures in mesh ontology and their application to information retrieval on medline. Master's thesis (2005)
9. Jiang, J.J., Conrath, D.W.: Semantic similarity based on corpus statistics and lexical taxonomy. arXiv preprint cmp-lg/9709008 (1997)
10. Leacock, C., Chodorow, M.: Combining local context and WordNet similarity for word sense identification. WordNet Electron. Lexical Database **49**(2), 265–283 (1998)
11. Lin, D., et al.: An information-theoretic definition of similarity. In: ICML, vol. 98, pp. 296–304. Citeseer (1998)
12. Martı, S., Valls, A., SáNchez, D., et al.: Semantically-grounded construction of centroids for datasets with textual attributes. Knowl.-Based Syst. **35**, 160–172 (2012)
13. Nelson, S.J., Johnston, W.D., Humphreys, B.L.: Relationships in medical subject headings (MeSH). In: Bean, C.A., Green, R. (eds.) Relationships in the Organization of Knowledge. Information Science and Knowledge Management, vol. 2, pp. 171–184. Springer, Dordrecht (2001). doi:10.1007/978-94-015-9696-1_11

14. Patwardhan, S., Banerjee, S., Pedersen, T.: Using measures of semantic relatedness for word sense disambiguation. In: Gelbukh, A. (ed.) CICLing 2003. LNCS, vol. 2588, pp. 241–257. Springer, Heidelberg (2003). doi:10.1007/3-540-36456-0_24

15. Pedersen, T., Pakhomov, S.V., Patwardhan, S., Chute, C.G.: Measures of semantic similarity and relatedness in the biomedical domain. J. Biomed. Inform. 40(3), 288–299 (2007)

16. Petrakis, E.G., Varelas, G., Hliaoutakis, A., Raftopoulou, P.: X-similarity: computing semantic similarity between concepts from different ontologies. JDIM 4(4), 233–237 (2006)

17. Rada, R., Mili, H., Bicknell, E., Blettner, M.: Development and application of a metric on semantic nets. IEEE Trans. Syst. Man Cybern. 19(1), 17–30 (1989)

18. Resnik, P.: Using information content to evaluate semantic similarity in a taxonomy. arXiv preprint cmp-lg/9511007 (1995)

19. Rodríguez, M.A., Egenhofer, M.J.: Determining semantic similarity among entity classes from different ontologies. IEEE Trans. Knowl. Data Eng. 15(2), 442–456 (2003)

20. Sánchez, D., Batet, M.: A new model to compute the information content of concepts from taxonomic knowledge. Int. J. Semant. Web Info. Syst. (IJSWIS) 8(2), 34–50 (2012)

21. Sánchez, D., Batet, M.: A semantic similarity method based on information content exploiting multiple ontologies. Expert Syst. Appl. 40(4), 1393–1399 (2013)

22. Sánchez, D., Batet, M., Isern, D., Valls, A.: Ontology-based semantic similarity: a new feature-based approach. Expert Syst. Appl. 39(9), 7718–7728 (2012)

23. Sánchez, D., Solé-Ribalta, A., Batet, M., Serratosa, F.: Enabling semantic similarity estimation across multiple ontologies: an evaluation in the biomedical domain. J. Biomed. Inform. 45(1), 141–155 (2012)

24. Saruladha, K., Aghila, G., Bhuvaneswary, A.: Information content based semantic similarity for cross ontological concepts. Int. J. Eng. Sci. Tech. 3(6), 327–336 (2011)

25. Seco, N., Veale, T., Hayes, J.: An intrinsic information content metric for semantic similarity in WordNet. In: Proceedings of the 16th European Conference on Artificial Intelligence, pp. 1089–1090. IOS Press (2004)

26. Spackman, K.: SNOMED CT milestones: endorsements are added to already-impressive standards credentials. Healthc. Inf. Bus. Mag. info. Commun. Syst. 21(9), 54–56 (2004)

27. Sy, M.-F., Ranwez, S., Montmain, J., Regnault, A., Crampes, M., Ranwez, V.: User centered and ontology based information retrieval system for life sciences. BMC Bioinform. 13, S4 (2012)

28. Vicient, C., Sánchez, D., Moreno, A.: An automatic approach for ontology-based feature extraction from heterogeneous textualresources. Eng. Appl. Artif. Intell. 26(3), 1092–1106 (2013)

29. Wu, Z., Palmer, M.: Verbs semantics and lexical selection. In: Proceedings of the 32nd Annual Meeting on Association for Computational Linguistics, pp. 133–138 (1994)

Agent-Based Assistance in Ambient Assisted Living Through Reinforcement Learning and Semantic Technologies

(Short Paper)

Nicole Merkle[1]([✉]) and Stefan Zander[2]

[1] FZI Forschungszentrum Informatik am KIT, Information Process Engineering,
Haid-und-Neu-Str. 10-14, 76131 Karlsruhe, Germany
merkle@fzi.de
[2] Institute for Computer Science, University of Applied Sciences Darmstadt,
Schöfferstrasse 8B, 64295 Darmstadt, Germany
stefan.zander@h-da.de

Abstract. For impaired people, the conduction of certain daily life activities is problematic due to motoric and cognitive handicaps. For that reason, assistive agents in ambient assisted environments provide services that aim at supporting elderly and impaired people. However, these agents act in complex stochastic and indeterministic environments where the concrete effects of a performed action are usually unknown at design time. Furthermore, they have to perform varying tasks according to the user's context and needs, wherefore an agent has to be flexible and able to recognize required capabilities in a certain situation in order to provide adequate, unobtrusive assistance. Hence, an expressive representation framework is required that relates user-specific impairments to required agent capabilities. This work presents an approach which (a) describes and links user impairments and capabilities using the formal, model-theoretic semantics expressed in OWL2 DL ontologies, (b) computes optimal policies through Reinforcement Learning and propagates these in an agent network. The presented approach improves the collaborative, personalized and adequate assistance of assistive agents and tailors the agent-based services to the user's missing capabilities.

1 Introduction

While people are getting older, their age-related diseases are increasing. Especially, motoric and cognitive capabilities deteriorate over time. Hence, elderly people need compensating assistance in their daily life, especially if they intend to continue to live an independent and self-determined life at home. Usually, in Ambient Intelligent environments, software agents are responsible for executing certain tasks in order to adequately assist users w.r.t. their contexts and needs. For this reason, an agent utilizes certain sensing and acting devices in order to

© Springer International Publishing AG 2017
H. Panetto et al. (Eds.): OTM 2017 Conferences, Part II, LNCS 10574, pp. 180–188, 2017.
https://doi.org/10.1007/978-3-319-69459-7_12

make observations in the environment for recognizing user activities and initiating state changes. The linking and processing of different evidences enables an agent to infer the user's activity and make decisions about the required assistance in the moment of need. Moreover, an agent has to evaluate if it is able to perform a specific kind of assistance. Hence, an agent needs to be aware about the capabilities it provides. For instance, if an agent deduces that the user needs the capability to open a door, the agent has to check first if it has a policy[1] (e.g. perform the action *OpenDoor* if the current state is *DoorClosed*) available in order to solve the task and if this is the case, it has to check the availability of an adequate acting device (e.g. a release buzzer) which is able to open the appropriate door.

The main challenges and problems of agent-based environments are the complex configuration and programming of agents. An on-the-fly integration is mostly hindered because of the complexity of such systems. Moreover, the agents have to face varying context-dependent tasks and different target users with different characteristics and needs. Another problem is the heterogeneity of IoT platforms which requires an abstraction of devices and applications.

Considering these issues, we provide a holistic model which explicitly amalgamate agent-, user-, and device capabilities as well as user impairments and their linkage to actions and user intentions. Furthermore, an agent has to be able to process this model in order to decide its next task. Therefore, we provide a framework for setting up a semantic representation of the environment, consisting abstract and physical things of interest and relevance (e.g. devices, agents, users, actions, events, etc.). To address the mentioned requirements, we utilize the model-theoretic semantics of a knowledge representation system for the AAL domain, which describes and links all agent relevant things, in order to allow the agent to learn and evaluate its observations and to decide its next actions according to given policies.

Moreover, we apply—based on a semantic state-action-reward model—a Reinforcement Learning (RL) approach in order to enable software agents to compute optimal policies for certain tasks according to their observations. The RL approach helps to overcome the restrictions of domain specific rule specifications. The agent is enabled by the RL approach to compute unobvious strategies for performing assistive tasks in an optimal way.

2   Related Work

In the field of Ambient Assisted Living (AAL) and Ambient Intelligence, much emphasis was placed on approaches that develop or provide context-aware and adaptive services to impaired people—only recently by utilizing machine learning and semantic technologies (e.g. [5,6,9,12,13,15,16]). Furthermore, there are several activity models, which are applied in order to detect activities of the user and to provide appropriate assistance. The work of Liming et al. [2] deals with an ontology-based hybrid approach to model user activities in Smart Homes

---

[1] A policy is a strategy for performing the best possible action in a certain state.

combined with data-driven model learning. Their objective is to provide generic activity models for all users and then to create based on this incrementally a user specific activity model. Our previous work [3] proposes an approach for modelling a light-weight AAL ontology in order to link user intentions to device capabilities by using semantic Web technologies. This paper enhances our previous work by considering user impairments and agent capabilities and by integrating probabilistic and RL approaches in order to consider task specific policies and required capability patterns. Hyoungnyoun et al. [8] present an approach for personalizing a user model and determining repetitive user activities by incremental clustering and Bayesian Networks. Hence, they interpret wearable sensor networks. They state that the user model has to reflect the life pattern of the user. Their user model is based on probabilistic modeling of user activity patterns. However, the aforementioned approach has no shared conceptualization and is just applicable in the presented device setting.

However, all approaches do just partially or not consider the agents capabilities related to the user's capabilities and impairment as well as the computation of required capabilities. In our model-theoretic semantics, we model explicitly the correlation between user impairments, activities and required capabilities, which allows us to adjust the model-theoretic semantics of the agent to the user requirements. Furthermore, our work addresses the autonomous and mutual programming of agents as well as the support of domain experts by appropriate modelling tools. In the considered related work, we could not find the support of the whole system life-cycle (e.g. from programming, configuration to quality assurance). We close this gap with our approach and framework.

3   Approach

The life-cycle of the presented approach comprises different chronological steps. In the first step, the whole system with its components (Web of Things Server (WoT), Semantic MediaWiki (SMW)[2], Triple Store, SPARQL endpoint) is installed via the Docker[3] deployment tool. The Docker integration allows to deploy software components in a simplified and platform independent way. After the installation, the components are started via the Docker console. In the second step, a domain expert provides (a) a semantic representation of the environment and (b) models the tasks by using a SMW system. The mentioned semantics serve as a description framework for expressing significant aspects of the user's environment (e.g. device-states and actions), the user's impairments and the agent's profile, containing process steps which an agent requires to perform. The process steps are: requesting frequently task descriptions from the SMW and computing the task policy as well as transforming this task policy into a

---

[2] A content management platform for generating a light-weight RDF(S) representation of annotated wiki pages. See: https://www.semantic-mediawiki.org/wiki/Semantic_MediaWiki.

[3] https://www.docker.com/.

semantic representation, in order to publish it in the local SMW. The advantage of using SMW is that the domain expert does not need to know RDF(S) in detail. Via individual forms and templates—integrated in SMW—a domain expert is able to create and annotate instances in the environment in a simple and semantically profound way. In the second step, the agent instances are deployed and started on different hardware platforms (e.g. Smartphone, Tablet, Eye-Tracking Glass, PC, etc.). The agents are requesting their profile description from the SMW via a SPARQL endpoint in order to know their process algorithm to perform. The process algorithm is described in the agent's service profile by rules and actions to perform. Moreover, the profile provides the devices, which an agent shall subscribe for via the WoT server in order to get environmental states.

The agent requests the task description from the SMW and builds from it an internal state-action model in order to compute the optimal policy. Afterwards, the agent transform this policy into a semantic representation and publishes it via the local SMW, which is accessible by all agents. By this information, the agent requests the matching policy to its given task and performs the proposed actions in order to assist the user during an activity. Therefore, the actions are sent by the agent to the WoT server, which forwards them to IoT system adapter components. These components are implementations of IoT system specific protocols. The IoT system then forwards the appropriate device commands to the addressed devices.

The description of the approach is divided into three different parts. In the first part, we present the used RL approach which is used by the agent to compute optimal policies. In the second part, we introduce our information model and the theory upon which our model is created. The policy as well as the linkage to capabilities are autonomous generated by the agent itself, after the agent has computed the optimal policy of the assigned task. In the third part, we introduce the transformation of the computed policies into a semantic representation as well as the ability of other agents in the network to search and retrieve these policies.

3.1 Reinforcement Learning

Before an agent can compute a policy, it requires a task description containing a state-action-reward model with a determining goal state. A resulting policy can be considered as a mapping from environmental states to optimal actions. For every given state S of a task, an optimal action A is computed by the agent, so that the agent knows in every possible state which action to perform. An optimal action is in this context an action that leads to the highest expected reward value. This implies that the agent needs to iteratively compute the optimal action for every state until the value functions of each state-action pair is converging. The life-cycle of policy creation envisages at the beginning that a human domain expert creates a semantic task representation, which consists of a state-action-reward model. The states are described by device states while the actions are supported functionalities by appropriate devices and agents. The goal state is also a state and represents the final state, where the task ends.

The agent learns in several episodes which actions lead to the maximum expected reward. If some quality value functions are already computed, the agent is also able to exploit (decide) which action to choose. According to [10] the agent should alternate between exploration and exploitation, since the exploration allows to find unknown and unobvious constellations, which can maximize the future expected reward of the agent. However, in particular at the beginning of the learning phase, the exploration of actions is necessary. If the policy is found, the agent creates a semantic representation, which is discussed in Sect. 3.2. The policy is computed by means of a quality value function [10]. Equation 1 depicts the value function.

$$Q^*(S_t, A_t) = \sum_{t=0}^{T} r_{t+1} + \alpha * max(Q^*(S_{t+1}, A_{t+1})) \qquad 0 \leq \alpha \leq 1 \qquad (1)$$

The quality value of a state-action pair is computed by considering the reward for choosing the next action plus a discount factor alpha multiplied with the maximum expected quality value of the next state and the next related action. The alpha discount factor helps to balance the importance of the next states' quality value. By means of the quality value function, the agent is able to compute for every state of a task a policy.

3.2   Agent Semantics

The presented modelling approach builds on the previous work of [1], which describes the use of a model-theoretic semantics in order to express and infer capabilities of IoT[4] devices and compute complex capabilities in cyber physical systems. We enhance this approach by transferring it to the AAL and healthcare domain by introducing the linking of agent capabilities to IoT device capabilities, required capabilities by user activities and -preferences in order to infer required actions, which are conducted by the appropriate agent. Furthermore, we consider user capabilities and user impairments in order to compensate these impairments and decide where the user needs help and where he/she has sufficient capabilities to conduct herself the appropriate activity. In order to express the relations between the mentioned instances, we use description logics (OWL2 DL), as defined by the W3C[5]. The advantage of OWL2 DL is that its reasoning problems are decidable and good tool support (several APIs are available for processing the OWL2 DL language). Moreover, it has a sufficient semantic expressivity regarding the modelling of complex facts. The approach based on [1] uses in the T-Box *role restriction* and *role inclusion axioms* in order to describe IoT devices and their capabilities while making use of the full feature set provided by the DL Reasoner. The objective is that a reasoner is

---

[4] Internet of Things.
[5] Further details: https://www.w3.org/TR/2012/REC-owl2-profiles-20121211/.

able to derive by subsumptions implicit and complex capabilities of compound IoT devices. Taking this idea, we transfer it to agent- and user capabilities, but do also consider device capabilities, because an agent needs devices in order to act in an AAL environment. Furthermore, our approach considers user activities and describes them also by role restriction axioms. In the following axioms, we illustrate how this is performed.

Axiom 2 describes an activity named *EnteringRoom* by using role restriction axioms. In this example, *EnteringRoom* is defined by the property *requiresAction* with the property range *SwitchLightOn*, *OpenDoor* and *RegulateClimate* which means, that the activity needs three different actions in order to be successfully conducted.

$$\text{EnteringRoom} \equiv \exists \text{requiresAction.SwitchLightOn} \qquad (2)$$

Axiom 3 defines a user impairment named *DebilityOfSight* which requires a capability (e.g. *RegulateIllumination*) in order to be compensated by the appropriate agent and its controlled devices.

$$\text{DebilityOfSight} \equiv \exists \text{requiresCapability.RegulateIllumination} \qquad (3)$$

Axiom 4 shows that the *LightSherlock* agent provides the missing capability named (*RegulateIllumination*) and has the two purposes to darken or brighten the user space in order to regulate for the user–in her presence–the brightness in the rooms in order to compensate the user's debility of sight. In this way, we relate the agent capabilities to the user impairment and user activity.

$$\text{LightSherlock} \equiv \exists \text{providesCapability.RegulateIllumination} \qquad (4)$$

The *RegulateIllumination* capability is described by axiom 5 which states that the mentioned capability is related to two opposite actions.

$$\text{RegulateIllumination} \equiv \exists \text{consistsOfAction.SwitchLightOn}$$
$$\sqcup \ \exists \text{consistsOfAction.SwitchLightOff} \qquad (5)$$

Axiom 6 shows that the *Lamp* device supports both actions which are linked in the *RegulateIllumination* capability. For this reason, the agent can decide to apply the lamp in order to achieve its purpose to lighten the room. In this way, we connect all relevant information, and provide a semantic model view for the agent.

$$\text{Lamp} \equiv \exists \text{supportsAction.SwitchLightOn}$$
$$\sqcup \ \exists \text{supportsAction.SwitchLightOff} \qquad (6)$$

Furthermore, we use property chains to define required capabilities. Axiom 7 shows this by an example. The role *requiresCapability* is defined by the roles *conductsActivity*, *requiresAction* and *belongsToCapability*. This means, that a user (who is defined as the domain of the property) conducts a certain activity.

This activity requires some action and this action belongs to a suitable capability. This capability is required by an impairment and defined as the required capability, which the agent can deduce. By linking the impairment–via the role (*isRequiredBy*)–to the activity requirements, we assure that the user is not able to execute the activity without help. In this cases, the agent can intervene.

$$\text{conductsActivity} \circ \text{requiresAction} \circ \text{belongsToCapability} \circ \text{isRequiredBy}$$
$$\sqsubseteq \text{requiresCapability} \quad (7)$$

3.3   Automated "Semantification" and Retrieval of Agent Policies

We have seen in the previous sections, that the agent increases its experience and performance by rewards and by computing policies for given tasks. If these policies are computed, the next step of the agent is to transform this internal policy representation into a machine-understandable respectively semantic representation. Therefore, the agent maps the policies to semantic rules expressed in SWRL[6]. The general representation structure in SWRL is depicted in Eq. 8.

$$\text{State(StateX)} \sqcap \text{Action(ActionY)} \sqcap \text{Agent(?a)} \sqcap \text{IsInState(?a, StateX)}$$
$$\Rightarrow \text{optimalAction(StateX, ActionY)} \quad (8)$$

This rule structure implies that an optimal action is inferred by a given state. In order to publish the policy, the agent creates a new semantically annotated policy wiki page via the MediaWiki API[7]. The policy page is structured by a predefined wiki template and linked by the agent to already existing capability representations in the system. The agent retrieves the suitable capabilities from the SMW SPARQL endpoint by requesting capabilities filtered by policy actions. The policy representation is also purported by a wiki template. The agent uses this template in order to assign and implicitly annotate the appropriate capabilities and rules. Moreover, the agent links all policy states to the policy, in order to allow other agents to query in a certain state for the right policy. If the policy page is created and annotated respectively linked to other RDF(S) instances of the wiki, other agents are able to search for the new published policy via the SPARQL endpoint.

4   First Results and Conclusion

For the evaluation of our approach, we implemented a *Navigation* agent. The task is to navigate from any room to the outdoor area of the building. First, we provide a task description by SMW. There are five rooms, while the fifth room represents the outdoor area. Every room has a reward value of zero except the goal state. We start the RL agent, which computes as expected a policy representation of the task. The computation provides for every state-action pair

---

[6] https://www.w3.org/Submission/SWRL/.
[7] https://www.mediawiki.org/wiki/API:Main_page.

a quality value. Afterwards, the agent subscribes for all registered IoT devices, which are in our case just virtual sensors, providing appropriate environmental states. The agent gets informed by a simulator that it is in state *Room2*. Now it requests from the SMW the appropriate policy. The policy suggests to go to *Room3*. The agent performs an appropriate action. After the action the agent gets informed that it is in state *Room3*. According to the policy, the agent can now decide if it goes to *Room1* or *Room4*. The agent decides randomly out of the both possibilities for *Room4*. In *Room4* the agent deduces by the policy, that it has to go to *Room5*, which is the goal state. In the goal state the policy provides the rule for staying in the goal state. The evaluation has proven that the presented approach works for the presented *Navigation* use case. The future work is to improve and complement all components of the presented framework by additional components. The objective is to provide a framework, which can easily be used by everyone who requires such a collaborative agent-based system.

References

1. Zander, S., Merkle, N., Frank, M.: Enhancing the utilization of IoT devices using ontological semantics and reasoning. Proc. Comput. Sci. **98**, 87–90 (2016). 7th EUSPN 2016 Conference, UK
2. Liming, C., Nugent, C., Okeyo, G.: An ontology-based hybrid approach to activity modeling for smart homes. IEEE Trans. Hum.-Mach. Syst. **44**(1), 92–105 (2014)
3. Merkle, N., Zander, S.: Improving the utilization of AAL devices through semantic web technologies and web of things concepts. Proc. Comput. Sci. **98**, 290–297 (2016). 6th ICTH
4. Merkle, N., Kämpgen, B., Zander, S.: Self-service ambient intelligence using web of things technologies. In: Proceedings of the 1st SEMPER Workshop, Co-located with the 13th Extended Semantic Web Conference (ESWC) 2016, Greece, 29 May 2016, pp. 1–10 (2016)
5. Stavropoulos, T., et al.: The DemaWare Service-Oriented Platform for AAL of Patients with Dementia, Czech Republic (2014)
6. Stelios, A., et al.: Dem@Home: Ambient Intelligence for Clinical Support of People Living with Dementia, Greece, May 2016
7. Beetz, M., et al.: Towards automated models of activities of daily life. Technol. Disabil. **22**(1–2), 27–40 (2010). IOS Press
8. Hyoungnyoun, K., et al.: Adaptive modelling of a user's daily life with a wearable sensor network. In: Tenth IEEE International Symposium on Multimedia (2008)
9. Kurschl, W., et al.: Modeling situation-aware ambient assisted living systems for eldercare. In: Information Technology: New Generations (2009)
10. Sutton, R., Barto, A.: Introduction to Reinforcement Learning, 1st edn. MIT Press, Cambridge (1998)
11. Russell, S., Norvig, P.: Artificial Intelligence: A Modern Approach, 3rd edn. Prentice Hall Press, Upper Saddle River (2009)
12. Yin, G.Q., Bruckner, D.: Daily activity model for ambient assisted living. In: Camarinha-Matos, L.M. (ed.) DoCEIS 2011. IAICT, vol. 349, pp. 197–204. Springer, Heidelberg (2011). doi:10.1007/978-3-642-19170-1_22
13. Denil, M., Colmenarejo, S.G., Cabi, S., Saxton, D., de Freitas, N.: Programmable agents. CoRR abs/1706.06383 (2017)

14. Mnih, V., et al.: Playing ATARI with deep reinforcement learning. J. CoRR abs/1312.5602 (2013)
15. Dragoni, M., et al.: A semantic-enabled platform for supporting healthy lifestyles, pp. 315–322 (2017)
16. Bailoni, T., et al.: PerKApp: a context aware motivational system for healthier lifestyles, pp. 1–4 (2016)

On the Need for Applications Aware Adaptive Middleware in Real-Time RDF Data Analysis (Short Paper)

Zia Ush Shamszaman$^{(\boxtimes)}$ and Muhammad Intizar Ali

Insight Centre for Data Analytics, National University of Ireland Galway, Galway,
Republic of Ireland
{zia.shamszaman,ali.intizar}@insight-centre.org

Abstract. Nowadays a handful applications are designed to consume dynamic real-time continuous stream data from IoT, Social network, Smart sensors and more. Several RDF Stream Processing (RSP) engines are available to query those data streams. Application designers have the freedom to select the best available RSP engine based on their application requirements. However, this selection needs to be done at design time resulting in early bound rigid solutions that are unable to adapt to changing application requirements. In this paper, we have evaluated two most popular RSP engines to proof that adaptivity is required to bridge the gap between RSP engines and applications requirement. Then we propose an adaptive middleware to adapt to dynamic application requirements during run-time. Moreover, adaptive middleware includes input and output control, monitoring the status of the underlying RSP engines, as a result the adaptive middleware is essential when single or multiple instances of same type engines are available.

Keywords: RDF · Stream processing · Query processing · IoT/WoT · Semantic web · Linked data · Data analysis

1 Introduction

Over the last few years, several RSP engines have been proposed for efficient processing of RDF streams [2–5,8,9]. However, the most popular and deployable RSP engines i.e. CSPARQL, CQELS are affected and differ by multiple aspects, including the execution method, input data model, query language, operational semantics, output streaming operators, execution time, processing techniques Considerable manual efforts go into creating and tuning such diverse and dynamic engines. In addition to that, application requirements, input data and workload properties may change over time, often in unpredictable ways. As a result, a mechanism to adapt to this diversity is required in order to satisfy user with a best possible way in changing environments by considering the user requirements and available resources of RSP engines. Due to this diversity, it is

© Springer International Publishing AG 2017
H. Panetto et al. (Eds.): OTM 2017 Conferences, Part II, LNCS 10574, pp. 189–197, 2017.
https://doi.org/10.1007/978-3-319-69459-7_13

hard to select a single RSP engine for any particular application. Moreover, few of the RSP engines capabilities are limited due to the decision made during the design time of any RSP engine, which can not be changed during the lifespan of a query engine. Existing approaches to stream query processing mostly offer rigid solutions and lack the adaptability to accommodate the changing requirements of the applications and properties of the underlying data streams. To satisfy diverse applications requirements data streams need to be queried and processed in a uniform way and semantic technologies are known for their effectiveness to facilitate the integration of diverse data formats. Existing approaches are based on SPARQL-like query languages for retrieving and manipulating data stored in RDF format.

We understand that different RSP engines are proposed with different understanding of RDF stream and we also believe that combining their strengths under a single framework is essential from the users perspective. In this way, their differences can be used as combined strengths rather than competing with each other. Consequently, we propose an adaptive layer on top of existing RSP engines after analysing their strengths and weakness at a granular level to come up with an efficient solution under a single flexible framework. Summary of our main contributions are, (a) We identify key application requirements to highlight the need for adaptivity. (b) We categorise different parameters and features of RSP engines and showcase how dynamic changes in application requirements can be handled by adapting to different settings of RSP engine parameters. (c) We present a conceptual architecture of adaptive RSP engine, which can monitor application requirements and RSP engine performance in real-time and trigger adaptivity accordingly. (d) We evaluate the performance of existing two RSP engines to emphasize the need for adaptivity.

2 Formalization of RSP Features

We classify RSP features into two main categories, (i) *design-time features* include aspects such as input data model, query language, execution strategy, supported output streaming operators, and (ii) *run-time features* include aspects such as execution time, time window size, input stream rate and available resources. We define design time features in Table 1. Matchmaking is required to select the best RSP engine for a given set of requirements and a set of RSP features. We formalise this Matchmaking process in the Definition 1.

Notation. $A = \{a_i, i = 1, 2, ..., n\}$ is the set of application requirements. $C = \{c_j, j = 1, 2, ..., m\}$ is the set of RSP capabilities. $\xi = \{E_k : k = 1, ..., r\}$ is a set of engines. Each engine $E \in \xi$ has its set of RSP capabilities $C_E = \{c_j^E \in C : j = 1, ..., m\}$. $< a, c >$ iff an application requirement $a \in A$ is matched to a capability $c \in C$ (Table 2).

Definition 1. *Given a set of application requirements A. An engine $E \in \xi$ is defined as "the best-matched engine to A" iff:*

$$|f(A, E)| \geq |f(A, E_k)|, \forall E_k \in \xi$$

Table 1. Narrowed down design time RSP features and denotation

Features	Type	Description	Denotation
Execution	Time	follows a periodic schedule	X_d
	Data	when new stream data arrives	X_t
Output	I	includes only the new arrivals added to the relation	O_i
	R	includes the entire output relation in the stream	O_r
	D	includes only those which are deleted from the relation	O_d
ERN	Yes	notifies if there is no mapping relation	ERN_y
	No	do not provide notification	ERN_n

Table 2. Systems Comparison

Engine	Feature			
	Execution	Output	ERN	Query Language
CQELS	Data	I	No	CQELS Native
CSPARQL	Time	R	Yes	CSPARQL Native
SPARQLstream	Time	R, I, D	Yes	SPARQLstream Native

Legend: Execution=Query Execution strategy, Output=Output streaming operators, ERN=Empty Relation notification, Time=Time driven, Data=Data driven, I/R/D=I/R/D stream, No=Not supported, Yes=Supported

where $f(A, E) = \{< a, c >: a \in A, c \in C_E\}$, and $|f(A, E)|$ is cardinality of $f(A, E)$

3 Evaluation of RSP Engines Based on Formalized RSP Features

To proof the need of an adaptive middleware for RSP and focus the impact of different design-time and run-time RSP features we have conducted an empirical evaluation over two RSP engines i.e. CQELS and CSPARQL. We set-up our testbed using CityBench datasets [1], which are based on real datasets collected from the City of Aarhus, Denmark. CityBench also provides a set of queries, we have used three CityBench queries Query 1, Query 4 and Query 6, we named these three queries Query1 - Query3. We intend to evaluate RSP engines behaviour in terms of latency, breaking point and memory consumption of two RSP engines. All the experiment run on PC environment in a Mac book Pro, processor 2.6 GHz Intel Core i5, Memory 8 GB 1600 MHz DDR3.

effects of input stream rate on the performance of RSP engines, we intend to see the effect of different input stream rate in CQELS and CSPARQL in terms of latency and memory. The delay of input stream was set at different frequency in millisecond i.e. 200, 600, 800, 1000, 1200, 1400, 1600, 1800, 2000.

effects of concurrent queries on the performance of RSP engines, processing of concurrent queries in CQELS or CSPARQL demand more memory and include processing delay. Hence we have registered concurrent queries to both engines by duplicating the same query.

effects of time Window size in RSP engines, according to the design architecture CQELS and CSPARQL have two different time window mechanisms. In this experiment, we have used three different sizes of time window with concurrent queries and different input rates to observe the changes in memory and latency.

3.1   Latency and Breaking Points

Time difference between the input stream and output result stream at an RSP engine is defined as latency. We have investigated the latency by changing several run-time features i.e. varying the input stream rate, registering concurrent queries and changing the time window size. Three sets of windows have been used in the experiment i.e. default CityBench query's windows (3s and 20s), large windows (9s and 40s) and small (500 ms and 3s). Figures 1 to 3 depict the latency of Query1 for CQELS and CSPARQL. We observed that CSPARQL never responded when the input stream delay is 200 ms to 600 ms but CQELS seems consistent at lower input stream delay. However, in Fig. 2 both engines crossed each other at several points.

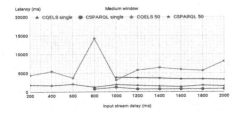

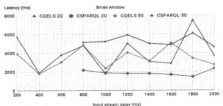

Fig. 1. Q1-various input delay with single and 50 concurrent queries

Fig. 2. Q1-various input delay with 20 and 50 concurrent queries

Large window causes the decrement in latency for both engines in Fig. 3 compare to the CityBench default windows in Fig. 1. However, with the 50 concurrent queries in Fig. 3 it takes a rise again. Figure 2 shows that small window increases the latency. It is worth mentioning here that we have used all three queries but both engines stop responding after a while.

We observe different critical points (breaking points), where given engines stop responding. Both engines stopped responding for Query2 and Query3 at certain points as shown in Fig. 4. This evaluation is helpful to figure out at what

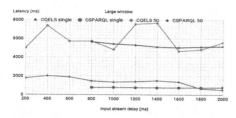

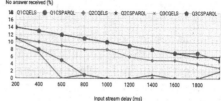

Fig. 3. Q1-various input stream delay with single and 50 concurrent queries

Fig. 4. Percentage chart of No results received with 100 concurrent queries

settings both engines stopped responding, especially when input rate is higher i.e. input delay is low like 200 ms to 800 ms, at that time both engines stopped responding and the trend is, misses fewer results towards higher input stream delay. However, CQELS never stopped responding on Query 1 and CSPARQL never responded on 200 ms, 400 ms and 600 ms input stream delay for all three queries.

3.2 Memory Consumption

Figures. 5(a) to 5(d) depicts memory consumption in both engines. CQELS consumed huge memory in Query1 compare to CSPARQL at every experiment because of multiple streams. However, CSPARQL consumes significantly less memory in Query1. Consequently, CQELS consume more memory when concurrent queries increase in number. Both engines memory consumption go higher with higher stream rate. However, CSPARQL did not consume less memory than CQELS with smaller window. Additionally, both engines consumed slightly more memory with a medium window than a large window.

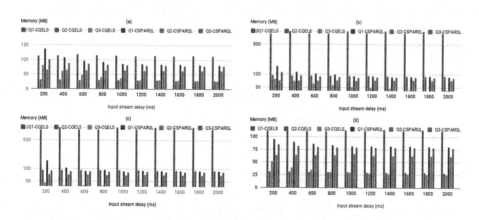

Fig. 5. Memory consumption at various input delay (a) large window with single query (b)large window with 50 concurrent queries (c) small window with 50 concurrent queries and (d) medium window with single query

4   Adaptive Middleware for RSP

Dynamic applications may require the different combination of RSP capabilities i.e. design time & runtime features. The present status of RSP engine (free resources) is crucial to select a suitable RSP engine. Hence, we consider present status, window size, concurrent queries and input stream rate as run-time features of RSP engines. We also intend to control input stream rate and output results rate, where it requires to keep the resource consumption as low as possible of the RSP engines. Additionally, engine level fail-over can also be handled by adaptive layer by monitoring underlying engines.

4.1   Adaptive Layer

Adaptive middleware analyses query before registering to an RSP engine by considering the status of underlying components and finds the best match RSP engine. Figure 6 shows the architecture of adaptive middleware.

Negotiator receives applications request and divides it into three different segments i.e. controlling input and output rate, select best match RSP engine and query handler.

Input stream handler calculates input stream rate, and selects data model i.e. point based or interval based. In the point based model data arrives at any time but in the interval based mode data arrives at a certain interval. This component sets stream pushing rate to RSP engine according to the negotiator's instruction.

Status monitor collects resource consumption of RSP engines periodically and on demand. When memory consumption goes beyond the threshold, status monitor immediately reports to the negotiator.

Matchmaker is the key component for finding a suitable engine by using an Algorithm 1 to select an RSP engine. It is a complex calculation as it requires both *design time features* and *run time features*. Initially, we consider *design time features*, because of their impact on output results [7,10] and RSP engine's status (free resources) as run time features.

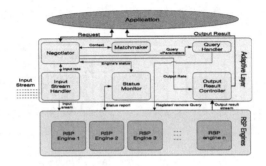

Fig. 6. Adaptive middleware architecture

Algorithm 1. Matchmaker

Input: Application's requirements, RSP Capabilities, Status and Threshold
Output: Selected RSP Engine

 1: **procedure** BESTMATCH(M)
 2:     $A \leftarrow getApplication requirements$
 3:     $C \leftarrow getRSPCapabilities$
 4:     $E \leftarrow getAvailableRSPengines$
 5:     $S \leftarrow getRSPenginesScore()$
 6:     **for** $A \in C$ **do**
 7:         $MatchedEnginesList \leftarrow getMatchedEngines(A, E)$
 8:         **for** $MatchedEnginesList \leq MatchedEngines$ **do**
 9:             $SelectedEngine = BestScoredEngine(E, S)$
10:         **end for**
11:     **end for**
12:     **return** $SelectedEngine$
13: **end procedure**

We use Multiple Criteria Decision Making (MCDM) [6] technique to select a suitable best match engine. Five design time and runtime features are considered, but other features may also be included without any fundamental changes. The features are numbered from 1 to 5. 1 = Execution 2 = Output operator 3 = Empty relation notification 4 = Input rate 5 = Memory consumptions. Given a query Qj in the adaptive layer, there is a set of candidate RSP engines $Ej = E1j, E2j, ...Enj$, that can be used. By merging all the design time and runtime features a matrix $F = (Fi, j; 1 <= i <= n, 1 <= j <= 5)$ is built, where each row Fj corresponds to a RSP engine Eij and each column corresponds to a feature dimension. Some features are positive i.e. choosing them to increase user satisfaction and also helps to occupy system resources without overloading. On the other hand, some features are negative i.e. the higher the value, the lower the quality. Hence, positive features are execution strategy, output operator, empty relation notification and negative features are Input rate, Memory consumption. Positive features are aligned to Eq. 1 and negative features are aligned to Eq. 2,

$$
V_{i,j} = \begin{cases} \frac{F_j^{max} - F_{i,j}}{F_j^{max} - F_j^{min}} & \text{if } F_j^{max} - F_j^{min} \neq 0 \\ 1 & \text{if } F_j^{max} - F_j^{min} = 0 \end{cases} \tag{1}
$$

$$
V_{i,j} = \begin{cases} \frac{F_{i,j} - F_j^{min}}{F_j^{max} - F_j^{min}} & \text{if } F_j^{max} - F_j^{min} \neq 0 \\ 1 & \text{if } F_j^{max} - F_j^{min} = 0 \end{cases} \tag{2}
$$

where F_j^{max} is the maximum value of a feature in the matrix F i.e. $F_j^{max} = Max(F_{i,j}), 1 \leq i \leq n$ while F_j^{max} is the minimal value of a feature in the matrix F i.e. $F_j^{min} = Min(F_{i,j}), 1 \leq i \leq n$ while F_j^{max}, Hence we set a matrix $V = (V_{i,j}; 1 \leq i \leq n, 1 \leq j \leq 5)$ in which each row V_j corresponds to a RSP engine $E_{i,j}$ while each column corresponds to a feature dimension.

Now we calculate the score of each RSP engines using the following formula, $Score(s_i) = \sum_{j=1}^{5}(V_{i,j} * W_j)$, where $W_j \in [0,1]$ and $\sum_{j=1}^{5} W_j = 1$. W_j represents the weight of criterion j. Users express their preferences according to the weight of each features. Finally based on the score the best suitable engine is selected.

Query handler currently, all RSP engines have different query languages, though they are extended from SPARQL but they differ from each other. If the received query language and the selected RSP engine does not match then query handler transform the query according to the selected RSP engine and then register the query to the RSP engine.

Output result controller controls the output stream rate according to the instruction from negotiator. If this module receives results faster than the application expects then it may discard some results, but it is always better to control the execution than control the output, that is why we introduce an input stream handler.

5   Conclusion

RSP engines support the development of smart applications with the capability of providing continuous query and continuous results, but current RSP engines lack adaptivity, which is a key solution for improving RSP engines results in diverse settings. We believe that addressing data and application requirements can make a difference in the correctness of RSP query results and systems resource utilisation. In this paper, we considered a few significant features of RSP engines for the matchmaking process, while this paper proves the necessity of adaptive layer by evaluating two RSP engines, but in future, we will further evaluate the performance of adaptive layer itself to compare the cost of adaptivity and monitoring against the performance gain for applications.

References

1. Ali, M.I., Gao, F., Mileo, A.: CityBench: A Configurable Benchmark to Evaluate RSP Engines Using Smart City Datasets. In: Arenas, M., et al. (eds.) ISWC 2015. LNCS, vol. 9367, pp. 374–389. Springer, Cham (2015). doi:10.1007/978-3-319-25010-6_25
2. Anicic, D., Fodor, P., Rudolph, S., Stojanovic, N.: EP-SPARQL: a unified language for event processing and stream reasoning. In: WWW Conference. ACM (2011)
3. Barbieri, D.F., Braga, D., Ceri, S., Della Valle, E., Grossniklaus, M.: C-sparql: Sparql for continuous querying. In: World Wide Web Conference (2009)
4. Bolles, A., Grawunder, M., Jacobi, J.: Streaming SPARQL - extending SPARQL to process data streams. In: Bechhofer, S., Hauswirth, M., Hoffmann, J., Koubarakis, M. (eds.) ESWC 2008. LNCS, vol. 5021, pp. 448–462. Springer, Heidelberg (2008). doi:10.1007/978-3-540-68234-9_34
5. Calbimonte, J.-P., Jeung, H., Corcho, O., Aberer, K.: Enabling query technologies for the semantic sensor web. In: IJSWIS (2012)
6. Fandel, G., Gal, T.: Multiple Criteria Decision Making. Springer, Heidelberg (1997)

7. Le-Phuoc, D., Dao-Tran, M., Pham, M.-D., Boncz, P., Eiter, T., Fink, M.: Linked stream data processing engines: facts and figures. In: Cudré-Mauroux, P., et al. (eds.) ISWC 2012. LNCS, vol. 7650, pp. 300–312. Springer, Heidelberg (2012). doi:10.1007/978-3-642-35173-0_20

8. Le-Phuoc, D., Dao-Tran, M., Xavier Parreira, J., Hauswirth, M.: A native and adaptive approach for unified processing of linked streams and linked data. In: Aroyo, L., et al. (eds.) ISWC 2011. LNCS, vol. 7031, pp. 370–388. Springer, Heidelberg (2011). doi:10.1007/978-3-642-25073-6_24

9. Shamszaman, Z.U.: Adaptive stream query processing approach for linked stream data: (extended abstract). In: Web Reasoning and Rule Systems (2014)

10. Zhang, Y., Duc, P.M., Corcho, O., Calbimonte, J.-P.: SRBench: A Streaming RDF/SPARQL Benchmark. In: Cudré-Mauroux, P., et al. (eds.) ISWC 2012. LNCS, vol. 7649, pp. 641–657. Springer, Heidelberg (2012). doi:10.1007/978-3-642-35176-1_40

Learning Probabilistic Relational Models Using an Ontology of Transformation Processes

Melanie Munch[1]([✉]), Pierre-Henri Wuillemin[2], Cristina Manfredotti[1],
Juliette Dibie[1], and Stephane Dervaux[1]

[1] UMR MIA-Paris, AgroParisTech, INRA, Universite Paris-Saclay,
75005 Paris, France
melanie.munch@agroparistech.fr
[2] Sorbonne Universites, UPMC, Univ Paris 06, CNRS UMR,
LIP6, 75005 Paris, France

Abstract. Probabilistic Relational Models (PRMs) extend Bayesian networks (BNs) with the notion of class of relational databases. Because of their richness, learning them is a difficult task. In this paper, we propose a method that learns a PRM from data using the semantic knowledge of an ontology describing these data in order to make the learning easier. To present our approach, we describe an implementation based on an ontology of transformation processes and compare its performance to that of a method that learns a PRM directly from data. We show that, even with small datasets, our approach of learning a PRM using an ontology is more efficient.

Keywords: Probabilistic relational model · Ontology · Learning

1 Introduction

Probabilistic Relational Models (PRMs) extend Bayesian networks (BNs) with the notion of class of relational databases. Thanks to the addition of the oriented-object concepts (e.g. class, instantiation, reference) they offer a new expressivity to BNs: they provide a qualitative description of the structure of complex domains while representing the quantitative information provided by the probability distribution. However, because of this richness, learning PRMs from data is a difficult task. This is due, on the one hand, to the learning of both the **high level structure** (i.e. classes and relations between them) and the **low level structure** (i.e. attributes and their probabilistic dependences); that leads us to deal with a two layers learning problem. On the other hand, their expressivity allows the modelization of systems with a small amount of data which increases the complexity of the learning task. These difficulties explain the complexity of determining the best structure among all the possible ones.

Ontologies are nowadays used as a common and standardized vocabulary for representing a domain (e.g. in life-science, geography). They organize and structure the knowledge in terms of concepts, relations between these concepts

© Springer International Publishing AG 2017
H. Panetto et al. (Eds.): OTM 2017 Conferences, Part II, LNCS 10574, pp. 198–215, 2017.
https://doi.org/10.1007/978-3-319-69459-7_14

and instances of these concepts [13]. The aim of this paper is to show that we can use the knowledge represented by an ontology to map the high level structure of PRMs easing, in this way, the learning of their probability distribution. We choose to use ontologies as opposed, for example, to relational databases, because in the future we are interested in modeling non-stationary domains and the structure of an ontology is more adaptable to changes in the domain than that of a relational database.

We present, in this paper, our approach of learning a PRM using an ontology. We propose to use the knowledge of an ontology, first, to define the high level structure (i.e. the **relational schema**) of a PRM and, then, to learn this PRM from data. Using ontology helps us by integrating the experts' knowledge to ease the learning in complex domains.

To illustrate our approach of learning a PRM using an ontology, we propose to use an ontology of transformation processes where a transformation process can be represented as a sequence of operations, receiving different inputs and designed to obtain a specific output. Such an ontology allows the representation of the knowledge of a complex domain with several interesting characteristics:

- it is **complex**, multiple operations can occur at the same time and are linked together; inputs and outputs are characterized at multiple scales (i.e. environment, population, cellular and molecular) and studied with different types of measurement (e.g. physiological, biochemical, genetic);
- data is **scarce**, due to the difficulty to obtain results, this imposes to gather information from various sources;
- it presents problems of **missing data** (e.g. a parameter is not controlled) and **missing values** (e.g. the process' instructions are not precise);
- even with complete information, it is still characterized by **uncertainty**, instruments used to take measurements during a transformation process are able to return only an estimation of the quantity observed because their calibration cannot be entirely defined and repeated from an experiment to another and some internal and uncontrollable parameters (from both devices or outside the experiment) can influence the final result.

This paper is organized as follows. In Sect. 2 we present the ontology of transformation processes used, PRMs and their existing learning methods. In Sect. 3 we describe our approach of learning a PRM using an ontology. In Sect. 4 we present preliminary results where the efficiency of our approach is evaluated through a comparison of its performance to that of a method that learns probabilistic models without ontology. We conclude in Sect. 5.

2 Backgrounds

2.1 The Ontology of Transformation Processes

To illustrate our approach, we propose to use the Process and Observation Ontology (PO^2) [9], written in OWL 2, designed to represent transformation processes.

A transformation process is denoted as a sequence of steps (i.e. operations), receiving different participants (i.e. inputs) and designed to obtain a specific product (i.e. output).

An ontology is a representation of the knowledge of a domain and is composed of two main components: the conceptual component where the concepts, relations between these concepts and axioms are defined and the instance component which contains the facts. The conceptual component of PO^2 contains the following three main parts (see Fig. 1):

- **Step part:** contains the concepts *step, itinerary* and *process*
- **Participant part:** contains the concepts *method, mixture* and *device*
- **Observation part:** contains the concepts *observation, scale, measure, sensor output* and *computed observation*

In this ontology, a *process* is a whole operation: processes that are the same share the same goal. A variation in one *process* is called *itinerary*. An *itinerary* is defined as a succession of different *steps* linked to each other: each *step* is associated to the one(s) following it according to a chronological order. A *step* is defined both by its duration and its participants, that can be a *method*, a *mixture* or a *material*. Participants are characterized by inner attributes defined by experimental conditions; moreover, a *mixture* is composed of different *products* that represent its composition. Finally, during each *step*, one or more *observations* can take place to make measurements of one participant: they are made using specific participants (independently of the other step's participants), and at a specific *scale*. They have for result a *sensor output* and/or a *computed observation*, each of them can have for value a function or a simple measure. A *measure* is characterized by either a quantity and a unit of measure or a symbolic concept and a measurement scale.

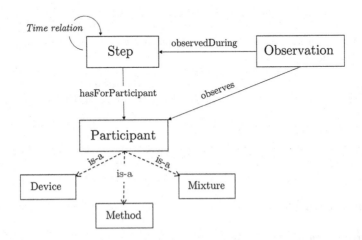

Fig. 1. Simplified schema of the conceptual component of PO^2. The ontology is divided in three main parts: **Step, Participant** and **Observation**. These parts interact to each other through semantic relations.

Each *step* is defined as a concept to which a set of descriptor concepts is linked: participants (i.e. devices, mixtures and methods) are concepts whose parameters are set *a priori*; observations are concepts whose parameters are measured during the step. Therefore there exists for each step a compartmentalization between the different domain's objects. Moreover, the *time relation* linking steps gives information about their relative time (inside the process and with other steps). The instance component of PO^2 allows one to represent different transformation processes by a succession of instances of steps and instances of their associated descriptors.

We introduce an example of a domain ontology about the micro-organisms stabilization transformation process denoted by PO^2_{stab}. Figure 2(a) gives an excerpt of the simplified conceptual component of PO^2_{stab} where there are 3 steps: *Fermentation*, *Culture* and *Stabilization* which are sub-concepts of the concepts *Step* and 2 attributes: *SugarQuantity* and *Temperature* which are sub-concepts of the concept *Attribute*. Figure 2(b) gives an excerpt of the simplified instance component of PO^2_{stab}. In this example, there are three instances of steps linked by a linear temporal dependency *Fermentation_1* that is before *Culture_1* that is before *Stabilization_1*. The instance *Fermentation_1* of the concept *Fermentation* has for participant *Mixture_1* (an instance of the concept *Mixture*) which has for sugar quantity (the instance *SugarQuantity_1* of the concept *Attribute*) the value: *2g*. Moreover, an observation (the instance *Observation_1* of the concept *Observation*) was made on the temperature (the instance *Temperature_1* of the concept *Attribute*) of *Mixture_1* which has for value: 5 °C.

2.2   Probabilistic Relational Models

Probabilistic Relational Models (PRMs) extend Bayesian networks (BNs) with the notion of class of relational databases. A BN is the representation of a joint probability over a set of random variables that uses a Directed Acyclic Graph (DAG) to encode probabilistic relations between variables (see Fig. 3(a)). However, in the case of numerous random variables with repetitive patterns (for instance different steps in the same transformation process), it cannot efficiently represent every probabilistic link.

PRMs extend the BN representation with a relational structure between potentially repeated fragments of BN called classes [15]. A **class** is defined as a DAG over a set of attributes. These attributes can be inner attributes or attributes from other classes referenced by so-called **reference slots**. The analysis of the BNs in Fig. 3(a) reveals two recurrent patterns, that can be translated into two interconnected classes $\mathcal{E}$ and $\mathcal{F}$, as presented in Fig. 3(b).

The high level structure of a PRM (i.e. its **relational schema**, see Fig. 3(b)) describes a set of classes C, associated with attributes $A(C)$ and reference slots $R(C)$. A slot chain is defined as a sequence of reference slots that allows one to put in relation attributes of objects that are indirectly related.

The probabilistic models are defined on the low level structure (i.e. at the class level) over the set of inner attributes, conditionally to the set of outer

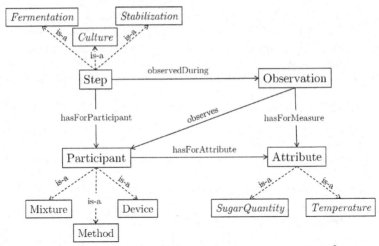

(a) Excerpt of the simplified conceptual component of PO^2_{stab}

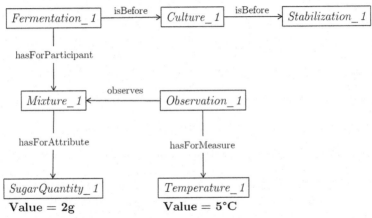

(b) Excerpt of the simplified instance component of PO^2_{stab}

Fig. 2. An example of a domain ontology about the micro-organisms stabilization transformation process: PO^2_{stab}

attributes and represent generic probabilistic relations inside the classes. This is the **relational model** of the PRM (see Fig. 3(c)).

Classes can be **instantiated** for each specific situation (see Fig. 3(d)). A system in a PRM provides a probability distribution over a set of instances of a relational schema [16]. PRMs define the high-level, qualitative description of the structure of the domain and the quantitative information given by the probability distribution [5].

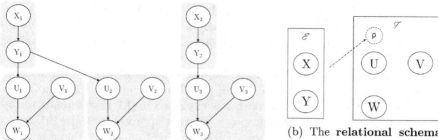

(a) An example of two BNs. The gray areas represent the repetitive patterns, but are not part of the BN specification.

(b) The **relational schema** of the PRM. It is composed of two connected classes E and F .

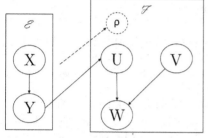

(c) The PRM **relational model**. Relational links between attributes were added to the relational schema in (b).

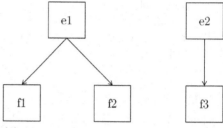

(d) A system for the PRM in (c). Instantiation of the classes of the PRM representing the BN in (a).

Fig. 3. BNs and PRMs: the analysis of the BN in (a) reveals two recurrent patterns, that can be translated into two interconnected classes $\mathscr{E}$ and $\mathscr{F}$ of a PRM (b) and (c). An equivalent system can, thus, be constructed through the instantiation of twice the class $\mathscr{E}$ and three times the class $\mathscr{F}$ (d).

2.3   Learning PRMs

PRM learning is composed of two different parts: **structure selection** and **parameter estimation**. Structure selection can be decomposed in two layers: a high level layer that organizes the knowledge under an entity-relation pattern (using classes and references); and a lower level layer that employs a graphical language to represent the probability distribution in a compact way by exploiting the probabilistic dependencies between the attributes. The **relational schema** is learned with the high level layer while the **relational model** is the final result of structure selection. Due to these multiple layers, the number of free parameters is high, and the target model is not unique: selecting one requires making subjective choices. Moreover, the richness of this tool allows us to represent new and complex systems where data can be scarce or incomplete. This can be another obstacle while learning PRMs, for example, in life science.

In [5] an algorithm based on an **heuristic search**, such as a greedy algorithm, is proposed to select the legal structure (i.e. a structure representing a coherent probability model) with the highest score. The score proposed has a decomposability property that helps to analyze small parts of structures, easing the search. Other score-based approaches have been equally proposed based on a relational extension [6].

On the contrary of heuristic search, **dependency analysis** tries to discover dependency relations from the data itself and then attempts to learn the structure. This constraint guided approach was exploited in [10] that extends to the relational context, or in [4] that proposes an exact approach to learn PRMs.

In this paper we propose to learn a PRM starting from its relational schema. This relational schema can be deduced from a relational database, however ontologies can also be used to define it. In fact, the notions of class in PRMs and of concepts in ontologies are very similar. We therefore propose to deduce the relational schema of a PRM from the concepts' structuration defined in the ontology's conceptual component.

The use of ontology has already been proposed for learning BNs [3,8] and Object Oriented Bayesian Networks [1]. An approach to define a relational schema from an ontology has been proposed in [11], but the task of learning PRMs using an ontology has not been addressed yet.

Indeed, once the structure of the relational schema is known, learning the relational model of a PRM can be compared to selecting the structure of a BN [6]. The main difference is that probabilistic dependences between attributes in the same class are forced to be identical: the PRM relational schema and the ontology's semantic knowledge give us patterns on which to learn.

3 Learning a PRM Using an Ontology

We present, in this section, our approach to learn PRMs using ontologies. We first present the relational schema mapping from the PO^2's conceptual component and then our **ON2PRM** algorithm.

3.1 Relational Schema Mapping

We briefly present our relational schema mapping from the PO^2's conceptual component that relies on the one proposed in [11]. Our mapping was motivated both by the description of transformation processes in PO^2 and the definition of *state* as explained in the theory of control and expert systems.

In the theory of control, a system can be described as a succession of *states* through time [14]. A state contains a set of every attributes that enables to describe the system. Observations can be made to evaluate these attributes: however, the act of observing is independent of the state itself. These definitions and the semantic representation of transformation processes defined in PO^2 allow us to define the following temporal dependences properties:

- **Observations can be longer in time than the states they are observing.** For instance, some measurement methods in biology are based on time dependent reactions; in this case, the result of observations can be physically obtained even if the step linked to these has ended before;

– **States influence the result of observations, but observations do not influence states' values.** From this property, we can deduce that observations cannot influence other observations.

In the relational schema, we therefore propose to define two classes built from the ontology's concepts defined in its conceptual component:

– The **Participant Class**, $\mathscr{P}$. It groups every *a priori* attributes: the attributes of the participant concepts mixtures, devices and methods (Fig. 1).
– The **Observation Class**, $\mathscr{O}$. It groups every measured attributes.

At each time step t, we instantiate these two classes: $\mathscr{P}_t$ and $\mathscr{O}_t$. We call *Step*, denoted by $\mathscr{S}_t$, the couple $\mathscr{P}_t$ and $\mathscr{O}_t$.

The temporal dependences properties introduced above can be formalized between the two classes $\mathscr{P}_t$ and $\mathscr{O}_t$ as the following **temporal dependences constraints**: $\mathscr{P}_t$ can have none or multiple $\mathscr{P}$ parents at time $t-1$ (that we call altogether $\mathscr{P}_{t-1}$), but always maximum one child at time $t+1$ ($\mathscr{P}_{t+1}$). $\mathscr{O}_t$ only depends on $\mathscr{P}_t$. To each $\mathscr{P}$ class an $\mathscr{O}$ class is linked. Through slot chain, each $\mathscr{P}_T$ class has access to every attributes of $\mathscr{P}_t$ with $t < T$, and each $\mathscr{O}_t$ has only access to the attributes of $\mathscr{P}_t$.

The relational schema mapped from the PO^2 ontology is represented in Fig. 4: the arrows represent the reference slots; given two classes $\mathscr{P}_t$ and $\mathscr{P}_{t-1}$, $\mathscr{P}_{t-1}$ o$\rightarrow$ $\mathscr{P}_t$ means that $\mathscr{P}_t$'attributes can depend on $\mathscr{P}_{t-1}$'s, attributes of $\mathscr{P}_{t-1}$ can be parents of attributes of $\mathscr{P}_t$. However, according to the temporal dependences constraints, attributes of $\mathscr{O}_{t-1}$ cannot be parents of attributes of $\mathscr{O}_t$.

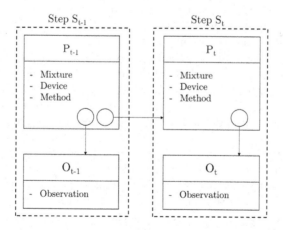

Fig. 4. Relational Schema mapped from the PO^2 ontology for two steps.

This relational schema has two interesting properties we use in the learning. First, it preserves the **compartmentalization** between the different steps and between participants in the process and observations about the process.

In the relational schema the attributes of an observation class only depend on the attributes of the participant class it is associated with. This will allow us to consider, while learning the relational model, only meaningful attributes, which are defined in the conceptual component of the ontology. In example of Fig. 2, we can deduce from the instance component of PO^2_{stab} that the *SugarQuantity* is an attribute of the mixture in the Participant class and the *Temperature* is an attribute of the Observation class. Moreover, we can deduce that these two attributes are specific to the Fermentation step.

Second, it preserves the integrity of the steps through time: a choice made at time t (i.e. the value of an attribute of $\mathscr{P}_t$) cannot influence an observation at time $t-1$. This leads us to define the **direction learning constraint** used in the learning presented below: if attributes are dependent in the instance component of PO^2, the learnt links between them can only have one direction. In example of Fig. 2, we can deduce from the instance component of PO^2_{stab} that the sugar quantity, an attribute of the mixture, can have an influence on the temperature, an observation attribute of the mixture. Moreover, considering that the fermentation step is before the culture step, the sugar quantity can also have an influence on the values of the attributes associated with the culture step.

In the next subsection, we present our algorithm for learning PRMs' relational models using its relational schema and the ontology PO^2.

3.2   Our ON2PRM Algorithm

The learning approach we propose for learning a PRM given its relational schema is very similar to a classic approach for learning BNs. However, we propose to use the semantic knowledge of the ontology: the concepts' structure defined in its conceptual component and the links between concepts defined in its instance component, as presented in the mapping defined above.

Let us consider a database D about a transformation process, where each attribute using concepts defined in the conceptual component of the ontology is represented (e.g. fermentation is a step according to the concepts' hierarchy of PO^2_{stab} as presented in Fig. 2). Following the **compartmentalization** property introduced in the relational schema, several sub-databases are created, during the learning from D, each containing the data of one step: only the attributes of this step (i.e. attributes from the $\mathscr{P}_t$ and the $\mathscr{O}_t$ classes) and their parents (i.e. attributes from the $\mathscr{P}_{t-1}$ class) are considered. This ensures that the organization between participant and observation is preserved. Afterwards, using the **direction** learning constraint, we force a learning order over the attributes of the same sub-database. This ensures that the temporal order between steps is preserved. However preserving organization and temporal order does not imply links existence but only that, if they exist, the orientation of the links is defined by the direction and the organization given. In example of Fig. 2, we can deduce from the instance component of PO^2_{stab} that the attribute quantity of sugar is included both in the fermentation $\mathscr{P}_t$ and the culture $\mathscr{P}_{t-1}$, while the temperature is only included in the fermentation $\mathscr{O}_t$ class.

We call ON2PRM(M) our algorithm of learning a PRM's relational model from an ontology where M is a learning method for Bayesian Networks that can be used to draw probabilistic dependencies between attributes from a database. For each step (e.g. the steps fermentation, culture and stabilization in Fig. 2), the ON2PRM(M) algorithm uses M over the attributes (e.g. the attribute quantity of sugar in the fermentation $\mathscr{P}_t$ and the culture $\mathscr{P}_{t-1}$ and the attribute temperature in the fermentation $\mathscr{O}_t$ class) following the established learning order to learn a small BN for each identified class of the PRM. Once every class has been learnt, the PRM relational model is defined and can be instantiated.

Input: ontology PO^2 + relational schema + database D + learning method M

Result: a PRM relational model

//the **for** loop is justified by the compartmentalization property of the relational schema

//the identification of the steps relies on the concepts and concepts' hierarchy defined in the conceptual component of PO^2

for *each step at time t* **do**

> //the identification of the attributes relies on the concepts and concepts' hierarchy defined in the conceptual component of PO^2 ;
> identify attributes for $\mathscr{P}_t$;
> identify attributes for $\mathscr{P}_{t-1}$;
> identify attributes for $\mathscr{O}_t$;
>
> create a **sub-database** from D from the identified attributes;
>
> //the **learning order** is defined from the instance component of PO^2 as defined in the direction constraint ;
> define the **learning order** ;
>
> learn a BN of a PRM class from **sub-database** + **learning order** + method M;

end

//the PRM relational model is the set of the PRM classes generated above, linked to each other following the PRM relational schema ;
create the PRM relational model ;

Algorithm 1. ON2PRM(M): Learning a PRM using an ontology

As explained in Sect. 2.2, the PRM relational model can be instantiated with data in D providing the system of the PRM. In the following we use the instantiated PRM to compare the performance of our approach to that of a method that learns BNs directly from data. We demonstrate that, thanks to the use of the semantic knowledge represented in an ontology, learning a PRM with an ontology is more efficient than learning without an ontology. We compare the performance of learning with our algorithm ON2PRM(M) to the performance of learning only with the method M.

4   Experiments

In order to validate our approach we propose to compare the performance of learning with and without ontology implementing two learning methods[1]:

- *Greedy Hill Climbing* algorithm with BIC score, denoted by $M1$;
- *Local Search with Tabu List* algorithm with BDeu score, denoted by $M2$.

The proposed experiment consists in comparing the instantiated PRM, learnt with ON2PRM from a database D of transformation processes and PO^2, with a BN learnt from D, using both methods $M1$ and $M2$. All our experiments were implemented using the PyAgrum Python library [7].

In order to have an experiment as generic as possible, we perform our learnings from several randomly generated databases D_i. We first present the generation of our test databases and then our results.

4.1   Databases Generation

The databases generation (1) generates PRM relational models representing transformation processes using the PRM relation schema of Fig. 4 and (2) builds the domain ontologies corresponding to the generated PRM relational models. We first present the generation of the PRM relational models, then the construction of the corresponding domain ontologies, finally the databases generation.

The PRM Relational Models Generation. One of our motivations to study transformation processes is their complexity (of which we presented the main characteristics in the introduction section. One process cannot therefore encompass alone the entire diversity spectrum of processes. We define five **process complexity degrees criteria** to randomly generate PRM relational models representing transformation processes as much as possible generic:

1. the number s of steps in a process;
2. the maximal number p of parallel steps, representing how many parents a step can have;
3. the number n of attributes in a class;
4. the number m of modalities for the attributes;
5. the number d of probabilistic dependencies an attribute may have.

The higher the process complexity degrees criteria are, the harder to learn the corresponding PRM relational models are. As a matter of fact, during the learning phase of a PRM relational model:

- a high number of steps induces more PRMs' classes to learn;
- a high number of parallel steps, attributes and probabilistic dependencies induces more possible links to draw;
- a high number of modalities induces a more difficult learning.

---

[1] These are two standard well known methods for learning BN. These and others methods can be found in [12].

In the following, we assume that the process complexity degrees criteria are better addressed by ON2PRM where the ontology semantic knowledge reduces the learning's complexity. Therefore, we argue that if the results of our approach outperforms that of a standard method for simple processes, it will have better results in learning more complex processes. Considering this assumption and to be as close as possible to the modelization of real transformation processes, we decided to fix two process complexity degrees criteria: $m = 2$ (i.e. binary attributes) and $d = 3$, and to have three criteria that vary: $s \in \{3, 5, 8\}$, $p \in \{1, 2, 3\}$ and $n \in \{2, 4\}$. This leads us to have 16 different configurations of possible processes, not considering the case $s = 3$ and $p = 3$ (i.e. a process composed of only three parallel steps without interaction) because it is not interesting.

With these 16 configurations, we can generate several different PRM relational models because of the possible relations between attributes. For example, links between steps are decided randomly, given s and p (see Fig. 5(a)); moreover, even inside the same class or sequence of two steps, links between attributes are decided randomly given n (see Fig. 5(b)). We generate 10 PRM relational models for each configuration, that corresponds to a total of 160 processes.

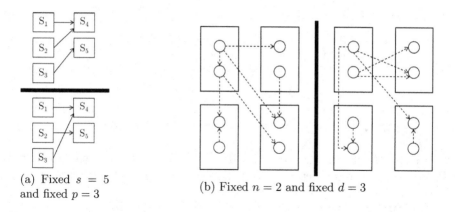

(a) Fixed $s = 5$ and fixed $p = 3$

(b) Fixed $n = 2$ and fixed $d = 3$

Fig. 5. Possible differences with fixed parameters

These PRM relational models will be on the one hand used to generate the test databases and on the other hand considered as the original models, i.e. the ground truths in the experiments' evaluation.

The Domain Ontologies Generation. In parallel to the generation of these PRM relational models, we build several domain ontologies, denoted by $PO^2_{dom_i}$, necessary for our ON2PRM learning algorithm. The domain ontology's generation is done using the same process complexity degrees criteria defined above. The number s of steps and the number n of attributes are used to create the conceptual component of each domain ontology $PO^2_{dom_i}$; the number p of parallel steps are used to create its instance component. For example, the domain ontology PO^2_{stab} of Fig. 2 has $s = 3$, $n = 2$ and $p = 1$.

Databases Generation. From each of the 160 PRM relational models, we generate 100 times four databases of different sizes as presented in Fig. 6: 100 databases of size 50, 100 databases of size 100, 100 databases of size 150 and 100 databases of size 200. The database size refers to the number of examples in it. We therefore generated $16 * 10 * 400 = 64\,000$ databases for experiments.

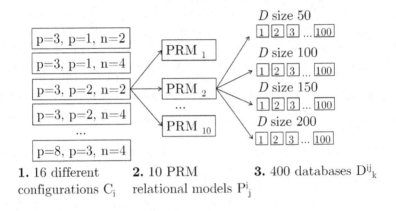

1. 16 different   2. 10 PRM   3. 400 databases D^{ij}_k
configurations C_i relational models P^i_j

Fig. 6. Databases generation

4.2   Results

We evaluate the performance of ON2PRM in learning the relational model of a PRM by comparing its performance with that of a method that learns probabilistic models without using an ontology.

The learning is performed on the $64\,000$ databases D_i generated above. We compare the instantiation of the relational models learnt by our algorithm ON2PRM using both learning methods $M1$ and $M2$, denoted by ON2PRM($M1$) and ON2PRM($M2$), with the BNs learnt by $M1$ and $M2$ alone. With the standard approach, the learning is done directly from the database.

The performance of our algorithm is evaluated using structural analysis (i.e. recall, precision and f-score scores) by comparing the structure of the graph learnt to the one of the ground truth. More precisely, let C_i, $i \in [1, 16]$, be one of the 16 possible process configurations, P^i_j, $i \in [1, 10]$, be one of the 10 PRM relational models generated from the configuration C_i and $D^{i,j}_k$, $k \in [1, 4]$, be one of the 400 databases generated from the PRM relational model P^i_j. The model (relational model or BN) learnt from the database $D^{i,j}_k$ with one of the four different methods ON2PRM($M1$), ON2PRM($M2$), $M1$ and $M2$ is compared with its ground truth P^i_j.

Let us notice that due to the semantic value added, edges orientation is crucial: that is why we consider the presence of arcs as well as their orientation while evaluating the performance.

Three different structural parameters were evaluated: **recall, precision** and **F-score** [2]. Recall is used to estimate the number of links found out of the total we have to find. Precision allows the estimation of the proportion of true links among the ones found. F-score is the average of recall and precision. In order to compute these parameters, we have to count the number of true positive and true negative (i.e. right learning), and false positive and false negative (i.e. wrong learning). These are defined following the heuristic reported in Table 1.

Table 1. Heuristic used to compare two BNs. TN: True negative. FN: False negative. TP: True positive. FP: False positive

Learned	Model		
	∅	→	←
∅	TN	FN	FN
→	FP	TP	FN
←	FP	FN	TP

Precision, recall and F-score are computed with the following equations:

$$\text{Recall} = \frac{TP}{TP + FN} \qquad \text{Precision} = \frac{TP}{TP + FP}$$

$$\text{F-score} = \frac{2 * Recall * Precision}{Recall + Precision}$$

In Table 2 we report the F-score of ON2PRM($M1$) compared with $M1$ and ON2PRM($M2$) compared with $M2$ on a database of size 50. In all cases, results with ontology are significantly better than without. This can be explained by the two properties of the ontology that are preserved by the relational schema (as explained in Sect. 3.1): both compartmentalization and the direction constraint drastically reduce the number of possibilities the method M can consider in the ON2PRM(M) algorithm.

Recall and precision are both as significant as F-score; however depending on the methods, performance varies. Precision tends to be, in fact, better with $M1$, while recall is better with $M2$. Since the difference between recall and precision for $M2$ is smaller than for $M1$, it explains why $M2$ has the best F-score. Table 3 shows performances' comparison between different databases' sizes (50, 100, 150 and 200) and between different process complexities (high-complexity processes (b) and low-complexity processes (a)). This score rises with the augmentation of the size of the database.

Even with few data a difference between the two learning approaches appears. Moreover while raising the size of the database, every score increases.

Table 2. Variation of F-score in function of different parameters tested with a database of size 50 with 100 repetitions: (mean [confidence interval 99%]). **bold**: highest value in column, *italic*: lowest value in column. *s*: number of steps, *p*: maximal number of parallel steps, *n*: number of attributes

s	p	n	ON2PRM($M1$)	ON2PROM($M2$)	$M1$	$M2$
3	1	2	**0.40** [0.03]	0.27 [0.03]	0.56 [0.03]	**0.33** [0.03]
		4	0.33 [0.02]	0.25 [0.02]	0.45 [0.02]	0.26 [0.02]
	2	2	*0.24* [0.03]	*0.17* [0.03]	0.43 [0.03]	0.26 [0.03]
		4	0.25 [0.01]	0.20 [0.01]	*0.38* [0.02]	0.24 [0.02]
5	1	2	**0.40** [0.02]	**0.29** [0.02]	0.54 [0.02]	0.27 [0.02]
		4	0.30 [0.01]	0.22 [0.01]	0.43 [0.01]	0.22 [0.01]
	2	2	0.37 [0.02]	0.27 [0.02]	0.54 [0.02]	0.27 [0.02]
		4	0.29 [0.01]	0.21 [0.01]	0.42 [0.01]	0.21 [0.01]
	3	2	0.37 [0.02]	0.28 [0.02]	0.52 [0.02]	0.27 [0.02]
		4	0.28 [0.01]	0.22 [0.01]	0.41 [0.01]	0.21 [0.01]
8	1	2	0.45 [0.01]	**0.29** [0.02]	**0.58** [0.01]	0.25 [0.01]
		4	0.31 [0.01]	0.21 [0.01]	0.43 [0.01]	*0.17* [0.01]
	2	2	0.37 [0.02]	0.25 [0.02]	0.52 [0.02]	0.22 [0.01]
		4	0.31 [0.01]	0.22 [0.01]	0.44 [0.01]	0.18 [0.01]
	3	2	0.34 [0.02]	0.24 [0.02]	0.52 [0.02]	0.22 [0.01]
		4	0.31 [0.01]	0.22 [0.01]	0.43 [0.01]	0.18 [0.01]

Table 3. Comparison of performances for recall, precision and F-score for $M1$ and $M2$ with different sizes of the database (mean [confidence interval 99%]).

Method	Length	Recall		Precision		Fscore	
		ON2PRM(M)	M	ON2PRM(M)	M	ON2PRM(M)	M
M1	50	0.26 [0.04]	0.16 [0.03]	0.95 [0.05]	0.81 [0.09]	0.4 [0.05]	0.27 [0.04]
	100	0.39 [0.04]	0.24 [0.04]	0.97 [0.02]	0.87 [0.07]	0.54 [0.05]	0.37 [0.05]
	150	0.47 [0.04]	0.28 [0.04]	0.97 [0.02]	0.86 [0.06]	0.62 [0.04]	0.41 [0.05]
	200	0.51 [0.04]	0.31 [0.04]	0.97 [0.02]	0.88 [0.06]	0.66 [0.04]	0.44 [0.05]
M2	50	0.44 [0.04]	0.27 [0.04]	0.82 [0.04]	0.46 [0.05]	0.56 [0.04]	0.33 [0.04]
	100	0.53 [0.04]	0.33 [0.04]	0.90 [0.03]	0.61 [0.06]	0.66 [0.04]	0.42 [0.05]
	150	0.57 [0.04]	0.38 [0.05]	0.92 [0.03]	0.69 [0.05]	0.70 [0.03]	0.48 [0.05]
	200	0.61 [0.04]	0.4 [0.04]	0.94 [0.02]	0.72 [0.05]	0.73 [0.03]	0.50 [0.05]

(a) Parameters of the process: s = 3, p = 1, n = 2

Method	Length	Recall		Precision		Fscore	
		ON2PRM(M)	M	ON2PRM(M)	M	ON2PRM(M)	M
M1	50	0.19 [0.01]	0.13 [0.01]	0.91 [0.02]	0.61 [0.03]	0.31 [0.02]	0.22 [0.01]
	100	0.29 [0.01]	0.21 [0.01]	0.93 [0.01]	0.73 [0.03]	0.44 [0.01]	0.33 [0.02]
	150	0.36 [0.01]	0.27 [0.01]	0.94 [0.01]	0.77 [0.02]	0.52 [0.02]	0.40 [0.02]
	200	0.42 [0.01]	0.32 [0.02]	0.94 [0.01]	0.8 [0.02]	0.58 [0.02]	0.46 [0.02]
M2	50	0.33 [0.02]	0.19 [0.01]	0.61 [0.02]	0.16 [0.01]	0.43 [0.02]	0.18 [0.01]
	100	0.42 [0.02]	0.26 [0.02]	0.78 [0.02]	0.32 [0.02]	0.54 [0.02]	0.29 [0.02]
	150	0.48 [0.02]	0.32 [0.02]	0.84 [0.02]	0.44 [0.02]	0.61 [0.02]	0.37 [0.02]
	200	0.52 [0.02]	0.36 [0.02]	0.87 [0.01]	0.52 [0.02]	0.65 [0.01]	0.42 [0.02]

(b) Parameters of the process: s = 8, p = 3, n = 4

In order to quantify and compare the performance of learning with ontology and without, we introduce the following ratio of the performances:

$$\text{ratio} = \frac{\text{performance with ON2PRM}}{\text{performance without ON2PRM}}$$

The more the ratio is above 1, the more the learning with the ON2PRM algorithm is efficient. We have used this value to compare the evolution of scores with processes complexity and the different complexity degrees defined (number of step s, number of parent p and number of attribute n).

Figure 7 illustrates the evolution of the ratio for two processes. The ratio is always above 1, but it is also decreasing with the augmentation of the database size. Depending on the methods this decrease can be narrower or wider: while $M1$ stays practically stable $M2$ drops faster. Moreover the ratio varies equally with the complexity for $M2$: ON2PRM efficiency is higher with a complex process.

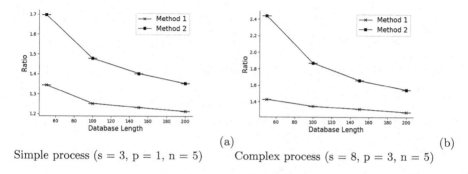

(a)   (b)

Simple process (s = 3, p = 1, n = 5)   Complex process (s = 8, p = 3, n = 5)

Fig. 7. Evolution of F-score ratio for two different processes with the database length

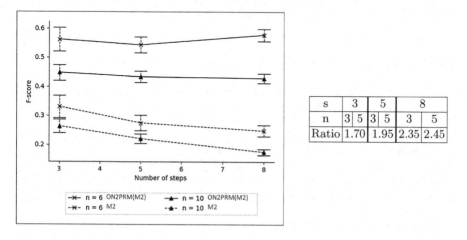

s	3		5		8	
n	3	5	3	5	3	5
Ratio	1.70	1.95	2.35		2.45	

Fig. 8. Evolution of F-score in function of n (p = 1, D size = 50) and ratio evolution in function of n and s for M2.

Figure 8 shows correlation between the number of attributes and the number of steps: the more a process is complex in terms of attributes, the lower is the F-score. Moreover, for the same number of attribute, the F-score decreases in case of the learning without ontology while it stays stable with ON2PRM. This explains that the ratio measure increases with the number s of steps (b).

5   Conclusion

In this paper, we presented an approach to learn a PRM by using expert knowledge extracted from an ontology. The advantage of our approach is that (i) the high level knowledge structure organization needed to construct the relational schema can be found directly in the ontology and (ii) this relational schema combined with the semantic knowledge represented by the ontology eases the learning afterward. Thanks to the addition of this semantic knowledge the learning's complexity is reduced and the learnt models are more meaningful than those learnt with a simple direct learning. In our experiments we demonstrated the efficiency of our approach compared to the one without prior knowledge, even in low-complexity processes or with few data.

In future works, we will generalize this approach to consider different temporal relations such as, for example, the time duration of the observations. In this paper, we identified the mapping between classes in the ontology and the PRM, and the temporal dependency properties from the concepts' structuration and the relations of an ontology, following the approach presented in [11]. A global framework of completely automated ON2PRM extraction from an ontology conceptual component and its instance component will face two main challenges: the discovery of oriented relations in ontologies (from data, from expert, etc.) and more complex relations between concepts in ontology and classes in PRM (more complex than a one-to-one mapping). We intend to address these issues by testing our approach and its limits on different ontologies.

References

1. Ben Ishak, M., Leray, P., Ben Amor, N.: Ontology-based generation of object oriented Bayesian networks. In: BMAW 2011, Spain, pp. 9–17 (2011). https://hal.archives-ouvertes.fr/hal-00644992
2. Manning, C.D., Raghavan, P., Schütze, H.: Introduction to Information Retrieval. Cambridge University Press, New York (2008)
3. Cutic, D., Gini, G.: Creating causal representations from ontologies and Bayesian networks. In: Arai, Gini, P.E. (eds.) Proceedings of Workshop NRF- IAS, Venice, Italy, pp. 1–12, 18–19th July 2014. http://home.deib.polimi.it/gini/papers/2014NFR.pdf
4. Ettouzi, N., Leray, P., Messaoud, M.B.: An exact approach to learning probabilistic relational model. In: Antonucci, A., Corani, G., Campos, C.P. (eds.) Proceedings of the Eighth International Conference on Probabilistic Graphical Models, pp. 171–182 (2016)

5. Friedman, N., Getoor, L., Koller, D., Pfeffer, A.: Learning probabilistic relational models. In: Dean, T. (ed.) Proceedings of the Sixteenth International Joint Conference on Artificial Intelligence, IJCAI 1999, Stockholm, Sweden, 31 July–6 August 1999, vol. 2, pp. 1300–1309, 1450 pages. Morgan Kaufmann (1999). http://ijcai.org/Proceedings/99-2/Papers/090.pdf

6. Getoor, L., Taskar, B.: Introduction to Statistical Relational Learning. Adaptive Computation and Machine Learning. The MIT Press, Cambridge (2007)

7. Gonzales, C., Torti, L., Wuillemin, P.H.: aGrUM: a graphical universal model framework. In: Proceedings of the 30th International Conference on Industrial Engineering, Other Applications of Applied Intelligent Systems, Arras, France, June 2017. https://hal.archives-ouvertes.fr/hal-01509651

8. Helsper, E.M., van der Gaag, L.C.: Building Bayesian networks through ontologies. In: Proceedings of the 15th Eureopean Conference on Artificial Intelligence, ECAI 2002, Lyon, France, July 2002, pp. 680–684 (2002)

9. Ibanescu, L., Dibie, J., Dervaux, S., Guichard, E., Raad, J.: PO^2 - a process and observation ontology in food science. Application to dairy gels. In: Garoufallou, E., Subirats Coll, I., Stellato, A., Greenberg, J. (eds.) MTSR 2016. CCIS, vol. 672, pp. 155–165. Springer, Cham (2016). doi:10.1007/978-3-319-49157-8_13

10. Li, X.-L., Zhou, Z.-H.: Structure learning of probabilistic relational models from incomplete relational data. In: Kok, J.N., Koronacki, J., Mantaras, R.L., Matwin, S., Mladenič, D., Skowron, A. (eds.) ECML 2007. LNCS, vol. 4701, pp. 214–225. Springer, Heidelberg (2007). doi:10.1007/978-3-540-74958-5_22

11. Manfredotti, C.E., Baudrit, C., Dibie-Barthélemy, J., Wuillemin, P.: Mapping ontology with probabilistic relational models. In: Fred, A.L.N., Dietz, J.L.G., Aveiro, D., Liu, K., Filipe, J. (eds.) KEOD 2015 - Proceedings of the International Conference on Knowledge Engineering and Ontology Development, part of the 7th International Joint Conference on Knowledge Discovery, Knowledge Engineering and Knowledge Management (IC3K 2015), Lisbon, Portugal, 12–14 November 2015, vol. 2, pp. 171–178. SciTePress (2015). https://doi.org/10.5220/0005590001710178

12. Neapolitan, R.E.: Learning Bayesian Networks. Prentice-Hall Inc., Upper Saddle River (2003)

13. Staab, S., Studer, R. (eds.): Handbook on Ontologies. IHIS. Springer, Heidelberg (2009). doi:10.1007/978-3-540-92673-3

14. Thrun, S., Burgard, W., Fox, D.: Probabilistic Robotics. Intelligent Robotics and Autonomous Agents. The MIT Press, Cambridge (2005)

15. Torti, L., Wuillemin, P.H., Gonzales, C.: Reinforcing the object-oriented aspect of probabilistic relational models. In: PGM 2010 - The Fifth European Workshop on Probabilistic Graphical Models, Helsinki, Finland, pp. 273–280, September 2010. https://hal.archives-ouvertes.fr/hal-00627823

16. Wuillemin, P., Torti, L.: Structured probabilistic inference. Int. J. Approx. Reason. **53**(7), 946–968 (2012). https://doi.org/10.1016/j.ijar.2012.04.004

ORDAIN: An Ontology for Trust Management in the Internet of Things

(Short Paper)

Kalliopi Kravari$^{(\boxtimes)}$ and Nick Bassiliades

Department of Informatics, Aristotle University of Thessaloniki,
Thessaloniki 54124, Greece
{kkravari,nbassili}@csd.auth.gr

Abstract. The Internet of Things is coming and it has the potential to change our daily life. Yet, such a large scaled environment needs a semantic background to achieve interoperability and knowledge diffusion. Furthermore, this open, distributed and heterogeneous environment raises important challenges, such as trustworthiness among the various types of devices and participants. Developing and sharing ontologies that support trust management models and applications would be an effective step in achieving semantic interoperability on a large scale. Currently, most of the ontologies and semantic description frameworks in the Internet of Things are either context-based or at an early stage. This paper reports on identifying and incorporating social and non-social parameters involved in the Internet of Things in a general-purpose ontology that will support trust management. This ontology will include among others data and semantics about trust principles, involved parties, characteristics of entities, rating parameters, rule-based mechanisms, confidence and dishonesty in the environment. Defining an ontology and using semantic descriptions for data related to trustworthiness issues will provide an important instrument in developing distributed trust (reputation) models.

Keywords: Ontologies · Semantics · Trust management

1 Introduction

The Internet of Things (IoT) aims to create a world where everyone and everything, called Things, will be connected, changing the way people live, work and communicate. Numerous research areas and applications is expected to benefit from this large scaled environment. Smart environment, living and healthcare are just a few cases [6]. Yet, this revolution has to be supported by an effortless diffusion of knowledge. Hence, promoting and applying semantic technologies to the IoT is vital for the needed interoperability. However, IoT raises challenges, such as intelligence and trustworthiness, due to its open and distributed nature which is combined with the enormous heterogeneity of things. The heterogeneity makes it difficult to standardize interaction and communication. The open and distributed environment allows malicious participants to pose a serious threat to the proper functioning of the network, harming its

© Springer International Publishing AG 2017
H. Panetto et al. (Eds.): OTM 2017 Conferences, Part II, LNCS 10574, pp. 216–223, 2017.
https://doi.org/10.1007/978-3-319-69459-7_15

credibility. Hence, Things acting in such an environment will have to make decisions about the degree of trust that can be invested, a vital but challenging task. [7, 8, 11]

Although there is no single accepted definition for trust, there is a wide range of proposed trust and reputation models [12]. This diversity, the context-based approaches and definition discrepancies lead to a need for a general-purpose ontology for trust. Such an ontology will improve knowledge reusability and diffusion, enabling inter-operability regardless of trust algorithms and/or mechanisms. Furthermore, it will support the design and development of novel approaches.

This paper reports on identifying and incorporating social and non-social parameters involved in the IoT in a general-purpose trust ontology, called ORDAIN. Non-social parameters are concepts related to establishment and maintenance of trust relationships. Usually, they can be found at most trust models and mechanisms. On the other hand, although, the IoT is not considered as a social network, studying the potential societal impacts and relationships objects and/or people is essential. In fact, research on the IoT is expected to shift from intelligent objects to objects with a real social consciousness. Hence, the social dimension of the IoT is currently an open research area [8, 10]. ORDAIN attempts to include data and semantics related to trust mainstream concepts and novel social approaches. The aim is to provide an instrument in developing distributed trust (reputation) models, which on their turn will allow Things to establish and maintain social relationships based on their experiences, preferences and requirements without complex underlying protocols.

2 Defining Trust and Reputation

A reference trust definition is provided by Dasgupta [2], according to him trust is a belief an agent has that the other party will do what it says it will (being honest and reliable) or reciprocate (being reciprocative for the common good of both), given an opportunity to defect to get higher payoffs. In other words, trust is generally defined as the expectation of competence and willingness to perform a given task. Yet, the involved parties are likely to be self-interested and might not always complete requested tasks. Moreover, given that the system is open, they can change their identity and re-enter, avoiding punishment for any past wrong doing. Since involved parties may be dishonest, reputation is a core element at trust establishment, in the sense that a better reputation can lead to greater trust. In general, reputation is the opinion of the public towards a party. Reputation allows parties to build trust, helping them to establish relationships that achieve mutual benefits [7].

Risk is a situation that involves exposure to danger or loss, since the probability of loss is usually non-zero. Hence, the amount of risk that a party may be willing to tolerate is directly proportional to the amount of trust that the party has in the other party. Finally, for purposes of better understanding consider a party A interacting with a party X; party A can evaluate the other party's performance, affecting its reputation. The evaluating party (A) is called truster whereas the evaluated party (X) is called trustee. After each interaction, the truster has to evaluate the abilities of the trustee according to some parameters, such as response time, validity or cooperation.

3 ORDAIN Ontology

3.1 Ontology Contents

The first step towards an ontology for trust management is to study and classify all concepts that affect reputation, the establishment and maintenance of trust between parties. Information sources, criteria, metrics and entities' roles are just a few of these. This subsection provides part of the reference taxonomy. This work is the result of a thorough literature review and previous work on reputation models [1, 4, 5, 7, 8, 12]. In order to elicit the requirements for such an ontology we compared available reputation models, extracting common concepts and relationships. Next, we studied IoT issues, such as the fact that devices are often not connected to the owners, and we tried to discard concepts that are or seem non applicable to IoT while we kept those that can be adopted even with some modifications.

Type of Trust
Trust can be distinguished in *communication, information, social* and *cognitive trust*. Communication trust studies uncertainties that cause low communication quality. Cognitive trust refers to truster's confidence or willingness to rely on trustee's competence. Social trust refers to entities' social relationships and how they affect trustworthiness, including metrics about influence, proximity, social ties and similarity.

Type of Control
There are two system types, *centralized* and *distributed*. A centralized approach identifies a central authority that observes, manages and controls the system. A distributed approach has no central authority. Centralized approaches, usually, lead to global reputation values whereas distributed approaches lead to personalized estimations.

Roles of Involved Parties
Parties may act as *Trusters, Trustees, Recommenders* or *Witnesses*. A witness provides reports based on personal previous experience whereas a Recommender usually propagates reports based on others' experience or observation.

Characteristics of Involved Parties
Each entity has its own unique characteristics. It is not possible to provide here an exhaustive list of the characteristics that might have an entity. Yet, the most common of them are trade relationships, occupation or type of service, club membership, etc.

Information Context
Contextual information is the means for a meaningful description of all available data, providing sufficient details about how parties interact. In the literature most cases refer to a *single* context domain whereas other more complex cases refer to *multiple*. Multiple context could be the result of multi-sourcing rating collection.

Information Sources
Collecting ratings in an open, distributed environment is not always easy. Possible sources are *direct experience* which is the result of an individual's personal interactions or *direct observation* where a party observes the interaction between two other parties

and records its opinion. Additionally there are cases of *indirect experience*, provided by witnesses and recommenders, called relational or social networks based trust. There is also another case, called *derived* information, which is obtained from sources that were not explicitly designed to be used as reference sources but act as such under specific circumstances. Finally, there is *prejudice*, which is a source that allows bootstrapping of trust and reputation when no other information is available.

Information Aggregation

Aggregation is the mechanism behind the estimation process. The *counting* category includes *summation, averaging, weighting* and *normalization*, considering reputation as single value. Other approaches consider reputation as a multiple *discrete* value, using qualitative values for the rating procedure, such as "Untrustworthy". Another aggregation category involves *probabilistic* approach that computes the likelihood of a hypothesis being correct. An improvement of this category, is the aggregation that uses logic. This is the case of rule-based mechanisms. There are approaches that use *fuzzy logic* or *defeasible logic*. Finally, there are the social approaches. They adopt principles mainly from *social graphs* and *peer-to-peer networks*.

Types of Evaluation

There are two evaluation approaches, the *holistic* and the *atomistic*. In the atomistic approach all past interactions are detailed described and taken into account. Some management systems in order to take into account more recent ratings, use weights and a time window. In the holistic cases, systems use summarized information rather than detailed reports in order to provide a single, overall trustworthiness estimation.

Evaluation Criteria

It is not possible to provide an exhaustive list of criteria. Besides, they are domain-specific. Yet, there are some of them that are frequently used in most models, e.g. *response time, validity, cooperation, competence, correctness* and *outcome feeling*.

Data Aging

Data aging is a technique that can reduce the available set of reports that have to be processed. Decaying information is the most common approach. It reduces the confidence and granularity of older rating reports as time passes. Another approach is to *discard* information after a specific time period or used-defined criteria.

Reward or Punishment

Self-interested entities are unwilling to sacrifice time and resources in order to contribute in a trust management system. Hence, there is a need for a motivation mechanism. To this end, there are two possible approaches, namely explicit *rules* and *incentives*. Rules force an entity to act only within a predefined manner. Incentives (or disincentives) motive or even guide entities by using rewards and/or punishments.

3.2   Ontology Implementation

The proposed ORDAIN ontology is an attempt to provide a reusable trust taxonomy and a tool that will support the development of novel trust management systems. It provides the necessary information that will clarify trust issues while new approaches in trust management, such as graph-based trust propagation, will promote research in the field. This section provides some information regarding the core implementation of the proposed ontology in OWL, using RDF/XML Syntax.

Involved Parties and Ratings

Involved parties, as discussed, can have any of the four potential roles: Truster, Trustee, Recommender and Witness. Yet, at a specific time point they comply only with one of them. As a result, the role classes, subclasses of class Entity, are disjoined in ORDAIN. Each of these classes is associated with a number of properties, such as those presented below for the Truster case. A Truster isInterestedIn a specific Trustee whereas it may requestsInformationFrom some Wintesses (Fig. 1). A Trustee could hadPreviouslyInteracted with a witness. If this witness isRequested InformationBy (inverse property with requestsInformationFrom) the aforementioned Truster will provideRating (range: Rating).

```
<owl:Class rdf:about="&ordain;Truster">
 <rdfs:subClassOf rdf:resource="&ordain;Entity"/>
 <rdfs:subClassOf><owl:Restriction>
   <owl:onProperty
                 rdf:resource="&ordain;isInterestedIn"/>
   <owl:someValuesFrom rdf:resource="&ordain;Trustee"/>
 </owl:Restriction></rdfs:subClassOf>
 <rdfs:subClassOf><owl:Restriction>
   <owl:onProperty
         rdf:resource="&ordain;requestsInformationFrom"/>
   <owl:someValuesFrom rdf:resource="&ordain;Witness"/>
 </owl:Restriction></rdfs:subClassOf>…</owl:Class>
```

Fig. 1. Part of Truster class' source code.

Ratings are core elements in reputation management. From a practical point of view, they include the evaluation data. A typical rating (Fig. 2) is in the form:

Rating [Truster, Trustee, TimeStamp, Evaluationcrite-$rion_1Value,…$, $Evaluationcriterion_nValue$, Confidence, Importance, TransactionValue] (1)

Aggregating Mechanism

Aggregation rating reports and trustworthiness values is perhaps the most difficult and challenging aspect of a trust management system. There are plenty of approaches while new are frequently proposed. ORDAIN includes each category as class with a number of subclasses and plenty of properties. For instance, a typical graph aggregation mechanism includes the following:

```
<owl:Class rdf:about="&ordain;Rating">
 <rdfs:subClassOf><owl:Restriction>
  <owl:onProperty rdf:resource="&ordain;byTruster"/>
  <owl:someValuesFrom rdf:resource="&ordain;Truster"/>
 </owl:Restriction></rdfs:subClassOf>
 <rdfs:subClassOf><owl:Restriction>
  <owl:onProperty rdf:resource="&ordain;forTrustee"/>
  <owl:someValuesFrom rdf:resource="&ordain;Trustee"/>
 </owl:Restriction></rdfs:subClassOf>
 <rdfs:subClassOf><owl:Restriction>
  <owl:onProperty
        rdf:resource="&ordain;hasEvaluationCriteria"/>
  <owl:someValuesFrom rdf:resource="&ordain;Criteria"/>
 </owl:Restriction></rdfs:subClassOf>
 <rdfs:subClassOf><owl:Restriction>
  <owl:onProperty rdf:resource="&ordain;hasConfidence"/>
  <owl:someValuesFrom><rdfs:Datatype>
    <owl:onDatatype rdf:resource="&xsd;float"/>…
  </rdfs:Datatype></owl:someValuesFrom>
 </owl:Restriction></rdfs:subClassOf>…</owl:Class>
```

Fig. 2. Part of Rating class' source code.

```
GraphAggregation [NumOfNodes, NumOfTies, connectedNodes,
NodeID, TieID, hasTieValue, isNodeEntity, hasNodeCharac-
teristics, NodeTrustworthiness] (2)
```

Actually, each of these elements/values, just like above, are associated with the class GraphAggregation, subclass of AggregationMechanism, with appropriate properties, such as hasNumOfNodes that refer to an integer number (rdf: resource="&xsd;integer").

Combining Information Sources

Combining different types of experience and, thus, available trustworthiness values is a really challenging task and, actually, an open research area. However, ORDAIN includes the TrustCombining class, with a number of subclasses (e.g. Weighted TrustCombining), that can be considered as a guideline.

```
TrustCombining [Trustee, Timestamp, SourceType₁Value, …,
SourceTypeₙValue, AggregationType] (3)
```

Beardly speaking, ORDAIN includes a variety of classes and properties that can enable a different degree of trust management simulation and implementation based on the domain specific needs.

4 Related Work

In the IoT a common agreement on ontological definitions is still an open research issue. Ontologies and semantic frameworks are either at an early stage providing just a few basic properties or they are defined in the context of different projects.

For instance, in [3] authors propose a service oriented ontology. They assume that trust can be directed towards either an agent, product or service. They propose an ontological representation of agent, service and product trust in the sense that an agent develop trust in an agent, product or service. In their approach there are three distinct domains, namely Agent Trust Ontology, Service Trust Ontology and Product Trust Ontology. Opposed to that limited approach, we provide a single general-purpose ontology that can be adopted in a variety of domains. However, we do acknowledge that services, being an important component, are involved in the IoT.

In [9] authors propose an ontology-based framework for information fusion, as a support system for human decision makers. They build their approach upon the concept of composite trust, consisting of four trust types, communication, information, social and cognitive trust. Based on the concept of multidimensional trust, they constructed a composite trust ontology framework, called ComTrustO, that embraces four trust ontologies, one for each trust type. Their approach, similarly to ours, acknowledges the need for comprehensive ontologies and identifies four trust types. However, they provide four domain specific ontologies rather than a general-purpose approach. Furthermore, our approach includes many other concepts, such as trust aggregation.

5 Conclusions

Internet of Things faces interoperability issues and challenges, due to its open, distributed, heterogeneous nature. This paper proposed an ontology for trust management in the IoT. This ontology was the result of a detailed study on trust management systems presented in the literature. The proposed approach is a general-purpose ontology that takes into account social and non-social features. Trust, reputation and risk were discussed while a taxonomy of concepts related to trust (reputation) management was reported. The key feature of the proposed ontology is that it captures the whole life-cycle of trust from the involved parties to decision mechanisms.

As for future directions, first of all, we plan to study further the proposed ontology in order to adopt any new concept or approach published in the literature. More technologies could be adopted for these purpose; machine learning techniques and user identity recognition and management being some of them. Another direction towards improving the proposed ontology is also to combine it with Semantic Web metadata for trust. Furthermore, we plan to evaluate it in order to report its added value as well as its weakness that will be subject of further improvement.

Acknowledgments. The Postdoctoral Research was implemented through an IKY scholarship funded by the "Strengthening Post-Academic Researchers/Researchers" Act from the resources of the OP "Human Resources Development, Education and Lifelong Learning" priority axis 6, 8, 9 and co-funded by The European Social Fund - the ESF and the Greek government.

References

1. Cho, J.-H., Chan, K., Adalı, S.: A survey on trust modeling. ACM Comput. Surv. **48**(2), 40 (2015). Article 28
2. Dasgupta, P.: Trust as a commodity. In: Gambetta, D. (ed.) Trust: Making and Breaking Cooperative Relations, pp. 49–72. Blackwell (2000)
3. Hussain, F.K., Chang, E., Dillon, T.S.: Trust ontology for service-oriented environment. In: IEEE International Conference on Computer Systems and Applications, pp. 320–325 (2006)
4. Kravari, K., Bassiliades, N.: DISARM: a social distributed agent reputation model based on defeasible logic. J. Syst. Softw. **117**, 130–152 (2016)
5. Kravari, K., Bassiliades, N.: HARM: a hybrid rule-based agent reputation model based on temporal defeasible logic. In: Bikakis, A., Giurca, A. (eds.) RuleML 2012. LNCS, vol. 7438, pp. 193–207. Springer, Heidelberg (2012). doi:10.1007/978-3-642-32689-9_15
6. Li, S., Da Xu, L., Zhao, S.: The internet of things: a survey. Inf. Syst. Front. **17**(2), 243–259 (2015)
7. Medić, A.: Survey of computer trust and reputation models – the literature overview. Int. J. Inf. Commun. Technol. Res. **2**(3), 254–275 (2012)
8. Nitti, M., Girau, R., Atzori, L.: Trustworthiness management in the social internet of things. IEEE Trans. Knowl. Data Eng. **26**(5), 1253–1266 (2014)
9. Oltramari, A., Cho, J.H.: ComTrustO: composite trust-based ontology framework for information and decision fusion. In: 18th International Conference on Information Fusion, pp. 542–549 (2015)
10. Ortiz, A.M., Hussein, D., Park, S., Han, S.N., Crespi, N.: The cluster between internet of things and social networks: Review and research challenges. IEEE Internet Things J. **1**(3), 206–215 (2014)
11. Whitmore, A., Agarwal, A., Da Xu, L.: The Internet of Things—a survey of topics and trends. Inf. Syst. Front. **17**(2), 261–274 (2015)
12. Yan, Z., Zhang, P., Vasilakos, A.V.: A survey on trust management for Internet of Things. J. Netw. Comput. Appl. **42**, 120–134 (2014)

APOPSIS: A Web-Based Platform for the Analysis of Structured Dialogues

Elisjana Ymeralli$^{(\boxtimes)}$, Giorgos Flouris, Theodore Patkos,
and Dimitris Plexousakis

FORTH-ICS, Institute of Computer Science,
N. Plastira 100 Vassilika Vouton, 700 13 Heraklion, Crete, Greece
{ymeralli,fgeo,patkos,dp}@ics.forth.gr
https://www.ics.forth.gr/isl

Abstract. A vast amount of opinions are surfacing on the Web but the lack of mechanisms for managing them leads to confusing and often chaotic dialogues. This creates the need for further semantic infrastructure and analysis of the views expressed in large-volume discussions. In this paper, we describe a web platform for modeling and analyzing argumentative discussions by offering different means of opinion analysis, allowing the participants to obtain a complete picture of the *validity*, the *justification* strength and the *acceptance* of each individual opinion. The system applies a semantic representation for modeling the user-generated arguments and their relations, a formal framework for evaluating the strength value of each argument and a collection of Machine Learning algorithms for the clustering of features and the extraction of association rules.

Keywords: Debating platforms · Opinion analysis · Association rules · K-means algorithm · Multi-aspect evaluation

1 Introduction

Social networks, debate platforms and forums have become major sources of knowledge sharing, interaction and collaboration among participants through the Web, where users express and share their opinions over a plethora of topics. Due to the lack of methods for analyzing and capturing the structure of the argumentative discussions in conjunction with the vast amount of available information encountered on the Web, users are often overwhelmed when trying to understand and make sense of the user-generated opinions. This makes the task of analyzing and identifying useful patterns of relationships among contributors and opinions difficult.

Amongst social networking platforms, debate portals are becoming increasingly popular in recent times (*e.g., Debate.org, Quora*). These applications provide features for collecting differing opinions related to goal-oriented topics of discussion where users exchange their views in the form of agreement or disagreement. People rebut to other user's posts by defending and justifying their

© Springer International Publishing AG 2017
H. Panetto et al. (Eds.): OTM 2017 Conferences, Part II, LNCS 10574, pp. 224–241, 2017.
https://doi.org/10.1007/978-3-319-69459-7_16

opinions with the purpose of persuading the audience. However, it is difficult and time consuming to browse through the useful opinions expressed within a dialogue when a large amount of comments is provided. Hence, it is essential and helpful to analyze the users opinions in a more comprehensible form, so that useful patterns of relations among online users can be identified and extracted from dialogues, helping contributors understand the dynamic flow of a community.

In this paper, we introduce and describe a web-based debating platform for modeling and analyzing online discussions. The system, called $APOPSIS^1$, motivates users to participate in goal-oriented topics of discussions by raising issues, posting ideas or solutions, posting comments in the form of (supporting or attacking) arguments and voting. Moreover, the platform offers the opportunity for a variety of groups of people to work and collaborate with each other with the goal of suggesting and sharing new ideas, regarding different open issues. Towards this contribution, the system can produce and extract useful conclusions and opinions expressed in a dialogue, that help sense-makers and expert users who wish to take advantage of the system, to understand the dynamics of social communities and make decisions on specific issues and problems.

Furthermore, our platform provides a range of functionalities which are presented next. An argumentation-based approach is applied for organising the conceptual components of a dialogue, based on the Issue-Based Information System (IBIS) model [1]. Then, a Semantic Web ontology is used for representing the conceptual components and their relations in the form of RDF^2 statements. Considering users' reactions (comments and votes), answers are evaluated through a general formal framework for computing the strength value of each argument, considering one or more aspects (*i.e. incorrect, irrelevant, insufficient*), as proposed in [2]. The main part of this work concerns the opinion analysis, where clustering and associations techniques are used for the clustering of features and the extraction of association rules, such as Expectation-Maximization (EM), K-means and Apriori algorithms, implemented in [3].

Example. Let's consider a city, where the City Council needs to take some important decisions about the city planning and the implementation of the city's actions, enabling residents to be involved in dialogues by expressing their agreements or disagreements and vote on other citizens' comments. The system aims at giving the opportunity to the citizens of the municipality to work together for designing policies for a municipality. A platform such as Apopsis can assist the City Council and decision-makers to understand and make decisions on specific problems and issues by identifying and extracting useful suggestions. As decisions made by the city council are rarely on a black-or-white basis, it is essential to identify the different trends and driving forces of the various groups that are formed, in order to try to accommodate as many of their needs as possible. *For instance, the system could identifies different groups of citizens (e.g., groups of civil society, other bodies) who share similar opinions on specific suggestions that can help the City Council decides on the effective governance of the city.*

---

[1] http://www.ics.forth.gr/isl/apopsis.
[2] http://www.w3.org/RDF/.

The rest of this paper is as follows. In Sect. 2, we present the relevant background followed by the related work. Section 3 introduces the ontology and continues with the methodology used for modeling and analyzing users' opinions. Section 4 presents the basic concepts and features of the web application. The last section draws some conclusions by emphasising the key points of our work and identifying issues for further research.

2 Background and Related Work

2.1 Computational Argumentation

Argumentation [4] is the research field dealing with the formal study of agreement and disagreement that people express with the goal of defending their opinions or convincing themselves and others. The theory of argumentation plays an important role in understanding, analysing, formalising and structuring both online and everyday human deliberation and discussions. This creates the need for effective formalizations and automated mechanisms that can model and valuate the users opinions. Some of the problems are approached by researchers, mostly in computational argumentation where well-defined frameworks are provided. Computational argumentation theories have found beneficial applications in the fields of Artificial Intelligence, decision-support systems and recently on the Social Web for facilitating online dialogues among multiple participants.

Many efforts have been proposed for the evaluation of arguments using a graded (numerical) acceptability ranking. Such works include the QuAD (Quantitative Argumentation Debate) [5] framework for quantifying the strength of opinions based on the aggregation of the strength of attacking and supporting arguments. Another framework that has attracted research attention is the SAAF (Social Abstract Argumentation Framework), proposed in [6]. SAAF incorporates a voting mechanism to calculate the strength of arguments by considering both votes and the attack strength that an opinion has received. The approach was later extended in [7], by incorporating also supporting relations among arguments, and a social voting for aggregating votes in order to identify the strongest answers.

2.2 Machine Learning

Machine Learning algorithms (ML) have been already an integral part of computing systems for exploratory data analysis. In this work, a collection of machine learning software is used for clustering analysis, including techniques for data pre-processing, clustering and association rules.

K-means Algorithm. One of the commonly used unsupervised learning algorithm for solving clustering problems. The algorithm aims to assign a set of data objects to clusters, in order to achieve a high intracluster similarity and a low inter cluster similarity. The K-means algorithm is adapted to many problem domains and can be applied to many fields, such as Marketing for product selling, Social Networks for online users behavior etc.

Expectation-Maximization (EM) Algorithm. An iterative probabilistic clustering algorithm that can be used as a pre-processing procedure of the K-means algorithm with the goal of deciding the optimal number of clusters that need to be generated for the clustering analysis.

Apriori Algorithm. An association rule mining algorithm that identifies frequent itemsets over a given set of observations. A prior knowledge of data is required in order to generate the next set of itemsets. Associations rules are useful for discovering interesting relationships among attributes of a dataset.

2.3 Existing Tools for Online Debates

Several online platforms have been developed to serve the need of modeling, evaluating and querying arguments in an informative and interactive way. A plethora of tools on the Web focus on user actions, allowing them to raise issues or ask questions about public concerns and post comments in support of or against a specific topic of interest.

Quaestio-it.com [7] is a web-based Q&A debating platform that offers an interactive way of engaging users in conversation regarding any question within the platform. The system provides a computational argumentation framework, called ESAAF (Extended Social Abstract Argumentation Framework), for modeling online discussions and identifying the strongest comments prevailing within debates. Best answers and arguments are highlighted and visualised as bubbles with their sizes indicating the participation rate. Compared to our system, Quaestio-it offers a comment-rating algorithm similar to ours in order to identify the most acceptable opinions, but does not consider clustering or any other types of dialogue analysis techniques for generating groups of related opinions. We use an evaluation algorithm that considers both arguments and votes strength score but more importantly, APOPSIS allows an automated opinion analysis that extracts useful conclusions to sense-makers in order to help them make sense of the discussions in social communities.

e-Dialogos [8] is a web application for open public debates that enables citizens to connect with other people and discuss problems related to the design and implementation of policies related to municipalities. Despite the fact that this system provides the ontological infrastructure for modeling discussions that are taking place in deliberations, they do not evaluate the users' answers, neither provide a voting mechanism for supporting or attacking other answers. Furthermore, the application provides a summarization form that displays all opinions shared during the deliberation process but there is no implementation of an opinion analysis algorithm. Our methodology can provide a reliable approach for evaluating and analyzing the users' answers by combining both argumentation theories and Machine Learning techniques to facilitate the users' reactions.

Many other social networks exist in the form of *question & answer* systems that are similarly related to Apopsis platform, such as *Quora*[3], *Answers*[4],

---

[3] http://www.quora.com.
[4] http://www.answers.com.

answerbag[5], *Yahoo Answers*[6] and *StackExchange*[7]. However, these systems lack some of the defining characteristics of a debating platform, such as the organization of opinions into pro and con, the formalization of well-defined methods for evaluating the strength value of each opinion individually and the analytical features (e.g., clustering) that would enable users to make better sense of the discussions and the opinions expressed within the dialogue.

3   A Methodology and Platform for Opinion Analysis

3.1   A Platform Methodology

Apopsis is a web-based debating platform that aims to motivate online users to participate on well-structured discussions by raising issues and posting ideas or comments that support or attack other opinions. The main goal of the system is to offer an automated opinion analysis that determines and extracts the most useful and strongest opinions expressed in dialogue, that help decision-makers understand the discussion exchange process.

In our platform, dialogues proceed in two different levels of discussions allowing the strongest arguments, based on their score value, to proceed in the next level of the dialogue with the help of moderators that ensure the quality of debate. Users may navigate amongst different dialogues existing within platform and debate on a particular topic of interest by providing their statements in the form of agreements and disagreements. Discussions are presented in the form of trees where subsequent levels of comments respond to the parent comment (position or argument). In the second phase of the dialogue, participants are not allowed to post new ideas (positions) but only provide positive or negative answers on existing positions. The nature of comment (positive or negative) is predefined by the system, allowing users to select whether support or attack an opinion. The system offers well structured dialogues and a voting mechanism for evaluating each argument considering one or mores aspects. Conversations are organized and represented by using the MACE-ontology for making the discussions available in the form of RDF statements and users' answers are evaluated based on a quantitative evaluation algorithm that takes into account both, arguments and votes. Then, a clustering analysis of opinions can be applied that aims to discover different trends (users and opinions), helping contributors to obtain a clear picture of the opinions expressed in social communities.

3.2   Knowledge Map Representation

The knowledge map is designed to structure the argumentation process by allowing five different types of elements: *issue, topic, position, pro-argument* and *con-argument* for facilitating online debates, see Table 1. Most of our concepts

---

[5] http://www.answerbag.com.

[6] http://www.answers.yahoo.com.

[7] http://www.stackexchange.com.

have their roots in the IBIS-style argumentation model [1] with slightly different semantics. In our approach, each type of element is represented as a node that denotes a specific meaning. For instance, an *issue* represents a question or statement that initialize a conversation where users can share and contribute ideas in a positive or negative way, a *topic* represents label or aspect of the issue matter where an argument is a response to specific topic related to that issue. Direct answers in the form of solutions or ideas to the initial point of conversation are considered as *positions* while the arguments that support or attack a position are defined as *pro-arguments* and *con-arguments* respectively. Each dialogue is represented as a directed graph with each node representing an argument and each directed edge indicating a support or attack relation. An overview of the knowledge map introduced in this work is given in Table 1.

Table 1. Knowledge map.

Node Element	Description	Stereotype
Issue	A question or statement that initialize a conversation.	
Topic	A label or tag that is closely related to issue.	
Position	A solution or idea that respond to the initial point of conversation.	
Pro-argument	The ability to support a position or another argument.	
Con-argument	The ability to attack a position or another argument.	

3.3 The MACE - Ontology Domain

The need for understanding and investigating how communities interact and argue in the context of specific domains, led many researchers in modeling ontological formalizations for structuring the information exchanged in these communities. We briefly mention them, which can model and represent online communities and relations related to online activity. The OPM (Opinion Mining Core Ontology) proposed by Softic and Hausenblas [10], describes concepts related to online discussions with the connection of two existing vocabulary, the SIOC (Semantically-Interlinked Online Communities) [9] and the SKOS (Simple Knowledge Organisation System Reference) [11]. Another important ontology that should be considered is the CiTO[8] ontology, which expresses some similar relation semantics (*e.g.*, *agrees with/support, disagrees with/attack*) with MACE[9] ontology by allowing a more complex set of interaction among users.

In Apopsis, we designed and implemented an RDF ontology (MACE) for organizing and representing online discussions and their relations that aims

---

[8] http://www.sparontologies.net/ontologies/cito/source.html.
[9] http://www.ics.forth.gr/isl/mace/.

to accommodate complicated opinions found in online debates by semantically querying and presenting them to the audience. The main goal of MACE-ontology is to represent dialogues and opinions through Semantic Web technologies, which are organised according to specific types of argumentation elements as shown in Table 1. Although many ontologies are applied in specific domains, our ontology is powerful and generic enough to represent the content produced and exchanged within the platform for online communities. The formal ontology (MACE) can also be accessable through the Apopsis platform. An overview of our ontology and its properties for making the content and their relations available in the form of RDF statements is given next.

3.3.1   A Taxonomy of Ontological Concepts and Properties

Our ontology consists of 12 classes and 27 properties. Properties that start with the prefix (P) are the main properties of ontology that describe the relations among classes hierarchies while properties that start with the prefix (RP) contain information that can be derived from the composition of other properties.

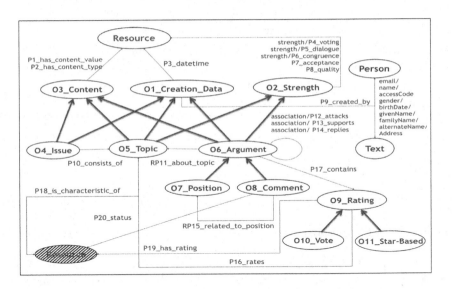

Fig. 1. The MACE - ontology domain.

A core idea behind the design of the ontology is to have top classes (*O3_content, O1_Creation_Date, O2_Strength*) that represent different function-alities. This way, the other entities (sub-classes and sub-properties) inherit features and functionalities from the top classes, as shown in Fig. 1. For instance, the Content class contains information about the data related to each object, and is a superclass of those entities that have content (Issue, Topic, Argument). Similarly, the Creation Data class contains information about the provenance of

each object, including datetime and author, and is also a superclass of the entities (Issue, Topic, Content). Finally, the Strength class comprises information
about the calculation of the score value of each argument and is a superclass of
(Topic. Argument). Considering the *O6_Argument* hierarchy, an argument can
be either a position or comment. Positions are not allowed to respond to other
positions (as they are, essentially, top-level comments), whereas arguments can
be positive or negative related to positions or other comments. We have used
and extended the Person class from the *schema.org*[10] ontology that describes
people and provides information about their profiles. Moreover, since the system
offers a voting mechanism, the ontology includes a class, named *O10_Vote* that
stores the votes (positive or negative) that an argument has gained.

All class entities are associated with other classes through appropriate properties. Some properties contained in this schema are more complex than others.
These properties are the *strength property* and the *association property*. The
strength property denotes the score calculation of arguments (comments and
positions) and is based on the calculation of five more relations: *voting, dialogue,
congruence, acceptance* and *quality*, as shown in Table 2.

Table 2. Dialogue properties.

Dialogue	Description
Voting	The value of the voting strength for both (acceptance and quality)
Dialogue	The overall dialogue strength by combining the arguments and votes strength
Congruence	The strength of an argument considering only the supporting votes of an argument, normalized by the attacks
Acceptance	Represents how acceptable an argument is, based on the strength of arguments that support or attack other arguments
Quality	Determines how well-justified the arguments are presented in dialogues

Similarly, the association property consists of three more relations including
the *P12_attack relation*, the *P13_support relation* and the *P14_replies* relation,
denoting an attack or support relation between two arguments that disagree or
agree with other arguments. The P14_replies property identifies a simple relation
between two arguments denoting that the argument of the property replies on
the object without denoting if it agrees or disagrees.

3.4   Evaluating Opinions

This work builds on a multi-dimensional framework (s-mDiCE) for estimating users' reactions (comments and votes), based on different metrics.

---

[10] http://schema.org/Person/.

The framework introduces interesting functions and properties that can guarantee an intuitive behavior for interlocutors who wish to react on social discussions by commenting or vote on other's answers. This quantitative algorithm is generic enough to capture the features of online communities on the Social Web and may benefit several platforms from debate portals to decision-making systems by providing a more reliable and effective approach for the score calculation of both, comments and votes. For a more comprehensive overview of the quantitative framework, we refer the reader to [2].

3.4.1   s-mDiCE Properties

The framework consists of a set of internal and generic functions that allows the evaluation of users' reactions (comments and votes). Each argument is characterized by two different values for a given strength, the *quality* (QUA) and *acceptance score* (ACC) of an argument in each aspect, based on the reactions (responses, positive and negative votes) related to that aspect (e.g., *incorrect, irrelevant, insufficient*). Moreover, the framework introduces concepts such as the *base score* and the *blank argument* metaphor. The notion of the base score is used for capturing the initial rating over users opinions, where the score value may change either positively or negatively through users arguments and votes. Votes are considered as arguments *(blank arguments)* without carrying any content on its own, rather share the content of the arguments they support or attack. The reliability of these opinions that can support or attack a target argument can be estimated by adding those opinions into supporting or attacking blank arguments, respectively.

A set of generic functions is introduced for calculating the acceptance and the quality score of an argument by considering the *dialogue strength*, and the *congruence strength*. Positive and negative votes are aggregated into a single strength score of votes while supporting and attacking arguments are aggregated into a strength score of arguments. In this approach, arguments have a greater impact than votes on a dialogue as they remain a strong belief for an opinion on a given aspect, asserted in order to add more information or explain better the opinion stated. The overall strength value of an argument is calculated by aggregating the strength score of votes and the strength score of (supporting and attacking) arguments on a given aspects.

4   Methodology: Analysing User-Generated Opinions

We propose an approach for analysing the user behavior in social communities, where a plethora of people express their opinions with the goal of defending their ideas or convincing other people by providing well-justified opinions. The ultimate goal of applying a clustering analysis is to offer different means of opinion analysis, allowing contributors to obtain a clear and complete picture of the validity, justification strength and the public acceptance of each opinion expressed in large-volume discussions. The rest of this section goes through all the features and implementation of the opinion analysis methodology.

User-Generated Opinions Estimation. In order to facilitate opinion analysis, we need to evaluate the users' arguments, according to their agreements and disagreements on a particular position. Different relations are defined which represent the users' viewpoints, considering: *support relation* (s), *attack relation* (a) and *unknown relation* (?), as shown in Table 3. Specifically, a relation (s) represents a positive answer to the target position (P), a relation (a) denotes a negative answer to the target position (P) and a missing value (?) relation represents the unknown answers of users to the target argument (P). Figure 2 shows how combinations of attack/support relations propagate across the dialogue.

Table 3. Combinations of support/attack relations.

Argument (**aj**)		Argument (**ai**)		Argument (**P**)
Support (s)	⟶	Support (s)	⟶	Support (s)
Support (s)	⟶	Attack (a)	⟶	Attack (a)
Attack (a)	⟶	Attack (a)	⟶	Missing Value (?)
Attack (a)	⟶	Support (s)	⟶	Missing Value (?)

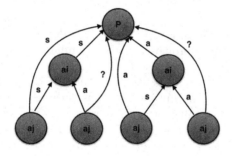

Fig. 2. Scenarios illustrating the argumentation graph of debates

4.1 Clustering Analysis

A key component of this work is the clustering analysis of users' opinions that aims to help users identify useful trends and patterns relationships among participants, towards the better sense-making of the dialogues and the opinion exchange process. Machine Learning algorithms are used for the clustering of features and the extraction of association rules. Based on users reactions and their profiles, both algorithms (K-means and Apriori) are used to identify users behavior in online discussions. Next, we present and describe all information needs introduced in this platform.

Sharing Similar Opinions with Specific Users. This opinion analysis aims at identifying trends of users whose opinions are closely related to particular authors. The input to this analysis is, (a) the topic of discussion that need to be determined and (b) a set of users, where $u = \{u(i),u(2),.....u(n)\}$. The output determines and extracts groups of users whose opinions are similar to particular authors $u(i)$. *For example, if we consider groups of users who participate and collaborate on a topic of discussion that concern the regeneration of the city center, then different trends of users are identified such as $\{Isabella,Karmen\}$ who share similar opinions with the predefined group of users $\{Kate,Brian\}$ on improving the city with new parks and intuitive arts, but they also disagree with the group of users $\{Mathew,Oliver,Andrew\}$ who do not suggest any regeneration for the city.*

Algorithm 1. Sharing similar opinions with specific users

---

input : Topic of discussion, set of authors.
output: Several groups of users, sharing similar opinions with specific authors.
begin
    1: Determine **K** clusters using EM algorithm
    2: Define groups of users based on positions, using **K-means** algorithm.
end

Same Profiles with Similar Opinions. This debate analysis aims at identifying and extracting user profiles that share similar views with other contributors involved in online discussions. The input to the analysis is (a) a subject of discussion that needs to be determined and (b) a set of positions, where $p = \{p(1),p(2),.....p(n)\}$. The output identifies several groups of user profiles. Each such group shares similar or dissimilar opinions considering other groups on particular positions. *For example, if there is a group of users who participate and suggest solutions on how to make a city more livable, then the debate analysis identifies and extraxts trends of user profiles where (single and young men) share similar opinions among them, concerning the regeneration of the new parks and arts within a city, but they express dissimilar opinions with groups of users who are (young, single and well-educated) and disagree on these measures fo improving the city.*

Same Users with Dissimilar Opinions. This type of analysis determines dissimilar opinions among participants who share similar profile characteristics. The input to this analysis is (a) the topic of discussion that need to be determined and (b) a set of positions, where $p = \{p(1),p(2),.....p(n)\}$. The output extracts several groups of users that share dissimilar opinions and similar profile characteristics with other groups on particular positions $p(i)$. *For example, if there exist groups of users who express their opinions in support of or against other's opinions on a topic of discussion of how to make the city more livable, then the opinion analysis identifies and extracts groups of users $\{Matthew,Oliver,Andrew\}$*

who are (young and single men) and share dissimilar views with the group of users {Brian,Kate,Isabella,Karmen} on regenerating the city with green parks and arts.

Similar Opinions Based on Different Profiles. This opinion analysis identifies relevant opinions expressed by users with different profiles characteristics. The input to this type of analysis is (a) the topic of discussion that need to be identified and (b) the profile characteristics that users choose in order to perform debate analysis. The output determines similar opinions that different users profiles share with each other on positions $p(i)$. *For example, if we consider groups of users who suggest and post new ideas on a subject of how to improve and generate the city, then the clustering analysis identifies similar opinions expressed by young women such as {Kate,Isabella,Karmen} that concern solutions on improving the city center with new green parks and arts.*

Algorithm 2. Same profiles with similar opinions/Same users with dissimilar opinions/Similar opinions based on different profiles

---

    input : Topic of discussion, [set of positions or different profile characteristics].
    output: Several groups of users or users profiles, sharing similar or dissimilar
             opinions
begin
    1: Determine **K** clusters using EM algorithm
    2: Define several groups of users, using **K-means** algorithm.
    3: **For each** cluster
    4:     Apply **Apriori** algorithm to identify similar profile characteristics or
           relevant opinions.
    5: **end**
end

Same Users with Similar Preferences. In this analysis, we need to identify trends of users that share not only similar opinions with other groups but also similar profile characteristics. The input of this analysis is the subject that need to be specified by users in order to perform debate analysis. The output identifies several groups of users who share similar preferences (opinions and profile characteristics). *For example, if there exist groups of users who express their agreements and disagreements on how to make a city more livable, then the opinion analysis determines groups of users {Brian,Kate,Isabella,Karmen} who are (young, single and well-educated) and share similar opinions on improving the city but they have dissimilar opinions with the group of users {Matthew,Oliver, Andrew} who do not support the regeneration of the city center.*

Different User Profiles with Similar/Dissimilar Opinions. This information need identifies several groups of users who can either have similar or dissimilar opinions to share with other groups. The input to this analysis is (a) the topic of discussion that need to be determined and (b) the type of relation

---

Algorithm 3. Same users with similar preferences

---
input : Topic of discussion.
output: Several groups of users, sharing similar opinions and profiles.
begin
   1: Determine **K** clusters using EM algorithm.
   2: Define groups of users based on positions, using **K-means** algorithm.
   3: **For each** cluster
   4:     Perform **Apriori** algorithm to identify similar opinions.
   5:     Perform **Apriori** algorithm to identify similar profile characteristic.
   6: **end**
end

---

which can be either similar or dissimilar opinions. The output determines different trends of groups with different profile characteristics, sharing either similar or dissimilar views with other groups found in opinion analysis. *For example, if there exist groups of users who contribute with each other on how to improve the city, then the output determines trends of users {Matthew,Oliver,Andrew} who share dissimilar viewpoints with the group of users {Kate,Isabella,Karmen} on suggesting solutions for the regeneration of the city. Both groups share different profile characteristics with each other.*

Algorithm 4. Different users profiles with similar/dissimilar opinions

---
input : Topic of discussion, a relation (similar or dissimilar opinions).
output: Several groups of users, sharing different profile characteristics and
         either similar or dissimilar opinions.
begin
   1: Determine **K** clusters using EM algorithm.
   2: Define groups of users based on their profiles, using **K-means** algorithm.
   3: **For each** cluster
   4:     Perform **K-means** algorithm to identify similar or dissimilar opinions.
   5: **end**
end

---

5   Apopsis Implementation and Architecture Details

5.1   Debating Functionality

The system enables users to interact and collaborate with other people by providing functionalities such as: *creating new topics of discussion, posting new positions* and *posting new arguments.* Users may navigate amongst existing dialogues or create new topics of interest in order to initiate a conversation. Within a dialogue, online users exchange opinions by posting comments in the form of supporting or attacking opinions on a particular topic of discussion.

When a disagreement takes place, users need to explicitly state the reason that disagree with others by justifying their arguments based on a particular aspect. In case of an agreement, the user does not need to justify their stance over a comment as it is considered that the user totally agrees with the content of the argument. Our platform is offering threaded representation style when it comes to represent online discussions. In Fig. 3, we provide a screenshot of a dialogue about how to make a city more livable where users participate and suggest their viewpoints for improving the society.

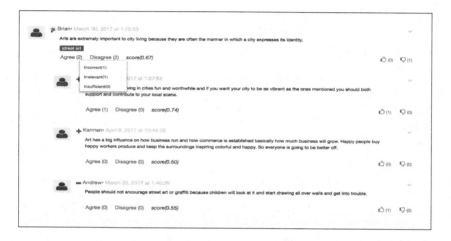

Fig. 3. Debating functionality.

5.2 Debating on Different Levels of Discussions

The platform offers different levels of discussions where contributors can participate and collaborate in different ways. At first, each dialogue is open for discussion where users can post comments (positions or arguments) and vote on existing arguments. The strongest positions and answers proceed in the next stage by a moderator, where users continues to post their positive or negative opinions and vote upon them but they are not allowed to post new positions on particular topic of discussion. The second level of dialogue is a necessary procedure in order for users to apply opinion analysis that are interested in.

5.3 Voting Mechanism

Our platform provides a voting mechanism by allowing users to express their positive or negative votes over arguments. Votes indicate users beliefs over particular answers and are reduced on the argument level of each discussion. When a negative vote takes place, users need to state the reason why they vote negatively, considering one or more aspects. In case of a positive vote, we consider

Fig. 4. Voting mechanism.

that the user who votes positively does totally agree with the argument and its content. The rating mechanism is presented in the form of *like or dislike* an opinion. Figure 4 presents a screenshot about voting on a particular level of argument.

5.4   Searching Mechanism

The system provides a searching functionality on topics, authors, positions and arguments (supporting or attacking arguments) for a particular subject of discussion. We designed and implemented several searching types of querying that enable users to extract information about different aspects of a dialogue. Each searching type has a different input and output, depending on the type of query that users of the system are interested at. *For example, the user u(i) can choose to query the arguments that can either support or attack other opinions, expressed by a specific author, such as Matthew, as shown in Fig. 5.*

Searching Results

Below are shown the results of your queries.The debate object here is mentioned to the elements of debate, which can be the arguments,positions,topics etc.

Debate Object	Author	Score	Topics
✚ I agree with you and also they facilitate face to face interaction and increase voter participation.	Matthew	0.50	increase life quality · academic research
✚ The current Social Security program will become insolvent by 2034 so a better system is urgently required	Matthew	0.50	personal accounts · low risk investments · reduced government workforce
━ Building new parks will not address our citys biggest issue which is growing inequity. In fact doing so might actually exacerbate this problem.	Matthew	0.45	small public plazas · parks and green spaces

Fig. 5. Searching functionality.

5.5   General Features

The platform offers several other features that enrich the system's usability. As a web system, it provides a logging and registration mechanism that allows online users to actively participate in online discussions. A logging mechanism was developed in order to store additional information about user profiles and their actions in each debate. Users' profile characteristics are of prime importance when a debate analysis takes place as they provide a rich description for different trends of users found in opinion analysis. Another feature is the group of users, named *moderators*, who ensure the quality of debates by proceeding the strongest arguments to the next level of discussion for later use in debate analysis. Within each debate, additional statistics information are provided about users' action over a topic of discussion.

5.6   Technical Information and Architecture

We designed and implemented a web-based platform as a web application project, using the platform Netbeans and the Tomcat server for its deployment. The front-end side is designed based on web technologies such as jQuery, HTML5, CSS3, JSP, JSON and AJAX requests while for the server-side we used the JAVA language to implement all features and functionalities of the web-system. We implemented the Apopsis system on a MacBook Pro (OS X Yosemite). The platform consists of three main tiers: the *Server Tier*, the *Client Tier* and the *Data Tier*, illustrated in Fig. 6.

Server Tier: Considering the back-end implementation of the system, we designed a web platform as part of a Java application, developed to provide a complete API of all provided features and capabilities. It was designed as a servlet Web-based service, using Java API and Semantic Web Standards such as RDF statements. An important aspect of the server side is that our platform uses WEKA's implementation methods and functionalities, used as libraries into the system.

Client Tier: The client-side environment constitutes the interface of our system which is designed using a bootstrap[11] template. It is implemented based on

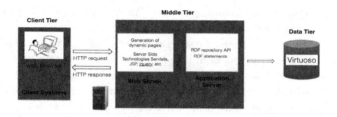

Fig. 6. Basic architecture of APOPSIS.

---

[11] http://colorlib.com/polygon/gentelella.

modern design patterns, following the basic user interface design principles. The client-side of the web application consumes the servers services as JSON data and provides representations using all the latest Web features that all modern browsers support. The Ajax Requests work as intermediary between client and server, orchestrating data exchanges.

Data Tier: The data-tier represents the Virtuoso repository, a cross-platform universal server that acts as a virtual database engine for combining the functionality of a traditional databases in a single system. The system was developed over jdbc Virtuoso provider, allowing users of Virtuoso to leverage the jdbc framework to modify, query, and reason with the Virtuoso triple store using Java language.

6 Conclusion and Future Work

A plethora of opinions are presented on the Web where users exchange their viewpoints and argue over different problems and issues, thereby raising the need for modeling well-structured arguments and analysing online user opinions. Well-defined formalisations are needed for evaluating and analysing the user-generated arguments. This work addresses the problem by introducing a web-based debating platform, called Apopsis, that facilitates and analyzes well-structured dialogues in support of social communities.

Apopsis is a web-based application for modeling and analyzing user opinions found in social discussions. Recent efforts have used quantitative methods to assess the credibility and acceptability of arguments, based on computational argumentation models [5]. Our platform uses a formal framework, named s-mDiCE [2], which offers intuitive methods for (numerical) evaluation of the argument's quality and acceptance, according to users' reactions (comments and votes). In Apopsis, argumentation models are used for organising the different argumentative elements of the dialogue, which are presented in the form of RDF statements through the MACE - ontology. An important aspect of this work is the opinion analysis where our methodology applies clustering and association techniques in order to extract useful trends and relations among users and different clusters of opinions. This way, decision-makers can obtain a clear picture of the *validity*, the *strength* value and the *acceptance* of the opinions expressed throughout the discussion exchange process. Moreover, the platform offers and implements many other functionalities that enrich the system usability.

Regarding future work, there are several aspects that are worth investigating. An important future direction would be to evaluate our system with real users and large datasets of discussions. Specifically, the process will include expert walk through evaluation of the prototype, adjustments based on expert evaluation results, user-based evaluation with UI experts and with real-users in a laboratory setting, and further adjustments of the prototype based on the results of the evaluation. Another important aspect would be the representation of the dialogue and the different clusters of opinions through different visualizations.

Finally, we plan to work on making the system interface more usable and intuitive in order to improve the engagement of users in conversation regarding any topic of discussion.

References

1. Kunz, W., Rittel, H.W.J.: Issues as Elements of Information Systems, vol. 131. Institute of Urban and Regional Development, University of California, Berkeley (1970)
2. Patkos, T., Flouris, G., Bikakis, A.: Symmetric multi-aspect evaluation of comments. In: ECAI 2016—22nd European Conference on Artificial Intelligence, The Netherlands, vol. 285. IOS Press (2016)
3. Machine Learning Group at the University of Waikato. http://www.cs.waikato.ac.nz/ml/weka/
4. Schneider, J., Groza, T., Passant, A.: A review of argumentation for the social semantic web. Semant. Web 4(2), 159–218 (2013)
5. Baroni, P., Romano, M., Toni, F., Aurisicchio, M., Bertanza, G.: Automatic evaluation of design alternatives with quantitative argumentation. Argument Comput. 6(1), 24–49 (2015)
6. Leite, J., Martins, J.: Social abstract argumentation. In: Twenty-Second International Joint Conference on Artificial Intelligence (2011)
7. Evripidou, V., Toni, F.: Quaestio-it.com: a social intelligent debating platform. J. Decis. Syst 23(3), 333–349 (2014)
8. Anadiotis, G., Alexopoulos, P., Mpaslis, K., Zosakis, A., Kafentzis, K., Kotis, K.: Facilitating dialogue - using semantic web technology for eParticipation. In: Aroyo, L., Antoniou, G., Hyvönen, E., ten Teije, A., Stuckenschmidt, H., Cabral, L., Tudorache, T. (eds.) ESWC 2010. LNCS, vol. 6088, pp. 258–272. Springer, Heidelberg (2010). doi:10.1007/978-3-642-13486-9_18
9. Passant, A., Bojārs, U., Breslin, J.G., Decker, S.: The SIOC project: semantically-interlinked online communities, from humans to machines. In: Padget, J., Artikis, A., Vasconcelos, W., Stathis, K., da Silva, V.T., Matson, E., Polleres, A. (eds.) COIN 2009. LNCS, vol. 6069, pp. 179–194. Springer, Heidelberg (2010). doi:10.1007/978-3-642-14962-7_12
10. Softic, S., Hausenblas, M.: Towards opinion mining through tracing discussions on the web. In: The 7th International Semantic Web Conference, p. 79. Citeseer (2008)
11. Miles, A., Bechhofer, S.: SKOS simple knowledge organization system reference (2009)

Identifying Opinion Drivers on Social Media

Anish Bhanushali$^{(\boxtimes)}$, Raksha Pavagada Subbanarasimha,
and Srinath Srinivasa

International Institute of Information Technology, 26/C, Electronics City,
Bangalore 560100, Karnataka, India
{anish.bhanushali,raksha.p.s}@iiitb.org, sri@iiitb.ac.in

Abstract. Social media is increasingly playing a central role in commercial and political strategies, making it an imperative to understand its dynamics. In our work, we propose a model of social media as a *"marketplace of opinions."* Online social media is a participatory medium where several vested interests invest their opinions on disparate issues, and actively seek to establish a narrative that yields them positive returns from the population. This paper focuses on the problem of identifying such potential "drivers" of opinions for a given topic on social media. The intention to drive opinions are characterized by the following observable parameters: (a) significant level of proactive interest in the issue, and (b) narrow focus in terms of their distribution of topics. We test this hypothesis by building a computational model over Twitter data. Since we are trying to detect an intentional entity (intention to drive opinions), we resort to human judgment as the benchmark, against which we compare the algorithm. Opinion drivers are also shown to reflect the topical distribution of the trend better than users with high activity or impact. Identifying opinion drivers helps us reduce a trending topic to its "signature" comprising of the set of its opinion-drivers and the opinions driven by them.

Keywords: Social media · Drivers · Opinion marketplace

1 Introduction

The participatory nature of social media is increasingly impacting corporate and political strategies, in addition to affecting the daily lives of individuals. Social media is rapidly becoming the platform of choice for commercial and political campaigns and is seen as a strategic tool for engaging with potential stakeholders.

While campaigns have always been a part of mainstream media, the participatory nature of social media makes the dynamics much more complex. In fact, on social media, there may not be a clear distinction between a campaign and an organically trending open discussion. Confirmation bias plays a major role in online discussions [10] and even routine discussions are replete with different parties trying to implicitly promote their point of view (POV), essentially making any trending topic, a collection of several micro-campaigns.

Given the dearth of understanding of such dynamics, there is a need for suitable models. In our work, we approach this problem by modeling social media

© Springer International Publishing AG 2017
H. Panetto et al. (Eds.): OTM 2017 Conferences, Part II, LNCS 10574, pp. 242–253, 2017.
https://doi.org/10.1007/978-3-319-69459-7_17

as a "marketplace of opinions." Social media is seen as a secondary marketplace, where different vested interests invest their *opinions* on issues that affect them, and actively promote their "investments" in order to gain greater acceptance for such opinions. The "market share" of such opinions usually bring concrete returns in the primary marketplace where the parties operate.

For instance, promoting positive opinions about one's product may result in greater sales. Similarly, promoting dissonance or distrust about an opponent in elections may bring concrete results in terms of vote shares.

Social media offers a relatively small and equal opportunity cost for different parties to invest and grow their opinions. This is in contrast to mainstream media, where viewers typically have very little opportunities to air their opinions. As a result, opinion dynamics on social media are much richer and variegated. Here, opinions not only clash and compete with one another, but compatible opinions often "team up" to collectively increase their impact. Such constellations of mutually compatible opinions sustained over time, become *narratives* which are powerful foundations that shape worldviews, policies and even the course of history.

In this paper we address a specific problem under the proposed opinion-marketplace paradigm. This is the problem of identifying opinion "drivers" for a trending topic – or user accounts that are intending to drive a specific opinion or viewpoint on the issue. Identifying opinion-drivers and their respective opinions, on a given trending topic helps us in reducing the trend down to two dimensions: *who are driving this trend*, and *in what directions are they driving it*.

Opinion drivers are characterized by their latent *intent* to drive opinions. While the intent cannot be directly observed, we posit that it manifests in the following observables:

1. Proactive interest in the trending topic, and
2. Consistency or a narrow focus in one's vocabulary.

We test this hypothesis on different trending topics on Twitter, by analyzing the information content in the topics expressed by participants. Opinion drivers are then identified by measuring the entropy in their expressed topics, tempered with their participation rate.

2   Related Literature

Understanding underlying intent in social media communications, has elicited a lot of research interest. Lampos et al. [9] measure users' intention to vote for a particular party during elections by focusing on users' vocabulary and their role (normal people, journalist, businessman etc.) as characteristic attributes for prediction.

A problem closely related to driver identification is the "expert" identification problem. Users who possess expertise in a topical area often exert great influence and also drive conversations on that topic along specific lines. Cognos [7] is an approach that uses Twitter lists for identifying topical experts. Users' membership in lists representing specific topics, is used as a criteria to rank users for that topic. Pal et al. [13] address this problem by clustering users into disjoint groups called authorities and non-authorities, based on a measure of "self-similarity" or topical consistency in users' tweets. Wagner et al. [15] use classic topic models like Latent Dirichlet Allocation (LDA) [2] to identify distributions of topic in set of tweets. A measure called Normalized Mutual Information (NMI) is then used to compare the entropy of topic distribution for a user with a ground truth obtained from Wefollow (now called about.me).

Another study by Cha et al. [4] has shown that by observing number of retweets, followers and mentions that a user has, we can determine whether the user is an influencer. Bakshy et al. [1] address the question of quantifying influence of a user. The main idea here is to give a score to each user based on the seed post done by that user and how much cascade it has generated. They observe bit.ly URLs in a tweet. If a user tweets some URL which was not posted earlier by any other users whom that user is following, then that URL post will be considered as one of the seed posts. A cascade of the seed post is defined as the number of users who spread the seed post by retweeting, mentioning and embedding link of seed post into their tweet.

A machine learning approach to identify campaigns is proposed by Ferrara et al. [6]. In this approach they construct 423 features including user network, user account property, timing, part-of-speech tags and sentiment. Since the proposed method uses supervised learning, it depends upon good training data. An unsupervised learning method to identify campaigns is proposed by Li et al. [12] based on Markov Random Fields. They build a network of activity bursts, users and URLs and compute a probability of every user being a promoter of the burst. In this method except the URL, other contents of tweets haven't been used.

Lee et al. [11] identify campaigns by building a network of tweets which share similar terms (called a "talking point"), after removing noise such as singleton tweets from the network. Then a clique detection algorithm is used to find sets of tweets that could potentially indicate a campaign. An improvement on this study has been done by Zhang et al. [16]. In this study they measure the entropy of only URLs posted by a user. Measuring entropy of URLs common between two accounts (say user A & B) and dividing it with the sum of individual entropy for URLs of users A and B gives the similarity score between two accounts. While Lee et al. [11] build a graph of tweets, Zhang et al. [16] build a graph of users and with their entropy-based similarity score as edge weight. From our observations, campaigns and opinion drivers do not just use URLs – their aim is to get greater acceptance for their opinions, which is why to detect opinion drivers, we need to focus on actual terms rather than just URLs.

Borge-Holthoefer et al. [3] analyze the dynamics of social media communication using symbolic transfer entropy of time series of information flow. They capture the characteristic time scale of the events on social media during various stages of the event. They compare the time series of magnitude of twitter posts across different locations, capturing both spatial and temporal aspects of the tweets. It was observed that in most of the cases characteristic time scale drop characterizes the beginning of a collective phenomenon. While the authors look at the magnitude and directional information flow of tweets, content of tweets and their intention has not been captured in this work. The work focuses on identifying the structural signature of a social phenomenon whereas we focus on identifying the semantic signature of a social media event.

Also, being a topical expert or influencer does not necessarily mean that the user is an opinion driver. Their influence score or expertise are mainly functions of their position in network rather than their intentions. This makes the driver identification problem, an open problem.

3   Driver Identification: Formal Model

We use Twitter as the social media of choice for identifying opinion drivers. However, the formal model for a driver is itself generic, and does not use any Twitter-specific elements in its formulation. The proposed model can be easily adapted to analyze trending activity on any other social media websites like Facebook, Reddit etc.

We start with a topic t representing a trending activity on the social media. A dataset D_t of the social media activity concerning topic t is extracted, represented as a set of posts. The term $users(D_t)$ represents the set of all users who have contributed to D_t, and the term $userposts(u)$ represents the set of all posts by user $u \in users(D_t)$. The topic t is represented as an abstract space characterized by a set of dimensions $(t_1, \ldots, t_m)$, each of which, represents a term that is relevant to t, where m is total number of terms in D_t. Each user $u \in users(D_t)$ is defined as a vector $\boldsymbol{u} = (f_1, \ldots, f_m)$, where each element f_i represents the frequency of term t_i as used by user u, measured in terms of the number of posts by u for which, t_i is relevant.

Activities of any user u, are compared against a hypothetical "null" user u_0, which is modeled based on expected levels of activity.

The support for a given term t_i, for a given user u, is the probability of u being the author of any post mentioning t_i:

$$support(u, t_i) = \frac{freq(t_i, u)}{|t_i(D_t)|} \tag{1}$$

Here, $freq(t_i, u)$ is the frequency of term t_i by user u and $t_i(D_t)$ is the set of all posts in D_t that mention t_i.

For the null user, support for each term is calculated as the expected support averaged over all users:

$$support(u_0, t_i) = \frac{1}{|users(D_t)|} \sum_{\forall u \in users(D_t)} support(u, t_i) \tag{2}$$

The "support vector" of user u, denoted as $support(\boldsymbol{u})$ is a vector showing the support of each term for user u. Our first evidence towards underlying intent to drive opinions is by observing how focused is this support vector, compared to the set of all terms describing t. This is formulated in terms of the *entropy* of this vector.

To calculate the entropy, we first reduce the support vector into a simplex, by normalizing the support values to make it into a probability vector:

$$p_u(t_i) = \frac{support(u, t_i)}{\sum\limits_{i=1}^{m} support(u, t_i)} \tag{3}$$

Entropy, which is the measure of spread or "scatteredness" of the vocabulary of user u, is calculated as follows:

$$H_u = -\sum_{i=1}^{m} p_u(t_i) \log_2 p_u(t_i) \tag{4}$$

Higher values of the entropy shows a scattered vocabulary, while low values of the entropy indicates greater focus in the vocabulary of the user. Entropy values for a given user, are compared against the "null" user u_0, whose entropy is calculated as the expected entropy – which in turn, is calculated from the expected support vector:

$$H_{u_0} = E(H_u) = -\sum_{i=1}^{m} p_{u_0}(t_i) \log_2 p_{u_0}(t_i) \tag{5}$$

In order to convert entropy values into an evidence score, we compute its complement, by comparing the entropy score against the maximum score obtained.

$$Evidence(u) = \frac{(\hat{H} + 1) - H_u}{\max\limits_{u}((\hat{H} + 1) - H_u)} \tag{6}$$

Here, $\hat{H}$ is the maximum value of entropy obtained from all the users in the corpus. A high value of the evidence score indicates that the vocabulary of the user is highly focused on a small set of terms, in comparison with the overall vocabulary used by the community for this topic.

The evidence score is by itself insufficient to conclude the intent to drive opinions. A casual user who has participated very minimally with just one post, would also show low values of entropy. Users who intend to drive opinions, would not only be focused in their vocabulary, but also display significant amounts of activity.

To measure this, we model a prior distribution of user activity manifested by an intent to drive opinions. The intention to drive opinions is modeled as a Dirichlet process $DP(B, \alpha)$, built around a base distribution B, based on observed activity distribution:

$$B(u) = \frac{|posts(u)|}{|D_t|} \tag{7}$$

where B is the base distribution and α is a positive real number called scaling parameter.

A Dirichlet process $DP(B, \alpha)$ generates a set of data elements $\{X_1, X_2, .., X_n\}$ such that for the nth data element [14]:

1. If $n = 1$, value of X_1 is drawn from base distribution B
2. If $n > 1$ then,
   (a) With probability $\frac{\alpha}{\alpha+n-1}$ value of X_n is drawn from base distribution B
   (b) Set $X_n = u$ with the probability $\frac{n_u}{\alpha+n-1}$ where n_u is the number of times u has been drawn in the past.

With N data elements thus generated, the probability of user u is defined as:

$$p(u) = \frac{n_u}{N} \tag{8}$$

Dirichlet processes are useful to model latent variables affecting observable phenomena. While the base distribution shows observable activity, the Dirichlet process uses relative variations in observable activity levels to estimate a prior probability of a latent intentional variable, driving the activity. After the completion of Dirichlet process all users in D_t are assigned new prior probability which represents their latent intention to drive an opinion. For our experiments, we have set $\alpha = 0.5$. This value is based on empirical verification of the stability of proposed prior values generated by the Dirichlet process across several synthetic generation runs, given the observed base distribution.

The prior distribution of the null user is given by a least-biased formulation of expected activity, by expecting all users to have the same levels of activity:

$$p(u_0) = \frac{1}{|users(D_t)|} \tag{9}$$

Given the above, the posterior likelihood that a given user is a driver of opinions, is given by:

$$L(u) = Evidence(u) \cdot p(u) \tag{10}$$

After calculating the posterior probability of all the users in the dataset, users are ranked in descending order of their posterior probability. Empirical examination of the distribution of $L(u)$ scores for several datasets, show that it is a very skewed distribution with steep changes in values from one user to the next. To classify the ranked list of users into drivers and non-drivers, we compute the pair-wise difference in scores between a given user and the previous one. The first point at which the differential is lesser than the expected differential score, we mark as the dividing point. We use the same technique to also mark a second dividing point to finally represent the dataset as either 2 or 3 classes.

4   Results and Evaluation

For the purpose of evaluating the model we considered Twitter data for several topics. A topic is either a keyword or a hashtag (For example "#jallikattu",

"USElections2016", "GlobalWarming" etc.). For each topic on an average 2488 tweets were fetched from Twitter. Before starting the process of identifying opinion drivers, the tweets were pre-processed. This pre-processing involved removal of stop-words, punctuations, urls, searched keyword or hashtag, whitespaces and converting all terms to lowercase. After the pre-processing step, with the set of terms thus obtained, opinion drivers score is computed for all users in the respective topic.

Some of the topics were known apriori to be promoted campaigns (#Ignis and #SpiceJetBigOrder), while others were organically trending topics. Intentionally promoted campaigns and organically trending topics were easily distinguishable, based on the location of their null user. Figure 1 shows the distribution of posterior scores (Eq. 10) and the position of the null user for an organically trending topic, while Fig. 2 shows the distribution of posterior scores and the position of the null user for a promoted campaign. The null user scores were found to be significantly inside the overall distribution for promoted campaigns, while the null user score is below the scores of other users in organically trending topics.

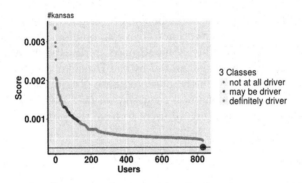

Fig. 1. Posterior probability score distribution for users of topic '#Kansas' divided in to 3 classes along with null user

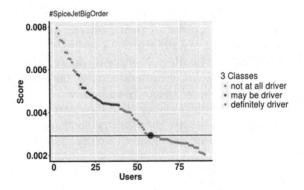

Fig. 2. Posterior Probability score distribution for users of topic '#SpiceJetBigOrder' divided in to 3 classes along with null user

In topics that are organically trending, the overall vocabulary is much richer, given high expected values of entropy, thus pushing the null user below.

Evaluation of scores was performed in two ways. In the first method, evaluators were described the definition of opinion drivers and evaluators were asked to classify users of various topics into three classes namely: "definitely driver" (clearly showing an intention to drive opinions), "may be driver" (potentially showing an intention to drive opinions) and "definitely not driver" (clearly showing no intention to drive opinions) based on the posts/tweets made by the respective user. In the second method, a 2-class classification scheme was used, where users were classified as either "driver"or "non-driver". In both the evaluation methods all users who have only one tweet in the dataset were removed from consideration.

Evaluation was performed by 19 human evaluators, largely coming from a graduate student pool, who were aware of and understood the problem. Each evaluator was provided with a list of 20 users from each topic chosen from five topics. The list provided to the evaluators comprised of users chosen from all classes with equal random probability.

Each evaluator classified 20 users across 5 topics, giving 100 values from each evaluator. Evaluation results form a matrix with dimensions 100×19. Rows represents users (twitter handles) and column represents evaluator scores. The modal value of the class in a row is chosen as the result of the evaluation for that twitter handle. The confidence of this evaluation is calibrated based on agreement across evaluators. This is calculated as:

$$Agreement_{user[i]} = \frac{count(mode(i))}{total\ number\ of\ evaluators} \tag{11}$$

Where *count* represents number of times the modal value is repeated in the list. Table 1 depicts the average agreement scores across all evaluators for both 3-class and 2-class models. Agreement scores were also computed by adding the algorithmically computed class as a column in the evaluation matrix. This is shown in rows 3 and 5 of Table 1 for both 2-class and 3-class models respectively. For 2 class evaluation, agreement score between the evaluators is 0.75 and agreement score between evaluators and the proposed model result is 0.78. For 3 class evaluation, agreement score between the evaluators is 0.60 and agreement score between evaluators and the proposed model result is 0.62. We interpret this to mean that even if the algorithm were to replace one of the human evaluators, it would result in little or no change in the overall consensus.

Algorithmic results were also calibrated for its precision and recall leading to the F1 measure. Precision was computed as a measure of the proportion of users identified as drivers by the algorithm, who were also identified as drivers by the evaluators. Recall was a measure of the proportion of the set of all users who were identified as drivers by the evaluators, who were also identified by the algorithm. The F1 score is the harmonic mean of precision and recall scores.

Finally, Cohen's kappa [5] was used to calibrate agreement between evaluation scores based on modal values of the class, and algorithmically computed

Table 1. Evaluation scores for different metrics

No. of classes	Metric	Score
2	F1 score	0.77
2	Agreement score (between evaluators)	0.75
2	Agreement score (between evaluators and algorithm result)	0.78
3	Agreement score (between evaluators)	0.60
3	Agreement score (between evaluators and algorithm result)	0.62
3	Cohen's kappa (quadratic weights)	0.67

class for each user. We obtained a score of 0.67 which indicates positive correlation between evaluator and algorithm scores.

Table 2. Kendall's correlation score comparing impact, activity and driver-score based ordering of terms with global ordering of terms based on frequency

Topic	Users with high impact	Users with high activity	Opinion drivers
Election	0.30	0.09	0.36
#TrumpGlobalGag	0.38	0.41	0.65
#Kansas	0.31	0.36	0.41
ABVP	0.31	0.51	0.58

Evaluating robustness of the model. We also asked how sensitive is the computed score to the specifics of the dataset and whether the driver scores would change with more or less data collected. The model is said to be robust, if it is impervious to minor perturbations in the evidence. To compute this, tweets of a topic were first sorted according to their time stamp. Driver scores were then iteratively computed, with each iteration reducing the data size by 3%. This process is repeated until data size is reduced to 45% of its original size. At each iteration top driver is noted. If the model is robust with respect to size of data, then the same user should be the top driver in all the iterations.

In our experiments on different topics, the top driver for the dataset remained intact on an average of 79% of the time across the iterations.

Comparison with user activity and user impact. We also address a hypothetical question whether opinion drivers can be trivially detected by the impact they have had on the discussion, or by their amount of activity. Since an opinion driver plans to drive collective opinions along specific directions, we compare opinion drivers and users with high activity or impact (based on retweets and likes), against collective opinion rankings. Collective opinion is modeled by a global ranking of terms, with respect to the number of tweets they appear in.

Similar term distributions are computed for each model. We first score users according to the respective model (activity, impact and driver). Terms used by these users are given an aggregated score by adding the respective model score every time a term is used by the user. Terms are then ranked based on this aggregated score, and compared with the global ranking using Kendall rank correlation coefficient [8].

Table 3. Signature of the narratives

Topic	Drivers	Top-1st word	Top-2nd word	Top-3rd word
#Ignis	Driver 1	{@nexaexperience}	{launch}	{electronation, ignis}
	Driver 2	{@nexaexperience}	{electronation}	{launch, looking, forward}
	Driver 3	{@nexaexperience}	{electronation, watch}	{launch, live}
ABVP	Driver 1	{#delhiuniversity}	{ramjas, college}	{umar, khalid}
	Driver 2	{#modiministry}	{ramjas, college}	{khalid, protest, clash}
	Driver 3	{well, done}	{traitors}	{medicine, long, due}
#Kansas	Driver 1	{feb, pm}	{#shooting}	{man}
	Driver 2	{#espn2, #tcu}	{#aclu}	{#doj}
	Driver 3	{kansas, #noplacelikeks}	{#ksbucketlist}	{city}

Table 2 shows the average Kendall rank correlation coefficient scores calculated for these three sets for different topics. It is evident that for all topics that were considered, rank ordering of terms by opinion drivers agreed more than that of users with high impact or activity. This indicates that opinion drivers represent the topical narrative more accurately.

Signature generation. The objective of driver identification is to reduce a trending topic into a "signature" representation – comprising of its drivers and the topics they represent. Table 3 shows top 3 keywords for the top 3 drivers for different topics. The user handles of the top three drivers have been anonymized for protecting their privacy.

The consistency of the top terms for drivers of the promoted campaign (#Ignis) is copious. Organically trending topics have drivers with different agenda, leading to different sets of top terms associated with them.

5   Conclusions

We conjecture that modeling social media as an opinion marketplace represents a significant leap in our understanding of its dynamics. This work represents preliminary research results in this direction of pursuit, where we reduce a trending topic on social media to a set of opinions and their drivers. This helps in better understanding of the latent intentional forces that are trending this topic. Reducing a trending topic into a set of directions and their drivers will be useful for several stakeholders – for government organizations to identify any suspicious activity, product companies to analyze and improve their products, Web science researchers to understand social media activities, social media enthusiasts to observe and participate in the direction they like. In future work, we plan to improve upon this model and extend this to characterize more elements of the opinion marketplace. We also plan to incorporate the temporal aspect of social media posts as a part of the model.

References

1. Bakshy, E., Hofman, J.M., Mason, W.A., Watts, D.J.: Everyone's an influencer: quantifying influence on Twitter. In: Proceedings of the Fourth ACM International Conference on Web Search and Data Mining, WSDM 2011, New York, NY, USA, pp. 65–74. ACM (2011)
2. Blei, D.M., Ng, A.Y., Jordan, M.I.: Latent dirichlet allocation. J. Mach. Learn. Res. **3**, 993–1022 (2003)
3. Borge-Holthoefer, J., Perra, N., Gonçalves, B., González-Bailón, S., Arenas, A., Moreno, Y., Vespignani, A.: The dynamics of information-driven coordination phenomena: a transfer entropy analysis. Sci. Adv. **2**(4), e1501158 (2016)
4. Cha, M., Haddadi, H., Benevenuto, F., Gummadi, K.P.: Measuring user influence in Twitter: the million follower fallacy. In: Proceedings of International AAAI Conference on Weblogs and Social Media, ICWSM 2010 (2010)
5. Cohen, J.: A coefficient of agreement for nominal scales. Educ. Psychol. Measur. **20**(1), 37–46 (1960)
6. Ferrara, E., Varol, O., Menczer, F., Flammini, A.: Detection of promoted social media campaigns. In: Tenth International AAAI Conference on Web and Social Media (2016)
7. Ghosh, S., Sharma, N., Benevenuto, F., Ganguly, N., Gummadi, K.: Cognos: crowdsourcing search for topic experts in microblogs. In: Proceedings of the 35th International ACM SIGIR Conference on Research and Development in Information Retrieval, SIGIR 2012, New York, NY, USA, pp. 575–590. ACM (2012)
8. Kendall, M.G.: A new measure of rank correlation. Biometrika **30**(1–2), 81 (1938)
9. Lampos, V., Preoţiuc-Pietro, D., Cohn, T.: A user-centric model of voting intention from social media. In: Proceedings of the 51st Annual Meeting of the Association for Computational Linguistics, ACL 2013, pp. 993–1003 (2013)
10. Lee, J.K., Choi, J., Kim, C., Kim, Y.: Social media, network heterogeneity, and opinion polarization. J. Commun. **64**(4), 702–722 (2014)
11. Lee, K., Caverlee, J., Cheng, Z., Sui, D.Z.: Content-driven detection of campaigns in social media. In: Proceedings of the 20th ACM International Conference on Information and Knowledge Management, CIKM 2011, New York, NY, USA, pp. 551–556 (2011). ACM

12. Li, H., Mukherjee, A., Liu, B., Kornfield, R., Emery, S.: Detecting campaign promoters on twitter using markov random fields. In: Proceedings of the 2014 IEEE International Conference on Data Mining, ICDM 2014, Washington, DC, USA, pp. 290–299. IEEE Computer Society (2014)

13. Pal, A., Counts, S.: Identifying topical authorities in microblogs. In: Proceedings of the Fourth ACM International Conference on Web Search and Data Mining, WSDM 2011, New York, NY, USA, pp. 45–54. ACM (2011)

14. Teh, Y.W.: Dirichlet Process, pp. 280–287. Springer, Boston (2010)

15. Wagner, C., Liao, V., Pirolli, P., Nelson, L., Strohmaier, M.: It's not in their tweets: modeling topical expertise of twitter users. In: Proceedings of the 2012 ASE/IEEE International Conference on Social Computing and 2012 ASE/IEEE International Conference on Privacy, Security, Risk and Trust, SOCIALCOM-PASSAT 2012, Washington, DC, USA, pp. 91–100. IEEE Computer Society (2012)

16. Zhang, X., Zhu, S., Liang, W.: Detecting spam and promoting campaigns in the twitter social network. In: 2012 IEEE 12th International Conference on Data Mining, ICDM 2012, pp. 1194–1199 (2012)

Representing Fashion Product Data with Schema.org: Approach and Use Cases

Alex Stolz[(✉)], Martin Hepp, and Aleksei Hemminger

Universität der Bundeswehr München, 85579 Neubiberg, Germany
{alex.stolz,martin.hepp,aleksei.hemminger}@unibw.de

Abstract. The last decade has seen a considerable increase of online shops for fashion goods. Technological advancements, improvements in logistics, and changes in buyer behavior have led to a dissemination of apparel goods and respective data on the Web. In numerous domains of knowledge management, ontologies have proven to be very useful for sharing meaning among organizations and individuals, and for inferencing. With schema.org, there already exists a collection of widely accepted Web vocabularies for the fields of gastronomy, accommodation, entertainment, sports, or products. Yet, schema.org still lacks a dedicated fashion ontology, which would allow for greater interoperability, higher visibility, and better comparison of fashion products on the Web. In this paper, we design and evaluate a Web ontology for garments as a compatible extension of schema.org. For our proposal, we take into account current best practices of Web ontology engineering, we formally evaluate our conceptual model, and we present practical use cases. We further contextualize our work by comparing our approach with state-of-the-art vocabularies for the fashion industry.

Keywords: Schema.org · Ontology engineering · Conceptual modeling · Fashion · Clothing · Garments · Apparel · E-business · E-commerce · Products · Product types ontology · DBPedia

1 Introduction

Online shopping constitutes an important application domain and sales channel for the fashion industry. Especially over the last decade, fashion e-commerce has gained a lot of traction. Technological advancements, improvements in logistics, and changes in buyer behavior have led to a dissemination of apparel goods and respective data on the Web [1]. According to [2], fashion-related products are globally the most popular category of online shopping. E-commerce statistics in the European Union [3] and the United States [4] further indicate a high online purchase volume in the apparel industry besides consumer electronics. In Germany, the e-commerce turnover for fashion and accessories has risen between 2007 and 2014 from 4.6% to 19.7% of the total annual turnover for this segment [1].

In numerous domains of knowledge management, ontologies have proven to be very useful for sharing meaning among organizations and individuals, and for inferencing. An ontology is generally defined as "a formal, explicit specification of a shared

© Springer International Publishing AG 2017
H. Panetto et al. (Eds.): OTM 2017 Conferences, Part II, LNCS 10574, pp. 254–272, 2017.
https://doi.org/10.1007/978-3-319-69459-7_18

conceptualization" [5], in other words it intends to share a common understanding of a domain across people and computers (cf. [5]). Because "an ontology is a concrete view of a particular domain" [6], ontology engineering often has to find a trade-off between an ontology's expressiveness (i.e. to maximize the coverage of the domain of discourse) and its general usefulness (i.e. to maximize overlap between different interpretations and thus optimize their reusability). A popular project that seeks to find conceptual models of domains that are accepted by large audiences on the Web is schema.org.

Schema.org is a joint initiative of the leading search engine operators Google, Yahoo!, Bing, and Yandex. The project was initiated in 2011 and it strives to compile a collection of commonly understood Web schemas (or vocabularies) under a single, consolidated namespace (i.e. http://schema.org/). These vocabularies allow to embed structured data into Web pages that can be used by search engines and other novel data-consuming applications. Over the years, the community has continuously improved the core schema and provided extensions for various domains such as creative works, events, health science, gastronomy, accommodation, entertainment, sports, products, or automobiles [7]. As a result, adoption of schema.org markup in Web pages has meanwhile taken off. However, schema.org still lacks a dedicated ontology for fashion goods, which would allow for greater interoperability, higher visibility, and better comparison of fashion products on the Web. In particular, fashion Web sites have currently no means to provide a detailed, machine-readable description of their products on the Web which would benefit novel applications.

The goal of this work is to design a light-weight ontology and modeling approach for the Web that covers the most important aspects of the apparel industry. We require our ontology to be light-weight for it can be easily incorporated into schema.org, it is useful for the majority of stakeholders, and still it can be easily extended using best practices provided by schema.org. Unlike classical ontologies intended for quite determinate contexts, the core problem with ontology engineering for the Web is that many of the potential stakeholders are unknown upfront (cf. [8]).

Having a fashion ontology within schema.org opens up a range of possible use cases. First and foremost, it allows to describe items within Web shops in a machine-friendly way so that data consumers like Google or Apple can easily grasp the structured content for novel applications. Furthermore, manufacturers can similarly benefit from marking up their products online and thus render them more visible for any kind of data-consuming applications (cf. [9]). This way even very sophisticated use cases are possible, e.g., with the upcoming Internet of Things devices such as washing machines and tumble dryers could autonomously query information about pieces of garment from the respective manufacturer Web sites and act intelligently upon by applying the correct care treatments.

The remainder of this paper is structured as follows: In Sect. 2 we provide context on clothing-related information and what the current challenges are. Section 3 describes our approach of engineering an ontology for garments that is compatible with schema.org. We evaluate our approach in Sect. 4. In Sect. 5 we compare our approach with important related works and give an outlook of how our ontology can be used and improved.

2 Overview of Clothing-Related Information

The main challenge of characterizing clothing-related information is the great variety of product types (trousers, shirts, shoes, dresses, underwear, accessories, etc.), the manifold manufacturers and other stakeholders, but also the versatility of textile materials. Besides the product-specific information for fashion goods, the most important dimensions related to clothing and accessories are garment sizes, colors, product variants with different garment sizes and colors, materials information, textile care recommendations, and certifications.

2.1 Garment Sizes

Garment sizes play a peculiar role in the fashion industry, as there exist several different standards for classifying sizes, including international ones (e.g. the European EN 13402 series of standards or the US clothing sizes standard) as well as systems that vary between individual countries (cf. [10]). Generally, garment sizes with these standards are based on anthropometry, i.e. measurements of human individuals (cf. [11]). Garment sizes can be grouped into size charts, i.e. "the artificial dividing of a range of measurements" [12]. Clustering sizes is both beneficial for manufacturers and vendors in production and retail (cf. [12]). Furthermore, size measures can differ across manufacturers. Nowadays it is no secret that garment sizes change over time, i.e. due to marketing policy and the like the physical sizes of clothing tend to increase with respect to the nominal sizes, also known as size inflation or vanity sizing [10]. Furthermore, size comparisons among clothing size standards are difficult because they often lack direct relationships to each other [10].

2.2 Colors

The color of a clothing is one of the most remarkable features of a clothing. Besides the type of clothing, it plays a critical role in receiving a first impression to an apparel good.

2.3 Product Variants

The fashion industry has to deal with a high diversity in the production of clothing. E. g., for the same piece of clothing there typically exist multiple configurations with different sizes (e.g. length, waist width) and/or colors.

2.4 Materials Information

The textile industry is characterized by different kinds of materials that can range from natural to synthetic fibers, and by a complex textile process chain (cf. [13]). Popular fibers that are used as materials for garments include Cotton, Wool, Silk, or Polyester. But even metal fibers can be found in clothing and accessories goods. For an overview of the kinds of materials in textile production, see e.g. [13].

2.5   Textile Care Recommendations

In order to prevent damages to clothing by improper handling, it is very common to sew care labels into textile goods. Such labels give care instructions using specific symbols for washing, bleaching, dry cleaning, ironing, and professional cleaning treatment (cf. [14]). The symbols and their meanings are defined by the International Association for Textile Care Labelling, GINETEX, and are standardized in the DIN EN ISO 3758 international standard [15]. This standard is commonly accepted in European countries. Other standards are available in other countries and regions, notably the Standard Guide to Care Symbols for Care Instructions on Textile Products [16] in the US.

2.6   Certifications

Certifications address sustainable production, usage, and disposal (cf. [17]). There exist a number of standards that impose requirements that are either industry-independent or target specific industrial branches. The Ecolabel Index[1] provides a directory of world-wide ecolabels from different industries. One of the most notable certification standard for the textile industry is the Global Organic Textile Standard (GOTS)[2], but also the OEKO-TEX® Standard 100[3] is widely known as a certification system for raw materials, intermediate products, and end products (cf. [17]).

3   Clothing Product Information with Schema.org

In this section, we discuss our modeling proposal and the design of a light-weight ontology for garments that relies on schema.org concepts and will possibly be included in one of the future releases of schema.org. In the course of this paper we will stick to the namespace prefix *cpi:* for Clothing Product Information in order to refer to concepts of our ontology. The rationale is to clearly highlight existing elements in schema.org that we can reuse and to discern them from the newly added ones. In the subsequent discussion, we assume the following prefix declarations without defining them repeatedly:

- cpi: http://www.ebusiness-unibw.org/ontologies/cpi/ns
- schema: http://schema.org/.

3.1   Method

Ontology experts have in the past contributed a large number of ontologies for many different domains. An overview of available ontologies is provided by popular online lookup services like Swoogle[4] or Linked Open Vocabularies[5]. Consequently, there

---

[1] http://www.ecolabelindex.com/ecolabels/.

[2] http://www.global-standard.org/certification.html.

[3] https://www.oeko-tex.com/en/business/business_home/business_home.xhtml.

[4] http://swoogle.umbc.edu/.

[5] http://lov.okfn.org/dataset/lov/.

have evolved best practices for the process of ontology engineering, consisting of (1) requirements analysis, (2) conceptualization, (3) implementation, and (4) evaluation [18]: The requirements analysis researches existing solutions, defines use cases, and formulates competency questions that describe a domain. It follows the modeling of the conceptual domain, defining classes, properties, and enumerated values, with a possible integration of existing sources. The implementation phase consists in formalizing the ontology model in a representation language. Finally, in the evaluation step the competency questions are typically answered automatically or semi-automatically (cf. [18]). Competency questions are a standard method in ontology engineering that help capture the scope and content of an ontology, and guide its form of evaluation (cf. [19]).

However, as our suggested schema.org extension for garments is intended for a rather large, unknown audience on the Web, the abovementioned, rigorous ontology design approach is replaced by a more practice-oriented approach that best possible meets the criteria and interpretations of the conceptual designers, but likewise of data publishers and data consumers on the Web with varying levels of semantics expertise (cf. [8]).

In the following, we formulate competency questions [19] for defining the scope of our extension proposal for schema.org. We base our competency questions on functional requirements that we derived from the general presentation of descriptions and features of textile goods in Web shops. Since commercial details of product offers can be readily supplied by schema.org (e.g., http://schema.org/Offer), we left out this aspect for our competency questions:

1. CQ: Which retrievable Web resources describe a
   - {footwear, clothing, accessories} product,
   - {by, with} a certain {manufacturer, brand name},
   - with a given garment size,
   - consisting of a combination of materials,
   - with one or more colors,
   - is designated for a particular group of customers,
   - has restrictions on textile care,
   - is certified by an organization or through a certification program, and
   - meets certain requirements on {properties, intervals for properties}.

2. CQ: Which clothing or accessories products are available?
3. CQ: How can we specify garments that contain at least 60% cotton?
4. CQ: Which types of certifications are available?
5. CQ: Which kinds of textile care recommendations do exist?
6. CQ: Which are the eligible customer groups?
7. CQ: How can the production origin (e.g. made in Germany) be specified?
8. CQ: How can variants of {footwear, clothing, accessories} products be expressed?
9. CQ: How can product identifiers be attached?
10. CQ: How can basic product information (e.g. images) be supplied?

In order to be able to answer these competency questions, the ontology must at least offer conceptual elements to represent basic product information (e.g. product types, descriptions, images), clothing-related information (e.g. garment sizes, materials), care recommendations (e.g. washing care instructions), and certifications (e.g. regarding sustainable production). In the upcoming subsections, we will illustrate how existing elements in schema.org and newly added concepts within our garment vocabulary allow to model all this information.

3.2 Representation of Basic Product Information

The domain that our vocabulary for garments shall interfere with is the product domain, since apparel items as they are offered by fashion Web shops are essentially products. We can classify any apparel good as a *schema:Product* and thereby take advantage of all concepts, individuals, and relationships related to this schema.org concept. This way clothing and accessories can immediately benefit from the full e-commerce capabilities within schema.org.

The core class that distinguishes clothing and accessories products from the generic product type is *cpi:ClothingAndAccessories*. It is defined as a subclass of *schema: Product*. A further and more granular distinction is possible by the definition of the two classes *cpi:Footwear* and *cpi:GarmentOrClothing*, which are subsumed by their common superclass *cpi:ClothingAndAccessories*. If we were to model accessories (e.g. scarf, hat, gloves), we would accordingly classify them under the class *cpi:ClothingAndAccessories*. With having all these concepts being subclasses of *schema:Product*, we can easily model basic product information for clothing products like

- name and description (*schema:name* and *schema:description*),
- product image and link (*schema:image* and *schema:url*),
- weight (*schema:weight* ($\rightarrow$ *schema:QuantitativeValue*)),
- manufacturer (*schema:manufacturer* ($\rightarrow$ *schema:Organization*)),
- (offer and) price information (*schema:offers* ($\rightarrow$ *schema:Offer*) and *schema:price*),
- brand (*schema:brand* ($\rightarrow$ *schema:Brand* or *schema:Organization*)),
- category (*schema:category*), and
- product identifiers, e.g. GTINs, MPNs, and identifiers like EPC/RFID (e.g. [20]) (*schema:gtin13*, *schema:mpn*, *schema:productID*, *schema:identifier*).

In addition, schema.org has a relatively simple yet powerful extension mechanism (*schema:additionalType* property) whereby products can be classified as more specific types. In our case, we can classify products as specific types of garments linking with *schema:additionalType* to corresponding classes in external ontologies such as the Product Types Ontology (PTO)[6]. The schema.org-specific *schema:additionalType* property corresponds to *rdf:type* in RDF data models.

```
foo:MyTShirt a cpi:GarmentOrClothing .
foo:MyTShirt schema:additionalType
<http://www.productontology.org/id/T-Shirt> .
```

---

[6] http://www.productontology.org/.

Ontologies for e-commerce like GoodRelations [21] and its derived counterpart in schema.org, typically favor the distinction between product models and product instances. While the product model denotes the blueprint of a product, a product instance represents the physical entity that can be sold, borrowed, owned, sought, or disposed. Although this is recommended practice, many data publishers will typically decide to model a product item using the more generic product type *schema:Product*. For the sake of simplicity, we also do this here wherever feasible.

3.3   Representation of Clothing-Related Information

Product-related information can well be covered by existing concepts in schema.org, as it was demonstrated in the previous subsection. However, schema.org is severely limited when it comes to describe clothing-related information. This subsection hence proposes the modeling of specific information related to garments.

Garment Sizes. Nowadays there exist many international and regional standards for garment sizes and partly incompatible interpretations of clothing sizes by various manufacturers. Because of this, garment sizes are complex to model in a correct way. For the purpose of our modeling approach, it is safe to neglect the comparability of different sizes from different standards or manufacturers when we specify garment sizes with respect to their standards or manufacturers. This processing step can be passed on to the data consumers.

Garment sizes in the Clothing Product Information ontology are attached to clothing and accessories products using a common property *cpi:clothingSize*, which likewise applies to (has as its domain) footwear, clothing, and accessories.

```
cpi:clothingSize (cpi:ClothingAndAccessories →
cpi:ClothingSize OR Text)
cpi:ClothingSize [schema:QualitativeValue]
   cpi:sizeStandard (→ schema:Organization OR Text OR
URL)
  schema:name
  schema:valueReference
```

For individual product sizes, we can indicate using the *schema:valueReference* property e.g. that "Size XXL for an Adidas sweater means width = 65 cm, length = 67.5 cm, and sleeve length = 67 cm". For brevity, we show only the reference to the width of the sweater. The unit code expects a unit of measurement code defined according to the UN/CEFACT Common Code standard [22].

```
adidas:SizeXXL a cpi:ClothingSize ;
  schema:name "XXL" ;
  cpi:sizeStandard <http://www.adidas.com/> ;
  schema:valueReference [ a schema:QuantitativeValue ;
    schema:name "width" ;
    schema:unitCode "CMT" ; # centimeters in UN/CEFACT
    schema:value 65.0
  ] .
```

Now we can use the size information to describe an instance of an Adidas sweater.

```
foo:AdidasSweater a cpi:GarmentOrClothing .
foo:AdidasSweater cpi:clothingSize adidas:SizeXXL .
```

Because clothing sizes are qualitative values, we can even consider to set up an ordering relation among sizes, e.g.

```
adidas:SizeM schema:greater adidas:SizeS .
adidas:SizeL schema:greater adidas:SizeM .
# etc.
```

The EN 13402 standard [23] part 3 defines range intervals and letter codes for garment sizes. In the following, we sketch briefly how we could model a size chart that uses codes (here with size "M").

```
en13402:SizeM a cpi:ClothingSize ;
  schema:name "M" ;
  # optional: designated for men (see subsection 3.6)
  schema:greater en13402:SizeS ;
  cpi:sizeStandard "EN 13402" ;
  schema:valueReference [ a schema:QuantitativeValue ;
    schema:name "chest girth" ;
    schema:unitCode "CMT" ; # centimeters in UN/CEFACT
    schema:minValue 94.0 ; schema:maxValue 102.0
  ] .
```

In our previous examples, we used the property *cpi:sizeStandard* to point at a standardization body or manufacturer, and we used *schema:greater/schema:lesser* properties to obtain an ordinal scale for garment sizes within manufacturers and size standards.

We opted for modeling the value and to refer to the size standards or size charts of manufacturers. We could have decided instead to materialize all possible standards as a Linked Open Data dataset. However, the problem with this approach is that manufacturers have different interpretations of sizes (i.e. facing the vanity sizing problem). Our approach is better from a schema.org point of view, as it passes on the processing task to the data consumer.

Colors. Schema.org has already a property to define the colors of products, namely *schema:color*. The expected value is a textual description, e.g. "red", "green", or "blue". We can thus take advantage of this property to describe the color of apparel goods. As we will see shortly, we can either add the color information right to the clothing product, or alternatively to the materials it is composed of, since materials are modeled as products too.

Materials. Schema.org defines a the property *schema:material* as "[a] material that something is made from, e.g. leather, wool, cotton, paper"[7]. We could blindly use this property and attach a material (essentially a *schema:Product*) to it, but then it would not be possible to provide granular descriptions on how much exactly of a material was incorporated. Additionally, it would be impossible to claim that a clothing contains at least/at most a certain amount of material, e.g. more than 80% merino wool. Instead we introduce a class *cpi:MaterialComponent* as a subclass of *schema:Product*. This way we can handle ternary relationships (cf. [21]) by introducing an intermediate node that specifies the information about the type of material and its relative share of a product.

```
cpi:MaterialComponent [schema:Product]
  schema:material (→ schema:Product OR Text OR URL)
  cpi:quantityOfMaterial (→ schema:QuantitativeValue)
    schema:value, schema:minValue, schema:maxValue
    schema:unitCode
```

Note that we use *schema:material* for linking from the product to the material component, and again from the material component to a material. This is by intention, because such a modeling approach allows to recursively decompose materials into their constituents in a transitive fashion.

To give an example, let us assume that we want to model a jacket made of 80% leather and 20% cotton using DBPedia[8] entities.

---

[7] http://schema.org/material.

[8] http://dbpedia.org/.

```
foo:Jacket a cpi:GarmentOrClothing .
foo:Jacket schema:material foo:Leather80, foo:Cotton20 .
foo:Leather80 a cpi:MaterialComponent ;
  schema:material <http://dbpedia.org/resource/Leather> ;
  cpi:quantityOfMaterial [ a schema:QuantitativeValue ;
    schema:value 80.0 ;
    schema:unitCode "P1" # percent in UN/CEFACT

  ] .
foo:Cotton20 a cpi:MaterialComponent ;
  schema:material <http://dbpedia.org/resource/Cotton> ;
  cpi:quantityOfMaterial [ a schema:QuantitativeValue ;
    schema:value 20.0 ;
    schema:unitCode "P1" # percent in UN/CEFACT
  ] .
```

Quite clearly, relying on the *schema:QuantitativeValue* class for modeling material quantities, we could even specify lower/upper limits on material compounds, e.g. to express that a clothing contains more than 25% recycled cotton.

Materials, as they are modeled as products, can carry a brand name, e.g. to specify that a windbreaker is made of GORE-TEX® material. Furthermore, treating materials as products invites to use the Product Types Ontology (PTO). However, entities in the Product Types Ontology are defined in the TBox modeling layer and not in the ABox layer, meaning that they describe classes rather than instances. In practice, it means that in order to take advantage of the Product Types Ontology we would have to create an enumerated list of materials that are instantiated product classes from the Product Types Ontology. An alternative way is to rely on DBPedia identifiers, which already represent instances, although are not defined as products.

We decided to create instances for a number of well-known fabrics. For additional context, we could easily link them with DBPedia using *owl:sameAs* links. Furthermore, DBPedia could be used to extend our current enumeration with additional materials.

Available Combinations. The fashion industry has to deal with a high diversity in the production of clothing. E.g., for the same piece of clothing there typically exist different sizes (e.g. length, waist width) and colors. When modeling such situations, we encourage to treat all possible combinations as variants. Schema.org already has a suitable property for that, namely *schema:isVariantOf*. The property allows to materialize variations of product models, i.e. all available configurations are represented explicitly with respect to a given base model.

```
foo:MyRedTShirt a cpi:GarmentOrClothing .
foo:MyRedTShirt schema:model foo:RedTShirtModel .
foo:RedTShirtModel schema:isVariantOf foo:BaseTShirt .
foo:RedTShirtModel schema:color "red" .
```

Currently, schema.org has no mechanism for rule-based generation of variants, but combinations can be generated in SPARQL 1.1 [24] using the VALUES clause. The SPARQL CONSTRUCT query below takes the VALUE clause to materialize all combinations of given garment sizes (S, M and L) and colors (red and blue) to a provided base model. The URIs for the new product and model entities are constructed using the BIND expression.

```
CONSTRUCT {
    ?product a cpi:GarmentOrClothing .
    ?product schema:model ?model .
    ?model schema:isVariantOf foo:BaseTShirt .
    ?product cpi:clothingSize ?size .
    ?product schema:color ?color .
} WHERE {
    foo:BaseTShirt a schema:ProductModel .
    VALUES (?size ?code) { (en13402:SizeS "S")
(en13402:SizeM "M") (en13402:SizeL "L") }
    VALUES ?color {"red" "blue"}
    BIND(IRI(CONCAT("http://foo.com/DerivedTShirtModel",
?code, ?color)) AS ?model)
    BIND(IRI(CONCAT("http://foo.com/DerivedTShirt", ?code,
?color)) AS ?product)
}
```

A generic mechanism for the modeling of variants of product models would be good for schema.org, but should not be clothing-specific. The Volkswagen Car Options Ontology[9] e.g., supports the modeling of configuration options for automobiles.

3.4   Representation of Textile Care Recommendations

Because of the vast diversity of textile care recommendations, it is not possible to cover the full range of standards and proprietary solutions for care instructions in our modeling. So we create an enumeration of textile care recommendations according to the ISO 3758 standard [15] and give further advice on extensions. There are two viable alternatives to model textile care instructions. One very light-weight approach would be to use the generic property-value mechanism[10] of schema.org. However, this approach does not use specific properties and is thus not very robust and gives a lot of freedom to the data publisher. We thus recommend a similar approach as for the garment sizes before, i.e. creating specific qualitative properties and values, and then to point at value references. This time, however, we also create more granular subtypes of the common type care instruction.

---

[9] http://www.volkswagen.co.uk/vocabularies/coo/ns.
[10] http://schema.org/PropertyValue.

```
cpi:careInstruction (cpi:ClothingAndAccessories →
cpi:CareInstruction OR Text)
cpi:CareInstruction [schema:QualitativeValue]
   cpi:careStandard (→ schema:Organization OR Text OR
URL)
   schema:valueReference
   schema:name
   schema:image
```

The subtypes of *cpi:CareInstruction* along with the number of instances defined in them are: *cpi:BleachingCare* (3), *cpi:DryingCare* (with subtypes *cpi:NaturalDrying* (8), *cpi:TumbleDrying* (3)), *cpi:IroningCare* (4), *cpi:ProfessionalCare* (9), *cpi: WashingCare* (11).

In the following, we present the basic description of a Jeans with the abovementioned care instructions.

```
foo:MyJeans a cpi:GarmentOrClothing ;
   schema:additionalType
<http://www.productontology.org/id/Jeans> ;
   cpi:careInstruction cpi:ColoredWash30,
cpi:LowTemperatureIron, cpi:LineDryingShade .
```

Textile care instructions are instantiated whilst providing a value reference, for instance the recommended maximal temperature for washing, and a link to the care symbol, if it is available.

```
cpi:FineWash30 a cpi:WashingCare ;
   schema:name "30 degrees Celsius fine wash" ;
   # optional symbol link via schema:image
   cpi:careStandard <http://www.ginetex.org/> ;
   schema:valueReference [ a schema:QuantitativeValue ;
      schema:name "Washing temperature" ;
      schema:unitCode "CEL" ; # Celsius in UN/CEFACT
      schema:value 30.0
   ] .
```

3.5   Representation of Certifications

For certifications, an enumerated list of predefined individuals for the most popular certifications or certification bodies is sufficient. In our schema, we define a property *cpi:certified* and link to it either enumerated values of the type *cpi:Certification*, which are subtypes of *schema:QualitativeValue*, or a certifying authority, URL (i.e., *schema: URL*), or Text (i.e., *schema:Text*). If the property points to a certification, the issuing certification body can be signaled via a URL.

```
cpi:certified (cpi:ClothingAndAccessories →
cpi:Certification OR schema:Organization OR Text OR URL)
cpi:Certification [schema:QualitativeValue]
  schema:name
  schema:url
```

The subsequent minimal example illustrates how certifications can be attached to clothing and accessories products (note that scarf is an accessory). We provided instances for the most popular certifications in the textile sector.

```
foo:MyEcologicalScarf a cpi:ClothingAndAccessories ;
  cpi:certified cpi:OekoTex100 .
cpi:OekoTex100 a cpi:Certification ;
  schema:name "OEKO-TEX® Standard 100" ;
  schema:url <https://www.oeko-tex.com/> .
```

3.6 Other Information

Besides the aforementioned core properties of clothing information, our proposal contains additional information, that we herewith briefly summarize:

- *cpi:countryOfOrigin* (*cpi:ClothingAndAccessories* → Text) with an ISO 3166-1 two-letter country code [25], e.g. to be able to express "made in Germany",
- *cpi:designatedFor* (*cpi:ClothingAndAccessories* OR *cpi:ClothingSize* → *cpi:Gender* OR Text) can be used to model designation for clothing, but also for garment sizes, e.g. to specify "for men",
- *cpi:fitDescription* (*cpi:ClothingAndAccessories* → Text) to attach a fit description like e.g. "regular fit".

Moreover, some existing properties of schema.org are very handy for modeling further information, namely

- *schema:additionalProperty*, the generic property-value mechanism that eases modeling properties of products that currently are not covered by our ontology,
- *schema:releaseDate* and *schema:productionDate* that allow to specify the variant release and production date of a specific clothing product,
- *schema:aggregateRating* and *schema:review* in order to link to ratings and reviews of clothing items as often used by Web shops.

3.7 Formalization and Results

We developed a first draft of our ontology proposal in the popular ontology editor Protégé. This tool allows to export the conceptual model in data formats (e.g. RDF/XML, Turtle, N-Triples) that are all compatible to each other. However, because it should be a schema.org-compatible extension for garments, we did the fine-tuning on a Notation 3 file in a text editor.

The design goal for the ontology was to keep it small, easy to grasp, and to reuse existing schema.org concepts, but still to be expressive enough for being useful to the majority of stakeholders. In total, our vocabulary consists of 15 classes, nine properties (two datatype properties and seven object properties), and 57 individuals. The latest version of our schema.org extension is available online[11] and provides besides the RDF data formats a human-readable documentation. Its current namespace is *cpi*, which stands for Clothing Product Information. We also performed some consistency and validation checks with Pellet reasoner[12] and Eyeball validator[13], and with the W3C RDF Validator service[14], which we all passed successfully.

4   Evaluation

We evaluate our conceptual model in the following two ways. First, we test its domain coverage and utility with real-world data from the Web. Second, we formally validate the functional requirements of our ontology with answering the competency questions.

Our first goal was to model real-world data with our approach and to show how all information categories can be represented. We selected for each category of garments, footwear, and accessories an arbitrary item from the Web. For that we visited Google Shopping[15] and entered the following three terms: "jeans" (garments), "running shoes" (footwear), and "hat" (accessories). We took the first vendor of the very first item that appeared on the result list. This way we got items from three different shops.

- Running shoes: https://www.goertz.de/nike-sneaker-juvenate-grau-hell-44365407/
- Jeans: https://www.limango-outlet.de/levi-s/jeans-501-regular-fit-in-dunkelblau-2-4904202
- Hat: http://www.stylefile.de/adidas-ac-bucket-hat-schwarz-fid-95779.html

As it can be seen in Fig. 1, the clothing-related product information varies between product types and to obtain a full coverage of all features would be infeasible for a light-weight ontology. However, with schema.org it is possible to model missing properties using the property-value mechanism. The examples of populating Web site content with clothing-related product information are published online[16].

At the beginning of Sect. 3 we have given competency questions to capture the domain and content of the planned conceptual design. In the following, we will translate the main questions into queries, and use SPARQL to answer them and hence formally evaluate our ontology.

---

[11] http://www.ebusiness-unibw.org/ontologies/cpi/.

[12] https://www.w3.org/2001/sw/wiki/Pellet.

[13] https://www.w3.org/2001/sw/wiki/Eyeball.

[14] https://www.w3.org/RDF/Validator/.

[15] https://www.google.de/shopping.

[16] http://www.ebusiness-unibw.org/ontologies/cpi/evaluation.html.

Produktinformation	Pflege & Material	Maße
• Verschluss: Knöpfe		• Unser Model (186cm, 103-81-95, Konfektionsgröße
• Gürtelschlaufen: ja	⌷ ★ ⊡ ⊟ ℗	50) trägt auf diesem Bild Größe W30/L32
• Tasche(n): 5-Pocket		
• Patch(es): Leder	• 99% Baumwolle	Hinweis: enthält nichttextile Teile tierischen Ursprungs.
• Detail(s): Zier- und Teilungsnähte	• 1% Elasthan	
• Passform: Regular fit		

Fig. 1. Jeans "501" - Regular fit - in Dunkelblau by Levi's, offered by Limango Outlet

Question 1: Return all white T-Shirts with size L.

```
SELECT ?product WHERE {
  ?product a cpi:GarmentOrClothing ;
    schema:additionalType
<http://www.productontology.org/id/T-Shirt> ;
    schema:color "white" ;
    cpi:clothingSize [ a cpi:ClothingSize; schema:name
"L" ] .
}
```

Question 2: What clothing or accessories products (explicitly) contain at least 60% cotton and are eligible for boiling wash and hot ironing?

```
SELECT ?product WHERE {
  ?product a ?type ;
    schema:material [ a cpi:MaterialComponent ;
      schema:material
<http://dbpedia.org/resource/Cotton> ;
      schema:quantityOfMaterial [ a sche-
ma:QuantitativeValue ;
        schema:value ?percentage ;
        schema:unitCode "P1" # percentage in UN/CEFACT
      ]
    ] ;
    cpi:careInstruction cpi:BoilingWash,
cpi:HighTemperatureIron .
  VALUES ?type {cpi:ClothingAndAccessories
cpi:GarmentOrClothing cpi:Footwear}
  FILTER(?percentage >= 60)
}
```

Question 3: Which entities for children clothing do exist, where do they come from, and what certification do they have?

```
SELECT ?product ?country ?certification WHERE {
  ?product a cpi:GarmentOrClothing ;
    cpi:designatedFor "children" ;
    cpi:countryOfOrigin ?country ;
    cpi:certified ?certification .
}
```

The recall of such queries might be limited in certain scenarios due to differing modeling patterns caused by the freedom of the data publisher to choose between entities and textual labels (e.g., material with a range of either *schema:Product*, Text, or URL).

5   Discussion and Conclusion

5.1   Related Work

As part of a project towards supporting the traceability of textile products in supply chains with Semantic Web technologies, a product ontology was specified that also models objects of the textile industry, including "fabrics, parts, dyes, yarns, mixing lots" [26]. The ontology utilizes subsumption hierarchies and also allows the modeling of certificates [26]. A more comprehensive, modular ontology for the whole fashion, textile, and clothing domain was suggested in [27].

In an attempt to represent knowledge about people and garments, a fashion advice ontology (Servive Fashion Ontology) was proposed and encoded in the Web Ontology Language OWL [28]. Its objective was to create personalized style recommendations by fitting garments to human profiles based on body measurements or facial features. The ontology comprises information about materials, garment features, and colors. The work in [29] proposes another approach to determine clothing similarities for fashion recommendation, with an ontology that covers clothing features as well as types of clothing by defining a taxonomy.

In the context of the OPDM project, a set of vertical product ontologies[17] as extensions of the GoodRelations ontology for e-commerce [21] were built, among others specific vocabularies for garments[18] and for shoes[19].

An approach similar to ours, namely to develop a clothing extension for schema. org, was proposed more recently by [30]. In contrast to many other approaches this proposal focuses less on the physical dimensions of clothing, but rather on subjective

---

[17] http://www.ebusiness-unibw.org/ontologies/opdm/.

[18] http://www.ebusiness-unibw.org/ontologies/opdm/garment.html.

[19] http://www.ebusiness-unibw.org/ontologies/opdm/shoe.html.

features of fashion goods through a subjective influence network model [30]. Finally, an ontology for garments was defined from a completely different angle, namely for the simulation of virtual dressing of garments [31].

5.2 Our Approach

In this paper, we have proposed a Web vocabulary for the domain of garments that is compatible with schema.org. It reuses large portions of schema.org and its light-weight extensions aim to be easy to grasp for the majority of stakeholders on the Web. As our Web schema aims to be compatible with schema.org and their intended user groups, our vocabulary was designed in a way that data publishers can pass the processing of information towards data consumers. E.g., if some information cannot be easily supplied by the data publisher (e.g. due to lack of granularity at the data source), it just suffices to attach the data with a textual description and let the consumer take care of lifting the data into the desired target structure.

As compared to existing ontologies for the textile domain like the OPDM garment ontology, our ontology is more designed towards the schema.org standard, but it also addresses clothing information often left out by other approaches, namely certifications and textile care recommendations. Furthermore, it is possible with our ontology to express the proportion of certain fabric materials contained in the clothing. Another unique characteristic is our reliance on and reuse of already existing properties and types of schema.org, the Product Types Ontology, and DBPedia.

5.3 Future Work

There are certainly some aspects and properties that have not yet sufficiently been addressed by our modeling proposal of garments, e.g. the degree of chemical substances in the clothing (cf. [32]). However, this aspect could intuitively be solved by the same modeling pattern that we used for materials.

In the course of this work, we gave a proposal on modeling garment sizes without attaching value to dimensions such as sleeve length, outer leg length, etc. While our recommendation for schema.org was to rely on standard measures with value references to this kind of quantitative dimensions, a vertical extension for garments could well complement them with specific properties. Yet another extension possibility, which we missed out in the context of this paper, is to provide information on the type of composition for materials, namely whether they were laminated, sealed, patched, etc., which would serve to testify the quality of clothing.

In summary, let us give a more prospective outlook for garments on the Web. We anticipate use cases like smart washing machines that can read RFID-tags patched into clothing and download care instructions from the manufacturer pages. Such care instructions can be populated in Web pages using schema.org and our garment extension.

References

1. KPMG: Fashion 2025: Studie zur Zukunft des Fashion-Markts in Deutschland (2015)
2. Nielsen: Share of internet users who have ever purchased products online as of November 2016, by Category. https://www.statista.com/statistics/276846/reach-of-top-online-retail-categories-worldwide/
3. Eurostat: E-commerce statistics for individuals: about two thirds of internet users in the EU shopped online in 2016. http://ec.europa.eu/eurostat/statistics-explained/index.php/E-commerce_statistics_for_individuals
4. eMarketer Inc.: Retail ecommerce set to keep a strong pace through 2017. https://www.emarketer.com/Article/Retail-Ecommerce-Set-Keep-Strong-Pace-Through-2017/1009836
5. Studer, R., Benjamins, R., Fensel, D.: Knowledge engineering: principles and methods. Data Knowl. Eng. **25**, 161–197 (1998)
6. Olivé, A.: Conceptual Modeling of Information Systems. Springer, Heidelberg (2007). doi:10.1007/978-3-540-39390-0
7. Schema.org: Organization of Schemas. http://schema.org/docs/schemas.html
8. Hepp, M.: From ontologies to web ontologies: lessons learned from conceptual modeling for the WWW (on Vimeo). https://vimeo.com/51152934
9. Stolz, A., Rodriguez-Castro, B., Hepp, M.: Using BMEcat catalogs as a lever for product master data on the semantic web. In: Cimiano, P., Corcho, O., Presutti, V., Hollink, L., Rudolph, S. (eds.) ESWC 2013. LNCS, vol. 7882, pp. 623–638. Springer, Heidelberg (2013). doi:10.1007/978-3-642-38288-8_42
10. Bogusławska-Bączek, M.: Analysis of the contemporary problem of garment sizes. In: 7th International Conference on Textile Science (TEXSCI 2010), Liberec, Czech Republic (2010)
11. Fryar, C.D., Gu, Q., Ogden, C.L.: Anthropometric reference data for children and adults: United States, 2007–2010. Vital Health Stat. **11**, 1–48 (2012)
12. Beazley, A.: Size and fit: the development of size charts for clothing - Part 3. J. Fash. Mark. Manag. Int. J. **3**, 66–84 (1999)
13. Cherif, C.: The textile process chain and classification of textile semi-finished products. In: Cherif, C. (ed.) Textile Materials for Lightweight Constructions, pp. 9–35. Springer, Heidelberg (2016). doi:10.1007/978-3-662-46341-3_2
14. Ginetex Switzerland: Textile care symbols. http://www.ginetex.net/files/pdf/gin_pfle_bro_ch_gb_web_rz.pdf
15. DIN Deutsches Institut für Normung: Textiles - Care labelling code using symbols (ISO 3758:2012); German version EN ISO 3758:2012 (2013)
16. ASTM D5489-14: Standard Guide for Care Symbols for Care Instructions on Textile Products, West Conshohocken, PA (2014)
17. Moore, S.B., Wentz, M.: Eco-labeling for textiles and apparel. In: Blackburn, R.S. (ed.) Sustainable Textiles: Life Cycle and Environmental Impact, pp. 214–230. Woodhead Publishing Limited, Oxford (2009)
18. Simperl, E., Tempich, C.: Exploring the economical aspects of ontology engineering. In: Staab, S., Studer, R. (eds.) Handbook on Ontologies. IHIS, pp. 337–358. Springer, Heidelberg (2009). doi:10.1007/978-3-540-92673-3_15
19. Uschold, M., Gruninger, M.: Ontologies: principles, methods and applications. Knowl. Eng. Rev. **11**, 93–136 (1996)
20. GS1 Germany: GS1-Standards für Fashion, Schuhe, Sport. https://www.gs1-germany.de/gs1-standards-fuer-fashion-schuhe-sport/

21. Hepp, M.: GoodRelations: an ontology for describing products and services offers on the web. In: Gangemi, A., Euzenat, J. (eds.) EKAW 2008. LNCS, vol. 5268, pp. 329–346. Springer, Heidelberg (2008). doi:10.1007/978-3-540-87696-0_29
22. United Nations Economic Commission for Europe: Recommendation No. 20: Codes for Units of Measure Used in International Trade (2006)
23. DIN Deutsches Institut für Normung: Size designation of clothes - Part 3: Body measurements and intervals; German version EN 13402-3:2013 (2014)
24. Harris, S., Seaborne, A.: SPARQL 1.1 Query Language. http://www.w3.org/TR/2013/REC-sparql11-query-20130321/
25. International Organization for Standardization: ISO 3166-1:2013: Codes for the Representation of Names of Countries and Their Subdivisions - Part 1: Country Codes (2013)
26. Alves, B., et al.: Fairtrace: applying semantic web tools and techniques to the textile traceability. In: Hammoudi, S., Cordeiro, J., Maciaszek, Leszek A., Filipe, J. (eds.) ICEIS 2013. LNBIP, vol. 190, pp. 68–84. Springer, Cham (2014). doi:10.1007/978-3-319-09492-2_5
27. Aimé, X., George, S., Hornung, J.: VetiVoc: a modular ontology for the fashion, textile and clothing domain. Appl. Ontol. 11, 1–28 (2016)
28. Vogiatzis, D., Pierrakos, D., Paliouras, G., Jenkyn-Jones, S., Possen, B.J.H.H.A.: Expert and community based style advice. Expert Syst. Appl. 39, 10647–10655 (2012)
29. Frejlichowski, D., Czapiewski, P., Hofman, R.: Finding similar clothes based on semantic description for the purpose of fashion recommender system. In: Nguyen, N.T., Trawiński, B., Fujita, H., Hong, T.-P. (eds.) ACIIDS 2016. LNCS, vol. 9621, pp. 13–22. Springer, Heidelberg (2016). doi:10.1007/978-3-662-49381-6_2
30. Bollacker, K., Díaz-Rodríguez, N., Li, X.: Beyond clothing ontologies: modeling fashion with subjective influence networks. In: Raykar, V.C., Klingenberg, B., Xu, H., Singh, R., Saha, A. (eds.) Machine Learning Meets Fashion KDD Workshop, pp. 1–7. ACM, San Francisco (2016)
31. Fuhrmann, A., Groß, C., Weber, A.: Ontologies for virtual garments. In: Proceedings of the Workshop Towards Semantic Virtual Environments (SVE 2005), Villars, Switzerland, pp. 101–109 (2005)
32. Kidmose Rytz, B., Sylvest, J., Brown, A.: Study on Labelling of Textile Products (2010)

Semantic Modeling and Inference with Episodic Organization for Managing Personal Digital Traces
(Short Paper)

Varvara Kalokyri, Alexander Borgida$^{(\boxtimes)}$, Amélie Marian, and Daniela Vianna

Department of Computer Science, Rutgers University, New Brunswick, NJ 08903, USA
{v.kalokyri,borgida,amelie,dvianna}@cs.rutgers.edu

Abstract. Many individuals generate a flood of personal digital traces (e.g., emails, social media posts, web searches, calendars) as a byproduct of their daily activities. To facilitate querying and to support natural retrospective and prospective memory of these, a key problem is to integrate them in some sensible manner. For this purpose, based on research in the cognitive sciences, we propose a conceptual modeling language whose novel features include (i) the super-properties "who, what, when, where, why, how" applied uniformly to both documents and autobiographic events; and (ii) the ability to describe prototypical plans ("scripts") for common everyday events, which in fact generate personal digital documents as traces. The scripts and wh-questions support the hierarchical organization and abstraction of the original data, thus helping end-users query it. We illustrate the use of our language through examples, provide formal semantics, and present an algorithm to recognize script instances.

Keywords: Personal digital traces · Conceptual model · Scripts · Plan recognition

1 Introduction

Our modern lives produce digital traces consisting of "personal digital documents" (PDDs) resulting from sources such as email, messaging, calendars, social media posts, web searches, purchase histories, GPS location data, etc.[1]

Our goal is to use this information to help users supplement their retrospective and prospective memory, as envisioned in the "personal memex" of Vannevar Bush [4]. In addition to the problem of user interfaces for such systems, the key difficulty is integrating the heterogeneous and disparate kinds of PDDs.

Traditional research on Personal Information Management (PIM) has been "object centric", in the sense that the programs were intended to identify objects and represent semantic relationships between them, so that "finding" was supported by associative search. We believe that this is in part because of the field's origins lie in supporting office work, where finding files and other office objects was the standard use-case.

In contrast, a core feature of our work is the existence and exploitation of PDDs from a wider variety of sources. Each individual information source may have its own

---

[1] We immediately acknowledge the sensitive nature of this information, and the very important privacy issues that they raise.

H. Panetto et al. (Eds.): OTM 2017 Conferences, Part II, LNCS 10574, pp. 273–280, 2017.
https://doi.org/10.1007/978-3-319-69459-7_19

natural (semi)structure (e.g., *from, to, date, subject, body* for an e-mail). What is needed is some way to integrate these "document schemas". One hypothesis embodied in our proposed conceptual model is that most information about a PDD can be fitted as answers to the questions *who, what, when, where, why* and *how* (the *w5h* questions), thereby providing a way of correlating the information on different kinds documents in a manner that is natural to humans. For example, the *who* property of an email include the sender and all the recipients, while the *who* property of a Facebook messenger/Google Hangouts conversation include the creator and the participants. Given this integration, one can provide relatively simple keyword search support along the "*w5h* dimensions".

A more significant novelty of our proposal, compared to traditional PIM, is motivated by the cognitive science literature (e.g., [15,16]), which shows that the intended use-cases are closely related to enhancing the user's *autobiographic memory*. This memory is centrally concerned with the *events* in one's life, which provide a narrative that connects the PDDs. For example, some emails concerning dinner, a confirmation of an OpenTable reservation, a Lyft receipt, and a credit-card payment, make much more sense as part of an episode of going out to dinner, *if* they have similar *when, where,* and *who* dimension values. Therefore, another hypothesis our work is that in developing a conceptual model for PDDs, one must make equal room for the modeling of events, both atomic actions and complex events. For the latter, we were inspired by the idea of *scripts* introduced by Schank and Abelson [13] for language understanding. These are stereotypical *plans* for common situations. Our current language for describing plans is based on Hierarchical Task Networks [7].

This paper illustrates the use of our proposed language through examples, provides a syntax and a formal semantics, and presents an algorithm to recognize script instances. It is a companion to the workshop paper [9] discussing some empirical results.

2 Related Work

We briefly mention here some of the many areas of related work.

The case for a unified logical data model for personal information has been made repeatedly in *PIM*, as has the use of ontologies (e.g., [11]) and semantic models like RDF(S) (e.g., [10]).

Since we are interested in representing (autobiographic) events and their instances, there is a vast literature on composite process and event representation spread across a wide variety of areas.

First, there are numerous formal process representation languages including program logics. Then there are the many graphical notations for describing real-world processes, such as Petri Nets, workflows and business process notations. These are all "prescriptive" in nature, while we are interested in "descriptive" formalisms that allow us to *recognize* script instances.

This leads us to several relatively closely related areas: *Activities of Daily Living, Ambient Intelligence, Behavior Recognition* and *LifeLogging*. A few relevant surveys are [2,12]. These areas separate two aspects: (i) the segmentation and recognition of atomic actions from (continuous) sensor/video data; (ii) the recognition of complex events composed of atomic ones.

A significant number of the approaches for complex event recognition are founded on probabilistic techniques, which rely on Machine Learning of process schemas followed by probabilistic inference. While data sets for this are easily generated for sensors, they are much harder to obtain in our case because of privacy issues, and because (as we shall see) personalized variants predominate [9].

There is extensive literature in the field of AI on *plan recognition*. A snapshot of this appears in [8], and references to it. As with complex event recognition, probability-based approaches using learning are frequent. We seem to have more in common with approaches that use plan libraries for recognition. An interesting approach is the work of Geib et al. [7], which is based on parsing hierarchical task networks (see Sect. 4), yet yields probabilistic results through "model counting".

The most important difference of our use cases from all the above approaches is that most of the digital traces we see are not part of any script, and a very large fraction of the plan steps in any particular instantiation of a script leave no trace ("missing actions")[2].

```
class DOCUMENT is a ENTITY {          class EMAIL is a DOCUMENT {
    hasPart :  set of ENTITY;             features:
    who :  set of PERSON                      threadId : STRING;
    what < hasPart: set of DOCUMENT;      properties:
    when :  set of TIME;                      from < who : PERSON;
    where :  set of LOCATION;                 to < who :  set of PERSON;
    why :  set of GOAL;        }              . . .
                                          actions
class SEND  is a ACTION {                      send : SEND
    sender < who: PERSON;                      forward : FORWARD
    recipients<who: set of PERSON;        constraints
    theme < what: DOCUMENT                     from = send.sender ;
    whenSent < when : TIME }                   send.whenSent < when;
                                              . . . }
```

Fig. 1. Specification of SEND and EMAIL classes

3   A Conceptual Model for Entities and Atomic Actions

Real-world entities. We start from a standard object-centered conceptual modeling language, whose fundamental notions include *individuals* (e.g., Calvin) that are related by *binary properties* (e.g., hasFriends). Individuals are grouped as instances of *classes* (e.g., PERSON). Classes specify restrictions on the range of values that properties can take for their instances, and at least whether they are functional (**set of** indicates that there is no upper bound on the number of fillers). Classes can be specialized into subclasses (e.g., RETIRED is a subclass of PERSON), during which new properties may be added, or existing properties may be restricted. Most importantly, properties can also be specialized into sub-properties (e.g., hasCloseFriends is a sub-property of hasFriends).

---

[2] The case study in [9] showed that 194 out of 316 episodes of eating out (61%) had *a single* PDD, corresponding to a single action in the plan associated with them.

We will use the notation `hasCloseFriends` < `hasFriends` to indicate such specialization relationships between properties.

Actions. We have argued that a key part of the conceptual model for PIM are events. We focus here on modeling primitive/atomic events, which we call *actions*. The most important part of describing actions is presenting their participants, the roles they play and restrictions on them. We illustrate this in Fig. 1, where the description of SEND includes properties for the important participants. So `sender` < `who` : PERSON is interpreted as saying that `sender` takes as value a PERSON instance, and is a sub-property of the `who` property.

Personal documents: We want to model PDDs in such a manner that *w5h* provide a unifying framework. For example, we will want to describe emails or reservations. Note however that there is no natural way to answer questions like "when?" or "who?" of such objects. (This is even more evident in the case of physical objects, like chairs say.) But if the object participates in an action, it can "derive" its *when* from the action. Thus one can ask when a message was sent, ... Therefore the model needs to express the natural properties of the PDD, connect to the actions involving it, and then assert the relation between their respective properties. One way to do this, inspired by our work on service description [3], is illustrated in Fig. 1, where EMAIL is connected to the SEND action through the `send` property, and then **constraints** equate the `to` property of the email to the `send.sender` property path, which passes thru the SEND action. This equation can be used to *infer* one path value when the other one is known.

We can now re-state our original hypothesis: a large subset of the properties of PDDs as well as of the actions, can be usefully viewed as specializations of *w5h* (*who, what, where, when, why, how*), when these are viewed as properties themselves. The result is the principled integration of heterogeneous object schemas we suggested.

Semantics and Inference. The above notation can be easily captured in UML, extended to support association hierarchies, which in turn can be translated to description logics (DLs) such as OWL[3]. (See [1] for an example translation.) DLs have precise formal semantics, and in fact languages such as OWL can express many more constraints, if needed. Several standard inferences are defined based on this semantics, and are supported by standard implementations of OWL, including *class inconsistency* (Is the specification of a class inconsistent, in the sense that it can never have any instances?) and *instance recognition* (Is an individual, with (partially) specified properties, necessarily an instance of some given class?).

One problematic feature of our language are **constraints**, called "complex role inclusions" in the DL literature. The general form we desire ($p_1 \sim q_1.q_2.\cdots.q_n$ where $\sim$ is $=, \sqsubseteq, \sqsupseteq$) is known to lead to undecidability of reasoning in most DLs. So such constraints can only be used to propagate information, using (epistemic) rules.

4   Conceptual Model for scripts

Our aim in using script instances is to *organize* PDDs, abstract out relevant information from them, and help humans access and make sense of episodes in their lives.

---

[3] https://www.w3.org/TR/owl2-overview/.

Therefore, relevant aspects of scripts include: (i) their goal (for purposes of human understanding; see [13, 16]); (ii) summary information of the participants in the plan, and other descriptive properties, especially *w5h* aspects; (iii) the hierarchical decomposition into sub-scripts and primitive actions (together with restrictions on their ordering), which describe how the script plan achieves its goal.

Our system will start with a a library of common, everyday scripts. Since this plan library will not be used to construct plans, only to recognize instances that have been enacted, we can consider only *plan skeletons*, which ignore issues like pre- and post-conditions for performing actions. Also, since our application for organizing PDDs does not require perfect execution of all episodes, the plans we describe can be *stereotypical*.

We start here from Hierarchical Task Networks (HTNs) [5], which are a classical expressive notation in AI for planning and plan recognition. To describe the "body" of a plan, which determines the set of valid atomic action sequences, a non-atomic script/goal T can be refined in one of two ways:

$T := (\textbf{Or } S_1, ..., S_n)$ T can be accomplished by achieving *one* of $S_1, ..., S_n$;
$T := (\textbf{And } S_1, ..., S_n(P))$ T can be accomplished by achieving *all* of $S_1, ..., S_n$, subject to the precedence constraints $i \prec j$ in P, requiring S_i to end before the start of S_j.

The following frequently occurring patterns that can be expanded into the above **And** and **Or** constructs with the use of the **no_op** action:

sequencing	$S_1 ; S_2$	$(\textbf{And } S_1, S_2 (\{1 \prec 2\}))$
iteration	$(\textbf{Loop } S)$	$T := (\textbf{Or } (S ; T) , \textbf{no\_op})$
optionality	$(\textbf{Optional } S)$	$(\textbf{Or } S, \textbf{no\_op})$

Future work concerns the addition of concurrency to the language.

Scripts also have properties, analogous to those of ordinary atomic actions and reminiscent of parameters/local variables in programs; and **constraints** relating these to the properties of their sub-scripts or actions mentioned in the body.

We illustrate these ideas by referring to Fig. 2, which describes the Eating_Out ("going out to eat") script.

```
class Eating_Out is a SCRIPT            (Optional MakeRstReservation);
locals:                                     AttendEatingOutEvent
whoAttended < who: set of PERSON        } // end Eating_Out
whereEating < where : EATERY
whenEating < when : TIME                class AttendEatingOut is a SCRIPT{
whatEaten < what : set of FOODS            body
purpose < why : GOAL                    GetToEatery ;
body                                     CheckIn ;
  InitiateGoingOut ;                     (Or    (OrderFood ; BeServed),
  (Loop ( Or DiscussWhenToEat,              SelfServeFood ) ;
  DiscussWhoWillEat,                     Eat ; Pay ; LeaveEatery      }
  DiscussWhereToEat ) )
```

Fig. 2. Definition of Eating_Out script and a sub-script

First, the values of whoAttended, whenEating, etc. describe and identify each script instance. Gathering the information into these properties provides the kind of higher level aggregation/organization of information in PDDs that we were looking for. Note that these properties are also organized along the *w5h* dimensions.

The body essentially describes the subgoals of the script plan. In this case, after an action initiating the idea of going out on this occasion, there are discussions about when, who, where (and hence what) to eat, which can be carried out in any order. Deciding when to eat in turn can be modeled by a script which shows exchanges of suggestions and discussions until agreement is reached.

As with regular classes, constraints like AttendEatingOut.whatEaten = what-Eaten, can be asserted to propagate information between a script and its sub-scripts, if we assume each subscript is uniquely named.

An instance of a script will have fillers for local properties, and an associated partially ordered set of sub-script and atomic action instances, which conform to the **body**. The major difference is that some participants and sub-scripts might be missing (because we may lack evidence for them in the form of PDDs).

Semantics. In the absence of conditions on states, the formal semantics of HTN interprets the above as a context-free-like grammar describing valid sequences of atomic tasks. An OR-decomposition corresponds to grammar rules $T \leftarrow S_1, T \leftarrow S_2, ...$. An AND-decomposition, when P is empty, corresponds to (exponentially many) rules for all permutations of $S_1, ..., S_n$. The above notation is powerful, because the task names can be used recursively, as in $S := (\textbf{And}\ (a, S, b), \{1 \prec 2, 2 \prec 3\})$.

5   Recognizing Script Instances from Documents

Let us review the context. We start with a model of the domain – classes for PDDs and scripts we are interested in. We then collect a large database of individual instances of PDD classes (the majority unrelated to any scripts). Our objective is to create episodes (instances of scripts) that the user was involved in, and relate them to documents. Algorithm 1 describes the steps for recognizing instances of script type $\mathscr{S}$.

We give further details of the algorithm steps, with reference to $\mathscr{S}$ = Eating_Out.

Algorithm 1. Algorithm for constructing instances of script class $\mathscr{S}$

1: $D :=$ documents indicating any potential instance of script class $\mathscr{S}$;
2: *Candidates* := $\emptyset$;
3: for all $d \in D$ do {
4:     *Candidates* += new instance c_d of script class $\mathscr{S}$, based on d;
5:     rate the strength of evidence for c_d;   }
6: repeat until no changes in *Candidates* {
7:     *MergeSet* := { $d \in Candidates$ such that there is sufficient corroboration that they refer
8:                 to the same real-world event };
9:     *Candidates* := (*Candidates* − *MergeSet*) $\cup$ {d':=combine(*MergeSet*)};
10:    rate the strength of evidence for d';
11:    use details of script $\mathscr{S}$ to look for additional documents that could be relevant to d';   }

Step 1: Retrieving documents D: First, create a list L of "trigger words/phrases", whose occurrence indicates that a document has something to do with an instance of script class $\mathscr{S}$. To find these words, look for goal sub-script(s)/action(s), and identify verbs that indicate an occurrence of it. E.g., for Eating_Out, the goal is AttendEating-Out, and indicative verbs are "eat" or "eat out". Next, generate a list of synonyms and hyponyms based on these verbs. To make this process replicable, we have used standard sources of synonyms and hyponyms like WordNet. In addition, one must also consider the *w5h* participants of these events by using resources like FrameNet [6], and again generate synonyms and hyponyms. For example, "restaurant" is a discriminating *where* value of "eat", which should be included in the search term list. The final list includes "breakfast", "lunch", "dinner", and "restaurant", plus hyponyms. A set of documents D is retrieved by searching for these terms, and then preprocessing them by (i) explicating/disambiguating information (e.g. terms like "today" or "Tuesday" are made absolute dates), (ii) performing entity resolution, (iii) grouping certain kinds of documents (e.g., related email threads/tweets).

Steps 4&5: Creating initial script instances c_d. Each retrieved document d results in a candidate instance c_d of $\mathscr{S}$, with some of its (sub)properties filled based on the document's *w5h* properties. For example, a restaurant credit card charge provides evidence for the *attendEatingOut* sub-script, together with information on its *when/whereEating- Occurred*, and one *whoAttended* value (the cardholder). We then assign a score $Score(c_d)$ to c_d based on the strength of the evidence that d manifests. This strength is based on the document type (e.g. a restaurant credit card charge is *stronger* evidence than an email), the location of keywords (e.g. in the subject of an email rather than body), or of the originator (e.g., the user being the sender rather than recipient).

Steps 7-10: Growing script instances from MergeSet. In order to combine multiple sources of evidence for the same script instance, $\mathscr{S}$ needs to specify *"keys"*: a rating of how *w5h* (sub)properties help identify instances. For the Eating_Out case, important keys are *when/whereEatingOccured* and, to a lesser extent, *who*. Each key-property can be assessed for similarity (e.g. time difference for *when*). Once two instances of $\mathscr{S}$ are judged sufficiently similar, they become merge candidates, and are combined by unioning their property fillers. The score for the merged instance is $1 - \prod_{s \in MergeSet}(1 - Score(s))$. This formula is Hooper's rule [14] for combining probabilistic evidence.

6 Summary

This paper addressed the problem of managing a database of personal digital traces. It introduced a semantic modeling language for entities, also used to describe digital documents and related activities. The novel feature of this language is organizing many object properties into hierarchies with *w5h* questions at the top. These help organize and unify the many heterogeneous data. The language was extended to support the representation of scripts, which are stereotypical plans with a mereological hierarchical structure — an idea motivated by research in the cognitive sciences [13, 16]. The *w5h* organization was continued. Script instances connect the PDDs generated by actions into meaningful episodes, and extract from them relevant summary information, in order to reconstruct autobiographic memories.

Instance recognition for scripts, in contrast to plans and complex events, is complicated by several factors: (1) The evidence for the occurrence of atomic actions in the form of PDDs is highly uncertain (in principle, an email can describe *any* task in the world!). (2) Most PDDs we collect are unrelated to any of the everyday scripts we foresee having in the library, so these must be ignored. (3) Most of the steps in any instantiation of a script do not leave digital traces. For this reason, the paper proposed a novel heuristic algorithm for recognizing the instances of a script, which was based on retrieving PDDs that contained systematically chosen keywords, and merging candidate instances.

A small case study involving the Eating_Out script [9] gave instance recognition precision ranging from 0.32% to 0.75% per user[4], showing that there are major differences between subjects and how they generate PDDs.

References

1. Berardi, D., Calvanese, D., De Giacomo, G.: Reasoning on UML class diagrams. Artif. Intell. **168**(1), 70–118 (2005)
2. Bikakis, A., Patkos, T., Antoniou, G., Plexousakis, D.: A survey of semantics-based approaches for context reasoning in ambient intelligence. In: Mühlhäuser, M., Ferscha, A., Aitenbichler, E. (eds.) AmI 2007. CCIS, vol. 11, pp. 14–23. Springer, Heidelberg (2008). doi:10.1007/978-3-540-85379-4_3
3. Borgida, A., Devanbu, P.T.: Adding more "DL" to IDL: towards more knowledgeable component inter-operability. In: Proceedings of ICSE 1999, pp. 378–387 (1999)
4. Bush, V.: As We May Think. The Atlantic Monthly, July 1945
5. Erol, K., Hendler, J., Nau, D.S.: HTN planning: complexity and expressivity. In: AAAI 1994, pp. 1123–1128 (1994)
6. Fillmore, C.J., Johnson, C.R., Petruck, M.R.: Background to framenet. Int. J. Lexicogr. **16**(3), 235–250 (2003)
7. Geib, C.W., Goldman, R.P.: A probabilistic plan recognition algorithm based on plan tree grammars. Artif. Intell. **173**(11), 1101–1132 (2009)
8. Goldman, R.P., Geib, C.W., Kautz, H.A., Asfour, T.: Plan recognition (dagstuhl seminar 11141). Dagstuhl Rep. **1**(4), 1–22 (2011)
9. Kalokyri, V., Borgida, A., Marian, A., Vianna, D.: Integration and exploration of connected personal digital traces. In: Proceedings of ExploreDB 2017 Workshop, pp. 1–6. ACM (2017)
10. Karger, D., et al.: Haystack: a general-purpose information management tool for end users based on semistructured data. In: Proceedings of CIDR 2005, pp. 13–26 (2005)
11. Katifori, V., Poggi, A., Scannapieco, M., Catarci, T., Ioannidis, Y.: OntoPIM: how to rely on a personal ontology for Personal Information Management. In: ISWC 2005 Workshop on The Semantic Desktop, pp. 258–262 (2005)
12. Rodríguez, N.D., Cuéllar, M.P., Lilius, J., Calvo-Flores, M.D.: A survey on ontologies for human behavior recognition. ACM Comput. Surv. **46**(4), 43 (2014)
13. Schank, R., Abelson, R.: Scripts, plans, and knowledge. In: IJCAI 1975, pp. 151–157 (1975)
14. Shafer, G.: The combination of evidence. Int. J. Intell. Syst. **1**(3), 155–179 (1986)
15. Tulving, E.: Episodic memory: from mind to brain. Annu. Rev. Psych. **53**, 1–25 (2002)
16. Williams, H.L., Conway, M.A., Cohen, G.: Autobiographical memory. In: Cohen, G., Conway, M.A. (eds.) Memory in The Real World, chap. 3, pp. 21–90. Psychology Press (2008)

---

[4] Low-end scores were due to factors such as absence of NLP and a couple sharing credit cards.

Linked Open Data for Linguists: Publishing the Hartmann von Aue-Portal in RDF

Alex Stolz[1(✉)], Martin Hepp[1], and Roy A. Boggs[2]

[1] Universität der Bundeswehr München, 85579 Neubiberg, Germany
{alex.stolz,martin.hepp}@unibw.de
[2] Florida Gulf Coast University, 10501 FGCU Blvd. South,
Fort Myers, FL, USA
rboggs@fgcu.edu

Abstract. The Hartmann von Aue-portal is a decade-long initiative to employ Web technology in order to support the study of the early German. It provides a comprehensive knowledge base on lexicographic and other aspects of the works of Hartmann von Aue, one of the key epic poets of Middle High German literature; namely lemmata, word forms, tagmemes, adverbs, and the like, including original contexts for entries. The portal is available for human users in the form of a Web application. Linked Open Data (LOD) is a recent approach in the evolution of Web technology that supports the publication of information on the Web in a way suitable for the intelligent consumption and processing of contents by *computers* instead of humans using Web browsers. In this paper, we study the use of modern LOD approaches for linguistics, describe the conversion of the complete Hartmann von Aue-portal into LOD, and show the usage for data-driven analyses via SPARQL queries and literate programming with Python.

Keywords: Web portal · Concordance · Computational linguistics · Middle High German · Poetry · Hartmann · LOD · RDF · SPARQL · Data-driven analysis · Python

1 Introduction

Hartmann von Aue was one of the key epic poets of Middle High German literature and his works are still studied by many students and researchers these days. The Hartmann von Aue-Portal[1] is a decade-long initiative to employ Web technology in order to support the study of Hartmann's poems in particular, and the early German in general. For this it provides a comprehensive knowledge base on lexicographic and other aspects related to Hartmann's works, including links to manuscripts, transcriptions, grammar and context equivalents, context dictionary, reverse word form lists, rhyme listing, and name register [1]. However, the online portal has a number of limitations that restrict the possibilities of linguistic research: (1) The data is only accessible by humans through a Web browser, (2) the databases are insufficiently replicated, i.e. the service has a single point of failure, and (3) the data cannot be easily combined with

---

[1] http://hvauep.uni-trier.de/ and http://www.fgcu.edu/rboggs/hartmann/HvAmain/HvAhome.asp.

© Springer International Publishing AG 2017
H. Panetto et al. (Eds.): OTM 2017 Conferences, Part II, LNCS 10574, pp. 281–299, 2017.
https://doi.org/10.1007/978-3-319-69459-7_20

other information for more sophisticated use. For this reason, the wealth of data available in the Hartmann von Aue-portal is practically inaccessible for modern methods of computational linguistics. In this paper, we describe an approach to convert the complete Hartmann von Aue-portal into Linked Open Data (e.g. [2, 3]) and thereby demonstrate how we can overcome those constraints.

1.1   Context

The interest in computational linguistics has recently experienced a renaissance due to enhanced (parallel) computing power, increased availability of data (texts), and sophisticated machine learning algorithms driven by the research efforts of huge companies such as IBM, Google, Apple, or Amazon. Nowadays, in the field of computational linguistics, applications like speech recognition and question answering are among the most prominent use cases [4].

Linked Open Data (LOD) [2] is a recent approach in the evolution of Web technology that supports the publication of information on the Web in a way suitable for the intelligent consumption and processing of contents by computers instead of humans using Web browsers. It has so far been adopted by a multitude of branches like governments, life sciences, media, social networks, geography, and linguistics.

The Hartmann von Aue-portal is a decade-long initiative by Roy A. Boggs and Kurt Gärtner. The portal provides Web access to a comprehensive knowledge base about the works of the epic poet Hartmann von Aue. It was mainly designed and developed for the purpose to support scholars in conducting Middle High German literature studies.

1.2   Problem Statement

The fact that the Hartmann von Aue-portal is openly available on the Web makes it an extremely useful information source for medieval literature studies. The portal conveniently collates and presents information related to works by Hartmann von Aue and thus improves accessibility. Nonetheless, a fundamental requirement that is currently missing is to make the valuable data in the portal more sustainable, accessible for machines, and reusable. The main challenges are:

1. **Ensure long-term access to the actual data, independent of the future of Web technology.** At the moment, the longstanding effort to create the portal depends on the benevolence of single organizations hosting and some few individuals running the portal. There is a risk that in the future nobody will feel responsible for keeping up and maintaining the service.
2. **Simplify reuse, in particular for data-driven approaches (data science for linguistics).** The portal is a walled garden with respect to machine-friendliness. All the data inside the portal cannot be easily used by researchers and practitioners for analysis.
3. **Simplify integration with other data sources.** The portal already offers references to external dictionaries. However, to make every data item (e.g. verse or word of a poem) more valuable, it must be possible to create contextual links to and from other data sources on the fly.

This paper addresses these challenges for the Hartmann von Aue-portal by asking the following research questions:

1. How can we adopt Linked Open Data for the Hartmann von Aue-portal to facilitate persistence of its digitized works regardless of organizations and individuals?
2. How can we unlock the data in the knowledge base portal for data-driven research?
3. How can the data be interlinked with other data sources and made accessible for innovative applications?

1.3 Relevance

The Hartmann von Aue-portal has become a focal point for Hartmann research. It started with a first version for "Der Arme Heinrich" [1] and grew to five poems and the lyrics by Hartmann von Aue, including a comprehensive number of lemmata and word forms. In the Mannheim Symposium of 1973, an interest group already met to discuss the issues and future work of computer-assisted studies of early German [5]. Roy A. Boggs shared there his experiences and problems he faced during the forthcoming construction of computer-generated concordances to the works of Hartmann von Aue [5]. In the mid-90s, Boggs and Gärtner prepared a first digital version of the works by Hartmann von Aue for compact discs, which later advanced into today's online portal. As of 2007, many individuals from various universities of seventeen countries had already contributed data and programming expertise to enhance the portal [1].

To get an understanding of the dimension of this project: There are nearly ten thousand lemmata in the whole corpus, and almost 140 thousands of tokens that needed to be lemmatized, translated into English and Modern High German, and assigned a context grammar – and, most importantly, all this was mainly crafted by a few humans.

Besides the considerable investments towards the development of the portal, linguists can expect several benefits from having such a dataset on the Web of Linked Open Data. Chiarcos et al. [6] mention five such benefits, namely structural interoperability, conceptual interoperability, query federation, ecosystem of formalisms and technologies, and linkage of resources from different data providers.

1.4 Organization of the Paper

The rest of this paper is structured as follows: In Sect. 2, we explain the idea of Linked Open Data. Section 3 describes the Hartmann von Aue-portal in greater detail. In Sect. 4, we outline our conversion approach of the Hartmann von Aue-portal into Linked Open Data. Section 5 evaluates our approach and finally, Sect. 6 compares related works with our work, discusses limitations, and concludes with an outlook.

2 Linked Open Data

Linked Open Data (LOD) was first proposed by Tim Berners-Lee, the inventor of the World Wide Web [2] and became very popular over the last ten years as a publishing paradigm for structured data on the Web. The core idea is to create typed links among distributed data sources [3], this way setting up a huge linked data space. In contrast to Linked Data which encompasses arbitrarily licensed data, Linked Open Data focuses on freely available data, i.e. linked data published using liberal licensing models.

Technically speaking, Linked Data is based on existing standards, namely the Hypertext Transfer Protocol (HTTP) for accessibility, Uniform Resource Identifiers (URIs) for identifying resources on the Web, and the Resource Description Framework (RDF) data model [3]. RDF uses triple notation to express binary relationships like *Hartmann von Aue* (subject) **is author of** (predicate) *Der arme Heinrich* (object). Each of the three resources in this triple are identified by HTTP URIs, and the data model can be encoded in various data formats, e.g. RDF/XML, Notation 3, JSON-LD, etc.

Linked Data allows a computer to reliably combine multiple statements into a consolidated graph, thereby gathering context information from various data sources and combining it into a sophisticated knowledge graph. This is what makes the Google Knowledge Graph [7] and related approaches extremely powerful.

In the context of linguistics, examples of popular deployments of lexical data sources as Linked Open Data are

- DBPedia [8] corresponds Wikipedia, the community encyclopedia on the Web, converted into RDF and published as Linked Open Data, and
- WordNet RDF (cf. [9]), a lexical database of English in RDF.

Other useful resources for linguists are consolidated into the so called Linguistic Linked Open Data (LLOD) cloud [10].

3   The Hartmann von Aue-Portal

The Hartmann von Aue-portal represents a comprehensive online collection of works by the medieval German poet Hartmann von Aue. It was designed and implemented by Roy A. Boggs in collaboration with Kurt Gärtner [11]. The project consists of knowledge bases related to the poems "Der arme Heinrich" (ah), "Erec" (er), "Gregorius" (gr), "Iwein" (iw), "Die Klage" (kl), and transcripts. In addition to that, it includes lyrics (ly) by Hartmann von Aue. All data is stored in relational databases and presented via a Web frontend.

3.1   Purpose and Current Status

Hartmann von Aue was "a major player of the first golden age of German literature" [1]. His works have been caught up by various famous German poets (e.g. Gerhart Hauptmann and Thomas Mann) and continue to be popular and widely enjoyed also today [1]. Consequently, the existence of a modern online portal is welcome to ease and foster literature studies about Hartmann von Aue.

The Hartmann von Aue online portal lifts the Hartmann von Aue knowledge bases onto the Web and makes them freely available for everyone. The portal aims to be a single point of reference both for students and scholars to learn about the works by Hartmann von Aue and medieval literature in general. By following a bottom-up approach scholars shall gradually become specialists [1], i.e. by starting one's studies from the critical editions (the transcriptions) and taking advantage of the additional resources provided by the portal.

The portal comprises several components, namely the aforementioned knowledge bases related to Hartmann's poems including links to manuscripts, transcriptions, grammar and context equivalents, context dictionaries, reverse word form lists, rhyme listings, and name registers [1]. The platform is available from different Web servers[2] and implemented in ASP/Access and a more mobile-friendly PHP/MySQL variant.

3.2 Data Model

To trigger a conversion of the Hartmann von Aue-portal into Linked Open Data, it was initially important to understand the underlying data model. In the course of this work, we examined the data model provided by the MS Access database of the ASP-based portal[3] hosted at Florida Gulf Coast University. The data model of the PHP/MySQL version[4] hosted at the University of Trier is conceptually very similar and the general approach should thus be no different.

For every poem, the corresponding data was stored in a relational database (MS Access) file. The common database schema that underlies each of these files is shown in the following entity-relationship (ER) diagram (Fig. 1). The upper part of the diagram represents the concepts specific to knowledge bases of the individual poems. We generalized them to Data, Text, and Notes. The actual naming pattern that was used for the database tables was each combination of {Ah, Er, Gr, Iw, Kl, ly} x {Data, Text, Notes}, i.e. AhData, AhText, AhNotes, ErData, etc.

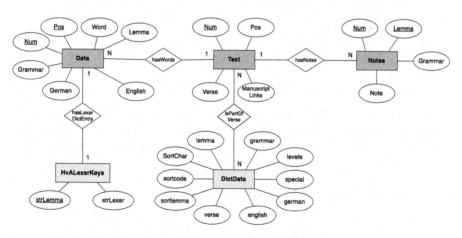

Fig. 1. Database schema of the Hartmann von Aue knowledge base portal

---

The knowledge bases in the Hartmann von Aue-portal are further accompanied by a context dictionary (DictData), where in a single huge database table every lemma is translated into English and Modern High German, and where references of lemmas to the positions of their occurrence within verses are made, also known as concordances.

Finally, a table (MHGDictKeys) was maintained where all lemmata in the texts are mapped into corresponding translations of an external Middle High German dictionary, e.g. the lemma "ritter" in the Hartmann von Aue-portal would link to its corresponding online dictionary entry http://woerterbuchnetz.de/Lexer/?lemid=LR01453.

3.3   Text Corpus Statistics

The Hartmann von Aue-portal has in total six knowledge bases, where five are poems and one are the lyrics. The cake diagram of Fig. 2 compares the sizes of the individual datasets[5]. In terms of the number of tokens (the total number of words), "Erec" is the longest poem with a proportion of 37.3%. Also in terms of lemmata (the number of distinct dictionary forms), "Erec" has the highest share with 30.2%. This clearly indicates that the richness of a text with respect to text length decreases in comparison to a shorter text, which is indeed what someone would intuitively expect.

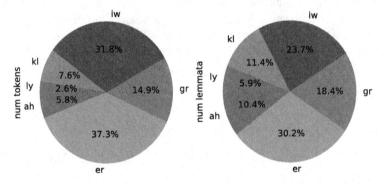

Fig. 2. Size comparisons across the Hartmann von Aue knowledge bases

Table 1 lists the absolute numbers of tokens and lemmata within the works by Hartmann von Aue. In total, the corpus contains 137,058 tokens and 9,757 lemmata. The type-token ratio (cf. [12]), a complexity comparison measure for similarly-sized texts, thus amounts to 137,058/9,757 = 14.05 in total.

Table 1. Statistics on tokens and lemmata in the Hartmann von Aue corpus

	ah	er	gr	iw	kl	ly	Total
# tokens	7,998	51,089	20,387	43,552	10,430	3,593	137,058
# lemmata	1,011	2,951	1,796	2,311	1,114	574	9,757
Type-token ratio	7.91	17.31	11.35	18.85	9.36	6.26	14.05

---

[5] The numbers and diagrams reported in this section were created from our conversion results of the Hartmann von Aue-portal.

4 Approach

In this section, we describe the main challenges of our approach, present our method for the conversion of the Hartmann von Aue knowledge base portal, outline the design and implementation of a URI scheme, define a vocabulary for semantically encoding the data, and eventually report on some of the conversion results.

4.1 Challenges

When creating a Linked Open Data representation for the Hartmann von Aue-portal, we were facing the following research challenges: (1) Reverse-engineering the existing data model, (2) ontology engineering including reuse of concepts, (3) data conversion, and (4) LOD deployment including URI schema definition, publication, and licensing.

4.2 Method

Since one of the authors of this paper played a crucial role in the creation of the Hartmann von Aue-portal, we are given unrestricted access to the relational database files of the project. Starting from there we analyze and reverse-engineer the data model.

As a next step we define a suitable ontology for our problem domain, i.e. a shared conceptualization that is formally and explicitly specified [13]. Ontology engineering typically involves finding and incorporating ontologies that concepts and relationships can be reused. In the linguistic domain, there exist multiple vocabularies that are eligible for modeling our linguistic knowledge base in RDF. However, such vocabularies can potentially add a lot of unnecessary complexity for our project, because (1) they are partly very sophisticated, and (2) they often serve different objectives. For example, the GOLD ontology [14] seems a promising candidate, but despite being a powerful vocabulary for descriptive linguistics (with 500 classes, 74 object properties, and 6 datatype properties) it still lacks some of the concepts that we needed, namely a specific concept "Verse" or a property "part of verse". Although it would be possible to reuse parts of an existing ontology like GOLD, we decided instead to create our own, light-weight schema suited to the relational schema of the Hartmann von Aue-portal. Still it would be an easy and worthwhile exercise to supply respective mappings to other, more widespread ontologies. For instance, the *lemon* model [15] primarily used for dictionaries could be used to model dictionary entries in the Hartmann von Aue-portal.

The ontology engineering is followed by defining mappings for the conversion of the knowledge base into Linked Open Data. Finally, we propose a URI scheme and publication approach considering best practices for publishing Linked Data.

4.3 Implementation

For the conversion of the Hartmann von Aue-portal into Linked Open Data, we developed a custom Python script. We could rather have used one of the available tools, e.g. Google Refine with RDF extension (for an overview of conversion tools, see https://www.w3.org/wiki/ConverterToRdf), but Python comes with RDFLib, a powerful library that can handle large RDF graphs and output data in a variety of syntaxes. From past conversions, we already knew that Python/RDFLib offers high scalability and allows for a seamless workflow integration. Parts of our approach materialized into an ontology and a set of knowledge bases that are available online[6].

Preliminary Work. We have exported tables in the Microsoft Access files (.mdb file extension) to comma-separated value (CSV) files using the script from https://github.com/brianb/mdbtools. E.g., to extract the table "lyData" from an Access database LyDataSet.mdb, we ran the following commands:

```
mdb-tables LyDataSet.mdb
mdb-export LyDataSet.mdb lyData > LyDataSet.lyData.csv
```

The first command allowed us to list all database tables from the database LyDataSet.mdb. The next instruction was used to export the table "lyData" to a CSV file.

Reverse-Engineering. The content within the CSV files corresponds the structure of the Access database tables. The database structure that we were able to re-engineer was already depicted in the ER diagram in Sect. 3.2.

Ontology Engineering. The ontology that we populated with data from the Hartmann von Aue knowledge base we designed from scratch. Our goal was to define an ontology that is light-weight but covers all the meaning in the Hartmann von Aue-portal. Figure 3 shows the main concepts of the vocabulary.

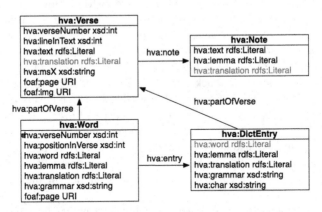

Fig. 3. Conceptual data model of the Hartmann von Aue Linked Data architecture (Some properties are conceptually useful but currently not required by the knowledge bases.)

The ontology consists of four classes (Verse, Word, Note, and DictEntry), three object properties, and 23 datatype properties. 14 datatype properties establish links to manuscript files (graphically represented by the property hva:msX). Every conceptual

---

[6] http://www.ebusiness-unibw.org/ontologies/hva/.

element of the ontology has a label and a short description in English. Our ontology is published online[7] and features a human-readable HTML documentation.

Conceptual Mapping. In the following, we highlight the main mappings from the relational data model to the ontology. The items with bold font face denote database tables, whereas the remaining lines are columns. To the left is the data structure of the database, the right side represents the corresponding ontological concepts in OWL:

What is missing here is that we used the MHGDictKeys database table to create foaf:page links for hva:Word-typed entities to externally defined lemmata. Using this array of mappings, it was possible to trigger fully-automated conversions (Table 2).

Table 2. Mapping of database tables to the ontology

Text → hva:Verse + URI(key = TextNum)
TextNum → hva:verseNumber (xsd:int)
TextVerse → hva:text (@gmh)
TextPos → hva:lineInText (xsd:int)
Data → hva:Word + URI(key1 = DataNum, key2 = DataPos)
DataNum → link via hva:partOfVerse to verse with URI(verseNumber = DataNum)
DataPos → hva:positionInVerse (xsd:int)
DataWord → hva:word (@gmh)
DataLemma → hva:lemma (@gmh)
DataGrammar → hva:grammar (xsd:string)
DataGerman → hva:translation @de
DataEnglish → hva:translation @en
Notes → hva:Note + BlankNode
NotesNum → reverse link via hva:note to verse w/URI(key = NotesNum)
NotesLemma → hva:lemma (@gmh)
NotesNote → hva:text (@lang)
DictData → (hva:Word hva:entry hva:DictEntry) + URI(key1 = lemma, key2 = levels)
lemma → hva:lemma(Word) (@gmh)
lemma → hva:lemma(DictEntry) (@gmh)
SortChar → hva:char(DictEntry) (xsd:string)
grammar → hva:grammar(DictEntry) (xsd:string)
german → hva:translation(DictEntry) (@de)
english → hva:translation(DictEntry) (@en)
verse → *verse number extraction* → link via hva:partOfVerse(DictEntry) to verse

URI Schema. Analogous to Web documents, entities on the Web of Data require globally unique keys to unambiguously identify and refer to them and thus to be able to create an integrated, giant graph of data. Linked Open Data relies on uniform resource

---

[7] http://www.ebusiness-unibw.org/ontologies/hva/ontology.html.

identifiers (URIs) to achieve this. We defined base URIs for the ontology (hereinafter referred to with the prefix hva:) and the respective knowledge bases as follows:

```
http://www.ebusiness-unibw.org/ontologies/hva/ontology#
http://www.ebusiness-unibw.org/ontologies/hva/{ns}#
```

The placeholder *ns* represents any of the namespaces *ah*, *er*, *gr*, *iw*, *kl*, *ly*, and *dict* corresponding to the datasets. Consequently, the RDF dataset of "Der arme Heinrich" (*ah*) is available at http://www.ebusiness-unibw.org/ontologies/hva/ah#. Under this common namespace, all entities that pertain to a specific dataset are subsumed. Their fragment identifiers are attached to the base URI. We specified the following URI patterns to mint URIs for instances of the different types of entities:

- Ontology: *Ontology*;
- Verse: *Verse-<verseNumber>*;
- Word: *Word-<verseNumber>-<positionInVerse>* for knowledge bases, and
- *Lemma- <lemma>* for the context dictionary;
- Note: For notes we only use blank nodes, i.e. resources without an identifier, as notes are always attached to verses and thus share their identity;
- DictEntry: Multiple dictionary entries are possible for every lemma, since the meanings can vary by context, thus *Lemma- <lemma>* and *Entry- <lemma>-<levels>*.

The placeholders within the URI patterns were derived from respective fields in the database tables. The only feasible way to create unique identifiers for lemmata was to encode the lemma text as part of the URI. Due to special letters in the lemma, this led to invalid URI strings, though. We resolved this by mapping those letters to ones that are accepted for URIs (e.g. ô to o). An alternative way is to create internationalized resource identifiers (IRIs) in place of URIs, which are supported by RDF 1.1 (cf. [16]).

Linked Data Deployment. We deployed the data on a hosted Web server under the previously outlined URI namespaces. According to Linked Open Data publishing best practices [17], we modeled the data in an application-independent way, we used cool HTTP URIs [18] to refer to objects, we used standard vocabularies where possible, we supplied basic metadata, we added an open license statement, and we provided human-readable descriptions and data access for machines. Machines can choose from a variety of data formats (e.g. RDF/XML, Notation 3, N-Triples), and the kind of syntax can be arranged via a simple content negotiation mechanism:

- **If** a client supplies a file extension, deliver that one (.rdf for RDF/XML, .n3 for Notation 3, .nt for N-Triples);
- **Else if** a client supplies HTTP Accept headers, try to deliver the respective media type. This mechanism is widely known as HTTP Content Negotiation: application/rdf+xml for RDF/XML, text/n3 for Notation 3, text/plain for N-Triples);

- **Else** deliver RDF/XML by default.

For example, we can request the N-Triples representation of "Der arme Heinrich" resolving the URI http://www.ebusiness-unibw.org/ontologies/hva/ah.nt, or alternatively using HTTP Content Negotiation [19] as exemplified by the curl command

```
curl -L -H "Accept: text/plain" http://www.ebusiness-
unibw.org/ontologies/hva/ah
```

With HTTP Content Negotiation, resolving the URI answers with an HTTP Redirect (status code 303) and an HTTP Location field in the response header. In order to immediately follow that link, the above command uses the "-L" parameter.

The datasets are deployed on an Apache Web server. An .htaccess file entails the configuration for the right delivery of the resources. The human-readable document is delivered if a client sends an HTTP Accept header containing text/html or application/xhtml+xml, or if it uses a User-Agent string of a popular Web browser. The human-readable representation with usage instructions on how to access the RDF data is generated during the conversion process.

Licensing. A simple way to assess the quality of Linked Open Data publishing is to refer to the five-star model[8]. One important requirement is an open license. We decided to make our data reusable under an open Creative Commons license[9].

Additional Design Decisions. In the Linked Open Data representation, we preserve the links to manuscript files relying on the msX columns in XXDataSet.XXText.csv of the database; e.g., AhMsARef="A197" in AhDataSet.AhText.csv creates the two links using foaf:page and foaf:img:

- http://hvauep.uni-trier.de/resources/armer/manuscripts/Ah_A/Ah_A_197.pdf
- http://hvauep.uni-trier.de/resources/armer/manuscripts/Ah_A/Ah_A_197[10]

In order to indicate the language of textual datatype properties, we assigned the language tag @gmh for Middle High German to textual descriptions (e.g. verses and words), @de for Modern High German, and @en for English.

The Hartmann von Aue-portal and the data structure of the relational database contains data entries especially for the reverse list view. However, there is no need to convert that, as we can easily obtain it using an appropriate SPARQL query (cf. [20]). The following SPARQL SELECT query obtains the reverse list for "Der arme Heinrich" sorted by beginning from the most frequent word:

---

[8] http://5stardata.info/en/.

[9] https://creativecommons.org/licenses/by-nc/4.0/.

[10] Host server must be configured to deliver images by default when the file extension is omitted.

```
PREFIX data: <http://www.ebusiness-
unibw.org/ontologies/hva/ah#>

SELECT ?word COUNT(?word) AS ?freq
WHERE {
 ?word_uri a hva:Word .
 ?word_uri hva:word ?word .
 ?word_uri rdfs:isDefinedBy data:Ontology .
}
ORDER BY DESC (?freq)
```

As it turns out from the query above, every entity has a reverse link to the ontology document it is defined by. This metadata allows for querying data within a specific knowledge base without the need to explicitly group sets of triples using named graphs, which is still not supported by all RDF storage engines.

4.4   Results

In this section, we report on some conversion results of the Hartmann von Aue-portal. We could successfully convert five poems, the lyrics, as well as the context dictionary. Furthermore, we have an RDF representation of the conceptual schema. The output format was Notation 3 (N3).

The execution time increases linearly with respect to the number of triples. On an early-2015 Macbook Air with a 2.2 GHz Intel Core i7 and 8 GB of 1600 MHz DDR3 RAM, the total execution time of the script was 7 min and 2 s. Table 3 below shows the execution times (averaged over three runs) and the number of triples obtained for each dataset. The biggest file in N3 syntax was produced for "Erec" and amounts to 29 MB.

Table 3. Quantitative comparison of conversion execution times and the number of triples

Dataset	Execution Time (s)	Number of Triples
Der arme Heinrich	26.25	142,584
Erec	150.41	847,892
Gregorius	59.95	334,647
Iwein	128.64	732,150
Die Klage	30.95	172,492
Lyrik	10.90	59,225
Context Dictionary	15.09	77,318
HvA Ontology	0.07	182
Total	422.27	2,366,490

5   Evaluation

In this chapter, we evaluate our approach. In particular, we aim to show the benefits of having the data from the Hartmann von Aue-portal readily available as Linked Open Data. Furthermore, we validate our conversion.

5.1   Formal Validation

We validated the completeness and consistency of our conversion. For this purpose, we drew a random sample of verses from each of the five knowledge bases related to the poems in the Hartmann von Aue-portal and tested rigorously whether they exist and are fully described in our converted datasets. The number of verses n are already given in the portal, so we could draw a sample of ten random numbers each in between one and n. We prepared two SPARQL ASK query templates, one for verses and one for words. We then populated these templates with the knowledge base namespaces and random verse numbers and executed the queries against a SPARQL endpoint that contains the RDF data of our conversion. As no exception was thrown by our evaluation script, it indicates that the conversion was complete and consistent for verses and individual words across all knowledge bases.

We further validated our published datasets using the Vafu and Vapour Linked Data validators[11]. These services perform several checks that assess the adherence of Linked Data deployments to current best practices. More precisely, they dereference a given URI requesting a human-readable HTML page, a machine-accessible RDF/XML document, and without content negotiation. The reported results were fine, but they would contain error messages if for example unexpected media types are returned or any redirects are not configured properly. We extended this validity check of Linked Open Data publishing with a validation of the RDF data. According to the online RDF Validator[12] by the World Wide Web Consortium (W3C), our datasets are described by valid RDF documents.

5.2   Usage for Data-Driven Research

In this part of the evaluation, we show how our data can be easily used for data analysis, and thus can prove its benefit for modern linguists.

Jupyter Notebook[13] (formerly IPython Notebook) is a browser-based integrated development environment that embodies the literate programming paradigm [21]. In a nutshell, documents are edited in the Web browser and can include live code in different programming languages (here we use Python), documentation, equations, and visualizations. Jupyter Notebook is extremely popular among data scientists.

With a Jupyter notebook, it is easy to implement a routine to manage the communication with a given SPARQL endpoint. The code snippet in Fig. 4 demonstrates

---

[11] http://vafu.redlink.io/ and http://linkeddata.uriburner.com:8000/vapour.

[12] https://www.w3.org/RDF/Validator/.

[13] http://jupyter.org/.

how a word frequency list (cf. [12]), a very popular method in computational linguistics, can be derived from a given corpus with only a few lines of Python code. Similarly, it would be easy to derive a concordance, often called key word in context (KWIC) (cf. [12]). The program code defines a SPARQL SELECT query that is executed against an endpoint, and the result is returned as CSV data. Alternatively, the result could be loaded into a Pandas dataframe for further data analysis with Python. In here we omit the discussion of any implementation details regarding endpoint communication.

```
# What are the ten most frequent words?
q = """
PREFIX hva: <http://www.ebusiness-unibw.org/ontologies/hva/ontology#>

SELECT ?lemma count(?lemma) as ?freq        "lemma","freq"
WHERE {                                      "dër",10465
    ?s a hva:Word .                          "ër",8115
    ?s hva:lemma ?lemma .                    "ich",4820
}                                            "und",3823
ORDER BY DESC(?freq)                         "sîn",3523
LIMIT 10                                     "daz",2496
"""                                          "ir",2443
result = query(q, "text/csv")                "si",2057
print result                                 "wësen",2022
                                             "dâr",2016
```

Fig. 4. Word frequency list in Python

By virtue of a Jupyter notebook we can further create a graphical plot of a rank-frequency profile of the Hartmann von Aue text corpus. Figure 5 displays the rank-frequency profiles for all datasets together with corresponding Zipf distributions starting at the highest-ranked word. The diagram confirms that text corpora in general follow a Zipf distribution, which also holds true for the Middle-High German corpus of Hartmann von Aue. The formula applied for drawing the Zipf distribution in Fig. 5 was

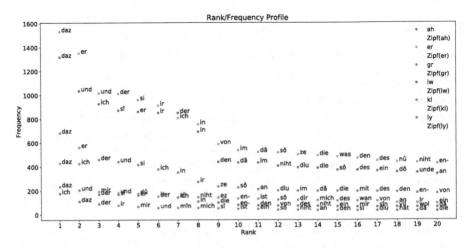

Fig. 5. Rank-frequency profile of the Hartmann von Aue text corpus and Zipf distributions with exponent 0.5

$$frequency_{dataset}\left(word_{pos=rank}\right) = \frac{1}{\sqrt{rank}} * frequency_{dataset}\left(word_{pos=1}\right) \qquad (1)$$

Quite obviously, there exist many more interesting examples for text statistics with Hartmann's works and data-driven research that computer linguists could come up with.

5.3   Interlinking of Hartmann von Aue Datasets with DBPedia

Linked Open Data provides a sophisticated means to perform information integration at Web scale, namely to combine structured data in otherwise isolated data silos using plain string comparison of entity URIs.

The SPARQL query language supports besides SELECT and ASK queries also CONSTRUCT queries that allow to create new data, i.e. to generate new triples in response to query results [22]. Query federation is a subquery mechanism to issue queries over multiple, remote SPARQL endpoints that results are then merged by the federated query processor [22]. The following federated SPARQL CONSTRUCT query (prefix declarations omitted) annotates any of our knowledge bases with titles and abstracts in multiple languages based on the respective data from DBPedia [8]. A similar approach of linking linguistic resources was described in [23].

```
CONSTRUCT {
  data:Ontology rdfs:label ?label .
  data:Ontology rdfs:comment ?abstract
}
WHERE {
  data:Ontology a owl:Ontology .
  {
    SERVICE <http://dbpedia.org/sparql> {
      dbpedia:Der_arme_Heinrich rdfs:label ?label ;
        dbo:abstract ?abstract .
    }
  }
}
```

And the resulting graph (omitting the text of the abstracts):

```
data:Ontology rdfs:label "Der arme Heinrich"@de, "Der
arme Heinrich"@en, "El pobre Enrique"@es,
"哀れなハインリヒ"@ja ;
  rdfs:comment "..."@en, "..."@es, "..."@ja .
```

5.4   Reverse-Engineering

In order to demonstrate the versatility of RDF data, we further set up two Web services that reverse-engineer the contents of the Hartmann von Aue-portal by relying on simple SPARQL queries. The first lists the contents of the poems along with their concordances, i.e. words along with their contexts, as well as notes and translations. The second one creates a name register, reverse list, lemma in context, and verses with notes.

Open questions of copyright clearance currently prevent us from publishing the full content of the poems at once, so for the moment we cannot publicly reveal the URIs of the respective services. Figure 6 outlines a screenshot of the second service.

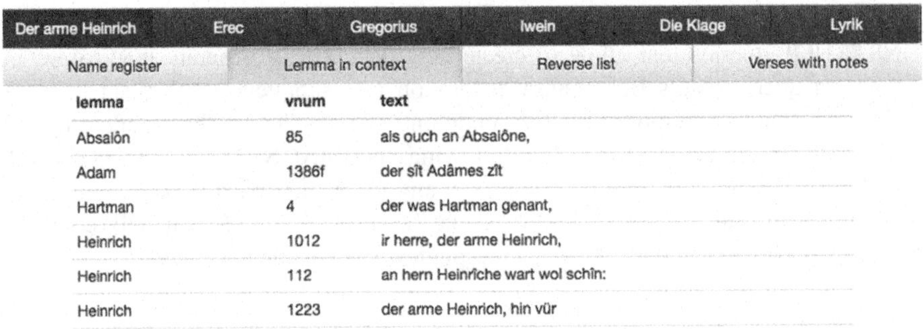

Der arme Heinrich	Erec	Gregorius	Iwein	Die Klage	Lyrik
Name register		Lemma in context		Reverse list	Verses with notes

lemma	vnum	text
Absalôn	85	als ouch an Absalône,
Adam	1386f	der sît Adâmes zît
Hartman	4	der was Hartman genant,
Heinrich	1012	ir herre, der arme Heinrich,
Heinrich	112	an hern Heinrîche wart wol schîn:
Heinrich	1223	der arme Heinrich, hin vûr

Fig. 6. Lemma in context view for "Der arme Heinrich" derived from LOD dataset

6   Discussion and Conclusion

In this section, we provide an overview of relevant related works and summarize our contributions and findings. We finally give a prospective outlook.

6.1   Related Work

Over the past ten years, much research work has been published under the umbrella of a Linguistic Linked Open Data cloud[14] [10], mainly put forward by the experts of the Open Linguistics Working Group[15] and others interested in computational linguistics. Many of the publications address the challenges of representing linguistic data sources at syntactical and conceptual levels in order to be able to render them interoperable.

Several corpora have been converted into RDF to become valuable input data sources for Linguistic Linked Open Data. Chiarcos et al. [6] distinguish between two important classes of linguistic data sources, namely lexical-semantic resources and annotated corpora. The first class includes ontologies, terminologies, and dictionaries, while the second covers textual data with linguistic annotations. With respect to

---

[14] http://linguistic-lod.org/.
[15] https://linguistics.okfn.org/.

dictionaries, WordNet (cf. [9]) is likely the most prominent converted dataset. Regarding annotated corpora, the Manually Annotated Sub-Corpus (MASC) of the Open American Corpus [24], or the German newspaper corpus NEGRA have been mentioned as primary examples [23].

Regarding conceptual interoperability between Linguistic Linked Data sets, a number of linguistic ontologies have been suggested. The Lexicon Model for Ontologies (*lemon*) [15] is a data model for the representation of lexicons in RDF. The Ontologies of Linguistic Annotation (OLiA) [25] forms a collection of shared vocabularies that contribute terminology for linguistic annotations [6]. The ontologies facilitate interoperability and information integration by acting as intermediate layers among linguistic ontologies like ISOcat [26] and GOLD [14], and defining mappings to them [6, 25].

The work in [27] is closely related to the ontology engineering task presented in this paper. The authors propose a poetic and metric ontology based on a digital system for Spanish medieval poetry. In [28], Chiarcos et al. present a linguistic Web portal that combines various linguistic resources for the aim to establish a powerful research tool. It collects information such as links, journals, databases, dictionaries, and diverse catalogs related to linguistics, e.g. bibliographic data. A planned future extension is to connect the portal with LLOD to make it even more valuable for users [28].

6.2   Our Contribution

In this work, we have described an automated conversion approach for the Hartmann von Aue-portal into RDF and its publication as Linked Open Data. We have built an ontology suitable for mapping the knowledge bases in the original relational database to RDF data. This involved also a sophisticated and robust URI schema design.

We claim that publishing the contents of the Hartmann von Aue-portal makes the data application-independent and thus more resilient against future changes of technology and organizational challenges. The data could be replicated as many times as desired. In our case, we have already created a first-stop, authoritative host for the ontology and RDF datasets of Hartmann's works. Furthermore, we have set up a SPARQL endpoint that again contains the very same data. Using SPARQL queries, we can build up custom services that reverse-engineer the contents of the Hartmann von Aue-portal.

In terms of data reuse and the ability for data-driven research in linguistics, we could show that unlocking the data enables powerful data analyses. With the human-centered Hartmann von Aue-portal, such advanced use cases are at least much more expensive.

With respect to linkage with other data sources, the current Hartmann von Aue-portal hampers innovative applications, because (a) the data is not available in structured form and (b) links are set up between documents rather than data items. With LOD and SPARQL, external data sources such as DBPedia can be integrated with ease.

6.3   Future Work

Within the scope of this paper, we decided to engineer our own light-weight ontology instead of referring to existing ontologies. A future extension consists in mapping our ontology to these candidate linguistic ontologies, such as GOLD [14], ISOcat [26], *lemon* [15], or OLiA [25]. Furthermore, links to a world's language reference such as Glottolog could be worthwhile (cf. [23]). In general, any improvements on the current data such as cleansing or augmentation can be made using simple SPARQL CONSTRUCT rules [22] or similar techniques.

One of the next steps is to prepare our RDF datasets for eligibility to register for the Linguistic Linked Open Data (LLOD) cloud [10]. This again would increase visibility and potentially attract linguists for doing more sophisticated data analysis with our data.

References

1. Boggs, R.A.: The Hartmann von Aue Internet (knowledge based) portal. an introduction and description. In: Fritz-Rößler, W. (ed.) Cristalin wort - Hartmann Studien, pp. 13–32. Lit. Verlag, Vienna (2007)
2. Berners-Lee, T.: Linked Data - Design Issues, http://www.w3.org/DesignIssues/LinkedData. html
3. Bizer, C., Heath, T., Berners-Lee, T.: Linked data - the story so far. Int. J. Semant. Web Inf. Syst. **5**, 1–22 (2009)
4. Jurafsky, D., Martin, J.H.: Speech and Language Processing: An Introduction to Natural Language Processing, Computational Linguistics, and Speech Recognition. Prentice Hall (2009)
5. Hirschmann, R., Lenders, W.: Computer-assisted study of early german: the mannheim symposium of 1973. Comput. Hum. **8**, 179–181 (1974)
6. Chiarcos, C., McCrae, J., Cimiano, P., Fellbaum, C.: Towards open data for linguistics: linguistic linked data. In: Oltramari, A., Vossen, P., Qin, L., Hovy, E. (eds.) New Trends of Research in Ontologies and Lexical Resources. Theory and Applications of Natural Language Processing, pp. 7–25. Springer, Heidelberg (2013). doi:10.1007/978-3-642-31782-8_2
7. Singhal, A.: Introducing the Knowledge Graph: Things, Not Strings, http://googleblog. blogspot.com/2012/05/introducing-knowledge-graph-things-not.html
8. Auer, S., Bizer, C., Kobilarov, G., Lehmann, J., Cyganiak, R., Ives, Z.: DBpedia: a nucleus for a web of open data. In: Aberer, K., et al. (eds.) ASWC/ISWC -2007. LNCS, vol. 4825, pp. 722–735. Springer, Heidelberg (2007). doi:10.1007/978-3-540-76298-0_52
9. McCrae, J., Fellbaum, C., Cimiano, P.: Publishing and linking WordNet using lemon and RDF. In: Proceedings of the 3rd Workshop Linked Data Linguistics (2014)
10. Chiarcos, C., Hellmann, S., Nordhoff, S.: Towards a linguistic linked open data cloud: the open linguistics working group. Trait. Autom. des Langues. **52**, 245–275 (2011)
11. Boggs, R.A., Gärtner, K.: Das Hartmann von Aue-Portal. Eine Internet-Plattform als Forschungsinstrument. Zeitschrift für Dtsch. Altertum und Dtsch. Lit. **134**, 134–137 (2005)
12. Adolphs, S.: Introducing Electronic Text Analysis: A Practical Guide for Language and Literary Studies. Routledge (2006)
13. Studer, R., Benjamins, R., Fensel, D.: Knowledge engineering: principles and methods. Data Knowl. Eng. **25**, 161–197 (1998)

14. Farrar, S., Langendoen, D.T.: A linguistic ontology for the semantic web. GLOT Int. **7**, 97–100 (2003)
15. Cimiano, P., McCrae, J.P., Buitelaar, P.: Lexicon Model for Ontologies: Community Report, 10 May 2016 (2016)
16. Cyganiak, R., Wood, D., Lanthaler, M.: RDF 1.1 Concepts and Abstract Syntax, http://www.w3.org/TR/2014/REC-rdf11-concepts-20140225/
17. Hyland, B., Atemezing, G., Villazón-Terrazas, B.: Best Practices for Publishing Linked Data, https://www.w3.org/TR/2014/NOTE-ld-bp-20140109/
18. Berners-Lee, T.: Cool URIs Don't Change, http://www.w3.org/Provider/Style/URI
19. Daigle, L., van Gulik, D., Iannella, R., Faltstrom, P.: URN Namespace Definition Mechanism, https://tools.ietf.org/rfc/rfc2611
20. Harris, S., Seaborne, A.: SPARQL 1.1 Query Language, http://www.w3.org/TR/2013/REC-sparql11-query-20130321/
21. Knuth, D.E.: Literate Programming. Comput. J. **27**, 97–111 (1984)
22. SPARQL 1.1 Overview, http://www.w3.org/TR/sparql11-overview/
23. Chiarcos, C., Hellmann, S., Nordhoff, S.: Linking linguistic resources: examples from the open linguistics working group. In: Chiarcos, C., Nordhoff, S., Hellmann, S. (eds.) Linked Data in Linguistics, pp. 201–216. Springer, Berlin Heidelberg (2012). doi:10.1007/978-3-642-28249-2_19
24. Ide, N., Baker, C., Fellbaum, C., Fillmore, C., Passonneau, R.: MASC: the manually annotated sub-corpus of american english. In: Proceedings of the Sixth International Conference on Language Resources and Evaluation (LREC 2008). European Language Resources Association (ELRA), Marrakech, Morocco (2008)
25. Chiarcos, C., Sukhareva, M.: OLiA – ontologies of linguistic annotation. Semant. Web. **6**, 379–386 (2015)
26. Snijders, M.K., Windhouwer, M., Wittenburg, P., Wright, S.E.: ISOcat: remodelling metadata for language resources. Int. J. Metadata, Semant. Ontol. **4**, 261 (2009)
27. González-Blanco García, E., del Río, M.G., Martínez Cantón, C.I.: Linked open data to represent multilingual poetry collections. A proposal to solve interoperability issues between poetic repertoires. In: Proceedings of the Fifth Workshop on Linked Data in Linguistics: Managing, Building and Using Linked Language Resources, Portorož, Slovenia, pp. 77–80 (2016)
28. Chiarcos, C., Fäth, C., Renner-Westermann, H., Abromeit, F., Dimitrova, V.: Lin|gu|is|tik: building the linguist's pathway to bibliographies, libraries, language resources and linked open data. In: Proceedings of the Tenth International Conference on Language Resources and Evaluation (LREC 2016). European Language Resources Association (ELRA), Paris, France (2016)

A Survey of Approaches to Representing SPARQL Variables in SQL Queries

Miloš Chaloupka$^{(\boxtimes)}$ and Martin Nečaský

Faculty of Mathematics and Physics, Charles University,
Malostranské nám. 25, 118 00 Praha 1, Czech Republic
{chaloupka,necasky}@ksi.mff.cuni.cz
http://www.ksi.mff.cuni.cz/

Abstract. RDF is a universal data model for publishing structured data on the Web. On the other hand, many structured data is stored in relational database systems. To support publishing data in the RDF model, it is essential to close the gap between the relational and RDF worlds. A virtual SPARQL endpoint over relational data is a promising approach to achieve that. To build a virtual SPARQL endpoint, we need to know how to translate SPARQL queries to corresponding SQL queries. There exist several approaches to such transformation. Most of them are focused on the processing of user-defined mapping. The user-defined mapping gives an user the ability to define a mapping of a stored relation data to almost any RDF representation. In this paper we focus on one of the core problems of the transformation: how to represent variables from a given SPARQL query in the corresponding SQL query. We survey variable representations from existing approaches; how the selected representation affects the soundness and performance of the whole transformation approach.

Keywords: RDB2RDF · SQL · R2RML · SPARQL · Relational to RDF mapping · SPARQL variable representation

1 Introduction

The RDF model is a universal and popular data model for publishing structured data on the Web. It enables to access and integrate data from different sources. It also provides a clear and standard way to data querying without the knowledge of information about data storage using the RDF query language SPARQL [10]. Usually, a publicly available web service called SPARQL endpoint is provided by a publisher to enable consumers to access the published data using SPARQL.

There already exist several database management systems which support the RDF model natively. For example, Marklogic[1], OpenLink Virtuoso[2] and Apache Jena[3]. However, it is much more common to store data in a relational database

---

[1] http://www.marklogic.com/, visited in July 2017.
[2] https://virtuoso.openlinksw.com/, visited in July 2017.
[3] http://jena.apache.org/, visited in July 2017.

H. Panetto et al. (Eds.): OTM 2017 Conferences, Part II, LNCS 10574, pp. 300–317, 2017.
https://doi.org/10.1007/978-3-319-69459-7_21

management system today. For example, on the ranking of database engines[4], the top 4 engines are relational. There is no system with native RDF support between the top 30 engines.

Therefore, it is essential to minimize the effort necessary to publish data stored in relational databases in the RDF model on the Web. A virtual SPARQL endpoint over a relational database can effectively achieve that. It allows for querying relational data without materializing the RDF triples beforehand. This is ensured by translating a given SPARQL query to a corresponding SQL query and transforming the relational result to its RDF equivalent. There are several research papers which describe the SPARQL to SQL transformation (see [4, 5, 14, 15]).

When translating a SPARQL query it is necessary to represent variables from the SPARQL query in the corresponding SQL query correctly. This part of the transformation is usually neglected in the existing literature although it usually affects the validity and efficiency of the approach. For example, one of the biggest differences between SPARQL and SQL query semantics is that a SPARQL variable does not have a defined type, although a SPARQL value has a type. So, the simplest representation where only a SPARQL value without type is present is not sound because, for example, it is not possible to correctly compare two values. Moreover, the simplest representation is not efficient because values may be constructed from several columns and comparing calculated values are usually slow and prevent the usage of indexes in the relational database.

Contribution. In this paper, we survey existing approaches to representing SPARQL variables in SQL queries. In particular, the paper has the following contributions

1. We specify the problem of representing SPARQL variables in SQL queries.
2. We provide an overview of 4 existing approaches: Ultrawrap [14], SparqlMap [15], Ontop [4] and EVI [5] in terms of SPARQL variable representation.
3. We evaluate the completeness, soundness and performance of the presented approaches. We show several problems with soundness of the approaches. We also show how the representation affects performance of each approach.

In Sect. 2, we describe the compared approaches. In Sect. 3, we explain our motivation why we focus on the variable representation. In Sect. 4, we specify the problem of representing SPARQL variables in SQL queries. In Sect. 5, we describe the representations used in the selected approaches. In Sect. 6, we evaluate the described representations.

2 Specification of Surveyed Approaches

In general, existing approaches differ in the way of using the underlying relational database. There are approaches which require a user-defined mapping to enable

---

[4] http://db-engines.com/en/ranking, visited in July 2017.

the creation of a virtual SPARQL endpoint over any relational structure. The mapping specifies how a given relational structure is mapped to a required RDF structure. The standardized language to define such mapping is R2RML [9]. Other approaches derive a mapping automatically [12]. These approaches do not allow the user to specify the required RDF structure in which relational data is published. The RDF structure is derived automatically from the structure of the underlying relational data. The rest of the approaches do not use any mapping at all. They only use the relational database system as an efficient storage engine. They prescribe a fixed relational database structure where they store RDF triples. Therefore, they do not enable to built a virtual SPARQL endpoint on top of an arbitrary relational structure and they cannot be used to publish an existing relational database in the RDF model on the Web.

In this paper, we focus only on the approaches which consider a user-defined mapping for publishing a given relational database in the RDF model on the Web through a virtual SPARQL endpoint. In the rest of this section we will list the selected approaches. Also, we will show how they transform the sample SPARQL query in Fig. 1(a).

Ultrawrap [14] is a simple solution for virtual SPARQL endpoints. It does not aim to optimize the queries because it completely relies on the efficiency of the underlying relational database. The sample SPARQL query is transformed by Ultrawrap to the SQL query in Fig. 1(b).

The authors of **SparqlMap** [15] proposed a solution which uses a set of SQL columns to represent a single SPARQL variable. This enables to resolve, e.g. the problem with typing variables. The sample SPARQL query is transformed by SparqlMap to the SQL query in Fig. 1(c).

The **Ontop** [3,4,13] system seems to be actively used and from the found tools it has the most active community. The sample SPARQL query is transformed by Ontop to the SQL query in Fig. 1(d).

Last but not least, we include to the survey our approach **EVI** [5]. The sample SPARQL query is transformed by EVI to the SQL query in Fig. 1(e).

We selected only the approaches which transform a SPARQL query into a single SQL query so no significant in-memory processing is needed to merge multiple results together. Therefore, for example D2R [8] is not evaluated.

3   Motivation

The main feature of a virtual SPARQL endpoint is rewriting a given SPARQL query into a corresponding SQL query which may be executed on the underlying relational database. The rewriting mechanism has to take into account a user-mapping of the relational schema of the underlying database into the required RDF structure.

One of the biggest differences between SPARQL and SQL query semantics is that a SPARQL variable does not have a defined type. It may contain values of various types. SPARQL (see [10]) declares rules for working with values of different types. Such rules are necessary for correct ordering and value comparisons.

```
SELECT ?paperId, ?title, ?author
WHERE {
  ?paperId a x:paper;
    x:name ?title;
    x:author ?authorId.
  ?authorId x:name ?author
}
```

(a) Sample SPARQL query

```
WITH triples AS (
  SELECT 'x:person/' + a.id AS s, a.id AS s_id,
    'rdf:type' AS p, 'x:person' AS o, NULL AS o_id
  FROM authors AS a WHERE a.id IS NOT NULL
  UNION ALL
  SELECT 'x:person/' + a.id AS s, a.id AS s_id,
    'x:name' AS p, a.name AS o, NULL AS o_id
  FROM
    authors AS a WHERE a.id IS NOT NULL AND
    a.name IS NOT NULL
  UNION ALL
  ...
)
SELECT
  t1.s as paperId, t2.o as title, t4.o as author
FROM
  triples AS t1, triples AS t2,
  triples AS t3, triples AS t4
WHERE
  t1.p = 'rdf:type' AND t1.o = 'x:paper' AND
  t1.s_id = t2.s_id AND t2.p = 'x:name' AND
  t1.s_id = t3.s_id AND t3.p = 'x:author' AND
  t4.s_id = t3.o_id AND t4.p = 'x:name'
```

(b) Ultrawrap

```
SELECT
  1 AS paperId_type,
  1 AS paperId_datatype,
  NULL AS paperId_text,
  NULL AS paperId_num,
  NULL AS paperId_bool,
  NULL AS paperId_time,
  2 AS paperId_reslength,
  'x:paper/' AS paperId_res_1,
  paper.id AS paperId_res_2,
  2 AS title_type,
  1 AS title_datatype,
  paper.title AS title_text,
  NULL AS title_num,
  NULL AS title_bool,
  NULL AS title_time,
  0 AS title_reslength,
  2 AS author_type,
  1 AS author_datatype,
  author.name AS author_text,
  NULL AS author_num,
  NULL AS author_bool,
  NULL AS author_time,
  0 AS author_reslength
FROM
  paper,
  author
WHERE paper.authorId = author.id
```

(c) SparqlMap

```
SELECT * FROM (
  SELECT
    1 AS paperIdQuestType,
    NULL AS paperIdLang
    'x:paper/' + safe_iri(QVIEW1.id)
      AS paperId,
    3 AS titleQuestType,
    NULL AS titleLang,
    QVIEW1.title AS title,
    3 AS authorQuestType,
    NULL AS authorLang,
    QVIEW2.name AS author
  FROM
    paper AS QVIEW1,
    author AS QVIEW2
  WHERE QVIEW1.authorId = QVIEW2.id
) SUB_QVIEW
```

(d) Ontop

```
SELECT
  paper.id as paperId,
  paper.title as title,
  author.name as author
FROM paper, author
WHERE
  paper.authorId = author.id
```

(e) EVI

Fig. 1. Sample SPARQL query transformations

Consequently, a SPARQL variable value comprises not only the actual value but also its type (or possibly a language tag).

The mapping adds complexity to the transformation algorithm. Even a simple SPARQL query may result in a large SQL query with joins and unions. For such query, it is essential that the variable representation allows efficient evaluation of joins. Moreover, the mapping may cause that a single SPARQL value is constructed from more than one column in a relational table.

Despite the importance of a correct representation of variables the current approaches neglect this problem. There is currently no general ground describing and comparing the used representations. Figure 1 shows that different approaches produce different queries for the same SPARQL query. Even the results of the queries are different although they have to represent the same SPARQL result. The main difference between the relational queries in the sample is how the SPARQL variables are represented.

In this paper, we will focus purely on the variable representation. We will describe and compare the variable representation in the existing approaches regarding efficiency and validity.

4   Problem Specification

In this paper, we focus only on the approaches which consider a user-defined mapping for publishing a given relational database in the RDF model on the Web through a virtual SPARQL endpoint. Moreover, we selected only the approaches which transform a SPARQL query into a single SQL query. Inputs of such transformation algorithm are a SPARQL query and a mapping from relational structure to RDF. The algorithm has to translate the SPARQL query to a corresponding SQL query, execute the SQL query and then translate the returned SQL result to a SPARQL result. The returned SPARQL result has to be the same as if the input mapping was firstly applied on the relational structure to create an RDF dataset and then the input SPARQL query was evaluated over such RDF dataset.

In this section, we will describe a subset of all issues which have to be handled in transformation algorithms. We focus only on the issues where a chosen variable representation matters. For simplicity, we use the term `column` when we speak about relational columns and the term `variable` when we speak about the SPARQL variable.

Type of a Value. The basic issue is that the representation needs to hold the type of the value. It is needed not only to correctly interpret a result of the SQL query evaluation as a result of SPARQL query evaluation. It is important even when creating the SQL query. For example, when comparing two values the type should be taken into account. Concretely, three literals `"10"`, `"10"^^xs:string` and `"10"^^xs:integer` have the same value but are not equal because their type differs. Moreover, we need to distinguish between RDF types `literal` and `IRI` because they are treated differently.

Type of a Relational Column. Relational columns are typed. If the value of a SPARQL variable is represented by a single relational column then it is needed to convert the SPARQL values to some relational type able to hold value of any SPARQL value. This is shown in Fig. 2.

```
SELECT
    CAST(authors.age AS nvarchar(MAX)) AS misc
FROM authors
UNION ALL
SELECT
    CAST(authors.birthDate AS nvarchar(MAX)) AS misc
FROM papers
```

Fig. 2. Casting values to common type

The problem with the approach shown in Fig. 2 is that SPARQL values stored in column `misc` can be hardly ordered using the SQL ORDER BY clause. For example, values `"41"^^xs:integer` and `"5"^^xs:integer` will have incorrect order because they are ordered as strings. According to SPARQL, they should be ordered as numerics.

Equality of Two Values. RDF defines rules declaring whether two values (terms) are equal or not (see [11]). For IRIs and blank nodes, it is simple. Two IRIs are equal if their value is equal. Two blank nodes are equal if their identifiers are equal. An IRI is never equal to blank node nor to a literal; a blank node is never equal to literal too. Two literals are equal if their value is equal, their type is equal (or both literals do not have a type defined), and the language is equal (or both literals do not have the language defined).

Equality is used for example when evaluating joins. A sample is shown in Fig. 3. In the $Join(M_1, M_2)$ the solution mapping with `"John"` is not part of the join result because the value in the solution mapping M_2 has a type and in the solution mapping M_1 has no type. Similarly, in the $Join(M_1, M_3)$ the solution mapping with `"Zack"` is not part of the join result because the `age` value in the solution mapping M_1 has type `xs:integer` and in the solution mapping M_3 it has type `xs:float`.

$$M_1 = \{(?name \leftarrow "John", ?age \leftarrow "41"^\wedge{}^\wedge xs{:}integer), (?name \leftarrow "Zack", ?age \leftarrow "53"^\wedge{}^\wedge xs{:}integer)\}$$
$$M_2 = \{(?name \leftarrow "John"^\wedge{}^\wedge xs{:}string), (?name \leftarrow "Zack")\}$$
$$M_3 = \{(?age \leftarrow "41"^\wedge{}^\wedge xs{:}integer), (?age \leftarrow "53"^\wedge{}^\wedge xs{:}float)\}$$

$$Join(M_1, M_2) = \{(?name \leftarrow "Zack", ?age \leftarrow "53"^\wedge{}^\wedge xs{:}integer)\}$$
$$Join(M_1, M_3) = \{(?name \leftarrow "John", ?age \leftarrow "41"^\wedge{}^\wedge xs{:}integer)\}$$

Fig. 3. Equality - sample join

Comparison of Two Values. In SPARQL there is another set of rules to compare two values. These rules are used to evaluate SPARQL functions and operators. For example, when we compare two variables on their equality using the FILTER operator (FILTER ?var1 = ?var2) then a different set of rules applies than the ones used for values equality.

The rules are shown on Table 1 (the detailed list is available in [10]). According to the rules, even if the RDF-terms are not equal, their comparison in FILTER may still be evaluated as equal. This can be seen in the sample on Table 2. Although the values of $?age_2$, $?age_3$ and $?age_4$ are different RDF terms it is true that $?age_2 = ?age_3 = ?age_4$, because they are numerically equal.

Value Ordering. There are also additional rules for value ordering (see [10]). When the ORDER BY clause is used over a variable then the values should be ordered as follows (from lowest to highest):

- No value assigned (unbound)
- Blank nodes - order between two blank nodes is undefined
- IRIs - ordered by comparing them as simple literals
- RDF literals

Table 1. Comparison rules

Rule #	Operator	Operands type	Comparison type
1	A=B, A>B, A<B	numeric	Numeric comparison
2	A=B, A>B, A<B	simple literal	String comparison
3	A=B, A>B, A<B	xs:string	String comparison
4	A=B, A>B, A<B	xs:boolean	Boolean comparison of values (true is bigger than false)
5	A=B, A>B, A<B	xs:dateTime	Date time comparison
6	A=B	other	Values equality

Table 2. Comparison samples

Comparison	Result	Rule used
$?name_1 = ?name_2$	✗	6 - $?name_1$ is simple literal, $?name_2$ is xs:string
$?name_2 = ?name_3$	✗	3 - Value is different
$?age_1 = ?age_2$	✗	6 - $?age_1$ is simple literal, $?age_2$ is xs:integer
$?age_2 = ?age_3$	✓	1 - Both values are numeric and the value is numerically equal
$?age_2 = ?age_4$	✓	1 - Both values are numeric and the value is numerically equal

$?name_1 \leftarrow$ "John", $?name_2 \leftarrow$ "John"^^xs:string, $?name_3 \leftarrow$ "Zack"^^xs:string, $?age_1 \leftarrow$ "41", $?age_2 \leftarrow$ "41"^^xs:integer, $?age_3 \leftarrow$ "41"^^xs:float, $?age_4 \leftarrow$ "041"^^xs:integer.

- Ordered by the comparison if it is defined for the values
- If one value is a simple literal and the second one has type $xs:string$ then the simple literal is lower

The ordering of the values from Table 2 is:

- $?name_1 > ?age_1$ - both values are simple literals
- $?name_{2,3} > ?name_1$, $?name_{2,3} > ?age_1$ - the values of $?name_2$ and $?name_3$ are $xs:string$ so they have to be after plain literal, after $?name_1$ and $?age_1$
- $?name_3 > ?name_2$ - both values are $xs:string$ and the value of $?name_3$ is bigger than $?name_2$
- $?age_{2,3,4} \overset{?}{<>} ?age_{2,3,4}$ - the values are compared as equal so the order is undefined
- $?age_{2,3,4} \overset{?}{<>} ?name_{1,2,3}$, $?age_{2,3,4} \overset{?}{<>} ?age_1$ - the orders between $xs:integer$ and simple literal and between $xs:integer$ and $xs:string$ are undefined

There are many orderings undefined, so the values can be ordered as: $?age_1 < ?age_4 < ?name_1 < ?age_2 < ?name_2 < ?name_3 < ?age_3$ or as: $?age_2 < ?age_3 < ?age_4 < ?age_1 < ?name_1 < ?name_2 < ?name_3$.

The Complexity Added by RDB2RDF Mapping. The mapping from relational database to RDF representation may contain advanced rules for generating RDF terms from relational data. For example, the R2RML language (see [9]) provides three options:

- Constant value - the value does not depend on anything from the relational dataset, the value is always the same.
- Column value - the value is taken from a single column. It is possible to declare a type (if omitted the type is automatically detected) and language.
- Templated value - the value is taken from one or more columns. The value is then concatenated using the defined template which may contain some hard-coded parts.

In case of templated IRI, all column values need to be converted into IRI-safe version. That means that all characters which are not alphabetical characters, digits, '-', '.', '_', '~' and some non-ASCII characters need to be encoded as defined in [9].

Even without IRI-safe conversion, the templated values affects the efficiency of joins. The queries in Fig. 4 return the same result. However, the second one is executed faster because the condition is evaluated directly over columns. Moreover, there can be indexes on the columns so the difference could be drastic. So, the variable representation affects the evaluation efficiency, because it matters whether the variable representation allows the query to process joins directly on relational columns instead over some calculated value.

```
SELECT foo                          SELECT 'start' + foo
FROM                                FROM
   (SELECT                             (SELECT
      'start' + table1.col AS foo          table1.col AS foo
   FROM table1) AS V1,                 FROM table1) AS V1,
   (SELECT                             (SELECT
      'start' + table2.col AS foo          table2.col AS foo
   FROM table2) AS V2                  FROM table2) AS V2
WHERE                               WHERE
   V1.foo = V2.foo                     V1.foo = V2.foo
```

(a) Join of templated values   (b) Join of column values

Fig. 4. Join of templated values

5   Approaches to Representing Variables

In this section we will describe individual approaches to the variable representation and discuss their advantages and disadvantages. In some cases we will demonstrate that the representation cannot provide valid results.

Simple Representation. The simplest representation uses a single relational string column in an SQL query to represent a SPARQL variable. The actual value of the SPARQL variable is stored in this relational column. This representation is used in Ultrawrap (see [14]). For example, a variable `author` is represented using a single column `author`.

Typed Simple Representation. The simple representation can be extended to have an extra column(s) to represent a type and a language. One string column contains the actual value of the variable. And additional columns represent the type and the language of the RDF-term. This representation is used for example in Ontop (see [4]). For example, a variable `author` is represented using a column `author` (which contains the actual value of RDF term), a column `authorQuestType` (which contains the type - the type is not there directly, but it contains an identifier of the type) and a column `authorLang` (which contains the language).

Representation with Fixed Column Set. An additional extension is to represent the actual value of a SPARQL variable by several columns in a SQL query to cover all needed types. As mentioned in Sect. 4 we need to handle numeric, string, boolean and date time literals differently. So, a variable can be represented using several columns - representing type, and typed columns for numeric value, string value, boolean value and date time value. A sample is shown in Fig. 5.

SparqlMap. A variant of the representation with fixed column set is the representation used in SparqlMap. The motivation behind this approach is to handle joins over templated IRIs efficiently [15].

```
3 AS foo_type,              2 AS foo_type,              1 AS foo_type,
1 AS foo_dataType,          NULL AS foo_dataType,       NULL AS foo_dataType,
NULL AS foo_text,           foo.label AS foo_text,      'x:foo/' + foo.id AS foo_text,
NULL AS foo_num,            NULL AS foo_num,            NULL AS foo_num,
NULL AS foo_bool,           NULL AS foo_bool,           NULL AS foo_bool,
NULL AS foo_time            NULL AS foo_time            NULL AS foo_time
```

(a) Int value   (b) Simple literal   (c) IRI

Fig. 5. Variable representation with fixed column set

For templated IRIs it is possible to split the value into multiple columns representing individual parts of the IRI. It is needed to align the templates to define the parts so two IRI values are equal if and only if all parts are equal. In common cases it is possible to align the templates in a way that no concatenation and replacement is needed. Therefore it is possible to perform joins on the table columns directly instead of on concatenations. A sample representation of a single variable is shown in Fig. 6. The IRI value is in this case represented using the paperId_reslength, paperId_res_1 and paperId_res_2 columns. In the previous approach, the value will be stored as a concatenation in the paperId_text column.

```
1 AS paperId_type,
1 AS paperId_datatype,
NULL AS paperId_text,
NULL AS paperId_num,
NULL AS paperId_bool,
NULL AS paperId_time,
2 AS paperId_reslength,
'x:paper/' AS paperId_res_1,
paper.id AS paperId_res_2
```

Fig. 6. Representation of a variable in SparqlMap

Dynamic Representation. There is also a completely different approach. A representation which does not use a fixed set of columns to represent a single variable. Instead of that, it is just querying the columns needed. This approach is used in EVI [5].

The example in Fig. 7 shows the representation of a single variable (the same as in Fig. 2). The value is represented by two columns c1, c2 containing the values and the column s used to decide which column should be used. It works in a way that the transformation algorithm not only produces a relational query but also a mapping from relational columns to the SPARQL solution mapping. So, the RDF-term is not completely present in the relational query. It just contains the data needed to reconstruct the value when translating the relational result into the SPARQL result.

The created mapping from relational columns to the SPARQL solution mapping is also used when translating the query. For example, when processing FILTER and JOIN clause, it is needed to use relational CASE clause to handle individual variants of the value.

```
SELECT
    authors.age AS c1, NULL AS c2, 1 AS s
FROM authors
UNION ALL
SELECT
    NULL AS c1, authors.birthDate AS c2, 2 AS s
FROM papers
```

Fig. 7. Dynamic representation

6 Evaluation

We evaluate the approaches from several perspectives - validity and performance perspective. Validity is evaluated from the completeness and soundness perspective. An approach is complete if it transforms any syntactically correct SPARQL query into a syntactically correct SQL query. An approach is sound when a SPARQL query and its corresponding transformed SQL query are semantically equal, i.e. they produce the same results on any input.

6.1 Completeness

All evaluated SPARQL variable representations are complete. The ability to transform a SPARQL query to a SQL query does not depend on the chosen variable representation. Any representation only implies how the individual SPARQL clauses and operators can be transformed into SQL. However, the variable representation may require some specific workarounds. For illustration, we will describe two issues where a straightforward transformation cannot be used but workarounds are available.

For example, to use the SQL SUM aggregation function, it is needed to use it over numeric column. That is problematic for the approaches where the variable value is represented in a SQL string column. In that case the proper workaround would be to use SUM(TRY_CAST(col AS numeric(38,8)))[5].

The approaches where multiple SQL columns are used to represent a single variable may have an issue with SQL operators which work only over a single operand. For example MIN operator. In that case, it is not possible to use it at all. It is needed to follow the SPARQL definition - it is defined as the first value from values sorted in the ascending order [10].

6.2 Soundness

In the Sect. 4 we have described the issues related to the variable representation. From that list, the following ones affect the soundness of transformation (for detailed description of the issues see the Sect. 4):

---

[5] It works because TRY_CAST returns NULL for values which cannot be casted to numeric type.

- **Type** of a value. A SPARQL variable value is not only the actual value but also the value type. Without a type, it is not possible to properly reconstruct SPARQL values from an SQL query result.
- **Equality** of two values. Two values are equal if they have the same type and the same actual value.
- **Comparison** of two values. In contrast to equality, for comparisons there are additional rules for specific types like strings, numerics, date and time and booleans.
- **Ordering**. The ordering uses the comparison logic but there are additional rules affecting how the values should be ordered.

The Simple variable representation does not contain the type information. Therefore, it is not possible to distinguish between various literal types; it is not even possible to make a distinction between literal and IRI.

It is possible to encode the value type and the actual value in a single column - both represented in a string column. However, it would be challenging to be able both test equality and compare the values in the FILTER or ORDER BY clause. Moreover, we consider this representation as the simplest one (in a way how it was used in Ultrawrap approach), so the type is not included.

The Typed Simple variable representation adds the type information. Therefore, the execution results of transformed SQL queries using this variable representation contain the type information. When checking equality of two values it is possible to use the type so even the equality works properly. On the other hand, the actual value is still represented as a single string column which prevents the correct comparison of two values. Moreover, without comparison it is not possible to handle ordering correctly. Except the types support, the problems are the same as with simple representation.

The Fixed Column Set variable representation extends the typed simple variable representation by adding multiple columns to represent the actual values. That is exactly designed to properly handle the comparison. Moreover, it also solves the ordering issue.

To check the equality we can just compare all the columns. It is just needed to take care of NULL values: NULL=NULL has to be evaluated to true. To compare values we will need to handle the type information carefully. That is the reason why there are two columns representing type: *_type and *_dataType. The *_type column contains identifier of a type. However, it is needed that this identifier is the same for all numeric types and the same for all date times. The distinction between individual types is maintained by the *_dataType column. The sample is shown in Fig. 8. In the sample we have chosen the *_type column values as: 0 - blank node, 1 - IRI, 2 - simple literal, 3 - numeric, 4 - string, 5 - boolean, 6 - datetime and 7 and higher for other types. The first three conditions check whether the comparison operator is even defined. If it is not defined then

```
WHERE (
  foo1_type = foo2_type AND
  foo1_type > 1 AND
  foo1_type < 7 AND
  ((foo1_text IS NULL AND foo2_text IS NULL) OR
    (foo1_text>foo2_text)) AND
  ((foo1_num IS NULL AND foo2_num IS NULL) OR
    (foo1_num>foo2_num)) AND
  ((foo1_bool IS NULL AND foo2_bool IS NULL) OR
    (foo1_bool>foo2_bool)) AND
  ((foo1_time IS NULL AND foo2_time IS NULL) OR
    (foo1_time>foo2_time))
)
```

Fig. 8. Comparison of two values (`FILTER ?foo1 > ?foo2`)

the filter is excluding the row from result - see [10]. If the operator is defined then it clearly compares the value on one of the four comparisons while on the other ones both values are NULL.

The ORDER clause is even simpler. The way how we have chosen the *_type column values helps us because it provides the ordering according to the variable type. It is only needed to translate ordering of a single variable to ordering of all columns representing the variable except the *_dataType column. For example, SPARQL clause ORDER BY ?foo can be translated to ORDER BY foo_type, foo_text, foo_num, foo_bool, foo_time. It works, because ordering by foo_type ensures the ordering of values with different type. Moreover, if two RDF-terms are comparable then their values are stored in the same column so the ordering is performed on that column. Otherwise the ordering is undefined so we do not care about the result.

The SparqlMap variable representation extends the fixed column set to improve efficiency of joins over IRI values. The IRI value is represented by several columns, where every column contain a part of a represented IRI value.

However, the ordering of such IRIs representation is complicated. It is not possible to correctly order the IRI parts, because they do not have a fixed length. As an example, an IRI composed from three parts `"prefix\"`, `col1`, `"\suffix"` and an IRI composed from two parts `"prefix\"`, `col2` are not easily comparable. The ordering of individual parts is different than the ordering of the concatenated values. Therefore, it is needed to order the concatenation of the parts.

The Dynamic variable representation uses a completely different approach. It depends on the implementation but the dynamic representation allows to construct the query exactly according to the SPARQL language rules.

The relational columns do not represent the exact RDF-term value, it is needed to handle DISTINCT and REDUCT clauses with care. Usually, the representation is altered in a way, that distinct relational rows represent distinct SPARQL solution mappings.

The similar situation is for ORDER BY and aggregation clauses. In that case, it is needed to alter the transformation to a similar form like in the fixed column representation. The advantage of this approach is that it is needed to produce only the needed columns in a form that the output will be valid. Moreover, the altered representation can be used only for the transformation of solution modifier (ordering, aggregation, DISTINCT and REDUCT clauses). For example, the IRI value can be represented in a single column for the ORDER BY clause while in the subquery the variable is represented directly by columns. Therefore, both efficient join and valid ordering can be achieved.

Summary. The soundness summary is shown on Table 3. The approaches which are sound are the **fixed column set**, the **SparqlMap** and the **dynamic** variable representations.

Table 3. The soundness of approaches

Representation\Issue	Type	Equality	Comparison	Literals ordering	Ordering
Simple	✗	✗	✗	✗	✗
Typed simple	✓	✓	✗	✗	✗
Fixed column set	✓	✓	✓	✓	✓
SparqlMap	✓	✓	✓	✓	✓
Dynamic	✓	✓	✓	✓	✓

6.3 Performance

The performance evaluation was executed on a Windows Server 2012 R2 running on Intel Xeon E5-2673 v3 with 4 GB of RAM. As the relational database, the Microsoft SQL Server 2014 was used.

As the relational dataset was selected the dataset from Berlin SPARQL Benchmark [1]. The Berlin SPARQL Benchmark is built around a use case in which products with various features and categories are offered by several vendors and consumers have posted reviews about products. The dataset was generated using the Berlin SPARQL Benchmark data generator [2] with scale factor of 284 826 - that means approximately 100 M triples after mapping to RDF. The used mapping definition maps the relational dataset to the RDF dataset used in the Berlin SPARQL benchmark.

Figure 9 shows the queries used for the evaluation. To evaluate only the variable representation, we have the same relational query pattern for all of them. In the join queries, every join operand is transformed into a subquery. These subqueries are then joined. In the case of filter query and ordering, the whole query is transformed into a single subquery (without self-joins), and then the filter or ordering is applied.

The performance results are shown in Fig. 10. It can be seen that there are huge differences between the selected approaches. Evaluation of joins can be almost thirty-times faster when comparing the slowest and fastest approach.

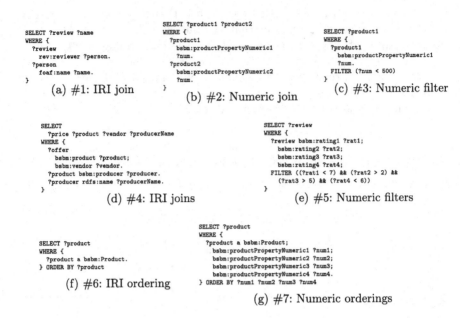

```
SELECT ?review ?name
WHERE {
  ?review
    rev:reviewer ?person.
  ?person
    foaf:name ?name.
}
```
(a) #1: IRI join

```
SELECT ?product1 ?product2
WHERE {
  ?product1
    bsbm:productPropertyNumeric1
    ?num.
  ?product2
    bsbm:productPropertyNumeric2
    ?num.
}
```
(b) #2: Numeric join

```
SELECT ?product1
WHERE {
  ?product1
    bsbm:productPropertyNumeric1
    ?num.
  FILTER (?num < 500)
}
```
(c) #3: Numeric filter

```
SELECT
  ?price ?product ?vendor ?producerName
WHERE {
  ?offer
    bsbm:product ?product;
    bsbm:vendor ?vendor.
  ?product bsbm:producer ?producer.
  ?producer rdfs:name ?producerName.
}
```
(d) #4: IRI joins

```
SELECT ?review
WHERE {
  ?review bsbm:rating1 ?rat1;
    bsbm:rating2 ?rat2;
    bsbm:rating3 ?rat3;
    bsbm:rating4 ?rat4;
  FILTER ((?rat1 < 7) && (?rat2 > 2) &&
    (?rat3 > 5) && (?rat4 < 6))
}
```
(e) #5: Numeric filters

```
SELECT ?product
WHERE {
  ?product a bsbm:Product.
} ORDER BY ?product
```
(f) #6: IRI ordering

```
SELECT ?product
WHERE {
  ?product a bsbm:Product;
    bsbm:productPropertyNumeric1 ?num1;
    bsbm:productPropertyNumeric2 ?num2;
    bsbm:productPropertyNumeric3 ?num3;
    bsbm:productPropertyNumeric4 ?num4.
} ORDER BY ?num1 ?num2 ?num3 ?num4
```
(g) #7: Numeric orderings

Fig. 9. Performance evaluation queries

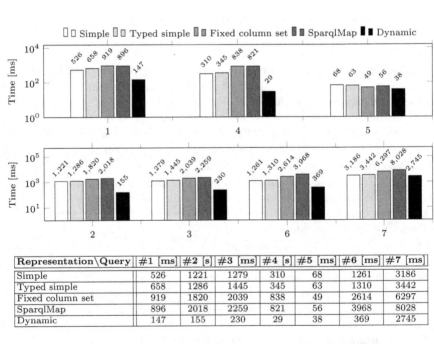

Representation\Query	#1 [ms]	#2 [s]	#3 [ms]	#4 [s]	#5 [ms]	#6 [ms]	#7 [ms]
Simple	526	1221	1279	310	68	1261	3186
Typed simple	658	1286	1445	345	63	1310	3442
Fixed column set	919	1820	2039	838	49	2614	6297
SparqlMap	896	2018	2259	821	56	3968	8028
Dynamic	147	155	230	29	38	369	2745

Fig. 10. The query performance (time of a query execution)

It can be seen that the dynamic approach is the best variant in all measured cases. The comparison of the dynamic and the simple approach shows that it is important to use joins directly on columns. On the other hand, according to the comparison of the fixed column set and the simple approach, it is important to keep the join conditions as simple as possible. The complexity of join conditions is the pain point of the performance of the SparqlMap approach. The join conditions are so complex that the performance difference for IRI joins (query #4) between fixed column set and SparqlMap approach is minimal. Moreover, the complex expressions are causing that these both approaches are the slowest ones.

7 Related Work

In the recent years, the topic of the virtual SPARQL endpoint over relational database becomes more and more important.

The D2R Platform is usually mentioned as the state-of-the-art tool [8]. It was one of the first working solutions supporting user-defined mapping. The tool transforms the SPARQL query into multiple SQL queries and then processes the returned rows in memory to create the final result of the SPARQL query. Because of that, it has a poor performance and sometimes it is not able to process the query at all. We have not included this approach in the survey because it does not transform a SPARQL query into a single corresponding SQL query. Therefore, most of the mentioned issues do not apply in their case. However, this approach is probably the first one with user-defined mapping. Their custom mapping language was later used to create the R2RML language.

Another approach was proposed by a team around Chebotko et al. [6,7]. Their approach uses the relational database only as an efficient storage. Chebotko defines a table where individual RDF triples are stored. The table has three columns: subject, predicate and object. A given SPARQL query is transformed to a single SQL query respecting this structure. The columns defined by the resulting SQL query are mapped one-to-one to the variable values in the SPARQL result.

Ultrawrap [14] is a solution based on Chebotko's work [6,7]. It considers an user-defined mapping. The mapping is used to define an SQL "view" with three columns: subject, predicate and object. Chebotko's algorithm is then evaluated on this view. A sample query is shown in Fig. 1(b).

The authors of SparqlMap have created a different approach [15]. They aimed to create a single unified queries. That means without an SQL "view" used in Ultrawrap. The main idea is that for every basic graph pattern in a given SPARQL query it generates a subquery. Such subquery returns all possible solution mappings[6] for the particular basic graph pattern. Then relational optimizations are performed, so the SQL query gets more compact and can be evaluated

---

[6] The term solution mapping is defined in [10]. Simply said it is a mapping from variables to RDF-terms.

more efficiently. The SparqlMap uses a set of SQL columns to represent a single SPARQL variable. A sample query is shown in Fig. 1(c).

Lately, the Ontop system has been extended with a support of creating virtual SPARQL endpoints [3,4,13]. Ontop uses defined mapping to annotate ontology. Such ontology is then used to generate queries. The Ontop seems to be actively used and from the found tools it has the most active community. A sample query is shown in Fig. 1(d).

Our approach EVI uses enhanced SPARQL algebra [5]. A SPARQL query is first transformed to an expression in our enhanced SPARQL algebra where every basic graph pattern is annotated with a mapping. On such algebra, it is possible to perform optimizations on the SPARQL query. The resulting algebraic expression is optimized and the optimized expression is transformed into a corresponding SQL query. Thanks to the annotation, the transformation is straightforward. A sample query is shown in Fig. 1(e).

8 Conclusions

In this paper, we have shown the importance of the variable representation to the overall validity and performance of the virtual SPARQL endpoint. Although this problem is often neglected, an incorrectly chosen representation may cause that the whole approach will not be efficient or valid.

We have discussed the validity and efficiency of different representations. We have found that the enhancements of fixed column set variable representation done in the SparqlMap approach have only negligible positive effect. Moreover, the enhancements complicate correct IRI ordering and they reduce the performance in comparison to fixed column set variable representation in all other cases than the IRI join(s).

The disadvantage of fixed column set representation is that the conditions are much more complicated than necessary. On the other hand, the advantage of the fixed column variable representation is that the condition has fixed complexity and the usage of such representation is simple.

When using the dynamic representation, it is harder to generate the query. Every condition may be different; it has to be produced according to the context. We have shown that the dynamic representation is valid and efficient. Actually, the difference is so huge that it seems to be essential to use the dynamic representation to implement efficient virtual SPARQL endpoint.

Acknowledgments. This work was supported by the Charles University in Prague, project GA UK No. #158215 and by the Czech Science Foundation (GAČR), grant number 16-09713S.

References

1. Bizer,   C.,   Schultz,   A.:   The   Berlin   SPARQL   benchmark. Int. J. Semant. Web Inf. Syst. **5**(2), 1–24 (2009). http://dblp.uni-trier.de/db/journals/ijswis/ijswis5.html#BizerS09

2. Bizer, C., Schultz, A.: Berlin SPARQL benchmark (BSBM) (2011). http://wifo5-03.informatik.uni-mannheim.de/bizer/berlinsparqlbenchmark/. Accessed Mar 2017

3. Brüggemann, S., Bereta, K., Xiao, G., Koubarakis, M.: Ontology-based data access for maritime security. In: Sack, H., Blomqvist, E., d'Aquin, M., Ghidini, C., Ponzetto, S.P., Lange, C. (eds.) ESWC 2016. LNCS, vol. 9678, pp. 741–757. Springer, Cham (2016). doi:10.1007/978-3-319-34129-3_45

4. Calvanese, D., Cogrel, B., Komla-Ebri, S., Kontchakov, R., Lanti, D., Rezk, M., Rodriguez-Muro, M., Xiao, G.: Ontop: answering SPARQL queries over relational databases. Semant. Web 8(3), 471–487 (2017). https://doi.org/10.3233/SW-160217

5. Chaloupka, M., Nečaský, M.: Efficient SPARQL to SQL translation with user defined mapping. In: Ngonga Ngomo, A.-C., Křemen, P. (eds.) KESW 2016. CCIS, vol. 649, pp. 215–229. Springer, Cham (2016). doi:10.1007/978-3-319-45880-9_17

6. Chebotko, A., Lu, S., Fotouhi, F.: Semantics preserving SPARQL-to-SQL translation. Data Knowl. Eng. 68(10), 973–1000 (2009)

7. Chebotko, A., Lu, S., Jamil, H.M., Foutouhi, F.: Semantics preserving SPARQL-to-SQL query translation for optional graph patterns. Wayne State University, Technical report, November 2006

8. Cyganiak, R.: D2RQ. Accessing relational databases as virtual RDF Graphs. http://d2rq.org/. Accessed Mar 2017

9. Das, S., Cyganiak, R., Sundara, S.: R2RML: RDB to RDF mapping language. W3C recommendation, W3C. http://www.w3.org/TR/2012/REC-r2rml-20120927/

10. Harris, S., Seaborne, A.: SPARQL 1.1 query language. W3C recommendation, W3C, March 2013. http://www.w3.org/TR/2013/REC-sparql11-query-20130321/

11. Lanthaler, M., Wood, D., Cyganiak, R.: RDF 1.1 concepts and abstract syntax. W3C recommendation, W3C, September 2012. http://www.w3.org/TR/2014/REC-rdf11-concepts-20140225/

12. Prud'hommeaux, E., Arenas, M., Bertails, A., Sequeda, J.: A direct mapping of relational data to RDF. W3C recommendation, W3C, September 2012. http://www.w3.org/TR/2012/REC-rdb-direct-mapping-20120927/

13. Rodríguez-Muro, M., Rezk, M.: Efficient SPARQL-to-SQL with R2RML mappings. Web Semant. Sci. Serv. Agents. World Wide Web 33(1), 141–169 (2015)

14. Sequeda, J., Miranker, D.P.: Ultrawrap: SPARQL execution on relational data. Web Semant. Sci. Serv. Agents. World Wide Web 22, 19–39 (2013)

15. Unbehauen, J., Stadler, C., Auer, S.: Accessing relational data on the web with SparqlMap. In: Takeda, H., Qu, Y., Mizoguchi, R., Kitamura, Y. (eds.) JIST 2012. LNCS, vol. 7774, pp. 65–80. Springer, Heidelberg (2013). doi:10.1007/978-3-642-37996-3_5

Semantic OLAP Patterns: Elements of Reusable Business Analytics

Christoph G. Schuetz$^{(\boxtimes)}$, Simon Schausberger, Ilko Kovacic, and Michael Schrefl

Johannes Kepler University Linz, Linz, Austria
{schuetz,schausberger,kovacic,schrefl}@dke.uni-linz.ac.at

Abstract. Online analytical processing (OLAP) allows domain experts to gain insights into a subject of analysis. Domain experts are often casual users who interact with OLAP systems using standardized reports covering most of the domain experts' information needs. Analytical questions not answered by standardized reports must be posed as ad hoc queries. Casual users, however, are typically not familiar with OLAP data models and query languages, preferring to formulate questions in business terms. Experience from industrial research projects shows that ad hoc queries frequently follow certain patterns which can be leveraged to provide assistance to domain experts. For example, queries in many domains focus on the relationships between a set of interest and a set of comparison. This paper proposes a pattern definition framework which allows for a machine-readable representation of recurring, domain-independent patterns in OLAP. Semantic web technologies serve for the definition of OLAP patterns as well as the data models and business terms used to instantiate the patterns. Ad hoc query composition then amounts to selecting an appropriate pattern and instantiating that pattern by reference to semantic predicates that encode business terms. Pattern instances eventually translate into a target language, e.g., SQL.

Keywords: Design patterns · Multidimensional model · Ad hoc queries · Semantic web technologies · Business terms

1 Introduction

Online analytical processing (OLAP) is a common approach to selecting and analyzing structured data (see [20] for further information). Typically, OLAP systems represent data using multidimensional models which can be realized as relational OLAP (ROLAP) or stored in another, highly optimized physical format. Logically, a multidimensional model consist of cubes with hierarchically-ordered levels of granularity, containing multiple data points. Each data point of a cube represents a fact, an occurrence of a business event of interest, quantified by several measures. Data points may then be selected based on different selection criteria (slice and dice) as well as viewed at different granularities to

© Springer International Publishing AG 2017
H. Panetto et al. (Eds.): OTM 2017 Conferences, Part II, LNCS 10574, pp. 318–336, 2017.
https://doi.org/10.1007/978-3-319-69459-7_22

obtain aggregated measures (roll up). Query languages, e.g., SQL or MDX, serve analysts to write the analytical queries in an ad hoc manner.

The ad hoc composition of analytical queries constitutes a challenge for domain experts with little to no experience with OLAP models and query languages. A common solution to this problem sees the provisioning of standardized reports on different subjects, which covers most of the domain experts' information needs in day-to-day analysis situations [6, p. 19]. Sometimes referred to as "ad hoc report navigation" [7, p. 2], analysts are handed input forms to customize reports within certain boundaries, e.g., by filtering the data. The number of such reports often lies in the hundreds [6, p. 21]. Even so, standardized reports do not provide answers to all analytical questions, covering only about 60–80% of the information needs [6, p. 19]. Therefore, some questions must be answered by ad hoc queries. In order to write these queries, domain experts require assistance.

Experience from our industrial research projects semCockpit [14] and agriProKnow[1] shows the existence of patterns of OLAP queries across different domains which can be leveraged to provide assistance to domain experts. Experts in different domains – e.g., analysts at health insurance providers, (research) veterinarians, and dairy herd managers – frequently perform various kinds of comparisons on multiple subjects between groups of facts. While the concrete measures and compared groups differ, the essence of the comparisons varies little across domains and subjects. The systematic gathering and formal representation of these patterns then allows analysts to select an appropriate pattern for instantiation. In the course of pattern instantiation, analysts provide values for specified formal pattern elements: elements from the data model, predicates which encode business terms, and pre-defined calculated measures. The same pattern can be reused for different domains and subjects.

The contribution of this paper is twofold. First, this paper identifies common patterns in OLAP. The presented set of OLAP patterns, however, consists only of a few selected representative patterns which also exist in more specialized variants, e.g., with a specific focus on temporal and spatial aspects. The presented, nonexhaustive set of OLAP patterns then serves to motivate and illustrate the presented approach to identify, represent, and instantiate OLAP patterns. Hence, the main contribution is a method and framework for pattern representation and instantiation based on semantic web technologies. Through the machine-readable representation of OLAP patterns using semantic web technologies, we obtain *semantic* OLAP (semOLAP) patterns which can be algorithmically translated into a target query language. The Resource Description Framework (RDF) serves as technological fundamental. The RDF representation formalizes the textual description of the identified OLAP patterns. Shared conceptualizations of business terms and calculated measures serve to instantiate semOLAP patterns, and can also be linked to existing domain ontologies and (semantic) web resources.

The remainder of this paper is organized as follows. Section 2 presents common patterns in OLAP. Section 3 proposes an approach to pattern representation.

---

[1] http://www.agriProKnow.com/

Section 4 briefly discusses pattern instantiation. Section 5 provides an evaluation of the presented approach. Section 6 reviews related work. The paper concludes with a summary and an outlook on future work.

2 Common Patterns in OLAP

Originally, the notion of pattern has been defined as a rule covering the relation between problem, context, and solution while considering possible constraints: A pattern provides instructions to creating a design artifact that solves a recurring problem in a given environment (see [1]). The Gang of Four (GoF) introduced the notion of pattern to object-orientated software design [10], identifying patterns as common communications vocabulary fostering reuse in the software design process. Similarly, organizational patterns [5] have been identified in order to facilitate the organization of the software design process, eventually leading to agile software development. In data analysis, however, the notion of patterns rarely exists. A proposal for patterns of data analysis in the statistical domain [19] aims at facilitating common data exploration tasks, e.g., detection of outliers and comparison of proportions. A common trait of patterns across domains is their discovery in a practical context rather than being invented [9, p. 8].

We define the notion of OLAP pattern as instructions to composing a query that satisfies the information need in a common analysis situation; the instructions address domain experts and are independent of physical data model and query language. We propose a description form for OLAP patterns (Table 1), adapted from common pattern forms [4], which consists of the pattern name as well as a description of analysis situation, solution, and structure, followed by a domain-specific example. In the following, we present a nonexhaustive set of common OLAP patterns that we identified in the course of industrial research projects. We distinguish *basic patterns* and *comparative patterns*, with a focus on comparative patterns. The examples are taken from the agriProKnow project, which investigates the benefits of data analysis in dairy farming, but similar examples are also found in other domains with different dimensions and measures.

Table 1. Description form for OLAP patterns

Name	What is a distinct name for the pattern?
Situation	What is the analysis situation that this pattern should be applied in?
Solution	What are the steps to follow?
Structure	What are the model elements that participate in the pattern?
Example	What is an example of the pattern's use in a specific domain?

2.1 Basic Patterns

The most elementary OLAP queries generalize multidimensional aggregation queries [18] which commonly target one or more measures from a single class of facts (i.e., cube), grouped along the cube's dimension hierarchies, optionally with filters applied. Multidimensional aggregation queries refrain from relating facts through comparison, i.e., comparing a set of interest with a set of comparison. Hence, a multidimensional aggregation query refers to a fact class along with the examined dimensions, grouping properties (dimensional attributes from the examined dimensions), and measures, combined with joins over dimensional and non-dimensional attributes [18]. In addition, a user may specify filtering criteria (slice and dice conditions) based on the grouping properties (dimensional attributes) and selected non-dimensional attributes, or selected measures.

The most important group of basic patterns are generalized multidimensional aggregation queries with spatial, temporal, and other semantic aggregations. An example query from dairy farming is the retrieval of the average blood sugar level (measure) in blood samples (fact class) of young and underweight cows (slice conditions) located in a specific farm site in Upper Austria in February 2017 (dice conditions). The terms "young" and "underweight" with respect to cows can be defined as shared, possibly parameterized (see Sect. 3) predicates. Another group of basic patterns is cumulative aggregation, e.g., the year-to-date milk yield of a specific farm site. Due to the simplicity of basic patterns, we focus on the more advanced comparative patterns.

2.2 Comparative Patterns

A comparative pattern describes a comparison between a *set of interest* (SI) and a *set of comparison* (SC). The SI refers to a set of facts which represents the state to be analyzed. The SC refers to a set of facts which represents a state that the SI is compared against. Both sets are characterized by a fact class (business event of interest), dimensions and dimension attributes as grouping properties, and measures. For each set, possible slice and dice conditions (selection criteria) further define membership conditions.

The *formal pattern elements* constitute a pattern's structure in the abstract description of the pattern. Analysts supply arguments for the formal pattern elements in order to instantiate the pattern. The pattern instance can then be algorithmically translated into an executable query (see Sect. 4). For example, an instantiation of a comparative pattern could see the SI to be defined as the lactations (fact class, business events) with milk yield (measure) for Simmental cattle (dice condition), the SC as the lactations with milk yield for all cows. Depending on the pattern, constraints apply to the relationships between formal pattern elements, restricting the actual values the formal pattern elements may assume. For example, each measure supplied as argument must apply to the respective fact class in the multidimensional model (milk yield must be a measure for lactations), argument slice and dice conditions refer only to attributes and

Table 2. Description of the set-to-base comparison pattern

Name	Set-to-base comparison
Situation	Compare SI and SC with the same grouping properties and same measures for the same type of business events (fact class). Both sets have common selection criteria but the SI represents a subset of the SC with additional selection criteria
Solution	The fact class, dimensions, grouping properties, measures, and selection criteria for the base set (SC) must be specified. The set of interest is specified by extending the selection criteria of the base set. All other parts of the definition of the SI are taken from the base set
Structure	SI: 1 fact class, 1..* selection criteria, 1..* dimensions, 1..* grouping properties, 1..* measures SC: analogous to SI + 1..* additional selection criteria JOIN: over common grouping properties
Example	Compare the amount of milk produced (measure) in January 2017 (selection criterion in the time dimension), per animal and farm site (grouping properties) (SC) with the overall amount of milk produced with a fat value above five percent (additional selection criteria) (SI)

measures supplied as arguments during instantiation (the breed, e.g., Simmental, is an attribute of the animal dimension).

The distinguishing factors of comparative patterns are as follows. First, the key distinguishing factor is the number and type of elements that are shared by SI and SC. Two groups of comparative patterns can hence be distinguished. If SI and SC refer to the same fact class then the pattern is a homogenous comparative pattern, otherwise the pattern is a heterogenous comparative pattern. Besides the fact class, other pattern elements may also either be shared or specified independently for each set, which allows for further distinction. Another distinguishing factor is the join condition for SI and SC. In the identified patterns there are generally two ways of joining SI and SC: Either the sets are joined over common dimension attributes or by an analyst-supplied join condition. In the following, we describe comparative patterns: set-to-base comparison, homogeneous independent-set comparison, and heterogeneous independent-set comparison.

Set-to-base comparison (Table 2) refers to a homogeneous comparative pattern where the SC (base) shares all elements with the SI and the SI has additional selection criteria (slice and dice conditions): The SI is a subset of SC. From a statistical point of view, set-to-base comparison yields inappropriate results[2] but users often demand such a comparison. A statistically correct variant of the set-to-base pattern employs as SC the complement-base, i.e., the base set without the members of the SI.

Homogeneous independent-set comparison (Table 3) refers to a homogeneous comparative pattern where the SI and the SC share all elements except the slice

---

[2] In that case, the base set includes the set of interest, which distorts the results.

Table 3. Description of the homogeneous independent-set comparison pattern

Name	Homogeneous independent-set comparison
Situation	Compare SI and SC with the same grouping criteria, same measures for the same fact class. Both sets have different selection criteria and are subsets of the same fact class. In addition, calculated measures can be derived based on the previously selected measures. The comparison result can be further restricted by additional result filters referring to the measures and grouping properties of SI and SC
Solution	The fact class, dimensions, grouping properties, measures, and common selection criteria for SI and SC must be specified. For both sets, further selection criteria must be specified independently. After the definition of both sets, calculated measures and further result filters can be defined which are derived from the sets
Structure	SI: 1 fact class, 1..* selection criteria, 1..* dimensions, 1..* grouping properties, 1..* measures SC: analogous to SI but with different, independently defined selection criteria JOIN: over common grouping properties RESULT: 1..* comparative measures, 1..* result filter
Example	Calculate the delta (comparative measure) between milk produced in January 2017 (SI) and milk produced in February 2017 (SC) per animal and farm site. Show only positive delta values (result filter)

and dice conditions: The SI and SC are different subsets of the same cube with the same structure, i.e., the same measures and grouping properties as well as selected non-dimensional attributes. The SI and SC are thus independent sets sharing no selection criteria but the common structure renders these sets directly comparable, allowing for the calculation of calculated measures over the sets. SI and SC are joined over common dimension attributes. A similar pattern allows for the definition of user-defined join conditions for SI and SC.

Heterogeneous independent-set comparison (Table 4) refers to a heterogeneous comparative pattern where the SI and SC potentially share no elements at all; most importantly, the compared classes of facts are different. The pattern shows two independent sets next to each other which are potentially at different levels of granularity. In that default variant, no calculated measure is derived from the result. Then, the result of a pattern instantiation really just amounts to a side-by-side display of two measures at different levels of granularity. Obviously, comparison of measures at the same granularity is preferable. In many cases, however, some data are just not available at the required level of detail. Domain experts still often wish to view these measures side by side in order to compare the values. An extended variant of the pattern allows for the possibility to derive calculated measures from the result.

Table 4. Description of the heterogenous independent-set comparison pattern

Name	Heterogeneous independent-set comparison
Situation	Compare SI and SC regarding to two different business events (fact classes). The SI and SC are independently defined through specification of fact class, dimensions, grouping properties, selection criteria, and measures. Therefore, the join condition is user-defined, as the SI and SC share no common traits. The comparison result can be further restricted by additional result filters referring to the measures and grouping properties of SI and SC
Solution	The fact classes, dimensions, grouping properties, measures, and selection criteria for the sets (SI and SC) must be specified independently. After the definition of both sets, one or more join conditions must be specified. Further result filters can be defined which are derived from the sets
Structure	SI: 1 fact class, 1..* selection criteria, 1..* dimensions, 1..* grouping properties, 1..* measures SC: analogous to SI but independently defined JOIN: over specified conditions RESULT: 1..* result filter
Example	Show the amount of milk produced in July 2017 per animal, farm site, and day (SI) next to the quantity of food given the day before on that farm site in July 2017 (SC)

3   Pattern Representation

An RDF representation formalizes the textual pattern descriptions: We obtain semantic OLAP (semOLAP) patterns. Each formal pattern element translates into an RDF property. Pattern instantiations are specified using these RDF properties to reference multidimensional model elements, calculated measures, and business terms. Patterns have a language-dependent expression that allows for automatic translation into an executable query (see Sect. 3.2).

3.1   Definition of Patterns, Predicates, and Calculated Measures

The RDF representation of OLAP patterns links a multidimensional model, represented using the RDF Data Cube (QB) [21] vocabulary and its extension QB4OLAP [8], to the pattern representation. The RDF representation of OLAP patterns itself is defined in a pattern language vocabulary with prefix pl.

The RDF representation of an OLAP pattern includes data properties for a textual description according to the form in Table 1 as well as object properties describing the formal parameters. Consider, for example, the RDF representation of homogeneous independent-set comparison (Listing 1). The textual description of the pattern is stated using the properties pl:name, pl:situation, pl:solution, pl:structure, and pl:example (Lines 2–6). The formal pattern elements of the pattern are stated using the `pl:hasElement` property (Lines 7–10). The `pl:result` property serves to define a pattern element as the result of a pattern. For example, the `compMeasure` formal element is a result of homogeneous independent-set comparison (Line 15). An element may also be renamed before inclusion in the result, e.g., by specifying a prefix (Lines 11–14).

Listing 1. RDF representation of homogeneous independent-set comparison

```
1  : homogeneous IndependentSetComparison  a  pl: Pattern ;
2     pl:name "Homogeneous  independent-set  comparison"@en ;
3     pl:situation  "Compare SI and SC with ..."@en ;
4     pl:solution  "The fact class , dimensions , ..."@en ;
5     pl:structure  "SI: 1 fact class , ..."@en ;
6     pl:example  "Calculate the delta ..."@en ;
7     pl:hasElement  :factClass ,  :factClassSlice ,
8        :measure ,  :dimension ,  :dimensionAttribute ,
9        :setOfInterest ,  :setOfComparison ,  :siSlice ,  :scSlice ,
10       :compMeasure ,  :resultSlice ;
11    pl:result  [
12       pl:element  :measure ;
13       pl:elementPrefix  "SI_"
14    ] ;
15    pl:result  :compMeasure ;
16    pl:result  :dimensionAttribute .
```

Listing 2. RDF representation of formal pattern elements for comparative patterns

```
1  :setOfInterest  a  pl:PatternElementGroup ;
2     pl:elementGroupAlias  "SI".
3  :setOfComparison  a  pl:PatternElementGroup ;
4     pl:elementGroupAlias  "SC".
5  :factClass  a  pl:PatternElement ;
6     rdfs:range  pl:Fact ;
7     pl:multiplicity  pl:One.
8  :factClassSlice  a  pl:PatternElement ;
9     rdfs:range  pl:FactPredicate ;
10    pl:multiplicity  pl:OneOrMore.
11 :measure  a  pl:PatternElement ;
12    rdfs:range  qb:MeasureProperty ;
13    pl:multiplicity  pl:OneOrMore.
```

A semOLAP pattern's formal elements are further described by RDF properties. Listing 2 shows an example definition of formal pattern elements for comparative patterns. There are two different types of formal pattern elements: atomic formal elements and pattern element groups. A pattern element group (`pl:PatternElementGroup`) consists of other formal pattern elements (Fig. 1). A pattern element group possibly has an alias assigned (`pl:elementGroupAlias`);

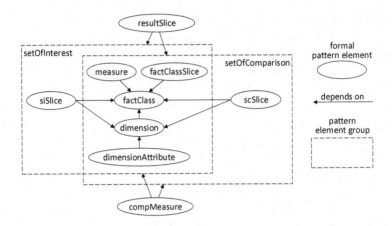

Fig. 1. Relationships between elements of homogeneous independent-set comparison

the group may in expressions be referred to by its alias. The second type of elements are atomic formal pattern elements (Lines 5–13). Each atomic element (pl:PatternElement) is an object property that serves for the description of pattern instances. On instantiation, depending on the pl:multiplicity, one or more values have to be supplied for each of these properties. The rdfs:range property defines the type which the provided values must be instances of.

A semOLAP pattern's formal elements are related to each other. Figure 1 illustrates the relationships between the formal elements of the homogeneous independent-set comparison. This comparison consists of the two pattern element groups setOfInterest and setOfComparison. These pattern element groups comprise formal pattern elements which depend on each other, e.g., a pattern element group's dimension as well as other elements either depend directly or indirectly on its fact class. Comparative measures and result filters do not depend directly on the fact class. Instead, these pattern elements depend on grouping properties and the result measures. The dependencies are important for code generation (see Sect. 3.2) and the user interface (see Sect. 5).

Business terms are represented as fact predicates, dimension predicates, and group predicates (Listing 3). The classes pl:ObjectCalcMeasure and pl:Group-CalcMeasure represent types of calculated measures. The aggregation function of an pl:ObjectCalcMeasure targets all entries within a group whereas a pl:GroupCalcMeasure depends on existing measures of the pattern element groups SI and SC. Examples of measure descriptions are provided in Listing 3. The pl:ObjectCalcMeasure (Lines 22–25) defines the structure of the sumMilk-Yield measure covering a language specific description and a reference to relevant concept in the AGROVOC ontology. The comparative measure delta-MilkYield in (Lines 26–35) represents a pl:groupCalcMeasure by defining the measures of the pattern element groups.

Listing 3. Example predicates and calculated measures

```
1  :highFat a pl:FactPredicate;
2    rdfs:comment "Restricts to fat content over "@en;
3    rdfs:seeAlso agrovoc:c_4360;
4    rdfs:isDefinedBy agri:highFat;
5    pl:uses agri:fatContent.
6  :monthJanuary a pl:DimensionPredicate;
7    rdfs:comment "Month is January"@en;
8    pl:usesDimension [
9      pl:usedDimLvl agri:date_;
10     pl:usedDimAttributes agri:month
11   ].
12 :DaytoDayBefore a pl:GroupPredicate;
13   rdfs:comment "Matches date from set of interest to
         previous day in set of comparison"@en;
14   pl:usesGroup [
15     pl:usedGroup  :setOfInterest;
16     pl:usedGroupAttribute agri:date_
17   ];
18   pl:usesGroup [
19     pl:usedGroup  :setOfComparison;
20     pl:usedGroupAttribute agri:date_
21   ].
22 :sumMilkYield a pl:ObjectCalcMeasure;
23   rdfs:comment "Summed-up milk yield"@en;
24   rdfs:seeAlso agrovoc:c_15998;
25   pl:uses agri:milkYield.
26 :deltaMilkYield a pl:GroupCalcMeasure;
27   rdfs:comment "Calculates the delta between sumMilkYield
         in SI and sumMilkYield in SC"@en;
28   pl:usesGroup [
29     pl:usedGroup  :setOfInterest;
30     pl:usedGroupAttribute  :sumMilkYield
31   ];
32   pl:usesGroup [
33     pl:usedGroup  :setOfComparison;
34     pl:usedGroupAttribute  :sumMilkYield
35   ].
```

3.2 Pattern-Based Query Writing

Pattern-based query writing relies on *pattern expressions*. A pattern expression
defines a template that realizes a specific pattern in a target language, with placeholders that correspond to the formal pattern elements. Language-specific templates allow to use the target language to full capacity, e.g., analytical functions,
built-in functions, stored procedures, and features of system-specific dialects.
In the RDF representation, the pl:expression property, with a string range,
associates a pattern with its pattern expression. Upon instantiation, the placeholders in the template are replaced with the actual pattern elements supplied
as arguments for the pattern's formal elements (see Sect. 4).

Listing 4. Grammar of the pattern expression language for SQL in EBNF notation

```
 1 sqlTemplate      = SQLTEXT | patternFunction ,
 2                    {SQLTEXT | patternFunction};
 3 patternFunction  = singleton | commaList | andList |
 4                    joinList | joinConditon ;
 5 singleton        = '!E', baseElement | asClause, '!E';
 6 commaList        = '!CL', baseElement | asClause, '!CL';
 7 andList          = '!AL', baseElement, '!AL';
 8 joinList         = '!JL', elementRole, '!JL';
 9 joinConditon     = '!JC',
10                    baseElement, '=' , baseElement ,
11                    '!JC';
12 asClause         = '![',
13                    baseElement, ['AS', prefix | SQLTEXT],
14                    '!]';
15 baseElement      = [tablePrefix], elementRole;
16 tablePrefix      = 'table' | SQLTEXT , '.';
17 elementRole      = [prefix], '<', ID, [attribute], '>';
18 prefix           = SQLTEXT , '!+';
19 attribute        = '.' , ID;
20 ID               = '_' | letter | digit ,
21                    {letter | digit | '_' | '-'}.
22 SQLTEXT          = '"', character , {character}, '"';
```

A pattern expression is defined using an *expression language* specific to the target query language. Listing 4 defines, in EBNF notation [13], the grammar of a pattern expression language for the translation of patterns into executable SQL queries. Expressions in that language consist largely of SQLTEXT, i.e., arbitrary SQL code[3], enclosed in double quotation marks ("). When the pattern expression is parsed, SQLTEXT goes in verbatim form into the resulting query. The ID element serves to identify formal pattern elements or their attributes. In a pattern expression, the ID of a pattern element corresponds, by convention, to the element's IRI without the prefix. The prefix element is used to indicate a string that is put before an element's name in the resulting query. The elementRole serves to reference formal pattern elements and their attributes. If the elementRole does not reference a formal pattern element but an element's attribute, e.g., the name or expression (in case of calculated measures and predicates), a second ID follows the first ID, separated by a dot. Elements may be assigned an alias (asClause). A baseElement has an elementRole and may have a tablePrefix that consists either of the "table" string or an SQLTEXT followed by a dot. In the final query, a "table" string as tablePrefix is replaced with the name of the table associated with the respective elementRole in the RDF representation of the multidimensional model. Since a pattern instantiation may have multiple actual pattern elements for each formal element, multiple actual pattern

---

[3] For simplicity, the grammar defines SQLTEXT as a string of characters.

elements may correspond to a single `elementRole` in the pattern expression. A `commaList` serves to indicate lists of values seperated by comma, an `andList` to indicate lists seperated by "AND". The `singleton` clause indicates that only one value is expected. Upon parsing a `joinList`, the actual pattern elements that correspond to the formal pattern element of the `elementRole` are treated as tables and corresponding inner join clauses are created. The `joinCondition` serves to explicitly define a join condition referring to base elements.

Listing 5 shows a pattern expression for homogeneous independent-set comparison using the expression language as defined in Listing 4. The query template consists of two subqueries: the first to retrieve the SI, the other to retrieve the SC. Both subqueries refer to the same formal element `measure` (Lines 11 and 17) in the `SELECT` clause. An instantiation may provide multiple actual elements for `measure`, thus multiple actual elements may correspond to that same `elementRole`. The expression of each actual element passed for `measure` in an instantiation goes into a comma-separated list in each subquery's `SELECT` clause. In the result projection, the measures are prefixed (Lines 6 and 7), thus mirroring the definition in Listing 1. In the `GROUP BY` clauses, the `table` attribute of the dimension attributes is referred to, the value of which is derived after instantiation at transformation time.

Listing 5. Pattern expression for homogeneous independent-set comparison

```
1  "WITH fact AS (
2    SELECT * FROM"!E <factClass> !E"fact
3    WHERE"!CL <factClassSlice.expression> !CL
4  ")"
5  "SELECT"!CL"SI".<dimensionAttribute> !CL","
6          !CL ![ "SI".<measure> AS"SI_"!+ !] !CL","
7          !CL ![ "SC".<measure> AS"SC_"!+ !] !CL","
8          !CL ![ <compMeasure.expression> !] !CL
9  "FROM ( "
10   "SELECT"!CL table.<dimensionAttribute> !CL","
11             !CL ![ <measure.expression> !] !CL
12   "FROM fact"!JL <dimension> !JL
13   "WHERE"!AL <siSlice.expression> !AL
14   "GROUP BY"!CL table.<dimensionAttribute> !CL
15  ") SI JOIN ("
16   "SELECT"!CL table.<dimensionAttribute>!CL","
17             !CL ![ <measure.expression> !] !CL
18   "FROM fact"!JL <dimension> !JL
19   "WHERE"!AL <scSlice.expression> !AL
20   "GROUP BY"!CL table.<dimensionAttribute> !CL
21  ") SC ON"!JC
22   "SI".<dimensionAttribute> =
23   "SC".<dimensionAttribute>
24  !JC
25  WHERE" !AL <resultSlice.expression> !AL
```

Listing 6. Example predicate expressions and calculated measures

```
1 :highFat pl:expression "fatContent > 5".
2 :monthJanuary pl:expression "date_.month = 1".
3 :matchDayToDayBefore
4   pl:expression "SI.date_ - 1 = SC.date_".
5 :sumMilkYield pl:expression "SUM(milkYield)".
6 :deltaMilkYield pl:expression
7   "SI.sumMilkYield - SC.sumMilkYield".
```

Listing 6 shows examples of predicate expressions and expressions for calculated measures. The simple SQL expressions can be directly inserted into the query template. The `highFat` predicate refers to a measure (`fatContent`). The `monthJanuary` predicate refers to a dimensional attribute (`month`); notice the mandatory use of the dimension name in expressions of dimension predicates. The `matchDayToDayBefore` predicate relates the day in the SC with the previous day in the SI; notice the use of group aliases. The `sumMilkYield` measure performs summarization of the `milkYield` measure during roll-up operations, and `deltaMilkYield` is a comparative measure. The definition of expressions for calculated measures and predicates adapts and extends the BIRD approach [16] to reference modeling for data analysis.

4   Pattern Instantiation

A pattern's formal pattern elements become object properties of the resource that represents a pattern instance. These properties represent actual pattern elements and refer to elements from the multidimensional model, predicates, or calculated measures. For example, when instantiating homogeneous independent-set comparison, the pattern elements defined in Listing 1, Lines 7–10, must be provided as arguments. Now consider a pattern instance to compare the sum of milk yields by animal and farm site in January and February in the year 2017 by calculating the change in milk yields (the "delta"), but only where that change is positive. Then, the pattern instance provides the `milk` fact as value for the `factClass` property and the `dateIn2017` predicate for `factClassSlice`. The pattern instance further provides the `sumMilkYield` calculated measure for `measure` along with the `animalDim`, `farmSiteDim`, and `dateDim` dimensions for `dimension`. The `animal` and `farmSite` dimension attributes serve as grouping properties. Furthermore, the `siSlice` property, restricting SI, refers to the `monthJanuary` predicate, and the `scSlice` property, restricting SC, refers to the `monthFebruary` predicate. Finally, the pattern instance provides as `compMeasure` property the `deltaMilkYield` calculated measure which compares the milk yield values of SI and SC, and as `resultSlice` the `deltaMilkYieldIsPositive` predicate in order to show only positive values of the comparative measure.

Upon pattern instance execution, the actual pattern elements are placed into the corresponding gaps in the pattern expression. The automatic translation of a pattern instance into an executable query in a target language requires retrieval of additional properties of the pattern instance's actual elements from the data model. For example, in order to translate the pattern instance to compare the sum of milk yields by animal and farm site in January and February in the year 2017 into an executable SQL query, the summMilkYield measure's expression must first be retrieved from the RDF representation using SPARQL queries since the corresponding pattern expression refers to the expression property of the measure formal pattern element (Listing 5, Lines 11 and 17). Derivation rules stored in RDF format encode knowledge about the retrieval of pattern element properties as strings containing SPARQL query fragments. The template is then executed resolving all pattern expression language element by providing the queried properties.

Relationships between formal pattern elements constrain pattern instantiation. For example, the scSlice element depends on factClass and dimension. Therefore, an instantiation may employ as actual value for scSlice only predicates that refer to measures in the corresponding fact class or attributes of the corresponding dimensions. The constraints must be considered when instantiating a pattern, otherwise the resulting query is invalid.

In order to increase flexibility of the semOLAP approach, we propose dynamic predicates which allow for ad hoc definition of simple predicates during pattern instantiation. The concept can be extended in order to support the dynamic definition of join conditions as well. Hence, analysts select an attribute (measure or dimensional attribute), a comparison operator, and a concrete value for the comparison. Dynamic predicates are represented by blank nodes and only valid within the pattern instance which they are defined for. For query execution, dynamic predicates are translated into SQL expressions.

An instantiation of a pattern that binds all of the pattern's formal elements with an actual pattern element is an executable pattern instance. But, it is also possible to create a non-executable pattern instance by providing actual elements as argument values only for a subset of the pattern's formal elements. When instantiating such a partial pattern instance, or pattern instance template, analysts provide only the missing values. That way, analysts may first select a fact class before selecting a partially instantiated pattern that already includes, for the previously selected fact class, the typically-relevant measures, dimensional elements, grouping properties, and selection criteria.

5   Evaluation

The purpose of semOLAP patterns is to support casual users with composing ad hoc queries to retrieve knowledge not covered by standardized reports. A prevalent approach to solving the same underlying issue of facilitating the composition of complex queries is *reuse of queries*. For example, QUERYAID [17], as a representative system, also aims to facilitate the composition of complex queries for casual users by storing, accessing, and reusing existing queries. QUERYAID's object-orientated approach that distinguishes queries for and by reuse differs from the pattern-based approach. The queries for reuse are specified by identifying concrete domain- and language-specific queries and transforming the queries to a canonical form, allowing for the identification of syntactically different but semantically equivalent query representations. QUERYAID suggests queries and join-conditions based on users' query formulation by determining semantically related terms in the respective query descriptions. The user is able to browse through a repository of queries which are related to each other over specialization and generalization hierarchies and to select or combine suitable ones. The approach of reusing queries, however, faces multiple drawbacks. First, the composition of queries from scratch is not supported since it always depends on preceding similar queries. Previous findings show that reusing and modifying similar queries instead of composing them from scratch is more error prone and provides less accurate results [2]. Supporting query reuse requires a repository, classification, and a sophisticated search mechanism as well as mechanisms to mitigate cognitive factors. Anchoring and adjustment bias have been identified as key factors leading to inappropriate results [2]. Anchoring refers to the fact that users tend to specify queries close to the initial query whereas the adjustment bias describes the failure of modifying the query in a way to fit the current information demand [2]. Errors and unnecessary functionality are thereby taken over from the initial query [2]. The QUERYAID system does not mitigate neither anchoring nor adjustment bias. Query reuse lowers the composition time but the probability of producing correct queries that satisfy the information need decreases [2]. The semOLAP patterns can also positively affect composition time, since query formulation as well as query translation and writing are supported. The (correct) satisfaction of the information need can also be fostered by semOLAP patterns since business terms close to natural language and description of the patterns are provided.

We developed a prototype application to instantiate patterns and manage pattern-related elements such as calculated measures and predicates. The prototype comprises three components: a ROLAP system, a triple store, and a web application. A ROLAP system covering business events of the agricultural domain is used to store the instance data. A triple store is used to persist the RDF multidimensional model, all patterns, and related RDF data, e.g., pattern instances, calculated measures, and predicates. The web application covers the user interface and the pattern engine. The user interface is implemented using the interaction flow modeling language (IFML)[4] and the WebRatio[5] platform which facilitates model driven development of data centric applications. Based on the pattern to be instantiated, the prototype dynamically generates a user interface. The relations of the pattern's formal pattern elements are used to calculate a corresponding graph. The edges of the graph are derived from the relations of the formal pattern elements (see Fig. 1) and the vertices represent the formal pattern elements themselves. The graph further represents the dependencies between the formal pattern elements, thus allowing to determine a navigation path for the user interface. Considering these dependencies, a wizard guides the user through the pattern instantiation process. For each formal pattern element, potential

---

[4] http://www.omg.org/ifml/.
[5] http://www.webratio.com.

input values are provided, which are derived from the pattern element's range and existing dependencies. The dependencies are represented as SPARQL snippets specifying retrieval of possible input values from the QB and QB4OLAP model. These snippets can be reused across patterns since dependencies between specific formal pattern elements are recurring. Yet unknown patterns with not considered dependencies between specific formal pattern elements can be instantiated with as little effort as defining the missing snippets. If values for all formal pattern elements are specified then an executable pattern instance is generated by the pattern engine. Otherwise the result is a non-executable partial pattern instance. The executable pattern instance is then used by the pattern engine to generate executable language specific queries, i.e., SQL queries which are used to query the underlying ROLAP system.

In course of the agriProKnow project, which aims to create a data analysis platform for precision dairy farming to increase efficiency of dairy production, we employ semOLAP patterns and the described user interface prototype. Domain experts, such as (research) veterinarians, employ the user interface for semOLAP patterns to extract knowledge from dairy farming data. Therefore, predicates representing agricultural business terms (see Listing 3) and corresponding calculated measures (see Listing 3) are available. Initial experience so far suggests that semOLAP patterns and the implemented user interface are valuable tools for domain experts and enable them to compose ad hoc queries without expert database knowledge. Further studies for evaluation of semOLAP patterns and the developed user interface are planned for future work.

6 Related Work

Pattern-based approaches to design [1] are found in many domains. In software engineering, a framework of object-oriented design patterns [10] facilitates bottom-up design of software systems. Software design patterns are interrelated through generalization, specialization, and composition relationships (see [10]). Design patterns, therefore, act as building blocks that foster reuse and maintainability throughout the design process. In statistics, data analysis patterns have been defined [19] to describe the methodology of exploratory data analysis (EDA). These data analysis patterns aim at guiding users through EDA tasks, focusing on graphical representation and helpful transformations of the explored data. The semOLAP patterns, on the other hand, help users to answer questions which arise in specific analysis situations. Therefore, EDA cannot be considered a focus of semOLAP patterns: EDA focuses on the exploration of data without former hypothesis whereas users in a specific analysis situation already have a question in mind. In conceptual modeling, patterns of conceptual models relevant across different domains have been identified [9,12]. In order to uncover the foundations of a modeling problem, modelers invest considerable effort. By reusing proven concepts, the effort to define conceptual models can be drastically reduced [9]. Furthermore, standardized modeling fosters readability, robustness against changes, and maintainability [12, p. 4]. Conceptual patterns should be seen as suggestions and not prescriptions acting as starting points in the modeling process [9, p. 12]. Similar to common patterns in conceptual modeling [9,12], the semOLAP patterns have emerged from experience gained in various industrial research projects. Likewise, semOLAP patterns can be seen as starting point for OLAP, providing analysts with suggestions to follow.

Providing support for query composition is a research area with a broad tradition. Various approaches exist to supporting the composition of multidimensional queries. Frequently, a visual interface is employed for query composition, sparing the user the intricacies of the underlying multidimensional data model. For example, the Semantic Data Warehouse Model (SDWM) [3] visualizes the elements of the multidimensional model as graphical representations by encoding the semantics using different shapes, colors, contours, and formats. The focus of SDWM is the representation of measures, following the idea that a potential user is primarily interested in measures, hence, using them as the starting point of the analysis. Patterns in the traditional sense [1,10,19] are not identified in SDWM, which instead classifies measures into atomic and complex measures. Atomic measures in SDWM can be compared to simple multidimensional aggregations using restrictions to values, which is captured in our approach by the basic semOLAP patterns. Complex measures in SDWM are limited to measure ratios relying on independent sets, subsets and base sets, or further dimension level conditions. The semOLAP patterns avoid these limitations by supporting arbitrary comparisons through abstraction of a measure-centric view to task-related patterns. Furthermore, the semOLAP patterns consider comparisons between different fact classes, which is not mentioned in SDWM. Apart from the visualization of query interfaces, recommender systems provide users with suitable queries for particular analysis situations. Common recommender systems [11,15] typically suggest queries to a user based on the user's previous query behavior (content-based) or based on the behavior of similar users (collaborative). Those similar users can be determined by comparing the user profiles, the query situations, or the currently investigated differences [11]. Rather than suggesting historical queries, recommender systems could be used to suggest the underlying semOLAP pattern beneath those queries and provide pattern-completion by predicting or suggesting multidimensional objects.

7   Summary and Future Work

We have proposed a pattern-based approach to facilitate ad hoc query composition from scratch in OLAP systems. The main contribution is a framework covering the semantic representation of patterns, the translation into a target query language, and supporting dynamic generation of a user interface. The applicability of the framework is demonstrated by its successful use in the agriProKnow project. Future work will cover the definition of further pattern expression languages, e.g., to support translation of semOLAP pattern instances to MDX. Furthermore, the introduced semOLAP patterns should be extended and distinguished according to temporal, spatial, and semantic aspects. The findings will then be evaluated in a study with casual users across domains.

Acknowledgments. The research reported in this work was funded by the Austrian Federal Ministry of Transport, Innovation and Technology (BMVIT) under program "Production of the Future", Grant No. 848610.

References

1. Alexander, C., Ishikawa, S., Silverstein, M., i Ramió, J.R., Jacobson, M., Fiksdahl-King, I.: A Pattern Language. Oxford University Press, New York (1977)
2. Allen, G., Parsons, J.: Is query reuse potentially harmful? anchoring and adjustment in adapting existing database queries. Inf. Syst. Res. **21**(1), 56–77 (2010)
3. Böhnlein, M., Ulbrich-vom Ende, A., Plaha, M.: Visual specification of multidimensional queries based on a semantic data model. In: von Maur, E., Winter, R. (eds.) Vom Data Warehouse zum Corporate Knowledge Center, pp. 379–397. Physica, Heidelberg (2002). doi:10.1007/978-3-642-57491-7_22
4. Coplien, J.O.: Software patterns. SIGS Management Briefings Series (1996)
5. Coplien, J.O., Harrison, N.B.: Organizational Patterns of Agile Software Development. Pearson Prentice Hall, London (2005)
6. Eckerson, W.W.: Pervasive business intelligence: techniques and technologies to deploy BI on an enterprise scale. TDWI Best Practices Report (2008)
7. Eckerson, W.W.: TDWI checklist report: Self-service BI. TDWI Research (2009)
8. Etcheverry, L., Vaisman, A.A.: QB4OLAP: a new vocabulary for OLAP cubes on the semantic web. In: Proceedings of the Third International Conference on Consuming Linked Data, pp. 27–38 (2012)
9. Fowler, M.: Analysis Patterns: Reusable Object Models. Addison-Wesley Professional, Reading (1997)
10. Gamma, E., Helm, R., Johnson, R., Vlissides, J.: Design patterns: abstraction and reuse of object-oriented design. In: Nierstrasz, O.M. (ed.) ECOOP 1993. LNCS, vol. 707, pp. 406–431. Springer, Heidelberg (1993). doi:10.1007/3-540-47910-4_21
11. Giacometti, A., Marcel, P., Negre, E., Soulet, A.: Query recommendations for OLAP discovery driven analysis. In: Proceedings of the ACM Twelfth International Workshop on Data Warehousing and OLAP, pp. 81–88. ACM (2009)
12. Hay, D.C.: Data Model Patterns: Conventions of Thought. Dorset House, New York (1996)
13. ISO/IEC: 14977: 1996 (E) - information technology - syntactic metalanguage - Extended BNF. International Standard (1996)
14. Neuböck, T., Neumayr, B., Schrefl, M., Schütz, C.: Ontology-driven business intelligence for comparative data analysis. In: Zimányi, E. (ed.) eBISS 2013. LNBIP, vol. 172, pp. 77–120. Springer, Cham (2014). doi:10.1007/978-3-319-05461-2_3
15. Sapia, C.: On modeling and predicting query behavior in OLAP systems. In: Gatziu, S., Jeusfeld, M.A., Staudt, M., Vassiliou, Y. (eds.) DMDW 1999. CEUR Workshop Proceedings, vol. 19. CEUR-WS.org (1999)
16. Schuetz, C.G., Neumayr, B., Schrefl, M., Neuböck, T.: Reference modeling for data analysis: the BIRD approach. Int. J. Coop. Inf. Syst. **25**(2), 309–344 (2016)
17. Seriai, A., Oussalah, C.: A reuse based object-oriented framework towards flexible formulation of complex queries. In: Larsen, H.L., Andreasen, T., Christiansen, H., Kacprzyk, J., Zadrożny, S. (eds.) Flexible Query Answering Systems. Advances in Soft Computing, vol. 7, pp. 128–137. Physica, Heidelberg (2001). doi:10.1007/978-3-7908-1834-5_12

18. Theodoratos, D.: Exploiting hierarchical clustering in evaluating multidimensional aggregation queries. In: Proceedings of the 6th ACM International Workshop on Data Warehousing and OLAP, pp. 63–70 (2003)
19. Unwin, A.: Patterns of data analysis. J. Korean Stat. Soc. **30**(2), 219–230 (2001)
20. Vaisman, A., Zimányi, E.: Data Warehouse Systems. Springer, Berlin (2014)
21. W3C: The RDF Data Cube Vocabulary - W3C Recommendation 16 January 2014. https://www.w3.org/TR/2014/REC-vocab-data-cube-20140116/

An Extensible Ontology Modeling Approach Using Post Coordinated Expressions for Semantic Provenance in Biomedical Research

Joshua Valdez[1], Michael Rueschman[2], Matthew Kim[2],
Sara Arabyarmohammadi[1], Susan Redline[2], and Satya S. Sahoo[1(✉)]

[1] Institute for Computational Biology and Electrical Engineering
and Computer Science Department, Case Western Reserve University,
Cleveland, OH, USA
{joshua.valdez, sara.arabyarmohammadi,
satya.sahoo}@case.edu
[2] Department of Medicine, Brigham and Women's Hospital
and Beth Israel Deaconess Medical Center, Harvard University,
Boston, MA, USA
{mrueschman,mikim,sredline1}@bwh.harvard.edu

Abstract. Provenance metadata describing the source or origin of data is critical to verify and validate results of scientific experiments. Indeed, reproducibility of scientific studies is rapidly gaining significant attention in the research community, for example biomedical and healthcare research. To address this challenge in the biomedical research domain, we have developed the Provenance for Clinical and Healthcare Research (ProvCaRe) using World Wide Web Consortium (W3C) PROV specifications, including the PROV Ontology (PROV-O). In the ProvCaRe project, we are extending PROV-O to create a formal model of provenance information that is necessary for scientific reproducibility and replication in biomedical research. However, there are several challenges associated with the development of the ProvCaRe ontology, including: (1) *Ontology engineering*: modeling all biomedical provenance-related terms in an ontology has undefined scope and is not feasible before the release of the ontology; (2) *Redundancy*: there are a large number of existing biomedical ontologies that already model relevant biomedical terms; and (3) *Ontology maintenance*: adding or deleting terms from a large ontology is error prone and it will be difficult to maintain the ontology over time. Therefore, in contrast to modeling all classes and properties in an ontology before deployment (also called *precoordination*), we propose the "ProvCaRe Compositional Grammar Syntax" to model ontology classes *on-demand* (also called *postcoordination*). The compositional grammar syntax allows us to re-use existing biomedical ontology classes and compose provenance-specific terms that extend PROV-O classes and properties. We demonstrate the application of this approach in the ProvCaRe ontology and the use of the ontology in the development of the ProvCaRe knowledgebase that consists of more than 38 million provenance triples automatically extracted from 384,802 published research articles using a text processing workflow.

© Springer International Publishing AG 2017
H. Panetto et al. (Eds.): OTM 2017 Conferences, Part II, LNCS 10574, pp. 337–352, 2017.
https://doi.org/10.1007/978-3-319-69459-7_23

Keywords: Precoordinated and postcoordinated expression · Ontology engineering · Provenance metadata · W3C PROV specification · ProvCaRe semantic provenance

1 Introduction

Scientific reproducibility is critical for ensuring validation of research results, scientific fidelity, and enabling the advancement of science through rigorous design of experiments [1, 2]. Therefore, the increasing adoption of data-driven research techniques, for example use of Big data in biomedical and healthcare research for better understanding of disease mechanism and drug discovery, has led to greater focus on scientific reproducibility [3, 4]. Multiple guidelines and best practices have been developed to ensure transparent reporting of scientific results that can be successfully replicated. For example, the US National Institutes of Health (NIH) has announced the "Reproducibility and Rigor" guidelines that requires biomedical researchers to provide contextual information for transparent reporting of research studies [5]. This contextual information describing the origin or source of data is called provenance metadata. Provenance metadata has been extensively studied in computer science for reproducibility in workflow systems and tracing data in relational database systems [6–8]. By leveraging provenance metadata in biomedical and healthcare research, researchers will have improved ability to collaborate, share data, identify "best practices" for reproducible scientific research [9]. In addition to scientific reproducibility, provenance metadata is also essential for evaluating data quality and computing trust value [10, 11]. Due to its application in a variety of domains, provenance has been modeled using multiple approaches, for example the Open Provenance Model (OPM) represented causal relationship between different provenance terms [12]. Similarly, the Provenir ontology used Semantic Web technologies, including the Web Ontology Language (OWL) [13] to incorporate partonomy, causal, transformation, and other categories of relationships to accurately represent provenance metadata [14].

The World Wide Web Consortium (W3C) provenance working group used various properties of these provenance modeling approaches to define the PROV specifications as a common representation model in 2013 [15]. The W3C PROV specifications consist of the PROV Data Model (PROV-DM), [15] a formal representation of the data model using description logic-based Web Ontology Language (OWL2) called PROV Ontology (PROV-O) [16], and a set of constraints to define "valid" provenance representations called PROV Constraints [17]. In particular, the W3C PROV Ontology was developed as an upper-level reference ontology that can be extended to model domain-specific provenance terms while ensuring interoperability through the use of a common set of ontology classes and properties [16]. We have extended the W3C PROV specifications to define a new provenance framework called Provenance for Clinical and Healthcare (ProvCaRe) to support scientific reproducibility in biomedical and healthcare domain. The ProvCaRe framework defines a reference model consisting of three categories of provenance metadata terms that we have identified as necessary for scientific reproducibility in biomedical research:

1. **Study Method:** The design of the research study in terms of study design, sampling, randomization technique and interventions (in experiments), data collection approach, and data analysis techniques (e.g., statistical models) are examples of provenance metadata describing Study Method;
2. **Study Tools:** The different instruments and their parameter values that are used to record and analyze data in a research study is the second essential component of the ProvCaRe framework. For example, the strength of the magnet used in a Magnetic Resonance Imaging (MRI) instrument is important provenance information that will allow other researchers to use an equivalent MRI machine to replicate the findings of the original experiment; and
3. **Study Data:** The provenance metadata describing the contextual information about the data elements used in a scientific experiment, for example drug information, demography information of participants, and timestamp values, are necessary to allow other researchers to replicate a given experiment.

Given the vast scope of biomedical and healthcare research, we initiated the development of the ProvCaRe framework using sleep medicine research as an example domain to identify, extract and analyze provenance information associated with research studies. We are using data from one of the largest repositories of sleep medicine studies at the National Sleep Research Resource (NSRR), which is working to release data from more than 40,000 sleep studies collected from 36,000 participants [18]. The NSRR project is an example of biomedical Big Data and it aims to allow researchers to validate results of previous studies using larger datasets from greater number of research studies and also facilitate the development of data-driven techniques in sleep medicine. Therefore, the systematic characterization of provenance metadata describing the research studies that involve analysis of NSRR datasets will not only demonstrate the role of provenance in scientific reproducibility, but also demonstrate the scalability of the ProvCaRe framework. Our objective is to model the provenance metadata extracted from NSRR related research studies using the "subject → predicate → object" triple model of W3C Resource Description Framework (RDF) [19]. The provenance terms used to construct the RDF provenance graphs are being modeled in the ProvCaRe ontology, which extends the W3C PROV Ontology and the resulting provenance graphs also conform to the PROV specifications [16].

The W3C PROV Ontology consists of three "core" classes, namely prov: Entity[1], prov:Activity, and prov:Agent, with nine "core" object properties, namely prov:wasGeneratedBy, prov:wasDerivedFrom, prov:wasAttributedTo, prov:startedAtTime, prov:used, prov:wasInformedBy, prov:endedAtTime, prov:wasAssociatedWith, and prov:actedOnBehalfOf [16]. Figure 1 shows the PROV-O schema with an illustrative representation of provenance metadata in sleep medicine research. The Entity class represents any physical, digital, or conceptual information resource. The Activity class represents information resources that occur over a period of time and Agent class represents information resources that are associated with Activity, Entity or have some responsibility related to another Agent. The object properties are used to link the

---

[1] prov represents the http://www.w3.org/ns/prov#namespace.

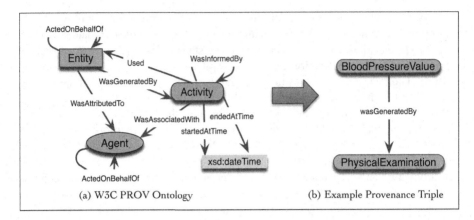

(a) W3C PROV Ontology   (b) Example Provenance Triple

Fig. 1. (a) The three primary classes and object properties of the W3C PROV Ontology (PROV-O). (b) An example showing representation of provenance information using RDF subject, predicate, and object structure.

provenance terms, for example blood pressure value (Entity) was generated during (wasGeneratedBy) a physical exam of the patient (Activity). These "core" classes and properties together with other PROV Ontology terms (as described in the PROV-O specifications [16]) are being extended to model biomedical domain-specific provenance information in the ProvCaRe framework. The use of PROV-O as the upper-level ontology will facilitate interoperability among provenance applications that conform to the PROV specifications. However, a key challenge for the ProvCaRe ontology is ensuring comprehensive coverage of the potentially hundreds of thousands of biomedical domain-specific terms in a single provenance ontology using *precoordinated* class expressions. Precoordinated class expressions are ontology constructs that already "built-in" in an ontology before the ontology is deployed or used (a detailed description of precoordination is presented in work by Rector et al. in [20]).

The biomedical domain covers a wide range of disciplines, including respiratory disease, neurology, and cardiovascular research, among others, and therefore it is almost impossible for a single ontology to model the relevant terms with consistency in a reasonable amount of time. In addition, there are more than 500 biomedical ontologies already available from the National Center for Biomedical Ontologies (NCBO) that represent a variety of biomedical terms at different levels of granularity and detail [21]. For example, the Human Phenotype Ontology (HPO) models abnormal phenotypes in human diseases and it covers different aspects of these abnormalities, including the mechanism for inheritance of these abnormalities, their onset and clinical course, and different categories of the abnormalities [22]. HPO includes more than 10,000 classes with many of the classes mapped to other biomedical ontologies. Similarly, the Systematize Nomenclature of Medicine Clinical Terms (SNOMED CT) is being developed as a comprehensive ontology for diseases and phenotypes with a large number of clinical terms modeled in the ontology [23]. The US edition of SNOMED CT 2015 version includes 300,000 concepts with more than 103,000 classes representing clinical findings. Therefore, it is intuitive to re-use these large numbers of existing ontology classes in the ProvCaRe project to model domain-specific termi-nology instead of re-creating the terms in the ProvCaRe ontology. The re-use of existing ontology classes also conforms to the ontology engineering best practice and facilitates interoperability across ontology-driven applications [24].

2 Background and Related Work: Use of Postcoordination in Biomedical Ontologies

Formal modeling of attributes related to the design and template of clinical and basic research studies have led to the development of multiple ontologies in biomedical research. The Ontology for Clinical Research (OCRe) was developed as part of a comprehensive effort to model protocols used in clinical research studies, including a classification of study designs, the plan components of the study protocols, and con-cepts describing statistical data analysis methods [25]. The OCRe project defines multiple attributes to represent various aspects of a research study, including the sampling method, number of participant groups, and whether a study is a longitudinal cohort or cross-sectional study. The OCRe project also developed a data annotation approach called Eligibility Rule Grammar and Ontology (ERGO), which extracts structured information regarding eligibility criteria used to identify participants in research studies [26]. A formal description of eligibility criteria is important metadata information that can be used by other researchers to replicate a biomedical or healthcare study. Similar to OCRe, the Ontology for Biomedical Investigations (OBI) models various attributes of basic sciences experiments, for example in genomics, proteomics, and parasite research domains [27]. The classes in OBI broadly represent five cate-gories of entities, including the objects used in experiment called material entity, activities in experiments such as planned processes, the data related to an experiment called information entities, different roles of participants in experiments, and instru-ments. OBI has been used in annotation of multiple biomedical databases, for example the Eukaryotic Pathogen Database and the Immune Epitope Database.

In contrast to OCRe and OBI, SNOMED CT is a model of clinical terminology organized into 19 top level concepts, for example clinical findings, procedure, speci-men, and body structure. These terms are linked to each other using attributes or properties, for example causative agent, associated morphology, and finding site. To address the challenging requirement of modeling extremely large variety of concepts from different biomedical disciplines, SNOMED CT uses two approaches to represent terms: (1) *Precoordinated Expressions*, and (2) *Postcoordinated Expressions*. Preco-ordinated expressions in SNOMED CT consist of a single class and are modeled in one of the 19 class hierarchies, for example Sleep disorder (ID: C0851578) is mod-eled as a subclass in the hierarchy of the top-level concept Clinical finding (ID: C0037088). However, it is almost impossible to model all possible attributes of a disease, which may evolve as new biomedical discoveries are made or additional clinical details that were not considered before and need to be modeled in context of a specific application. To address this challenge, SNOMED CT uses post coordination expressions to represent new terms by combining more than one SNOMED CT term using a set of rules defined in the SNOMED CT compositional grammar specification

[23]. For example, the post coordinated expression | hip joint | : | laterality | = | right | represents right hip joint using the classes hip joint and right together with the property laterality. We propose to use a similar approach to model provenance information for the biomedical domain through creation of postcoordinated expressions using classes from the ProvCaRe ontology together with classes from existing biomedical ontologies. We describe the details of our approach in the next section.

3   Modeling Provenance Metadata in ProvCaRe Ontology Using Postcoordinated Expressions

The ProvCaRe framework models the provenance description of a scientific study that may enhance the ability to replicate the study by researchers in other institutions or groups. The three objectives of the ProvCaRe framework are: (1) Create a biomedical domain-specific provenance ontology, (2) Extract provenance metadata from published biomedical articles and generate provenance graphs for analysis, and (3) Develop a provenance knowledgebase for users to search and identify research studies that can be replicated to validate important results or design new experiment studies. The three components of the ProvCaRe framework, namely Study method, data, and tools, were developed in close collaboration with a data manager working on the NSRR project. A data manager is responsible for working with researchers to identify the data needed to replicate results from previous studies and extract data for new research studies. Therefore, they are ideally placed to identify provenance information required for scientific reproducibility.

As described in Sect. 1, the three components of the ProvCaRe framework represents three essential provenance information types corresponding to the method used to conduct the research study (*Study Method*), the data used in the study as well as results generated from the study (*Study Data*), and details of the instruments used in the study (*Study Instrument*). The provenance information corresponding to these three components can be modeled in detail, which is important to accurately capture the contextual information necessary for replicating previous studies. For example, the Study Method term can be further subdivided into three categories of: (a) Study Design, (b) Study Data Collection Method, and (c) Data Analysis Method. Similarly, the Data Analysis Method can be further extended to model various categories of statistical data analysis methods, for example inferential or descriptive statistics. We use sleep medicine as an example domain with data from the NSRR project to define the ProvCaRe postcoordination-based ontology modeling approach and demonstrate the applicability of the ProvCaRe ontology.

3.1   Provenance Metadata in Sleep Medicine Research

NSRR is the largest repository of publicly available research sleep medicine data and it includes data from research studies that are representative of a wide variety of topics, for example cardiovascular diseases, neurocognitive functions, and metabolic disorders related to sleep disorders. Therefore, it is well suited to develop the ProvCaRe framework in terms of scalability and it is representative of the complexity of the biomedical research domain. Biomedical research studies are often defined in terms of the well-known *Population, Intervention, Comparison, Outcome,* and *Time* (PICO(T)) model to represent different aspects of a research study [28]. Many of the terms are modeled in SNOMED CT. However, the PICO(T) model does not include many of the critical provenance terms that are necessary to reproduce results generated from a scientific study. For example, the PICO(T) model does not represent provenance terms corresponding to the data analysis method (e.g., statistical model used to derive study results), instruments used to record data (e.g., type of blood pressure instrument used in the example research study). To address this issue, we extended the W3C PROV ontology in the ProvCaRe project to model provenance information corresponding to the three aspects of a scientific study, namely Study Tools, Method, and Data.

We extended the PROV-O class `Entity` to model `provcare:StudyData`, which includes the `provcare:StudyOutcome`, `provcare:ComparisonData`, and `provcare:StudyPopulation` corresponding to the PICO(T) components described earlier. The `provcare` namespace refers to the http://www.case.edu/ProvCa Re/provcare. The class `StudyDesign` represents three categories of biomedical research studies, namely `provcare:FactorialStudy`, `provcare:Inter-ventionalStudy`, and `provcare:ObservationalStudy`. The `Facto-rialStudy` class is similar to study design class modeled in OBI. The `Study Design` class is a subclass of `prov:Plan` class, which also has `StudyCon-straint` as a subclass. The `StudyConstraint` represents inclusion and exclusion criteria that are applied to select participants to be recruited into a research study. The ProvCaRe ontology also models multiple classes related to data analysis method as subclass of the `provcare:StudyMethod` class, which is modeled as a subclass of the `prov:Activity` class.

The `provcare:DataAnalysisMethod` class has multiple subclasses, including `provcare:MissingDataProtocol` and `provcare:Statisti-calMethod` that model different aspects of data analysis in research studies. The ontology models the two primary categories of statistical analysis methods, namely descriptive analysis and inferential analysis as subclasses of the `Statisti-calMethod` class. The ontology also models additional classes representing specific types of statistical analysis techniques as subclasses of the descriptive and inferential analysis classes. The `provcare:StudyInstrument` class is modeled as subclass of `prov:Agent` class and it models instruments used in research studies according to their function and modality. The `StudyInstrument` class includes electrophysio-logical signal recording instruments (e.g., Electrocardiograph and Electroencephalo-gram), imaging tools (e.g., MRI), and also software tools (e.g., statistical package R or SAS) as its subclasses. In addition to these ontology classes, the ProvCaRe ontology also extends the PROV-O properties to link ontology classes with appropriate relations. For example, we use OWL2 class-level restriction feature to assert that a `provcare:ResearchStudy` has different types of data collection methods (`provcare:DataCollectionMethod`), such as `provcare:BaselineDataCollection Method` representing data collected at start of the study and `provcare:Followup DataCollectionMethod` representing data collected at subsequent time intervals,

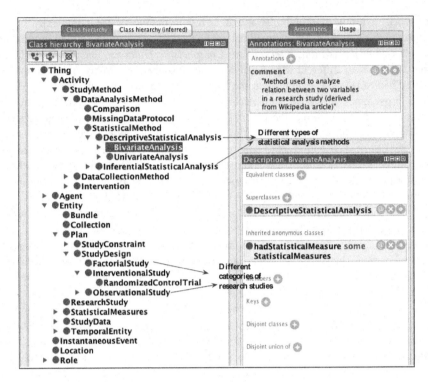

Fig. 2. A screenshot of the ProvCaRe ontology class hierarchy showing different components of the provenance metadata representation framework modeled in the ProvCaRe project

using restriction defined on the object property `provcare:hadDataCollec-tionMethod` (Fig. 2).

In addition to object properties, the ProvCaRe ontology models additional metadata information about the ontology classes using the RDF(S) annotation properties, for example `rdfs:label`, `rdfs:seeAlso`, and custom properties such as `synonym`. These metadata properties allow provenance applications, such as the ProvCaRe natural language processing (NLP) tool to effectively use the ProvCaRe ontology for entity extraction from biomedical literature. Figure 3 illustrates the class hierarchy of the ProvCaRe ontology. The ProvCaRe ontology provides the required set of precoordinated terms to represent provenance information corresponding to Study Method, Data, and Tools. However, as we discussed earlier we need a well-defined mechanism to create new postcoordinated expressions to represent provenance information and we describe the compositional grammar developed for the ProvCaRe framework in the next section.

3.2   Compositional Grammar Syntax in the ProvCaRe Ontology

We extend and adapt the postcoordination compositional grammar syntax defined to create SNOMED CT expressions to the ProvCaRe framework using classes and

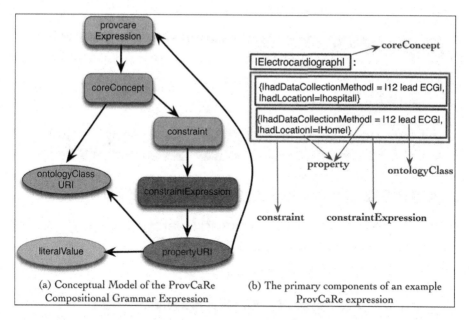

(a) Conceptual Model of the ProvCaRe
Compositional Grammar Expression

(b) The primary components of an example
ProvCaRe expression

Fig. 3. The conceptual model of the postcoordination expression compositional grammar used in ProvCaRe and an illustrative example are shown.

properties defined in the ProvCaRe ontology and existing biomedical ontologies, for example ontologies listed in NCBO [21]. A ProvCaRe postcoordinated expression consists of a single ontology class, which is the "core" provenance concept of the expression, and a set of properties as well as their values that qualify the core concept. The properties and the associated values may be defined either in the ProvCaRe ontology or NCBO listed ontologies (this ensures that the corresponding ontologies are publicly available) or the values may be RDF literal values (e.g., XML Schema data type). Figure 4 illustrates the conceptual view of the ProvCaRe postcoordination expression syntax. Each ProvCaRe postcoordinated expression is a triple structure with the form "class-property-expression", where the expression is a recursive structure consisting of either a single class or another expression and the "|" symbol is used as start and end delimiters of the terms (similar to the SNOMED CT compositional grammar syntax). The expression refines the core concept with additional values defined for properties and the corresponding concept represented by the expression is a subclass of the core concept.

We use the Augmented Backus-Naur Form (ABNF) [29] to define the ProvCaRe postcoordination expression syntax, which is described in Table 1.

Four categories of postcoordinated expressions can be composed using the compositional grammar in the ProvCaRe framework:

1. **Multi-Concept Expression:** Two or more ontology classes can be combined together using the "+" symbol to form a new concept, which is interpreted to be a

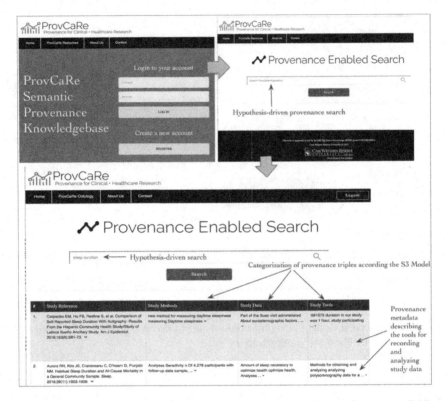

Fig. 4. The user interface of the ProvCaRe semantic provenance knowledgebase available at: https://provcare.case.edu/

subclass of the two original classes. For example, a data analysis method may involve both `provcare:CorrelationAnalysis` and `provcare:CovarianceAnalysis`, which can be represented using the expression[2]: | CorrelationAnalysis| + |CovarianceAnalysis|.

2. **Concept with Constraints Defined Over Properties:** A provenance term can be refined using additional constraints defined over a ProvCaRe or other ontology property. For example, it is important to record the provenance of blood pressure values of a research study participant in terms of the procedure. This provenance information can be modeled using an expression that combines ProvCaRe ontology and SNOMED CT terms: |Electrocardiograph|: |hadDataCollectionMethod| = |12 lead ECG| (ECG has SNOMED CT ID: C0180600 and 12 lead ECG has SNOMED CT ID: C0430456). A core concept can also include multiple constraints defined using multiple properties. For example, |Electroencephalogram|: |hadStudyInstrument| = |Scalp electrode cap|, |hadLocation| = hospital| expressions represent the provenance

---

[2] The namespace for the terms used in the expressions are not repeated for brevity.

Table 1. The specification of the ProvCaRe postcoordination expression syntax with explanatory description.

	Syntax expression	Description
1.	`provcareExpression = subexpression`	A provcare post coordinated expression consists of subExpressions
2.	`subExpression = coreConcept [":" constraint]`	A subExpression consists of a core provenance concept, which is refined through use of constraints, which may consist of multiple constraintExpressions. A subExpression is a subclass of the coreConcept
3.	`coreConcept = "|" ontologyClassURI "|"`	The coreConcept is a provenance ontology class defined in the ProvCaRe ontology
4.	`ontologyClassURI = nonPipe * (* nonPipe)`	The ontology class is listed using the concept ID or namespace aware URI. An ontology class URI may consist of any UTF-8 character except pipe "|" and conform to the URI specification as defined in the W3C URI specifications
5.	`constraint = (constraintExpression) * ([","] constraintExpression)`	A constraint consists of one or more constraint expressions that are optionally grouped together into a subunit separated by comma
6.	`constraintExpression = ["{"] propertyURI "=" ontologyclassURI/ "("constraintExpression ")"["}"] *constraintExpression`	A constraintExpression consists of an ontology property with an ontology class as value or a constraintExpression as value (for nested expressions) followed by additional constraints
7.	`propertyURI = nonPipe * (*nonPipe)`	The ontology property is listed using the concept ID or namespace aware URI. An ontology property URI may consist of any UTF-8 character except pipe "|" and conform to the URI specification as defined in the W3C URI specifications

information of an EEG in terms of the instrument used to record it and the location of the recording. The two properties `hasStudyInstrument` and `hadLocation` are ProvCaRe and PROV ontology terms respectively.

3. **Concepts with Constraints Defined Over Property Groups:** The ProvCaRe compositional grammar allows grouping of multiple properties into a subunit to reduce ambiguity regarding the ordering of the constraints using an approach that is similar to the SNOMED CT compositional grammar. For example, two ECG recordings may have been conducted at two different locations, which can be represented using the expression |Electrocardiograph|: {|hadDataCollectionMethod| = |12 lead ECG|, |hadLocation| = |hospital|}, {| hadDataCollectionMethod| = |12 lead ECG|, |hadLocation| = | Home|}. The curly braces group two or more properties together to allow humans and software tools to correctly parse the ordering of the constraints in an expression. Similar to the SNOMED CT compositional grammar, the comma between the two subunits is optional.

4. **Concepts with Nested Constraints:** As discussed earlier, the ProvCaRe postcoordinated expressions use a recursive definition, which allows the value in the triple structure of the expressions to be another expression. For example, a research study may use two statistical data analysis techniques that can be represented using the following expression: |ResearchStudy|: |hadDataAnalysisMethod| = (| CorrelationAnalysis| + |CovarianceAnalysis|). The nested structure may also be constructed using multiple properties, for example |ResearchStudy|: |hadDataAnalysisMethod| = (|StatisticalMethod|: |hadStatisticalMeasure| = |CentralTendencyMeasure|).

It is important to note that unlike the SNOMED CT compositional grammar, which allows interpretation of postcoordinated expressions as equivalent or subclass of a given class, the ProvCaRe postcoordinated expressions are interpreted only as subclass of the core concept.

4   Results

We describe the two-fold results of the ProvCaRe project: (1) we demonstrate the practical use of postcoordinated expression in the ProvCaRe ontology using an example research study published by O'Connor et al. [32]; and (2) we demonstrate the effectiveness of the ProvCaRe ontology in the extraction of 38 million provenance triples from 384,802 published articles.

Application of postcoordinated expression: The research study by O'Connor et al. [32] is classified as an `ObservationalStudy` (modeled as subclass of `StudyDesign` in the ProvCaRe ontology). Using the ProvCaRe compositional grammar, we model the data analysis related provenance information using ontology classes and properties modeled in the ProvCaRe ontology and existing biomedical ontologies. For example, the postcoordinated expression |ResearchStudy|: |hadDataAnalysisMethod| = |multivariate regression|, |hadSoftwareTool| = |SAS|, uses the terms `multivariate regression` and `SAS` modeled in the Bilingual

Ontology of Alzheimer's Disease and Related Diseases (ONTOAD) and Software Ontology (CWO) (listed in NCBO). Similarly, the population selected for the research study can be characterized using the following expression: |ResearchStudy|: {| hadStudyConstraint| = (|StudyExclusionCriterion|: |hadPrescription| = |Antihypertensive medication|)}. This expression represents important provenance metadata describing the constraints used to identify participants for the research study and is essential for other researchers who aim to replicate this study. The method used to collect the data in the research study can also be represented using postcoordinated expression: |ResearchStudy|: {|hadDataCollection Method| = |BaselineDataCollection|, |hadDataCollectionMethod| = (|FollowupDataCollection|: |hadTemporalAttribute| = |5 years|)}.

It is important to note that provenance-related postcoordinated expressions need to be created often by domain experts with little or no experience in ontology engineering practices. Therefore, development of a visual user interface form can significantly help domain experts to create valid postcoordinated expression. The ProvCaRe compositional grammar syntax supports the development of a form-based user input template that uses the property values as "widgets" and the corresponding ontology classes as "values". For example, hadDataAnalysisMethod can have a drop-down menu with list of ontology classes corresponding to DataAnalyisMethod or its subclasses. A similar approach is often used in development of ontology-driven user interface applications. The rules defined in the ProvCaRe compositional grammar syntax also supports systematic parsing of the postcoordinated expression, which can be used by provenance applications for validation, querying, and interpretation of research studies annotated with ProvCaRe postcoordinated expressions. The postcoordination-based modeling approach is also extensible as the new provenance-specific terms are modeled in the ProvCaRe ontology, for example detailed representation of how missing data is handled in research studies, and new biomedical ontologies are released through NCBO. This is an important feature of the proposed approach as the ProvCaRe project extracts and analyses provenance metadata information from additional biomedical domains, such as neurological disorders and lung cancer, as part of our ongoing and future work.

Creation of the ProvCaRe Semantic Provenance Knowledgebase: The extraction of structured data from free text is a significant challenge and this has been a focus of extensive research in computer science using statistical machine learning as well as rule-based techniques [30]. The use of Semantic Web techniques especially using ontologies as reference knowledge model has been an effective approach for natural language processing (NLP) [31]. However, we are not aware of any previous work that use ontologies for extracting provenance metadata from unstructured text. To extract and analyze the provenance information from published biomedical research studies, we have developed a novel Natural Language Processing (NLP) workflow using the ProvCaRe ontology [9]. Using the ontology-enabled NLP workflow, we have processed and extracted provenance information from 384,802 published articles describing biomedical research studies (the articles are available from the National Center for Biomedical Informatics PubMed resource, https://www.ncbi.nlm.nih.gov/pubmed/). We extracted more than 38 million provenance triples from these published articles by using the ProvCaRe ontology for named entity recognition (NER) and predicate identification. These provenance triples are available for querying and analysis in the ProvCaRe semantic provenance knowledgebase, which can be accessed at: https://provcare.case.edu/ (Fig. 4 shows the details of the ProvCaRe knowledgebase).

As far as we know, the ProvCaRe knowledgebase with 38 million provenance triples is one of the largest real world dataset of biomedical provenance information available to the research community for querying and analysis. The knowledgebase supports querying using two approaches: (1) users can use a "hypothesis-driven" query approach to search for previous research studies corresponding to a given hypothesis and view the provenance metadata associated with each of these studies for scientific reproducibility; and (2) use the provenance information of previous studies to design new experiments with rigorous protocols for ensuring transparent reporting as well as supporting reproducibility. As shown in Fig. 5, the provenance triples extracted from the published articles are classified into one of three categories of provenance metadata defined in the ProvCaRe S3 model. Table 2 lists the distribution of provenance triples in each of the three categories.

The distribution of provenance triples in Table 2 demonstrates that provenance

Table 2. The number and distribution of provenance triples in the ProvCaRe knowledgebase according to the S3 model

	Distribution of Provenance Triples (total: 38.47 million provenance triples)		
	Study methods	Study data	Study instruments
Total number of triples	12,212,129	15,361,311	10,905,018
Percent distribution of triples	32%	40%	28%

metadata describing the method and data of research experiments is well-described in published articles, however there is limited provenance metadata describing the instruments used in research studies. This highlights an important limitation of published articles describing research studies as the instruments used in an experiment and the parameters used to record experiment data are essential for reproducibility of scientific results. We believe new guidelines and best practices, for example the NIH Rigor and Reproducibility guidelines can help address these issues in transparent reporting of new research experiments.

5   Conclusions and Future Work

Our work was motivated by the need to represent provenance metadata information describing research studies in a variety of biomedical domains for scientific reproducibility. With the known limitations of modeling large number of classes and properties in a single ontology using precoordinated modeling approach, we extended and adapted the SNOMED CT compositional grammar syntax to create ProvCaRe postcoordinated expressions. The ProvCaRe postcoordinated expressions use provenance-specific classes and properties defined in the ProvCaRe ontology and re-uses terms from existing biomedical ontologies to represent provenance metadata. The ProvCaRe ontology extends the W3C PROV ontology to represent three core provenance terms: Study Method, Data, and Tools. We define the ProvCaRe compositional grammar syntax using ABNF notation and define four categories of postcoordinated expressions that can be created to represent provenance information. We demonstrate the application of the ProvCaRe postcoordinated expressions in modeling the provenance information associated with a research study and the use of the ProvCaRe ontology in the creation of one of the largest semantic provenance knowledgebase with more than 38 million provenance triples.

Acknowledgement. This work is supported in part by the National Institutes of Biomedical Imaging and Bioengineering (NIBIB) Big Data to Knowledge (BD2K) grant (1U01EB020955) NSF grant# 1636850

References

1. Collins, F.S., Tabak, L.A.: Policy: NIH plans to enhance reproducibility. Nature **505**, 612–613 (2014)
2. Landis, S.C., Amara, S.G., Asadullah, K., et al.: A call for transparent reporting to optimize the predictive value of preclinical research. Nature **490**(7419), 187–191 (2012)
3. Redline, S., Dean III, D., Sanders, M.H.: Entering the era of "Big Data": getting our metrics right. SLEEP **36**(4), 465–469 (2013)
4. Baker, M.: 1,500 scientists lift the lid on reproducibility. Nature **533**(7604), 452–454 (2016)
5. NIH: Principles and Guidelines for Reporting Preclinical Research (2016). https://www.nih.gov/research-training/rigor-reproducibility/principles-guidelines-reporting-preclinical-research. Accessed 20 July 2017
6. Buneman, P., Davidson, S.: Data provenance - the foundation of data quality (2010)
7. Goble, C.: Position statement: musings on provenance, workflow and (semantic web) annotations for bioinformatics. In: Workshop on Data Derivation and Provenance, Chicago (2002)
8. Sahoo, S.S., Sheth, A., Henson, C.: Semantic provenance for escience: managing the deluge of scientific data. IEEE Internet Comput. **12**(4), 46–54 (2008)
9. Valdez, J., Kim, M., Rueschman, M., Socrates, V., Redline, S., Sahoo, S.S.: ProvCaRe semantic provenance knowledgebase: evaluating scientific reproducibility of research studies. Presented at the American Medical Informatics Association (AMIA) Annual Conference, Washington DC (2017)
10. Zhao, J., Goble, C., Stevens, R., Turi, D.: Mining Taverna's semantic web of provenance. J. Concurr. Comput. Practice Exp. **20**(5), 463–472 (2008)
11. Simmhan, Y.L., Plale, A.B., Gannon, A.D.: A survey of data provenance in e-science. SIGMOD Rec. **34**(3), 31–36 (2005)
12. Moreau, L., Clifford, B., Freire, J., et al.: The open provenance model core specification (v1.1). Future Gener. Comput. Syst. **27**(6), 743–756 (2010)
13. Hitzler, P., Krötzsch, M., Parsia, B., Patel-Schneider, P.F., Rudolph, S.: OWL 2 Web Ontology Language Primer. In: W3C Recommendation. World Wide Web Consortium W3C (2009)

14. Sahoo, S.S., Sheth, A.: Provenir ontology: towards a framework for eScience provenance management. Presented at the Microsoft eScience Workshop, Pittsburgh, USA, October 2009
15. Moreau, L., Missier, P.: PROV Data Model (PROV-DM). In: W3C Recommendation. World Wide Web Consortium W3C (2013)
16. Lebo, T., Sahoo, S.S., McGuinness, D.; PROV-O: the PROV ontology. In: W3C Recommendation. World Wide Web Consortium W3C (2013)
17. Cheney, J., Missier, P., Moreau, L.: Constraints of the PROV data model. In: W3C Recommendation. World Wide Web Consortium W3C (2013)
18. Dean, D.A., Goldberger, A.L., Mueller, R., Kim, M., et al.: Scaling up scientific discovery in sleep medicine: the National Sleep Research Resource. SLEEP **39**(5), 1151–1164 (2016)
19. Cyganiak, R., Wood, D., Lanthaler, M.: RDF 1.1 concepts and abstract syntax. In: W3C Recommendation, World Wide Web Consortium (W3C) (2014)
20. Rector, A., Luigi, I.: Lexically suggest, logically define: quality assurance of the use of qualifiers and expected results of post-coordination in SNOMED CT. J. Biomed. Inform. **45**(2), 199–209 (2012)
21. Musen, M.A., Noy, N.F., Shah, N.H., Whetzel, P.L., Chute, C.G., Story, M.A., Smith, B.: NCBO team: The national center for biomedical ontology. J. Am. Med. Inform. Assoc. **19**(2), 190–195 (2012)
22. Köhler, S., Doelken, S.C., Mungall, C.J., et al.: The human phenotype ontology project: linking molecular biology and disease through phenotype data. Nucleic Acids Res. **42**, 966–974 (2014). Database Issue
23. Giannangelo, K., Fenton, S.: SNOMED CT survey: an assessment of implementation in EMR/EHR applications. Perspect Health Inf. Manag. **5**, 7 (2008)
24. Bodenreider, O., Stevens, R.: Bio-ontologies: current trends and future directions. Brief. Bioinform. **7**(3), 256–274 (2006)
25. Sim, I., Tu, S.W., Carini, S., Lehmann, H.P., Pollock, B.H., Peleg, M., Wittkowski, K.M.: The ontology of clinical research (OCRe): an informatics foundation for the science of clinical research. J. Biomed. Inform. **52**, 78–91 (2014)
26. Tu, S.W., Peleg, M., Carini, S., Bobak, M., Ross, J., Rubin, D., Sim, I.: A practical method for transforming free-text eligibility criteria into computable criteria. J. Biomed. Inform. **44**(2), 239–250 (2011)
27. Bandrowski, A., Brinkman, R., Brochhausen, M., et al.: The ontology for biomedical investigations. Plos One **11**(4), e0154556 (2016)
28. Huang, X., Lin, J., Demner-Fushman, D.: Evaluation of PICO as a knowledge representation for clinical questions. Presented at the AMIA Annual Symposium Proceedings (2006)
29. Overell, P.: Augmented BNF for Syntax Specifications: ABNF. https://tools.ietf.org/html/rfc5234. Accessed 20 Aug 2017
30. Hearst, M.A.: Untangling text data mining. In: 37th the Association for Computational Linguistics on Computational Linguistics meeting, pp. 3–10 (1999)
31. Rindflesch, T.C., Pakhomov, S.V., Fiszman, M., Kilicoglu, H., Sanchez, V.R.: Medical facts to support inferencing in natural language processing. Presented at the AMIA Annual Symposium Proceedings (2005)
32. O'Connor, G.T., Caffo, B., Newman, A.B., Quan, S.F., Rapoport, D.M., Redline, S., Resnick, H.E., Samet, J., Shahar, E.: Prospective study of sleep-disordered breathing and hypertension: the sleep heart health study. Am. J. Respir. Crit. Care Med. **179**(12), 1159–1164 (2009)

Complete Semantics to Empower Touristic Service Providers

Zaenal Akbar$^{(\boxtimes)}$, Elias Kärle, Oleksandra Panasiuk, Umutcan Şimşek, Ioan Toma, and Dieter Fensel

Semantic Technology Institute (STI) Innsbruck, University of Innsbruck, Innsbruck, Austria
{zaenal.akbar,elias.kaerle,oleksandra.panasiuk,umutcan.simsek, ioan.toma,dieter.fensel}@sti2.at

Abstract. The tourism industry has a significant impact on the world's economy, contributes 10.2% of the world's gross domestic product in 2016. It becomes a very competitive industry, where having a strong online presence is an essential aspect for business success. To achieve this goal, the proper usage of latest Web technologies, particularly schema.org annotations is crucial. In this paper, we present our effort to improve the online visibility of touristic service providers in the region of Tyrol, Austria, by creating and deploying a substantial amount of semantic annotations according to schema.org, a widely used vocabulary for structured data on the Web. We started our work from Tourismusverband (TVB) Mayrhofen-Hippach and all touristic service providers in the Mayrhofen-Hippach region and applied the same approach to other TVBs and regions, as well as other use cases. The rationale for doing this is straightforward. Having schema.org annotations enables search engines to understand the content better, and provide better results for end users, as well as enables various intelligent applications to utilize them. As a direct consequence, the region of Tyrol and its touristic service increase their online visibility and decrease the dependency on intermediaries, i.e. Online Travel Agency (OTA).

Keywords: Semantic annotations · Schema.org · Touristic service providers

1 Introduction

The tourism and leisure industry contributes significantly to the economic development of the region of Tyrol, Austria. With around 60,000 employees (25% of the full-time workplaces in the region were created in this industry), it generates sales approximately 8.4 billion Euros. In the tourism year 2015/2016, 11.5 million guests were arrived, generated 47.6 million overnight stays. The direct value

---

Completeness is something that can never be achieved. Therefore we think it is a proper goal to target our ambitions.

© Springer International Publishing AG 2017
H. Panetto et al. (Eds.): OTM 2017 Conferences, Part II, LNCS 10574, pp. 353–370, 2017.
https://doi.org/10.1007/978-3-319-69459-7_24

added of the industry to the region is 17.5%, higher than other regions such as Upper Austria (3.1%), Vienna (1.6%), or national level (5.3%)[1].

The TVB Mayrhofen-Hippach[2] is the tourism board of the Mayrhofen-Hippach region situated in Zillertal, Tyrol, Austria. It is the organization responsible for the marketing of the entire Mayrhofen-Hippach region and its members. As with all touristic service providers, it faces the challenge of achieving the highest visibility possible in search engines, and at the same time, they need to be present in various communication channels which are constantly growing. Website, for example, bridges the tourism organizations and tourists directly and plays roles at different stages of tourists decision making process [13]. Information quality, responsiveness, visual appearance, personalization are a few example of key factors for influencing website effectiveness. Specifically for the Alps region, the regional tourism boards have been enhanced their websites qualities significantly in various dimensions, not only information quality but also the adoption of new technologies including a few web standards and interactive maps [19].

But it is still challenging for the tourism sector, especially in the region of Tyrol, to provide useful content that could help potential guests to make a reservation decision directly as well as to be accessible by machine, i.e. semantically annotated [14]. In Austria national scope, most of the touristic service providers have not or minimally use the semantic annotations technology [20]. And most of the existing annotations of touristic service providers, especially hotels, were performed incorrectly [15]. This situation is critical for the industry because the use of semantic annotation such as schema.org[3] could increase a typical hotel website visibility by 20% [10]. More than just for increasing online visibility on search engines, a semantically annotated content of touristic service providers within a region could contribute to the tourism information system of the region. For example, enabling data query from distributed sources, topical or location-based data integration, matching of service providers and requesters, as well as transactional web services for tourists [18]. And we believe that the seamless interoperability among organizations which is still an issue in the tourism industry [21] can be solved by semantic web technologies including semantic annotations.

A substantial amount of semantically annotated content (possibly as complete as possible from every touristic service providers and touristic related information sources) could support every intelligent, machine processing decision making for the industry. Search engines such as Google consume annotated content and present it in a more interesting way visually such as stars for ratings instead of text, a structured layout for events, carousels for recipes[4]. With those richer search results, content annotation approach outperforms the conventional search engine optimization techniques. Annotated content also could help organizations to semi-automatically disseminate content to multiple online communication channels [2], reducing human efforts to manually collecting, curating content from different sources before distributing them to multiple channels. Most recently, semantically annotated content will be consumed by intelligent applications such as chatbot and personal digital assistant to automatically provide users with precise and personalized information.

This paper describes our systematic approach on annotating tourism information available on the region of Tyrol, Austria, started with the TVB Mayrhofen-Hippach website using semantic annotations, more precisely schema.org. The main goal of our cooperation, therefore, is to improve the online visibility of the region by enriching their content with machine processable data. Comparing to our similar efforts before, where we have been annotating various individual service providers such as hotel [10], this is by far the biggest effort regarding covered information sources, the number of produced annotation, types of annotations as well as how often an update need to be performed. More precisely the contributions of this paper are as follows: (1) an approach to **automatically generate semantic annotations** of dynamic data based on data APIs, as well as manually generate semantic annotations for the static data[5] (2) a method to **link the semantic annotations with the content** and (3) a regularly updated **schema.org annotations**[6] generated using a mixed approach i.e. automatically as well as manually. The rest of this paper is organized as follows. Section 2 describes related approaches that aim to address the creation of semantic annotations at large scale. Section 3 describes our methodology for identifying information that needs to be annotated and what types of annotations will be provided. Section 4 shows our implementation on annotating the tourist board Mayrhofen-Hippach website with schema.org, including its current results. Section 6 outlines our ongoing work on intelligently utilizing the obtained annotations, and finally, Sect. 7 concludes our paper and describes our future work.

2   Motivation and Related Work

The development in the mobile computing and artificial intelligence is leading the way to the development of a new layer on top of the web, so-called "headless web"[7], where the presentation of the web pages loses significance and publishing semantically described structured data becomes more important than ever. Among many vocabularies for embedding semantic data into the webpages, schema.org comes to the fore as a de-facto standard. Schema.org offers set of vocabularies that facilitate the publication of structured data on the web and it

---

[5] Dynamic data change rather frequently e.g. hotel offers and events, while static data change very rarely, e.g. contact information of a hotel.

[6] Per March 29, 2017, we have generated 1,6 Million (1,567,254 to be precise) triples of annotations for TVB Mayrhofen-Hippach.

[7] https://paul.kinlan.me/the-headless-web/.

has been evolved rapidly since its introduction in 2011. The success of schema.org can be measured by its adoption rate. The results of the Web Data Commons crawl in October 2016 shows that the web pages with triples are 39% of the overall crawled web pages which is 8% higher than the previous year[8]. The increased support from Content Management Systems (CMS) for schema.org as well as the support from other third-party software for tasks like event management have a great impact on this wide adoption [12].

When it comes to the tourism sector, the scene is quite different. Although the amount of schema.org annotations increase among the touristic service providers, the annotations usually come not directly from the lodging business' website (e.g. a hotel website), but from an entity like an online travel agency. Moreover, the annotations are usually incorrect or incomplete (e.g. missing values for important properties such as address) [15]. Even though CMS helps to publish a significant amount of structured content on the web, especially for the tourism sector, there is still a lot of data stored in the databases of proprietary software and served with an API. The publication of such data (e.g. events, offers from a hotel) described with schema.org carries a great importance in terms of online visibility and e-commerce in the headless web. Additionally, it will also contribute to their visibility on the search engines through features like Rich Snippets [11].

Given the results of the aforementioned analysis, this endeavor is challenging for two reasons: (a) there is a big development effort required from various parties to generate and publish structured data based on the existing internal data, (b) the lack of know-how of the touristic service providers and software producers in tourism field in terms of mapping internal metadata to correct schema.org types and properties. To tackle this challenge, we provide a solution that requires minimal development effort and know-how for the touristic service providers.

From a syntactic point of view, there are various ways to include schema.org into web pages, namely Microdata, RDFa, and JSON-LD. Microdata and RDFa have been around for many years and gained widespread usage [6]. Unlike Microdata and RDFa, JSON-LD does not require the annotations to be directly embedded in HTML markup blocks where the content reside, but it can be placed anywhere in the source of the web page in script tags. This is one of the main reasons we adopt JSON-LD for our implementation since it brings an advantage for dynamic injection of the semantic annotation to the web pages. The annotations can be prepared and hosted externally and be embedded on demand straightforwardly. There is a major effort from the semantic web and linked data community for generating semantic annotations based on unstructured text. These efforts are mainly focused on creating annotations in RDFa or Microdata format via editor interface. The approaches they adopt vary in terms of automation (e.g. usage of NLP techniques for named entity recognition and entity linking until some level). A comprehensive survey of semantic content authoring approaches and tools can be found in [17].

The generation of semantic annotation based on dynamic data served in a structured way (e.g. relational databases, web APIs) is critical considering the

---

[8] For detailed statistics: http://webdatacommons.org/structureddata/.

volume of the data. This publication method mainly requires mappings from the metadata of the data source to a vocabulary such as schema.org. Triplify [4] uses an SQL-based lightweight mapping approach to create RDF out of relational data. D2RQ [5] operates with a declarative mapping language and creates virtual RDF graphs on top of relational databases. RML [8] provides a mapping language and a processor for mapping data from various sources including but not limited to XML files, web APIs, relational databases and CSV files to RDF. All of the aforementioned techniques can be used for creating JSON-LD since it is an RDF serialization format.

Our main contributions are a holistic methodology and a proof of concept for analyzing and mapping static and dynamic data to schema.org and create semantic annotations to enrich touristic service providers' content on the web. For the static data, we create the annotations manually, to ensure high accuracy and domain coverage. For the dynamic data, we create mappings based on the domain analysis and generate automated annotations externally, which makes the deployment to the web pages feasible, since it does not require a major software development effort on the touristic service providers' side.

3 Methodology

In this section, we describe our methodology to annotate the TVB Mayrhofen-Hippach website, the starting point of our effort to annotate touristic service providers in the region of Tyrol, Austria completely. Our methodology comprises three essential activities: (i) data sources and format analysis, (ii) information modeling, and (iii) domain specification definition.

3.1 Data Sources and Format Analysis

Information available on the TVB Mayrhofen-Hippach website are originating from external and internal sources, where the external data came mostly from Feratel[9].

Feratel. Feratel offers a destination management system[10] for services related to tourism and travel industry. The system provides information about accommodation, packages, events, etc., including a real-time service to check room availability as well as to perform booking action. The system is widely used by service providers such as hotels to manage their booking system, as well as by TVBs to market a region. The system can be accessed through a web API so-called Deskline Standard Interface (DSI) [9] which serves data in an Extensible Markup Language (XML) format[11].

---

[9] http://www.feratel.at/.
[10] http://www.feratel.at/en/solutions/feratel-destination/.
[11] https://en.wikipedia.org/wiki/XML.

TVB Itself. The website also contains information which was created internally by the TVB itself by using a CMS. At the moment this study was performed, the TVB uses TYPO3[12] to manage their self-created content, where information will be entered in the backend and will be presented in a semi-structured format. In this case, TVB has full access to the CMS, so they can manage and define their content structure including installing extensions.

3.2   Information Modeling

After identifying the data sources and their formats as explained in the previous section, we start our next task to determine important concepts from every data source. For data originating from Feratel, we consulted the APIs documentation, XML responses, as well as how the data were presented in the relevant webpages. For self-created data, we visited each relevant webpage and performed analysis on it.

We started the information modeling by identifying the types of web pages and what kind of information are presented there. We went through of each webpage and analyzed what kind of data can be annotated. For example, the webpage Die Region[13] contains the information about the Mayrhofen-Hippach Holiday Region, its villages, and the latest news and events from the region. The webpage of the Mayrhofen region[14] contains the name of the region, url, location, information about the region, picture of it. This data are presented as text, image object, url. Another example is Globeseekers yoga event[15]. The web page contains the name of the event, its location, dates and time, description, information on prices, organizer, contact information and image.

Further, we defined what primary categories and subcategories could be chosen from the menu, submenu and web content. We made the list of categories and subcategories and selected the most important. We selected seven main categories: TVB Mayrhofen, Mayrhofen Hippach region, Ski Areas, Accommodations, Infrastructure, Events, Articles. For example, Mayrhofen Hippach region is the category, which includes the subcategories with the information of Mayrhofen, Ramsau, Schwendau, Brandberg, Ginzling, Hippach regions. Event is the category for different types of events and activities in TVB Mayrhofen Hippach, such as concerts, lectures, conferences, festivals, etc.

3.3   Domain Specification Definition

From the information modeling activities explained above, we obtained a list of identified information concepts (such as Place, News, Article, Event) including their attributes (such as name, location, start date, contact information). The next step will be to create a domain specification for every identified concept.

---

[12] https://en.wikipedia.org/wiki/TYPO3.
[13] http://www.mayrhofen.at/die-region/.
[14] http://www.mayrhofen.at/die-region/mayrhofen/.
[15] http://www.mayrhofen.at/en/events/detail/events/globeseekers-yoga/.

In this step, we selected the right class from schema.org which most adequately describes a concept. We searched for suitable schema.org classes for every concept, defined a selection of these properties, selected the range types and recursively repeated this process when structured types appear as ranges. For example, according to our domain analysis for the type Hotel, we selected the following properties: name, description, telephone, faxNumber, email, url, currenciesAccepted, address, aggregateRating, geo, makesOffer, and image. Each property has its range, e.g. Text, Url, DateTime, QuantitativeValue. Some elements can have external properties, and we considered them too. Address in schema.org has the range PostalAddress, where PostalAddress is defined by addressCountry, addressLocality, addressRegion, postalCode and streetAddress. Property makesOffer has the range Offer with the following properties: name, availability, itemOffered, priceSpecification. A subset of our domain specification is shown in Table 1.

Table 1. A subset of our domain specification

No.	Type	Property	Range type
1	Hotel	address	PostalAddress
		aggregateRating	AggregateRating
		currencyAccepted	Text
		description	Text
		geo	GeoCoordinates
		image	ImageObject
		makesOffer	Offer
		name	Text
		paymentAccepted	Text
		url	URL
3	PostalAddress	addressCountry	Text
		addressRegion	Text
		postalCode	Text
		...	
3	AggregateRating	ratingValue	Number
		reviewCount	Number
4	GeoCoordinates	latitude	Number
		longitude	Number

3.4 Discussion

We would like to outline a few important things we encountered during our analysis, information modeling, and defining domain specification. First, we worked with two different types of data: the static data and dynamic data. Static data refers to rarely changed information, meaning that once created then the information will stay as the original. Fall into this category are the information about the region and TVB itself, ski areas, press releases, and articles. On the other hand, dynamic data are changed regularly, for example, the price of an offer for an accommodation room could frequently be changed, and therefore the annotation should be updated regularly as well. Second, when selecting a class in schema.org, we tried to be as specific as possible. For example, information about concert should be annotated with MusicEvent (more specific type) instead of Event (generic type). And therefore, in our domain specification, we also consider the super and subclasses relationships between types.

4   Implementation

In this section, we explain our implementation on how to annotate content of the TVB Mayrhofen-Hippach website. Our implementation consists of a series of activities, starting from defining a mapping between data schema to classes and properties from the selected vocabulary. Next, we use the mapping to perform annotation automatically or manually, and finally, we attach annotations to the target website.

4.1   Automatic Annotation of Content

A significant part of the web content available on the TVB Mayrhofen-Hippach website is generated based on data made available by Feratel. Feratel provides information about events, accommodations, offers, and infrastructure for multiple regions in Austria including the Mayrhofen-Hippach region. We developed a software solution that automatically annotates events, accommodations, offers, and infrastructure in Mayrhofen-Hippach according to schema.org. Feratel data can be accessed through the DSI by accepting requests from a client, processes the request and produces responses, all in the format of XML according to a particular structure. To annotate every event, accommodations, offer, and infrastructure item coming from Feratel we take the following approach:

1. Define a mapping between the Feratel data types to the specification produced in the domain specification definition explained in the previous section.
2. Develop a software wrapper to communicate with the DSI, and consumes the mapping to produce annotations in a JavaScript Object Notation for Linked Data (JSON-LD)[16] format.

Table 2 shows the statistics of the data types mapping, where not all types available in Feratel can be mapped into schema.org. For example, for infrastructure, we were able to identify 67 types available in schema.org from 212 types provided by Feratel. There are various causes for the mapping deficits, such as: language differences, different conceptions and orientation for data types

---

[16] https://en.wikipedia.org/wiki/JSON-LD.

Table 2. Statistics of data types mapping

No	Description	Mapped data types	
		Feratel	Schema.org
1	Accommodation	19	6
2	Event	29	11
3	Infrastructure	212	67

providing by Feratel and schema.org. In German language there exist a lot of names which represent the same type of object, for example Gasthof, Pension, Aparthotel, Berghotel, Berggasthof are hotels based on their characteristics and features. Feratel, as mentioned, offers data related to the tourism and travel industry. They categorized data based on types of objects in Austria. This classification is on the one hand too general (e.g. the type of events Theater/Show/Tanz/Film/Kleinkunst), and on another hand too detailed (e.g. the types Fahrrad-Transport, Fahrrad-Werksttte, Fahrrad-Verleih, E-Bike). Schema.org provides the most common terms to annotate a great variety of entities on the web. But it is relatively young vocabulary and can't cover all content. That's why we had difficulty mapping some types from Feratel. For example, type SportsActivityLocation in schema.org contains only 9 subtypes, whereas in Feratel more than 40 are presented. Therefore for many types from Feratel we chose more generic classes from schema.org, such as: SportsActivityLocation, Local-Bussiness, TouristAtraction, Store, Event, CivicStructure and LodgingBusiness.

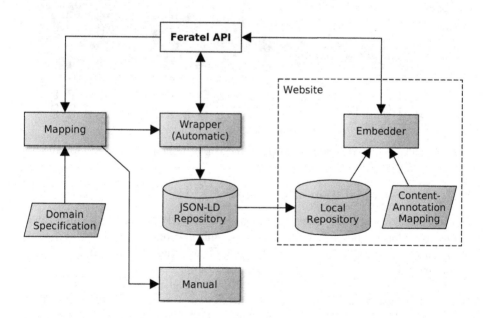

Fig. 1. Architecture of content annotation approach

Also, it is worth mentioning that in some cases, a data instance might inherit properties from multiples types, known as a multi-types entity.

Figure 1 shows the architecture of our approach for content annotations. For automatic annotation, the domain specifications will be aligned with the Feratel API specification to produce a mapping to be consumed by the wrapper. When defining the mapping, we started with our previous work [3], where we tried to map each XML element and attribute of Feratel data to class and property in the specifications. A plugin installed in the website server loads annotation in JSON-LD format from a repository and embeds it into the associated webpage identified by a mapping (detail explanation in Sect. 4.3). The wrapper runs on a daily basis, producing incremental updates. From about 1,6 million triples we produced per March 29, 2017, mostly dominated by information related to accommodation, as shown in Fig. 2. A full update needs to be performed whenever something changed in the domain specification, mapping, or wrapper implementation, e.g. when we supported multilingual annotation on 17.03.2017.

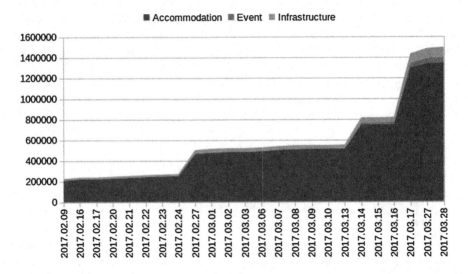

Fig. 2. The cumulative number of produced triples from automatic annotations

4.2   Manual Annotation of Content

Content not coming from content sources providing APIs, needs to be annotated manually. This is the case with ski areas, information about the Mayrhofen-Hippach region, press release articles, the TVB Mayrhofen-Hippach description and a few infrastructures which are not available in Feratel.

A conceptual analysis was performed before creating manually the annotations. For infrastructure for example, as its content is more complex in structure as all the other types of content available on Mayrhofen website, a large set of concepts and properties had to be considered. The conceptual analysis task enabled us to identify all the relevant information available in the Mayrhofen-Hippach holiday region and then clearly define the structure [1]. As shown in Fig. 3, we were able to produce around 8 thousand triples for this manually produced annotations, mostly for infrastructures and press release articles.

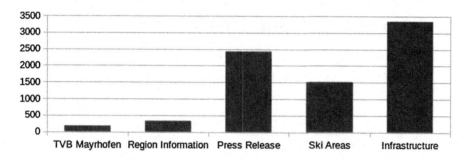

Fig. 3. Number of produced triples from manual annotations

We did not use any Natural Language Processing (NLP) technique for producing manual annotations due to time limitation. Finding the correct tuning parameters for an NLP algorithm requires data training which we do not have. Creating annotations manually was the best option, where an editor was used to guarantee annotations correctness and validity.

4.3   Linking of Content with Annotation

Once the annotations are created as described in the previous sections, what remains is to deploy them or in other words to link the annotations to the content which are available on TVB Mayrhofen website. This section describes our solution to achieve this goal. An important aspect is that the annotations and the content are available on different systems and are brought together via a deployment described in the rest of this section. Two core requirements needed to be fulfilled by our solution:

1. How to connect the content (which is in HTML) residing on the website with its annotation (JSON files) available on another server
2. How to embed the annotation (JSON files) into the content (HTML)

To fulfill the requirements, we designed our solution to separate the annotation process and embedding annotation to content process such that each process can be maintained without interfere with the other. In Fig. 1, the interlinking is done in the website server separately from the annotations processes. The separated processes will be performed as follows:

1. Annotation process produces all required annotations from Feratel API through automatic annotations (Sect. 4.1) as well as from manually generated annotations (Sect. 4.2). All annotations in JSON-LD format will be stored in a repository.

2. Embedder process which is installed as part of Content Management System (CMS) of the website loads a JSON-LD file from a local repository, where both repositories will be synchronized regularly. To identify which file should be loaded and embedded to a page, the Embedder reads a mapping between content and its associated annotation.

Table 3. An example of content to annotation mapping

Key	Value
0000	0a2346a9-3b05-4dc4-a056-1f32ccf05fe8.json
1111	0a19990c-b879-42ff-acc1-886d1ea59365.json
/meta/impressum	impressum.json
/service/kontackt	contact.json

As shown in Fig. 1, a plugin for Embedder will reside in the website (in this case in the CMS of TVB Mayrhofen website). When a request is received from a client, the plugin consults its mapping database, if a matching is found, then it will load the file from the local JSON-LD files repository and embed its content into the HTML response to the client. Items of the mapping database in the format of <key, value>, represent an association between a webpage and its annotation. In the current implementation, we have two types of association mapping: we use <Page-ID, Feratel-ID.json> and <Page-URL, Filename.json> for the data coming from Feratel and manually annotated respectively. "Page-ID" and "Page-URL" were obtained from the CMS and "Feratel-ID" from the Feratel API. A small fragment of the mapping is shown in Table 3, where an annotation will be identified either with an identification number for Feratel data and an URL for annotation generated from information on the website.

5 Result and Evaluation

In this section, we list and discuss the results of our work. After that, we explain the results of our qualitative and quantitative evaluations.

5.1 Results

As results, we were able to annotate numerous topics of information from the TVB Mayrhofen-Hippach website:

1. **Accommodation**, information related to accommodation including offered places such as a room that can be rented to stay for a given period.
2. **Event**, information about events that are happening at a particular time and location.
3. **Infrastructure**, information which is related to physical businesses or organization, including places that someone may find interesting (point of interest).

4. **Organization**, information which is related to the TVB itself, for example, its address and contact point, opening hours, etc.
5. **Press-release**, information which is related to news article or report including blog posting.
6. **Region**, information which is related to the region of Mayrhofen Hippach, for example, content about Mayrhofen and its villages, family holiday guides, winter guides, etc.
7. **Ski-area**, information which is related to ski-area such as ski-resort, ski-lift or slope which currently receives a minimal support in schema.org.

5.2 Evaluation

For qualitative evaluations, we monitored Google's search engine results especially the appearances of rich results. As shown in Fig. 4, Google Search was able to show events in a more structured way where the date of an event, as well as its location, will be additionally included in the search results. Google also produces richer and structured information for a hotel as shown in Fig. 5, where the rating and location for the hotel will be included in a rich card.

Food Festival Week : TVB Mayrhofen
www.mayrhofen.at/.../events/.../events/schmankerlwoche-schmanke... ▾ Translate this page
Information about the **Event**: Places: Hippach. Location: Hippach. Duration: 1 day. Additional **event** days: Mon, 03.07.2017 ICS-Download; Tue, 04.07.2017 ICS- ...

Mon, 3 Jul   Hippach, Mayrhofen, , AT

Fig. 4. Google Search rich-snippet for an event

Fig. 5. Google rich-card preview for a hotel

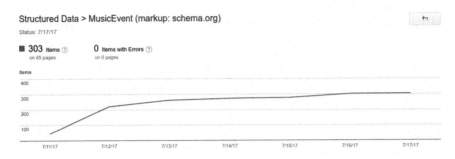

Fig. 6. Google search console for structured data of MusicEvent

For quantitative evaluations, we use the Google's Search Console[17] to measure a few aspects: (i) how long was required to detect the annotated pages, (ii) how often the annotated pages were crawled, (iii) how many errors were detected. As an example, we took the statistics of structured data of "MusicEvent" as shown in Fig. 6. From more than 300 detected items, 2 days were required for collecting about 220 items (status on 7/13/2017, compared to the status on 7/11/2017). Further, we used the "Last.Checked" history of Google's search console to measure how often the annotated pages were crawled. From all detected structured data of "LodgingBusiness", which were crawled for more than 50 times (status on 7/17/2017), the average of crawled frequency was 2 days. And lastly, there was no error detected.

6   Ongoing Work and Outlook

Besides the work described in this paper, we consider several more directions to continue and extend this topic. On the way to provide a holistic armamentarium for a semantic web contribution for touristic service providers, our ongoing and future work comprises of the tasks and ideas described below.

6.1   Schema.org 3.1 and Actions

Since May 2015 schema.org offers a new extension mechanism that facilitates the creation of specialized and/or deeper vocabularies based on the core vocabulary of schema.org. We have submitted an accommodation extension to schema.org that became an integrated part of schema.org 3.1 [16]. As future work we will update the set of annotations produced so far to be fully aligned and complete regarding with schema.org 3.1, particularly the hotel extension. Once accommodation offers and booking data are machine readable, a system that makes use of these data to enable automatic direct booking of offers can be established. The overall approach includes three main steps: (1) annotation of booking data, including data about room, offers, etc. This kind of data needs to be annotated

---

[17] https://www.google.com/webmasters/tools/.

with schema.org including the schema.org hotels extension as part of schema.org 3.1, (2) annotation of booking engines, meaning booking engines need to be annotated with schema.org in order to be found and understood and (3) implementation of an automated direct booking agent. As part of this last step a booking agent connects the booking data with the booking engine, crawls the booking data and the booking engines/endpoints and performs heuristic reasoning due to the fact that the booking data is usually not complete, approximately, and partially inaccurate. Our work so far has focused on the first step. As part of the current and future work we are tackling step 2 and 3 providing semantic annotations using schema.org actions and then designing and implementing an automated direct booking agent.

6.2 Chat Bots and Intelligent Personal Assistants (IPAs)

Since the introduction of chat bots on Facebook's F8 conference in April 2016, the topic has attracted a lot of attention from small and large companies alike. Big names in the software industry, including Amazon, Apple, Facebook, Google and Microsoft are developing their own solutions and are opening their APIs providing support for developers to build chat bots and personal assistants for their platforms. Tourism is a domain where chatbots and intelligent assistants have an immediate applicability from finding touristic service providers, their services and offers to booking/buying these items using new conversational, intuitive interfaces. As current work we are developing a chatbot for Mayrhofen-Hippach region. The bot is named as "Mayley", available as a Facebook Messenger bot as well inside a web widget deployable target website[18]. Mayley uses directly the Mayrhofen-Hippach region semantic annotations to create the appropriate answers given tourists natural language requests. Combined with user profile information and rules, Mayley delivers personalized content to its users. We are also working on using semantic annotations to update the set of entities and intents the chatbot understands.

6.3 Schema.org Annotation Generation Platform

The generation of schema.org annotations for web content is, especially for people with no background in programming languages or semantic technologies, not trivial. But often these are the people who should actually make use of schema.org annotations: content creators for enterprises or touristic service providers, event promoters or blog editors. Various different factors keep them from annotating their content on the web, broken down, three main challenges emerge: (1) what vocabulary to use, (2) how to create JSON-LD files and (3) how to import annotation files into websites. To tackle those three challenges we are working on a web platform which offers assistance in the whole annotation creation process. This platform should remove a big obstacle between the "normal" web content creator and the semantic web.

---

[18] https://www.facebook.com/MayleyMayrhofen.

6.4   Validation of Schema.org Annotated Data

A study about the use of schema.org in the hotel domain [15] showed, that a lot of content creators, enterprises and touristic service providers on the web want to use semantic annotations on their website, but are not able to do it in a correct way. So besides providing a solution to create and publish semantic annotations (mentioned in Sect. 6.3) we also work an a means to validate existing semantic annotations. There are several validators around on the web, with Google's Structured Data Testing Tool[19] leading the way, but those validators either only validate for syntactical correctness or do only very limited or biased semantic validation. The solution we work on is based on two different fundamentals. First of all the recommended or required vocabulary is defined. The second foundation is a set of rules to define correlation between schema.org properties and their ranges as well as the correlation between different properties. More information about the work on that idea can be found in [7]. This tool allows content creators on the web to not only generate some mandatory "meta-tags" but valid, high quality annotations which lead to reusable, high quality web content.

6.5   Touristic Knowledge Graph

With the idea of a touristic knowledge graph, we want to support tourists as well as touristic service providers and also provide a means for analyzing touristic developments over time. The knowledge graph comprises information about touristic services, the infrastructure of a region, points of interests but also information about arrivals of guests, events, weather data and other factors influencing tourism. This idea is in a very early phase and will be followed in the futures.

7   Conclusion

In this paper, we presented the work done to provide better online visibility for touristic service providers by using semantic technologies, particularly semantic annotations using schema.org. We used Mayrhofen-Hippach region as a pilot, created and deployed for the touristic services providers in this region a substantial amount of annotations. The annotations covered a wide variety of information topics including events, accommodations and accommodation offers, ski areas, the region, press release articles, the organization itself and a large variety of infrastructure information. Moreover, most of the annotations need to be regularly updated on a daily basis due to the dynamic changes in the data sources, for example, the price of an accommodation offer. The annotations were created in a mixed manner, automatically and manually, where the software tools for generating annotations automatically as well as tools to support the human users to create annotations are currently deployed for internal use only. After a few improvements, including integration with latest recent mapping languages

---

[19] https://search.google.com/structured-data/testing-tool.

such as RML, we will offer them as services to be used not only by TVBs but also by any organization that willing to annotate their webpages.

The same approach has been successfully applied to other TVBs, namely Seefeld[20] and Fügen[21] among others. We also applied the approach to some other use cases in the tourism industry, including ski schools, ski resorts, golf places. And currently, we are working to annotate an interactive map provided by General Solutions[22]. The map contains rich geo-related information such as hiking or biking routes, entry points for a route as well as point of interests along a route. Our ultimate goal is to be able to annotate all tourism relevant information in the region of Tyrol, Austria, not only to increase the online visibility of the region but also to enable intelligent applications to run on top of them.

Acknowledgements. This work was partially supported by the EU project EUTravel. We would like to thank Daniel Ackstaller, Daniel Eppacher, Christian Esswein, Omar Holzknecht, Philipp Kratzer, Jonas Stock, Johannes Strickner, Simon Targa, Sahin Ucar, Hannes Vieider, and Jakob Winder for their fruitful discussions and input.

References

1. Ackstaller, D., Akbar, Z., Eppacher, D., Esswein, C., Holzknecht, O., Kärle, E., Kratzer, P., Simsek, U., Stock, J., Strickner, J., Targa, S., Ucar, S., Vieider, H., Winder, J.: Semantic annotation for mayrhofen.at. Technical report, Semantic Technology Institute (STI) Innsbruck, August 2016. http://oc.sti2.at/results/white-papers/semantic-annotation-mayrhofenat
2. Akbar, Z., García, J.M., Toma, I., Fensel, D.: On using semantically-aware rules for efficient online communication. In: Bikakis, A., Fodor, P., Roman, D. (eds.) RuleML 2014. LNCS, vol. 8620, pp. 37–51. Springer, Cham (2014). doi:10.1007/978-3-319-09870-8_3
3. Akbar, Z., Toma, I.: Feratel content annotation with schema.org. Technical report, Semantic Technology Institute (STI) Innsbruck, February 2015. http://oc.sti2.at/results/white-papers/feratel-content-annotation-schemaorg
4. Auer, S., Dietzold, S., Lehmann, J., Hellmann, S., Aumueller, D.: Triplify: lightweight linked data publication from relational databases, WWW 2009, pp. 621–630. ACM, New York (2009)
5. Bizer, C.: D2RQ - treating Non-RDF databases as virtual RDF graphs. In: Proceedings of the 3rd International Semantic Web Conference (ISWC 2004) (2004)
6. Bizer, C., Eckert, K., Meusel, R., Mühleisen, H., Schuhmacher, M., Völker, J.: Deployment of RDFa, microdata, and microformats on the web – a quantitative analysis. In: Alani, H., Kagal, L., Fokoue, A., Groth, P., Biemann, C., Parreira, J.X., Aroyo, L., Noy, N., Welty, C., Janowicz, K. (eds.) ISWC 2013. LNCS, vol. 8219, pp. 17–32. Springer, Heidelberg (2013). doi:10.1007/978-3-642-41338-4_2
7. Şimşek, U., Kärle, E., Holzknecht, O., Fensel, D.: Domain specific semantic validation of schema.org annotations. In: Ershov, A.P. (ed.) Informatics Conference. Springer (2017, to appear)

---

[20] https://www.seefeld.com/.

[21] https://www.best-of-zillertal.at/.

[22] https://general-solutions.eu/.

8. Dimou, A., Vander Sande, M., Colpaert, P., Verborgh, R., Mannens, E., Van de Walle, R.: RML: a generic language for integrated RDF mappings of heterogeneous data. In: Proceedings of the 7th Workshop on Linked Data on the Web, April 2014

9. Ebner, C.: Deskline 3.0 standard interface (DSI) documentation (2016), ver. 1.0.67, 28 June 2016

10. Fensel, A., Akbar, Z., Toma, I., Fensel, D.: Bringing online visibility to hotels with Schema.org and multi-channel communication. In: Inversini, A., Schegg, R. (eds.) Information and Communication Technologies in Tourism 2016, pp. 3–16. Springer, Cham (2016). doi:10.1007/978-3-319-28231-2_1

11. Goel, K., Guha, R.V., Othar, H.: Introducing Rich Snippets (2009). https://webmasters.googleblog.com/2009/05/introducing-rich-snippets.html

12. Guha, R.V., Brickley, D., Macbeth, S.: Schema.org: evolution of structured data on the web. Commun. ACM **59**(2), 44–51 (2016)

13. Gupta, D.D., Utkarsh: Assessing the website effectiveness of top ten tourist attracting nations. Inf. Technol. Tourism **14**(2), 151–175 (2014)

14. Hepp, M., Siorpaes, K., Bachlechner, D.: Towards the semantic web in e-tourism: can annotation do the trick? In: Proceedings of the 14th European Conference on Information System (ECIS 2006), pp. 2362–2373 (2006)

15. Kärle, E., Fensel, A., Toma, I., Fensel, D.: Why are there more hotels in Tyrol than in Austria? Analyzing Schema.org usage in the hotel domain. In: Inversini, A., Schegg, R. (eds.) Information and Communication Technologies in Tourism 2016, pp. 99–112. Springer, Cham (2016). doi:10.1007/978-3-319-28231-2_8

16. Kärle, E., Simsek, U., Akbar, Z., Hepp, M., Fensel, D.: Extending the Schema.org vocabulary for more expressive accommodation annotations. In: Schegg, R., Stangl, B. (eds.) Information and Communication Technologies in Tourism 2017, pp. 31–41. Springer, Cham (2017). doi:10.1007/978-3-319-51168-9_3

17. Khalili, A., Auer, S.: User interfaces for semantic authoring of textual content: a systematic literature review. Web Semant. Sci. Serv. Agents World Wide Web **22**, 1–18 (2013)

18. Maedche, A., Staab, S.: Applying semantic web technologies for tourism information systems. In: Wöber, K., Frew, A., Hitz, M. (eds.) Proceedings of the 9th International Conference for Information and Communication Technologies in Tourism, ENTER 2002, Innsbruck, Austria, 22–25th January 2002. Springer, Vienna (2002)

19. Mich, L.: The website quality of the regional tourist boards in the Alps: ten years later. In: Xiang, Z., Tussyadiah, I. (eds.) Information and Communication Technologies in Tourism 2014, pp. 651–663. Springer, Cham (2013). doi:10.1007/978-3-319-03973-2_47

20. Stavrakantonakis, I., Toma, I., Fensel, A., Fensel, D.: Hotel websites, Web 2.0, Web 3.0 and online direct marketing: the case of Austria. In: Xiang, Z., Tussyadiah, I. (eds.) Information and Communication Technologies in Tourism 2014, pp. 665–677. Springer, Cham (2013). doi:10.1007/978-3-319-03973-2_48

21. Werthner, H., Alzua-Sorzabal, A., Cantoni, L., Dickinger, A., Gretzel, U., Jannach, D., Neidhardt, J., Pröll, B., Ricci, F., Scaglione, M., Stangl, B., Stock, O., Zanker, M.: Future research issues in IT and tourism. Inf. Technol. Tourism **15**(1), 1–15 (2015)

Distributed Holistic Clustering on Linked Data

Markus Nentwig[(✉)], Anika Groß, Maximilian Möller, and Erhard Rahm

Database Group, University of Leipzig, Leipzig, Germany
{nentwig,gross,rahm}@informatik.uni-leipzig.de,
m.moeller@studserv.uni-leipzig.de

Abstract. Link discovery is an active field of research to support data integration in the Web of Data. Due to the huge size and number of available data sources, efficient and effective link discovery is a very challenging task. Common pairwise link discovery approaches do not scale to many sources with very large entity sets. We propose a distributed holistic approach to link many data sources based on a clustering of entities that represent the same real-world object. Our approach provides a compact and fused representation of entities, and can identify errors in existing links as well as many new links. We support distributed execution, show scalability for large real-world data sets and evaluate our methods with respect to effectiveness and efficiency for two domains.

1 Introduction

Linking entities from various sources and domains is one of the crucial steps to support data integration in the Web of Data. A manual generation of links is very time-consuming and nearly infeasible for the large number of existing entities and data sources. As a consequence, there has been much research effort to develop link discovery (LD) frameworks [10] for automatic link generation. Platforms like `datahub.io` and `sameas.org` or repositories such as LinkLion [11] collect and provide large sets of links between numerous different knowledge sources. They can be reused to avoid an expensive re-determination of the links. It is particularly complex to ensure high link quality, i.e., the generation of correct and complete link sets. Existing link repositories cover only a small number of inter-source mappings and automatically generated links can be erroneous in many cases [2]. Despite the huge number of sources to be linked, most LD tools focus on a pairwise (binary) linking of sources. However, LD approaches need to scale for n-ary linking tasks as well as for an increasing number of entities and sources that are added to the Web of Data over time [13].

To address these shortcomings we recently proposed an approach to cluster linked data entities from multiple data sources into a holistic representation with unified properties [8]. The method combines entities that refer to the same real world object in one compact cluster instead of maintaining a high number of binary links for k sources. The approach is based on existing `owl:sameAs` links and can deal with entities of different semantic types as they occur in many sources (e. g., for geographical datasets, countries, cities, lakes). Input links are

© Springer International Publishing AG 2017
H. Panetto et al. (Eds.): OTM 2017 Conferences, Part II, LNCS 10574, pp. 371–382, 2017.
https://doi.org/10.1007/978-3-319-69459-7_25

checked for consistency and new links (e. g., for previously unconnected sources) are identified.

Considering the huge size and number of sources to be linked, scalability becomes a major issue. Linking and clustering approaches usually comprise complex operations such as similarity computations to identify similar entities or clusters. These complex work steps can often be parallelized in a distributed environment in order to reduce execution time significantly. Big Data frameworks like Apache Spark or Apache Flink [1] provide execution engines to process very large datasets in a distributed environment. With regard to the ever increasing amount of data that needs to be linked and integrated in typical big data processing workflows, it is essential to develop scalable solutions for link discovery and holistic entity clustering.

Herein we study a distributed holistic clustering approach. In contrast to the previous work, we support blocking strategies to reduce unnecessary comparisons and present a comprehensive evaluation for quality and efficiency on real-world data for different domains. An extended version with more details on Flink implementation and creation of a reference dataset can be found in [9]. We make the following contributions:

- We present a distributed holistic clustering approach for linked data to enable an effective and efficient clustering of large entity sets from many data sources.
- We evaluate the efficiency and effectiveness of the distributed holistic clustering for very large datasets with millions of entities from two domains.

We present the implementation of the distributed holistic clustering in Sect. 2. Then, we show evaluation results in Sect. 3. Finally, we discuss related work in Sect. 4 and conclude in Sect. 5.

2   Distributed Holistic Clustering Approach

In this section we outline the workflow and implementation for our distributed holistic clustering approach based on the big data stream and batch processing system Apache Flink [1]. Starting with a introduction to Apache Flink (Sect. 2.1), we present the transformation and adaptation of the holistic clustering approach towards a distributed processing workflow (Sect. 2.2).

2.1   Apache Flink and Gelly API

Apache Flink's batch processing provides the DataSet API and well-known dataset transformations like *filter, join, union, group-by* or *aggregations* (relational databases) and *map, flat-map* and *reduce* (MapReduce paradigm). Special in-memory, distributed data structures called DataSets store data within Flink programs. DataSets can be manipulated based on so called transformations that return a new DataSet. Some transformation operations make use of user-defined functions (UDFs) and allow for customized definitions how DataSet values need to be changed. We make use of the graph processing library (Gelly) in our

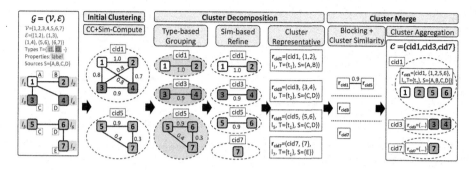

Fig. 1. Example clustering workflow.

holistic clustering workflow. In particular, we employ Gelly graphs containing a `DataSet<Vertex<K, VV>>` vertices and a `DataSet<Edge<K, EV>>` edges.

The complex data types `Vertex` and `Edge` are inherited from the Flink `Tuple` classes `Tuple2<K,VV>` (type K as vertex id, VV as vertex value), `Tuple3<K,K,EV>` (source vertex id, target vertex id (each type K) and EV as edge value), respectively. Operators like *join*, *filter* or *group-by* rely on tuple positions (starting from 0), e. g.,

```
vertices.join(edges).where(0).equalTo(1)
    .with((vertex, edge) -> new Tuple1<>(edge.getSimilarity))
    .filter(tuple -> tuple.f0 >= 0.9);
```

will join all edges with the vertices where the vertex id (position 0 in `vertices`) equals the target id of the edge (position 1 in `edges`) and returns the similarity value if the accompanied filter function is evaluated and returns true.

Besides the used graph data model we benefit from Flink's and Gelly's abstract graph processing operators like graph neighborhood aggregations or abstracted models for iterative computations. In particular, we will make use of the Flink delta iteration in different variations as discussed in the following sections.

2.2 Distributed Holistic Clustering

In this section, we will discuss the transformation and adaptation of the holistic clustering workflow towards a distributed processing workflow in Apache Flink. From a high-level perspective, we read input entities and links into a Gelly graph $\mathcal{G}$ with vertices $\mathcal{V}$ and edges $\mathcal{E}$ and apply a set of transformation operators to generate entity clusters $\mathcal{C}$. We illustrate the workflow steps using the running example in Fig. 1. There are six input edges $\mathcal{E}$ having optional similarity values and seven input vertices $\mathcal{V}$ further described by a label ($l_1, l_2, \ldots$), the originating data source S and colored dependent on their semantic type ($t1, t2$ or no type).

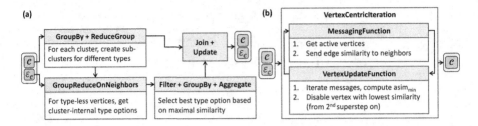

Fig. 2. Sub-workflows with operators for type-based grouping (a) and similarity-based refinement (b).

Preprocessing. During preprocessing we apply several user-defined functions on the input graph, e. g., to harmonize semantic type information, remove inconsistent edges and vertices and normalize the label property value. First, we compute similarities only for given input edges based on vertex property values. For each vertex, we carry out a consistency validation using grouping on adjacent vertices and associated edges, and remove neighbors with equal data sources (details in [8]). We omit the preprocessing in the example (Fig. 1) and directly start with the preprocessed input graph $\mathcal{G}$.

Initial Clustering. To determine initial clusters, we determine the connected components (CC) within $\mathcal{G}$ and assign a cluster id to each vertex. In the example, vertices 1-4 obtain cluster ids *cid1* and vertices 5-7 *cid5*. Intra-cluster edges are then generated within each cluster accompanied by a similarity computation based on properties such as a linguistic similarity on labels or normalized geographical distance.

Cluster Decomposition. *Type-based grouping* is the first part of the decomposition to split clusters into sub-components dependent on the compatibility of semantic types. Figure 2a shows the sequence of applied transformations and short descriptions. Within clusters, a ReduceGroup function assigns new cluster ids based on semantic types, e. g., in the example vertex 3 and 4 are separated from vertex 2. Vertices without type (like vertex 1) require a special handling. We apply GroupReduceOnNeighbors (a Gelly CoGroup function to handle neighboring vertices and edges) to produce tuples for vertices with missing semantic type, e. g., vertex 1 creates a (id,sim,type,cid) for each outgoing edge $((1,2),(1,3),(1,4))$, namely (1, 1.0, t_1,cid1) for edge $(1,2)$ and (1, 0.8, t_2, cid2) for edge $(1,3)$ and $(1,4)$. Grouping on the vertex id executes an aggregation function for each group to return the tuple with the highest similarity per vertex, which is (1, 1.0, t_1, cid1) for vertex 1, processed vertices update their cluster id accordingly (e. g., vertex 1 → *cid1*). The result of the type-based grouping is a set of clusters with intra-cluster edges.

Similarity-based Refinement. We further decompose clusters by removing non-similar entities from their cluster. We use a Gelly vertex-centric iteration using

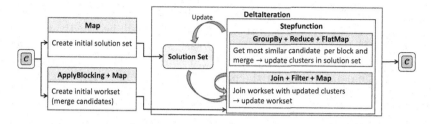

Fig. 3. Sequence of transformations for the cluster merge using Flink DeltaIteration

the core idea to iterate between a custom MessagingFunction and a VertexUpdateFunction (see Fig. 2b for details). In the first round, all vertices are active and send messages to all their neighbors. Messages are tuples containing the originating id, the edge similarity and an average edge similarity $asim$ over all incoming messages (0 in the first iteration). Starting with the second iteration, we illustrate the sent messages for vertex 7 in cluster $cid5$ in our example: vertex 5 sends (5, 0.4, 0.65) to 7, vertex 6 sends (6, 0.3, 0.6) to 7 and vertex 7 sends messages to 5 and 6, resulting in $asim = (0.4 + 0.3)/2 = 0.35$ for vertex 7. Now in each cluster the vertex with the lowest $asim$ will be deactivated (and is therefore excluded from the cluster) given that this $asim$ is below a certain similarity threshold. In Fig. 1, vertex 7 will be deactivated and isolated into cluster $cid7$. Vertices send only messages if they are updated and deactivated vertices never send messages again, therefore, iteration termination is guaranteed.

Finally, we create a unified *Cluster Representative* for each cluster based on contained entities. Aggregation of property values is used for covered data sources and semantic types as well as selection of best label or geographic coordinates, see Fig. 1.

Cluster Merge. The main operators for the merge phase are sketched in Fig. 3. The implemented DeltaIterate function iteratively combines highly similar clusters into larger ones. To avoid the quadratic complexity comparing all clusters, we employ blocking strategies to avoid unnecessary comparisons, e. g., standard blocking on properties like label, not comparing representatives with incompatible semantic type and check for already covered data sources. In our example in Fig. 1 three blocks are created by applying blocking strategies. Only r_{cid1} and r_{cid5} need to be compared, such that a triplet $(r_{cid1}, 0.9, r_{cid5})$ is created as a merge candidate.

The delta iteration starts with an initial solution set containing the previously determined clusters and an initial workset (merge candidates) as seen in Fig. 3. Each iteration updates the workset applying a custom step function to generate changes for the solution set. In detail, for each block the merge candidate with the highest similarity is selected using a custom Reduce function. For our running example, $(r_{cid1}, 0.9, r_{cid5})$ is the best candidate and is merged using a custom FlatMap function. The new cluster r_{cid1} contains combined values for properties

S, T and l (see Fig. 1). This will directly affect the cluster representatives in the solution set, and the already merged cluster r_{cid5} will be deactivated in the solution set. Merge candidates in the workset are adapted based on changed clusters within the iteration step (see Fig. 3 solution set). Again, triplets are discarded if the data sources for the participating clusters overlap or exceed the maximum possible number of covered sources.

The delta iteration ends when the workset is empty (true for our running example after the first iteration). Note that parts of the dataset will converge faster to a solution, when clusters can not be merged anymore. These parts will not be recomputed in following iterations, such that only smaller parts of the data will be handled.

3   Evaluation

In the following we evaluate our distributed holistic clustering approach w.r.t. effectiveness and efficiency for datasets from the geographic and music domains. We first describe details of the used datasets (Sect. 3.1). We then evaluate the effectiveness and efficiency of our approach (Sect. 3.2).

3.1   Datasets

We use five datasets of different sources from the music and geographic domains (Table 1). Datasets DS1 and DS3 are used to evaluate the quality of entity clusters generated by the distributed holistic clustering while DS2, DS4 and DS5 are used to analyze the efficiency and scalability (see Sect. 3.2).

We use two datasets (DS1, DS2) from the *Geographic Domain*, covering entities from the data sources DBpedia, GeoNames, NY Times, Freebase for DS1 and additionally LinkedGeoData for DS2. Entities for both datasets have been enriched with properties like entity label, semantic type and geographic coordinates by using SPARQL endpoints or REST APIs. DS1 is based on a subset of existing links provided by the OAEI 2011 Instance Matching Benchmark[1]. For DS1, clusters and links have been manually checked and create a novel reference dataset for multi-source clustering [9]. We provide this dataset covering

Table 1. Overview of evaluation datasets. Number of resulting clusters and deduced correct links are given for reference datasets.

Domain	Entity properties	Dataset	#entities	#sources	#correct links	#clusters
Geography	label, semantic type	DS1	3,054	4	4,391	820
	longitude, latitude	DS2	1,537,243	5	-	-
Music	artist, title, album	DS3	19,375	5	16,250	10,000
	year, length, language	DS4	1,937,500	5	1,624,503	1,000,000
	number	DS5	19,375,000	5	16,242,849	10,000,000

[1] http://oaei.ontologymatching.org/2011/instance/.

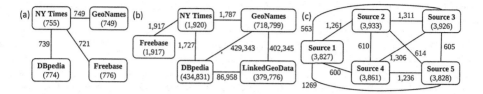

Fig. 4. Dataset structure for DS1 (a), DS2 (b) and DS3 (c) with number of entities and links.

the input dataset and the perfect cluster result as JSON files[2]. Dataset DS2 (Fig. 4b) originates from the link repository LinkLion [11]. We reuse about 1 Mio existing `owl:sameAs` links from LinkLion as input for the holistic cluster-ing. However, there is no reference dataset available to evaluate the quality of created clusters for dataset DS2. We use DS2 to evaluate the scalability of our approach for very large entity sets.

For the *Music Domain*, we use the publicly available Musicbrainz dataset covering artificially adapted entities to represent entities from five different data sources [4]. Every entry in the input dataset represents an audio recording and has properties like title, artist, album, year, language and length. The property values have been partially modified and omitted to generate a certain degree of unclean data and duplicate entities that need to be identified. Beside a set of artificially created duplicates, each dataset covers cluster ids from which links between entities, that refer to the same object, can be easily derived. DS3 will be used for quality evaluation, while DS4 and DS5 are used to analyze the scalability of the distributed holistic clustering.

3.2 Experimental Results

We now present evaluation results w.r.t. the quality of the determined clusters as well as the scalability of the distributed holistic clustering for the five datasets DS1-DS5.

Setup and Configurations. The experiments are carried out on a cluster with 16 workers (Intel Xeon E5-2430 6x 2.5 GHz, 48 GB RAM) operating on OpenSUSE 13.2 using Hadoop 2.6.0 and Flink 1.1.2. All experiments are carried out three times to determine the average execution time.

We created input links for DS1 using three different configurations (confs) - computing similarities based on JaroWinkler on the entity label; confs 2 and 3 additionally compute a normalized geographic distance similarity below a max-imum distance of 1358 km. Conf 1 applies a minimal similarity threshold of 0.9 for labels while confs 2 and 3 apply threshold 0.85 and 0.9 for the average label and geographic similarity, respectively.

---

[2] https://dbs.uni-leipzig.de/research/projects/linkdiscovery.

Table 2. Evaluation of cluster quality for geography dataset DS1 w.r.t. precision (P), recall (R) and F-measure (F1).

	Config 1			Config 2			Config 3		
	P	R	F1	P	R	F1	P	R	F1
Input links	0.933	0.806	**0.865**	0.964	0.938	0.951	0.981	0.799	0.881
Best (star1, star2)	0.863	0.844	0.853	0.963	0.941	**0.952**	0.951	0.838	0.891
Holistic	0.903	0.824	0.862	0.913	0.919	0.916	0.968	0.836	**0.897**

For the music dataset *DS3* we created input links using a soft TF/IDF implementation weighted on title (0.6), artist (0.3) and album (0.1) with a threshold of 0.35. *DS4* and *DS5* are used to show scalability, we simply create edges based on the cluster id from the perfect result by linking the first entity of each cluster with all its neighbors.

Quality. We analyze the achieved cluster quality for all datasets based on precision, recall and F-measure. The input links DS1 (see Fig. 4 a) are manually curated, therefore, they achieve a precision of 100%. However, missing links between lead to a recall of only 50%, resp. F-measure of 66.7%. With the holistic clustering approach, we achieve very good results w.r.t. recall (97.1%) while preserving a good precision (99.8%) resulting in the F-measure of 98.5%. This shows that we produce high-quality clusters based on existing input links thereby finding many new links.

However, as input mappings are not perfect in real-world situations, we used automatically generated input links ations (config 1-3) as described above. To evaluate the cluster quality, we further compare our results with the best configurations recently published results in [14]. Star clustering creates overlapping clusters, thus clusters may contain duplicates. Besides, star clustering does not create a compact cluster representation. Table 2 shows results w.r.t. the cluster quality for the computed input links, the best result of [14] and our approach. The distributed holistic clustering nearly retains the input link quality for config 1, while best(star1, star2) achieves slightly worst results. For config 2, the star2 implementation achieves a slightly better F-measure compared to the input mapping. For config 3 the holistic clustering improves the quality of the input mapping by 1.6% w.r.t. F-measure.

For the music domain, we evaluate the cluster quality for DS3 using a set of computed input links (see setup in Sect. 3.2). Overall, the quality of the input links is lower than for DS1. Due to strongly corrupted entities and more properties, DS3 is more difficult to handle. Applying the holistic clustering, we identify a quality improvement for both precision (89.0%) and recall (86.1%), resulting in a significant increase of F-measure by approx. 7% to 87.6% showing that our approach is able to handle such unclean data.

#workers		pre	dec	merge	total
DS2	1	312	668	351	1331
	2	164	367	268	799
	4	79	231	207	518
	8	45	130	186	361
	16	23	42	162	227
DS4	1	423	419	608	1450
	2	224	236	417	876
	4	121	123	301	545
	8	62	73	238	372
	16	40	35	237	312

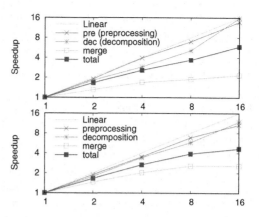

Fig. 5. DS2 and DS4 execution times (left) in seconds and speedup (DS2 top right, DS4 bottom right) for the single workflow phases and total workflow.

Overall, the holistic approach achieves competitive results although the DS1 dataset facilitates achieving relatively good input mappings making it difficult for any clustering approach to find additional or incorrect links.

Scalability. To evaluate the distributed holistic clustering w.r.t. efficiency and scalability, we determine the absolute execution times as well as the speedup for the very large geographic (DS2) and music datasets (DS4, DS5). To show scalability, we vary the number of Flink workers and use a parallelism equal to the number of workers.

Figure 5 show the achieved execution times for DS2 and DS4 for different phases of the clustering workflow and the overall workflow execution time. For each phase, an increased number of workers leads to improved execution times. The best improvement can be achieved for the preprocessing (pre) and decomposition (dec). The merge phase is by far more complex. While preprocessing and decomposition operate within connected components and clusters, the merge phase attempts to combine similar clusters based on the assignment in the blocking step and therefore can suffer from data skew problems for some blocks. These effects become also clear in Fig. 5 showing the speedup results compared to the linear optimum. Preprocessing and decomposition achieve nearly linear speedup, while the merge phase shows decreased speedup values. In total, we achieve a good speedup of 5.86 for the large geographic dataset DS2 and 4.65 for the large music dataset DS4. For the largest dataset DS5, we could determine results for two configurations: 8 workers could finish the complex task in 43,589 seconds, 16 workers finished after 24,722 seconds (reduced by factor ≈1.8).

Overall, the distributed holistic clustering achieves good execution times and moderate scalability results for very large entity sets. The approach is scalable for different data sources and employs a multi-source clustering instead of basic binary linking of two sources. The distributed implementation further allows to scale for a growing number of entities and data sources and is very useful for complex data integration scenarios in big data processing workflows.

4   Related Work

Link discovery (LD) has been widely investigated and there are many approaches and prototypes available as surveyed in [10]. Typical LD approaches apply binary linking methods for matching two data sources but lack efficient and effective methods for integrating entities from k different data sources to provide a holistic view for linked data. Some approaches enable distributed link discovery or for matching two data sources, e.g., Silk [6] and Limes [5] realized LD approaches based on MapReduce before distributed data processing frameworks like Spark or Flink became state of the art, therefore they are suffering from limitations of MapReduce like repeated data materialization and lack of iterations. They further focus on pairwise matching and do not support reuse of existing links sets.

While LD is driven by pairwise linking of data sources, support for multiple data sources can be found in related research areas. In [3] ontology concepts from multiple data sources are clustered based on topic forests for extracted keywords from concepts and their descriptions to determine matching concepts within groups of similar topics. In [7] a maximum-weighted graph matching and structural similarity computations are applied to concepts of multiple ontologies to find high quality alignments. However, these holistic ontology matching approaches do not focus on clustering and have limitations w.r.t. scalability for LD.

There are few LD approaches for linked data on multiple sources. Thalhammer et al. [15] present a pipeline for web data fusion using multiple data sources applying hierarchical clustering. The unsupervised LD approach Colibri [12] considers error detection for LD in multiple knowledge bases based on the transitivity, while clustering of entities is not the main focus. Both approaches do not realize distributed execution and have not been evaluated w.r.t. scalability. The work in [14] considers the implementation of existing clustering algorithms on top of Apache Flink for entity resolution of several data sources. The approach does not handle incorrect links or semantic type information and does not create a compact cluster representation.

5   Conclusion

We presented a distributed holistic clustering workflow for linked data using the distributed data processing framework Apache Flink using dataset transformations and user-specific Flink operators. Our approach is based on the reuse of existing links and is able to handle entities from various data sources. We presented comprehensive evaluation results for datasets from two domains with up to 20 million entities showing that the proposed approach can achieve a very high cluster quality. In particular, we were able to find many new correct links and could remove wrong links. The distributed execution in a parallel cluster environment resulted in very good execution times for the considered dataset sizes and good overall scalability results.

For future work, we plan to further improve the scalability of our approach, e. g., by realizing sophisticated blocking and load balancing methods for the complex cluster merge phase. We further plan the development and combination with an incremental clustering to support the addition of new entities and data sources, particularly to address the ongoing growth of the Web of Data.

Acknowledgments. This research was supported by the Deutsche Forschungsgemeinschaft (DFG) grant number RA 497/19-2.

References

1. Carbone, P., Katsifodimos, A., Ewen, S., Markl, V., Haridi, S., Tzoumas, K.: Apache flink™: stream and batch processing in a single engine. IEEE Data Eng. Bull. **38**(4), 28–38 (2015)
2. Faria, D., Jiménez-Ruiz, E., Pesquita, C., Santos, E., Couto, F.M.: Towards annotating potential incoherences in bioportal mappings. In: ISWC, pp. 17–32 (2014). doi:10.1007/978-3-319-11915-1_2
3. Grütze, T., Böhm, C., Naumann, F.: Holistic and scalable ontology alignment for linked open data. In: WWW2012 Workshop on Linked Data on the Web (2012)
4. Hildebrandt, K., Panse, F., Wilcke, N., Ritter, N.: Large-Scale data pollution with apache spark. IEEE Trans. Big Data **PP**(99), 1–1 (2017). doi:10.1109/TBDATA. 2016.2637378
5. Hillner, S., Ngonga Ngomo, A.C.: Parallelizing LIMES for large-scale link discovery. In: I-Semantics 2011, pp. 9–16. ACM, New York (2011). doi:10.1145/2063518. 2063520
6. Isele, R., Jentzsch, A., Bizer, C.: Silk Server - Adding missing Links while consuming Linked Data. In: Proceedings of the First International Workshop on Consuming Linked Data, CEUR Workshop Proceedings, vol. 665 (2010). CEUR-WS.org
7. Megdiche, I., Teste, O., Trojahn, C.: An extensible linear approach for holistic ontology matching. In: Groth, P., Simperl, E., Gray, A., Sabou, M., Krötzsch, M., Lecue, F., Flöck, F., Gil, Y. (eds.) ISWC 2016. LNCS, vol. 9981, pp. 393–410. Springer, Cham (2016). doi:10.1007/978-3-319-46523-4_24
8. Nentwig, M., Groß, A., Rahm, E.: Holistic entity clustering for linked data. In: Proceedings ICDM Workshops, pp. 194–201. IEEE (2016). doi:10.1109/ICDMW. 2016.0035
9. Nentwig, M., Groß, A., Möller, M., Rahm, E.: Distributed holistic clustering on linked data. CoRR abs/1708.09299 (2017)
10. Nentwig, M., Hartung, M., Ngomo, A.N., Rahm, E.: A survey of current link discovery frameworks. Semant Web **8**(3), 419–436 (2017). doi:10.3233/SW-150210
11. Nentwig, M., Soru, T., Ngonga Ngomo, A.-C., Rahm, E.: LinkLion: a link repository for the web of data. In: Presutti, V., Blomqvist, E., Troncy, R., Sack, H., Papadakis, I., Tordai, A. (eds.) ESWC 2014. LNCS, vol. 8798, pp. 439–443. Springer, Cham (2014). doi:10.1007/978-3-319-11955-7_63

12. Ngonga Ngomo, A.-C., Sherif, M.A., Lyko, K.: Unsupervised link discovery through knowledge base repair. In: Presutti, V., d'Amato, C., Gandon, F., d'Aquin, M., Staab, S., Tordai, A. (eds.) ESWC 2014. LNCS, vol. 8465, pp. 380–394. Springer, Cham (2014). doi:10.1007/978-3-319-07443-6_26

13. Rahm, E.: The case for holistic data integration. In: Pokorný, J., Ivanović, M., Thalheim, B., Šaloun, P. (eds.) ADBIS 2016. LNCS, vol. 9809, pp. 11–27. Springer, Cham (2016). doi:10.1007/978-3-319-44039-2_2

14. Saeedi, A., Peukert, E., Rahm, E.: Comparative evaluation of distributed clustering schemes for multi-source entity resolution. In: Kirikova, M., Nørvåg, K., Papadopoulos, G.A. (eds.) ADBIS 2017. LNCS, vol. 10509, pp. 278–293. Springer, Cham (2017). doi:10.1007/978-3-319-66917-5_19

15. Thalhammer, A., Thoma, S., Harth, A., Studer, R.: Entity-centric data fusion on the web. In: Proceedings of the 28th ACM Conference on Hypertext and Social Media. ACM (2017). doi:10.1145/3078714.3078717

Ontologies for Commitment-Based Smart Contracts

Joost de Kruijff[(⊠)] and Hans Weigand

Tilburg University, P.O. Box 90153, 5000 LE Tilburg, The Netherlands
{j.c.dekruijff,h.weigand}@uvt.nl

Abstract. Smart contracts gain rapid exposure since the inception of block-chain technology. Yet there is no unified ontology for smart contracts. Being categorized as coded contracts or substitutes of conventional legal contracts, there is a need to reduce the conceptual ambiguity of smart contracts. We applied enterprise ontology and model driven architectures to abstract smart contracts at the essential, infological and datalogical level to explain the system behind computation and platform independent smart contracts rather than its functional behavior. This conceptual paper introduces commitment-based smart contracts, in which a contract is viewed as a business exchange consisting of a set of reciprocal commitments. A smart contract ensures the automated execution of most of these commitments.

Keywords: Blockchain · Commitments · Commitment-Based smart contracts · Enterprise ontology · Model driven architecture

1 Introduction

Blockchain technology emerged as an alternative approach to payments, by using cryptographic methods to provide an alternative trust mechanism between two or more transacting parties. Blockchain is considered to be an unprecedented innovation in computer science as it enables a network of distributed nodes to reach consensus without having to resort to any form of central authority [1]. Blockchain also differs from traditional transaction systems with respect to how it irreversibly stores transaction data and executable contract code in a distributed ledger. In recent years, attention has increasingly shifted towards the core elements of blockchain itself and how its nature as a distributed ledger for transactions could be further leveraged, leading to the emergence of 'second-generation' blockchain platforms (e.g. Ethereum, Couterparty and Tendermint). These platforms extend blockchain's initial design to transfer value between actors by offering features for writing and executing so called *smart contracts* on the blockchain [2]. Smart contracts are coded contracts on the blockchain that automatically move digital assets according to arbitrary pre-specified rules [3], allowing parties that do not trust each other to transact safely within the context of a contract, without intermediation by third parties. In the event of contractual breaches or aborts, the blockchain automatically ensures that honest parties obtain commensurate compensation [4]. Objectives and principles for the design of smart contracts on the blockchain are derived from legal principles, economic theory and theories of reliable

© Springer International Publishing AG 2017
H. Panetto et al. (Eds.): OTM 2017 Conferences, Part II, LNCS 10574, pp. 383–398, 2017.
https://doi.org/10.1007/978-3-319-69459-7_26

and secure protocols [5]. The concept of smart contracts already emerged in the 1990s, but only gained exposure since the inception of blockchain technology. Today's smart contracts are coded using imperative programming languages, in mainstream programming languages (e.g. Javascript and Go for Tendermint) and non-mainstream procedural languages (e.g. Solidity for Ethereum). Representing contractual terms in code rather than natural language, could bring clarity and predictability to agreements, as a smart contract could then be tested against a set of material facts, allowing legal professionals on either side to know precisely how the contract would execute in every computationally-possible outcome [6].

Smart contracts receive plenty of exposure from the professional and research community. There has been an explosion of interdisciplinary research and experimentation, bundling legal, social, economic, cryptographic and even philosophical concerns into tokenized intellect. Yet there still seems to be a lack of a unified ontology for smart contracts [7], and the term "smart contract" still lacks a formal and settled definition. Smart contracts are occasionally defined as "autonomous machines", "automated contracts between parties stored on a blockchain", "any computation that takes place on a blockchain" or "algorithmic, self-executing and self-enforcing computer programs" [8]. Various debates about the nature of smart contracts are considered as contests between competing terminology. The different definitions can be categorized as follows: as a specific technology – code that is stored, verified and executed on a blockchain as *smart contract code*, or as a specific application of that technology as a complement or substitute for legal contracts called *smart legal contracts* [9]. If we accept the advantages for smart contracts, it is important to understand the ontology behind this technical and legal categorization. On the one hand, the ontology must be aligned with the conceptualization of business users, and on the other hand it should have a transparent operationalization. If the operationalization is not transparent, results become unpredictable.

Ontologies have been used to reduce conceptual ambiguities and inconsistencies in the blockchain domain [10]. Since most smart contract implementations focus on transactions with- or between enterprises, we apply enterprise ontology to leverage the collection of terms and natural language definitions relevant to these enterprise adopters. We use DEMO to model the ontologies at the essential (business), infological and datalogical perspective as DEMO has been proven to be a helpful methodology to formalize systems that are ambiguous, inconsistent or incomplete [11].

In line with Enterprise Ontology [11] and Business Ontologies like REA [12], we model smart contracts as a bundle of interrelated *commitments* among those parties who have signed it. According to [13], commitments come into being when an individual, links extraneous interests with a consistent line of activity by making a side bet. Side bets are often a consequence of the person's participation in social interactions. Commitments determine the robustness of a contract. For example, robustness is enhanced when a commitment to provide a service occurs with a commitment to resolve problems in cases where that service could not be delivered. A commitment-based approach to smart contracts is therefore considered as an interesting alternative for (deontic) logic-based smart contracts, that primarily build upon existing contract- and legal practice [14]. Our claim is that a commitment-based approach is not only aligned with business ontologies but also allows for a transparent implementation.

Due to the relative immaturity of the blockchain technology, the number of current real-world applications is still very limited. The evolution of digital platforms requires an approach with a combination of technological, economic and legal perspectives [8]. Our research objective is to make a theoretical contribution to scientific body of smart contract technology from a knowledge engineering perspective. Hereby we design smart contract ontologies abstracted at the essential, infological and datalogical layer conforming to the principles of enterprise ontology. These ontologies should help professionals involved in drafting and managing contracts to functionally design a platform independent commitment-based smart contract that eventually can be implemented on specific blockchain platform. We use a pragmatic approach in the sense that aim to describe the smart contract in its lifecycle and effects. Although we present an introduction to platform specific operationalizations of smart contracts, the actual automatic implementation is out of scope for this paper, as a conversion language for commitment-based smart contracts, required for defining infological to datalogical mapping rules, is set as a near future deliverable. For readability purposes, we use a running example to facilitate the explanation of the concept. The example consists of bed and breakfast platform AirBNB and a customer. The basic idea is that AirBNB (party) and the customer (counterparty) engage in a smart contract with each other, whereby AirBNB commits to provide a room upon payment and the customer commits to pay the room fee upon confirmation that a room is provided on payment.

This paper is structured as follows. Section 2 explores the concept of smart contracts, followed by the concept of commitment-based smart contracts in Sect. 3. The essential, infological and datalogical ontologies are presented and explained in Sect. 3, followed by a discussion and conclusion in Sects. 4 and 5.

2 Contract Robustness

A contract is a legally binding or valid agreement between two or more parties. The main objective of a contract is to fulfill a certain goal and to safeguard against undesirable outcomes, together referred to as contract robustness [6]. Contracts that are not robust may lead to transaction costs, expensive conflict resolution, or even a collapse of a transaction as a whole. Since the age of the internet, contracts in a business context have been rapidly digitalized. These electronic contracts are defined by law as contracts formed in the course of e-commerce by the interaction of two or more individuals using electronic means, such as e-mail, the interaction of an individual with an electronic agent, such as a computer program, or the interaction of at least two electronic agents that are programmed to recognize the existence of a contract. The Uniform Computer Information Transactions Act [15] provides rules regarding the formation, governance, and basic terms of an e-contract. Recently, smart contracts took over the spotlight, utilizing blockchain technology to exchange money, property, shares, or anything of value in a transparent, conflict-free way, while avoiding the services of a middleman [16]. Although smart contracts can theoretically exist without the blockchain, smart contracts are more powerful once stored on the blockchain due to its enhanced traceability and trust. A smart contract requires material- and formalizable inputs that can be measured and exchanged [15], like the delivery of a product or service. Transactions that are concerned with arbitrary inputs (e.g. emotions, opinions) or lack a social context (e.g. transactions that consist of only one actor) exist on their own merits and do not qualify as a smart contract from a business-level or social perspective. Once another actor gets involved (for example, as the requester of the transaction), this transaction could qualify as a smart contract.

Idelberger et al. [14] describes the contract lifecycle, which is considered to be the *defacto* standard for the mentioned contract types and has proven to be a powerful tool to decompose the process to draft and managing a contract. It consists of several phases, starting with the formation and negotiation (1) of the contract goals. To succeed this process, all participants have to coordinate and formalize both the content and process of the contract [15]. With regard to content, both the party and counterparty must have similar understanding of the contract by means of shared information or common ground, like mutual knowledge, mutual beliefs or mutual assumptions. [16] stated that such common ground cannot be properly updated without a process called grounding. Grounding requires purpose, what the parties try to accomplish, and a certain medium to fulfil this purpose, like a face-to-face conversation, a phone call, or as hypothesized in this paper; a smart contract. Once a contract is drafted, it is Notarized by a Notary or by using consensus technology and subsequently stored (2) in a physical vault or on a blockchain. Once stored, the contract is monitored and enforced (3) upon. Conventional contracts require more active monitoring due to the fact that actions related to the contract are not enforced automatically. For example, when tenants breach their commitment to pay for their room, they may still occupy the space as the action to remove them requires the decision and action by the landlord (e.g. to call the police), which is time consuming. Well-designed smart contracts have actions and consequences (like imposing a penalty) predefined as part of the contract formation, which automatically enforce and execute on either compliance or breach (e.g. using blockchain transactions connected to Internet of Things devices). A smart contract may not automate everything, as more analysis is needed to determine possible repercussions. Conventional contracts are potentially more vulnerable than smart contracts as they are non-enforceable and their free text form complicates analyzing their content in any automated way [6]. Since contract monitoring and enforcement are not directly aligned, traditional contracts are said to serve a mere ceremonial purpose [17]. The contract lifecycle is concluded by dispute resolution (4). [6] identified two types of dispute resolution for both traditional- and smart contracts: (A) adjudicative resolution, such as litigation or arbitration, where a judge, jury or arbitrator determines the outcome, and (B) consensual resolution, such as collaborative law, mediation, conciliation, or negotiation, where the parties attempt to reach agreement. There is insufficient jurisprudence to confirm whether or not the adjudicative dispute resolution clauses that are coded into a smart contract are legally valid and do not require being verified and decided upon in court. In these occasions, smart contracts will lead to consensual dispute resolution to re-evaluate common practice, as lawyers and clients discover what types of agreements and terms are best suited to code and which should be left to natural language and how to combine each to achieve the best of both worlds [18]. The uncertainty of the legal validity of smart contracts may impact the robustness of contracts [6, 14]. Smart contracts are therefore considered to be efficient and disruptive, but maybe not as a full replacement of conventional contracts (Table 1).

Table 1. Contract lifecycle automation between traditional- and smart contracts

Contract lifecycle	Traditional contract	Smart contract
Formation and negotiation	Via traditional media	Via traditional media or a smart contract
Notarization and storage	Manually notarized and stored in a vault	Notarized and stored on the blockchain
Monitoring and enforcement	Manual	Automated by the smart contract
Adjudicative dispute resolution	Manual	Automated by the smart contract
Consensual dispute resolution	Manual	Manual

As part of the enterprise ontology, [19] introduced the transaction layer model that can be related to the contract lifecycle. In this model, the success layer models the "happy path" of the transaction, consisting of initialization (where the parties agree on the commitment), execution, and evaluation (where the parties agree on the fulfilment of the commitments). This layer abstracts from all exceptions. When something goes wrong, e.g. a party wants to negotiate or cancel the transaction, the process switches to the contingency handling layer. When the normative framework in which the transaction takes place is challenged, the process switches to the discourse layer.

Table 2 shows that the contract lifecycle and the transaction model of Enterprise Ontology are closely aligned. Only the notarization and storage has no direct counterpart as Enterprise Ontology puts storage on a different ontological level.

Table 2. Contract lifecycle comparison

Contract lifecycle	Van Reijswoud & Dietz	
Formation and negotiation	Initialization	(success layer)
Notarization and storage	(datalogical)	
Monitoring and enforcement	Fulfilment	
Adjudicative dispute resolution	Contingency handling (dispute layer)	
Consensual dispute resolution	Norm formation (discourse layer)	

In theory, the entire contract lifecycle could be automated by the blockchain. Nevertheless, there is a lack of factual evidence on the effectuation of such a scenario. Therefore, this paper will focus on the two steps of the smart contract lifecycle that are actually automated on the blockchain: Notarization and Storage, Monitoring and Enforcement and Adjudicative Dispute Resolution (from now on Contingency Handling).

3 Designing a Smart Contract Ontology

We propose to adopt the distinction axiom of Enterprise Ontology as ontological basis for the separation of conceptual model and implementation of smart contracts.

The distinction axiom of enterprise ontology distinguishes three basic human abilities: *performa, informa*, and *forma* [11]. The performa ability concerns the bringing about of new, original things, directly or indirectly by communication. Communicative acts at the performa level are about evoking or evaluating commitment; these communicative acts are realized at the informa level by means of messages with some propositional content. The informa ability concerns the content aspects of communication and information. Production acts at the forma level are infological in nature, meaning that they reproduce, deduce, reason, compute, etc. information, abstracting from the form aspect. The forma ability concerns the form aspects of communication and information. Production acts at the forma level are datalogical in nature: they store, transmit, copy, destroy, etc. data.

The distinction axiom is highly relevant for the smart contracts. Following the three abilities, we distinguish three ontological layers (Table 3). We start with the description of the economic meaning of the essential smart contract transactions using the essential layer. This is the preferred level of initial specification for a smart contract as it abstracts from the implementation choices. We use the infological layer to describe the formal logic of smart contracts as effectuating an (immutable) open ledger system. This layer aims to abstract from the various implementations that exist today or in the future. Finally, we describe the datalogical layer that describes smart contract transactions at the technical level in terms of blocks and code.

Table 3. Layers of DEMO versus MDA

Enterprise ontology	Model driven architecture
Essential layer	Computation independent model
Infological layer	Platform independent models
Datalogical layer	Platform specific models

We combine enterprise ontology with Model Driven Architecture (MDA), as MDA is a framework based on UML and other industry standards for visualizing, storing, and exchanging software designs and models. Whereas DEMO emphasizes modeling capabilities, MDA is capable of (automatically) of mapping business models to models that can (automatically) be converted into formal models for implementation. MDA defines a system specification approach that separates the specification of business models (Computation Independent Model or CIMs), system functionality (Platform Independent Models or PIMs) from the specification of the implementation of that functionality on a specific technology platform (Platform Specific Models or PSMs). This approach offers multiple advantages: (1) CIMs represent an informal specification of the business functionality provided by a certain system [20], (2) PIMs provide formal specifications of the structure and functions of the system that abstract away technical details, (2) implementation on several platforms can be done from one PIM,

(3) system integration and interoperability can be anticipated and planned in the PIM but postpone to the PSMs, and (4) it is easier to validate the correctness of models [21].

In MDA, one of the key features is the notion of mapping. A mapping is a set of rules and techniques used to modify one model in order to get another model. Mapping is established through transformation rules [20]. Since all ontologies in our model are modeled using UML class diagrams, mapping is established between classes or groups of classes. An initial CIM to PIM mapping between the essential and infological layer is provided, as well as PIM to PSM mapping between the infological and datalogical layer. By combining DEMO with MDA, we aim to leverage on both the modeling strengths of DEMO and operationalization capabilities of MDA.

3.1 Essential Layer

The essential or business level is concerned with what is created directly or indirectly by communication. In the Language/Action Perspective [11], the key notion in communication is commitment as a social relationship based on shared understanding of what is right and what is true. Communicative acts typically establish or evaluate commitments. In a narrower sense, a commitment (promise, commissive) is about what an actor is bound to do (so what is right in a future situation). Such a commitment being agreed upon by two parties, formalized by a smart contract, is a one change in the social reality, as is the agreed upon fulfillment of that commitment.

An infological smart contract transaction, moving some value from one account to another represents a change in the social reality, e.g. transfer of ownership. Such a change is what we identify as the essential smart contract transaction. It might be objected that commitments represent the positive actions to be performed by the actors only. What about prohibitions? For instance, a music customer is not allowed to copy the music he can download. In some cases, like the music copying, there may be technical means to make the prohibited action impossible, but there are also actions of agents that are not under control. In these cases, what the parties should commit to, is to take the consequences (e.g. paying a penalty). Hereby, the prohibition – don't do A – is reformulated as a contingency commitment – IF <A> THEN <consequence>, where a transformation is described between deontic logic and dynamic logic [22]. In the example, the customer commits himself to be a penalty when he has made an illegal copy. Our claim is that all contract clauses can be similarly represented as commitments (validation of this claim is out of the scope of this paper).

Enterprise Ontology is not specific about the content of the change. In terms of MDA, the essential layer is similar to CIM as it does not present the details or structure of a system. In order to shape the business model for smart contracts, we combine Enterprise Ontology with the Business Ontology of REA. The REA model developed by Bill McCarthy [12] can be viewed as a domain ontology for accounting. REA intends to be the basis for integrated accounting information systems focused on representing increases and decreases of value within an organization or beyond. REA atomic constituents of processes are called economic events. Economic events are carried out by agents and affect a certain resource, like a (crypto) currency or physical good. The duality axiom says that provides and receives are always in balance. For instance, in an exchange process, some resources are transferred to or from an external agent in return for a monetary payment. In our smart contract ontology, we will use the term "transaction" to represent a transfer of some resource, and a process consists of two or more transactions in duality. In a commitment-based smart contract context, these transaction events are informally described as increment- and decrement commitments, whereby a commitment is labeled a future transaction instead of a transaction itself. A transaction is only triggered upon internal (e.g. by another commitment within the smart contract) or external events (Internet of Things trigger) in the future. These events may or may not happen. The actual event is represented by a transaction that is the result of a commitment being called into action. For example, a smart contract may contain reciprocal commitment by AirBNB and customer regarding check-in times. Non-compliance to these commitments result in a 10% penalty for the customer. This future event may never be called in cases where the customer complies to the agreed check-in times.

In the process to design an essential ontology for commitment based smart contracts, commitments are informally described in natural language and should also not have any technology-specific implementation information, in order to capture maximum business value. This model does not have to be exhaustive, as the mapping of concept to machine readable code occurs at the infological level. Table 4 shows an overview of the classes, depicted in the essential ontology in and Fig. 1.

Table 4. Essential ontology classes

Class	Explanation
Agent	An agent is individual or organization that controls resources and can initiate a transaction or commitment
Transaction	A (business) transaction is a change in the economic reality consisting of increment and decrement stock-flow events
Smart contract	A contract is an agreement between agents consisting of mutual commitments. A smart contract is a contract in which the commitment fulfillment is completely or partially performed automatically
Commitments	Commitments contain future events that are fulfilled with the execution of transactions
Increment commitment	An Increment Commitment is an event that increments the resources of an actor as part of the duality axioms
Decrement commitment	An Decrement Commitment is an event that decrements the resources of an actor as part of the duality axioms
Exchange process	An economic exchange is a process that changes the economic reality by means of exchange and consists of two or more transactions
Conversion process	An economic conversion is a process that changes the economic reality by means of conversion and consists of two or more transactions.
Resource	A resource is an asset with a certain economic value controlled by an agent

Commitment-based smart contracts are owned by multiple actors and are created through a transaction by an actor. Actors provide and receive transactions in order to interact with the smart contract through commitments that either increment or

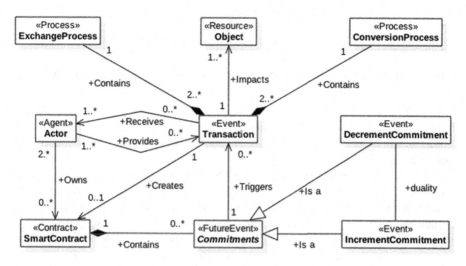

Fig. 1. Essential ontology

decrement a certain promise. For example, an AirBNB customer may transact with the smart contract by paying for a certain room. Since the increment and decrement commitments are in duality, the customer's payment triggers an event by the smart contract itself to reciprocate that commitment. Hereby a transaction is created to reserve a room for the paying customer.

3.2 Infological Layer

In the 1970s, Langefors was the first to make the important distinction between information (as knowledge) and data (as representation) [23]. When blockchain is described in the current literature as a "distributed ledger" [3, 26], this is an infological characterization that abstracts from transactions at the essential level. A transaction, in this ledger system, is not just a fulfillment of a reciprocal commitment, but a transfer of some value object (e.g. Bitcoin) that impact the state of the smart contract on the blockchain. A ledger consists of accounts (e.g. debit account), and this concept is indeed generic across the majority of blockchain providers that are part of this analysis. Accounts are not limited to have a (crypto) currency- balance or quantity, but may also refer to other types like stocks or a claim as mainchains other than Bitcoin (not taking sidechains into account) allow to register custom account types. Commitments at the infological layer are programmable functions that reflect promises made between the contract participants. They can also be represented by accounts. This implies that an essential transaction typically corresponds to multiple infological transactions, just like it corresponds in current general ledgers to multiple bookings (and not only the bank account or inventory account) (Table 5).

A commitment-based approach also represents a proposed business exchange: the party and counterparty include the considerations of the creditor and debtor. The robustness of their smart contract depends on how its commitments relate to the goals

Table 5. Mapping between the essential and infological layer

Essential transaction	Infological transaction
An essential transaction represents a change in the social reality, e.g. transfer of ownership	An infological transaction represents an infological transfer of some value object between accounts

of the contract parties [6]. This definition implies that a commitment-based smart contract should at least contain (1) goals, (2) commitments, (3) conditions and (4) actions. We have added (5) timing in order to force actions to actually execute. The purpose of a contract is to define the parties' responsibilities with respect to a desired scenario outcome to the level of detail necessary to make all parties comfortable with respect to the social relationship. Hereby, contract (1) goals, also known as 'articles' in conventional contracts [29], safeguard the value proposition that both parties rely on for business success. According to [28], a value proposition consists of first- order value objects (what is offered), but also second-order value objects (how it is offered). Second-order value objects are mostly intangible in nature and define the quality of the value transfer, like convenience or speed. This way it is possible to distinguish between the failure to deliver and the failure to meet a second-order value, e.g. when an e-commerce product is delivered but two days too late. In this case, one commitment is fulfilled, the other not. Once a commitment of any order is not met, contingency handling procedures may trigger other commitments (e.g. payment of a penalty) in order to reciprocate the other party. As a result, the goal is essentially the total outcome of a bundle of reciprocal commitments. (2) Commitments are commonly defined in the formation and negotiation phase of the contract lifecycle. As each contract goal consists of reciprocal commitments, for every commitment by one party, there should be a balancing commitment by the counterparty that is part of the smart contract transaction. It should be noted that not only the fulfillment of the commitments is an economic event (transaction) that transfers some value, but also providing a commitment that transfers some value (although it is potential). The negotiation/initialization phase is regarded to be a reciprocal exchange of commitments and each transfer as a blockchain transaction. (3) A condition determines whether or not an action related to a commitment is executed. Conditions verify the context or scope of the contract (for example: approved data sources for input, like the blockchain, internet, websites or only certain data providers.) and to determine if an actor lived by a commitment as part a contract. Conditions should only process automatically verifiable input. (4) Actions are events that are executed when a condition is true [29]. An action can be both internal (e.g. a blockchain transaction), or external (e.g. an IoT device trigger) and may be automated (transfer value) or manual (going to court). Actions depend on (5) timing, in order to allow for complex events that depend on timing constraints (e.g. only perform an action after 8 PM).

The infological ontology (Table 6 and Fig. 2) extends the infological blockchain ontology by [10] by including the elements that constitute to commitment-based smart contracts.

Table 6. Infological ontology classes

Class	Explanation
Ledger	A ledger maintains a continuously growing list of transaction records called blocks. Each block contains a timestamp and a link to a previous block
Account	An account sends and receives value to and from a transaction
Object	An object type is a custom stock or a claim (type) traded by an account via a transaction
Ledger	A ledger maintains a continuously growing list of transaction records called blocks. Each block contains a timestamp and a link to a previous block
State	A state refers to a variable that belongs to a smart contract and is stored on the blockchain
Transaction	A transaction represents the atomic inputs and outputs between accounts
Journal	A journal is list of transactions
Rules of engagement	Rules of engagement refer to the explicit rules to which a transaction should behave as specified in a smart contract
Default	A default is a definitional clause, which defines relevant concepts occurring in the contract
Clause	Clauses regulate the actions of the parties for contract performance
Goal	A goal safeguards the value proposition that both parties rely on for business success
Commitment	Commitments are explained as acts of binding oneself to a course of action
Condition	A condition determines whether or not an action related to a commitment is executed
Action	An action is an event that is executed when a condition is true
Timing	Timing enables actions to execute based on certain time constraints

In the context of our running example, infological transactions may either be a simple value transfers or may include smart contract code that contains the defaults and clause functions between AirBNB and the customer, explained as rules of engagement.

From an operational perspective, the infological ontology in Fig. 2 hints at a smart contract structure that could be applied through any opinionated blockchain platform (e.g. Ethereum). Like in conventional practices, contracts consist of finite sets of clauses. There are two types of these clauses: (1) definitional clauses, which define relevant concepts occurring in the contract, and (2) normative clauses, which regulate the actions of the parties for contract performance [29]. Commitment-based smart contracts represent definitive clauses and values as defaults, which are set by the constructor upon initialization, after the smart contract has been stored on the blockchain. Since commitment-based smart contracts require as many owners as participants (likely represented by a single public key), a required default is the public key of the owner of the smart contract, and may set other default values that outweighs values inside normative clauses. Normative clauses consist of commitments, that contain actions that may trigger other transactions based on conditions and time constraints.

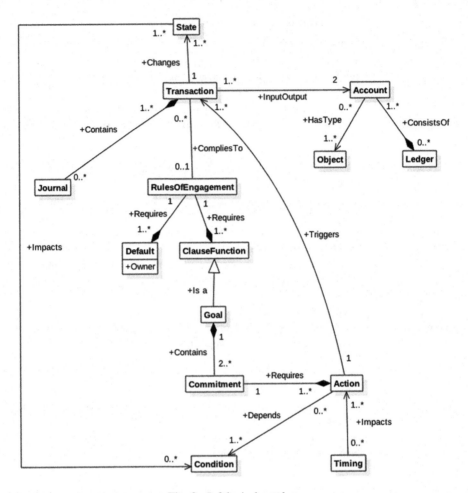

Fig. 2. Infological ontology

Finally, a delete function should provide insight in the legal provisions to terminate the commitment-based smart contract.

3.3 Datalogical Layer

Smart contracts are frequently explained in the context of technology as smart contract code that is executed on the blockchain. This technological context is positioned at the datalogical level, the level where smart contracts are created, mined or validated and stored in a blockchain. The datalogical level is the level where conceptual requirements from the infological level are translated into engineering means through a PIM to PSM transformation.

The construction of a datalogical ontology for smart contract uses the blockchain domain ontology as presented by [10], which provides a logical overview of the technique behind blockchain transactions, including smart contract transactions.

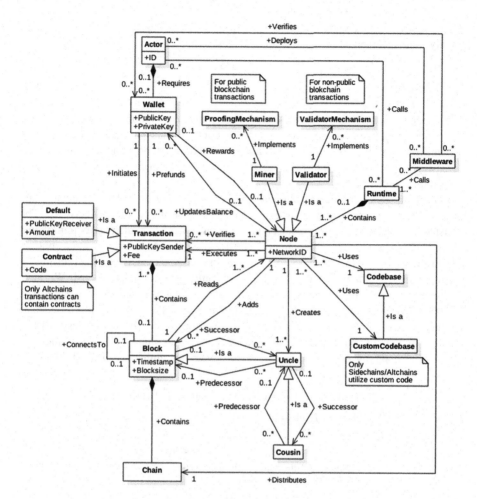

Fig. 3. Datalogical domain ontology (based on [10])

This datalogical system presented in Fig. 3, from initiation to transaction storage on the blockchain, is the datalogical characterization of an infological transaction.

Central to the datalogical domain ontology (Fig. 3) are the wallet, the transaction and the node. Each blockchain concept relies on these concepts, whereby a transaction can be either simple or contain smart contract code. We focus on the latter in this paper. Whenever AirBNB and a customer engage in a smart contract to transact, a datalogical transaction is initiated by a multi-owner wallet (owned by AirBNB and the customer) and presented to the network of nodes. This transaction consists of logic, which is the smart contract code that constitutes the infological contract goals, commitments, conditions, action and timing functions. Once mined and verified, the smart contract is stored on the blockchain. Interactions with this smart contract, through new transactions or queries, from inside or outside the blockchain ecosystem, occur through the nodes via runtimes or middleware, by means of API's and sockets, or via sidechains directly, which is our next research focus.

4 Discussion

This paper introduced commitment-based smart contracts and applied the concept of abstraction through enterprise ontology. The outcomes of this approach should increase the understandability of smart contracts by reducing conceptual ambiguity and increasing insight in how to design a smart contract from a business model all the way to an implementation.

The main idea behind commitment-based smart contracts is that it enhances the robustness of a proposed business exchange between a party and counterparty. The robustness of their smart contract depends on the extent to which its commitments relate to the goals of the smart contract. Robustness is important to safeguard a contract against the negative consequences (e.g. transaction costs, expensive conflict resolution and a collapse of a transaction as a whole) of contracts that are not robust. We have used a running example in this paper to elaborate the practical value of this approach.

In our search for an initial process to design and deploy a commitment-based smart contract, we combined DEMO's modeling capabilities with MDA in order to better translate functional requirements into commitment-based smart contracts implementations. The mapping between the ontological layers is summarized next (Table 7).

Table 7. Mapping between the ontological layers

Layer	UML class	Explanation
Essential	Transaction	An essential transaction represents a change in the social reality, e.g. transfer of ownership
Infological	Transaction	An infological transaction represents an infological transfer of some value object between accounts
Datalogical	Transaction	A datalogical transaction represents the entire workflow from initialization until fulfillment of a smart contract transaction on the blockchain

The main concept of this paper is that smart contracts are best informally described at the essential level. Hereby, all contract goals should contain sets of reciprocal commitments in order to capture maximum business value and robustness. This model is then mapped (CIM to PIM) to an infological smart contract design (UML class mapping) that has the capability to be machine readable and platform independent. The infological design subsequently formalizes the essential commitments into a smart contract structure and code in the form of default and normative clause functions. The infological model can be mapped to a datalogical implementation (PIM to PSM). Although out of scope of this paper, the infological model should be eventually be convertible into executable code using formal transformation rules, given some implementation platform. This would also enable end-to-end validation of the quality and value of the presented ontologies.

5 Conclusion

This paper introduces commitment-based smart contracts at the essential, infological and datalogical layer. The ontologies and mapping between the layers should provide a better understanding into designing smart contracts. It also can be used to support application development, as it suggests to specify a blockchain application on the business level first. At best, the ontologies help to generate a smart contract implementation automatically for any blockchain platform, with some design parameters to be set.

This paper aimed to provide a basic yet fundamental ontology for smart contracts. Besides plotting the running example to our model, we have not been able to do an extensive formal validation of the ontology yet, so the proposed model should be seen as an initial one. Validation is pivotal to test and improve the capability of the ontology to design smart contracts for any platform specific implementation without dogmatism.

The ontologies presented in this paper do not stop the need for further research on smart contracts, which are still relatively immature. In contrast, the current model may serve as a foundation for specific and formal smart contract syntax and semantics that translate the proposed smart contract structure into executable code.

References

1. Can blockchain technology send notaries on vacation for good? When trust in men is replaced by trust in math, Medium, May 2015
2. Delmolino, K., Arnett, M., Kosba, A., Miller, A., Shi, E.: Step by step towards creating a safe smart contract: lessons and insights from a cryptocurrency lab. In: Clark, J., Meiklejohn, S., Ryan, P.Y.A., Wallach, D., Brenner, M., Rohloff, K. (eds.) FC 2016. LNCS, vol. 9604, pp. 79–94. Springer, Heidelberg (2016). doi:10.1007/978-3-662-53357-4_6
3. Buterin, V.: A Next Generation Smart Contract & Decentralized Application Platform
4. Kosba, A., Miller, A., Shi, E., Wen, Z., Papamanthou, C.: Hawk: the blockchain model of cryptography and privacy-preserving smart contracts. In: 2016 IEEE Symposium on Security and Privacy, 22–26 May 2016
5. Szabo, N.: Formalizing and securing relationships on public networks. First Monday 2(9) (1997)
6. Chopra, A.K., Oren, N., Modgil, S., Desai, N., Miles, S., Luck, M., Singh, M.P.: Analyzing contract robustness through a model of commitments. In: Weyns, D., Gleizes, M.P. (eds.) AOSE 2010. LNCS, vol. 6788, pp. 17–36. Springer, Heidelberg (2011). doi:10.1007/978-3-642-22636-6_2
7. Hoskinson, C.: Thoughts on an ontology of smart contracts. IOHK, March 2017
8. Lauslahti, K., Mattila, J., Seppala, T.,: Smart Contracts – How will Blockchain Technology Affect Contractual Practices?, Work and Wealth in the Era of Digital Platforms, January 2017
9. Stark, J.: Making Sense of Blockchain Smart Contracts, June 2016
10. de Kruijff, J., Weigand, H.: Understanding the blockchain using enterprise ontology. In: Dubois, E., Pohl, K. (eds.) CAiSE 2017. LNCS, vol. 10253, pp. 29–43. Springer, Cham (2017). doi:10.1007/978-3-319-59536-8_3
11. Dietz, J.: Enterprise Ontology. Springer, New York (2006)

12. McCarthy W.: The REA accounting model: a generalized framework for accounting systems in a shared data environment. Acc. Rev. **LVII**(3) (1982)
13. Becker, H.: Notes on the concept of commitment. Am. J. Sociol. **66**(1), 32–40 (1960)
14. Idelberger, F., Governatori, G., Riveret, R., Sartor, G.: Evaluation of logic-based smart contracts for blockchain systems. In: International Symposium on Rules and Rule Markup Languages for the Semantic Web, RuleML 2016: Rule Technologies, Research, Tools, and Applications, pp 167–183 (2016)
15. Clark, H., Brennan, S.: Grounding in communication. In: Perspectives on Socially Shared Cognition, pp. 127–149 (1991)
16. Clark, H., Schaefer, E.: Collaborating on contributions to conversations. Lang. Cogn. Process. **2**(1), 19–41 (1987)
17. Norta, A.: Designing a smart-contract application layer for transacting decentralized autonomous organizations. In: Singh, M., Gupta, P.K., Tyagi, V., Sharma, A., Ören, T., Grosky, W. (eds.) ICACDS 2016. CCIS, vol. 721, pp. 595–604. Springer, Singapore (2017). doi:10.1007/978-981-10-5427-3_61
18. Stark, J.: How Close Are Smart Contracts to Impacting Real-World Law? Coindesk Opinion, 11 April 2016
19. Reijswoud, V.: The structure of business communication. Theory, model and application, Ph.D. dissertation, Technical university, Delft (1996)
20. Rhazali, Y., Hadi, Y., Mouloudi, A.: CIM to PIM transformation in MDA: from service-oriented business models to web-based design models. Int. J. Softw. Eng. Appl. **10**(4), 125–142 (2016)
21. Caplat, G., Sourrouille, J.L.: Model Mapping in MDA, Workshop in Software Model Engineering (WISME) (2002)
22. Wieringa, R., Meyer, J.-J., Weigand, H.: Specifying dynamic and deontic integrity constraints. Data Knowl. Eng. **4**(2), 157–189 (1989)
23. Goldkuhl, G.: Information as action and communication. In: The Infological Equation, Essays in honour of Langefors, B., Dahlbom, B. (eds.) Gothenburg Studies in Information Systems, Gothenburg University (1995). (also: Linkoping Univ. report LiTH-IDA-R-95-09)
24. McDonald, B.D.: The uniform computer information transactions act. Berkeley Technol. Law J. **16**(1), 461–484 (2001). Article 24
25. Waqar, H.: Fintricity: London Blockchain Meetup, December 2016
26. Pilkington, M.: Blockchain Technology: Principles and Applications, Research Handbook on Digital Transformations, September 2015
27. Baldwin, R., Cave, M.: Understanding Regulation: Theory, Strategy and Practice. Oxford University Press, Oxford (1999)
28. Weigand, H., Johannesson, P., Andersson, B., Bergholtz, M., Edirisuriya, A.: Strategic analysis using value modeling - the c3-value approach. In: Proceedings of 47th Hawaii International Conference on System Sciences. IEEE (2014)
29. Governatori, G., Rotolo, A.: Modelling contracts using RuleML. In: The Seventeenth Annual Conference on Legal Knowledge And Information Systems, pp. 141–150 (2004)

A Framework for User-Driven Mapping Discovery in Rich Spaces of Heterogeneous Data

Federica Mandreoli[(✉)]

University of Modena and Reggio Emilia, Modena, Italy
federica.mandreoli@unimore.it

Abstract. Data analysis in rich spaces of heterogeneous data sources is an increasingly common activity. Examples include exploratory data analysis and personal information management. Mapping specification is one of the key issues in this data management setting that answer to the need of a unified search over the full spectrum of relevant knowledge. Indeed, while users in data analytics are engaged in an open-ended interaction between data discovery and data orchestration, most of the solutions for mapping specification available so far are intended for expert users.

This paper proposes a general framework for a novel paradigm for user-driven mapping discovery where mapping specification is interactively driven by the information seeking activities of users and the exclusive role of mappings is to contribute to users satisfaction. The underlying key idea is that data semantics is in the eye of the consumers. Thus, we start from user queries which we try to satisfy in the dataspace. In this process of satisfaction, we often need to discover new mappings, to expose the user to the data thereby discovered for their feedback, and possibly continued towards user satisfaction.

The framework is made up of (a) a theoretical foundation where we formally introduce the notion of candidate mapping sets for a user query, and (b) an interactive and incremental algorithm that, given a user query, finds a candidate mapping set that satisfies the user. The algorithm incrementally builds the candidate mapping set by searching in the dataspace data samples and deriving mapping lattices that are explored to deliver mappings for user feedback. With the aim of fitting the user information need in a limited number of interactions, the algorithm provides for a multi-criteria selection strategy for candidate mapping sets. Finally, a proof of the correctness of the algorithm is provided in the paper.

Keywords: Mapping discovery · Pay-as-you-go information integration · Dataspace

This work is partially founded by the University of Modena and Reggio Emilia under the project "MapQS: mapping discovery and refinement driven by query samples in dataspace".

H. Panetto et al. (Eds.): OTM 2017 Conferences, Part II, LNCS 10574, pp. 399–417, 2017.
https://doi.org/10.1007/978-3-319-69459-7_27

1   Introduction

In the last few years, we are witnessing to the proliferation of data sources because of the ease of publishing data on the Web, the availability of data sharing services, and the diffusion of open data access policies. As a consequence, in many contemporary data analysis scenarios, users are faced with large collections of independent heterogeneous data sources with which they initially have limited understanding of the structure and semantics. Some example use cases of this scenario are: querying the web of linked data [8], Personal Information Management (PIM) systems [1], and exploratory data analysis [5].

The technical solutions for data sharing in this data management setting are by now quite mature, what remains fundamentally in querying over such rich spaces of data sources are issues of orchestrating the sources for querying, that is, aligning and exchanging data between sources, towards resolving an information seeking task. One of the key issues in this context is mapping specification. *Mappings* are declarative specifications that are used in information integration to spell out the relationship between a target data instance and possibly more than one source data instance [6].

Up to now, mapping specification has been almost exclusively considered from a data-centric perspective, often based on the availability of schemas, and primarily geared towards expert users [2,15]. Recently, some pay-as-you-go approaches for mapping specification have been proposed [3,9] that aims to reduce the up-front cost required to set up mappings among loosely connected data sources by involving end users in the mapping refinement and validation tasks.

In contrast to this view, analytics in the dataspace scenarios discussed above is often iterative and dynamic, where new query results inform and guide further analysis, in an open-ended interaction between discovery and data orchestration. At the center of this process lies ordinary users with their information needs. Therefore, as also evidenced in [4], mapping specification for such classes of users is even more compelling. At the same time, it is an arduous task, because of the complexity of the semantics mappings can express.

In this light, in this paper we propose a novel paradigm for mapping discovery where mapping specification is interactively driven by the information seeking activities of end users and the exclusive role of mappings is to contribute to users satisfaction. The underlying key idea is that data semantics is in the eye of consumers. Hence, we start from their primary visible expressions, the queries, which we try to satisfy in the dataspace. In this process of satisfaction, we often need to discover new mappings, to expose the user to the data thereby discovered for their feedback, and, possibly, continued reformulation of their queries.

For instance, when a data enthusiasts, such as a journalist, interact with an exploratory data analysis tool to achieve new and essential knowledge of the domain of interest, the satisfaction process can be initially supported by the discovery of mappings over jargon-free data objects that would bring easily understandable answers. Then, it will move toward more specialized data sources that could help the user to gain a deeper understanding of the domain of interest.

In this way, each user will have his/her own view of the dataspace that will evolve over time according to his/her level of knowledge and topics of interest. Similarly, in a PIM context the discovered mappings can reflect the personal interpretation of the issued queries. In our running example presented in Sect. 2, for instance, the user asks a dataspace made up of a social network and two knowledge graphs for the countries related to the movies his/her friends like. The delivered answers depend on what the user means by country, e.g. country of the movie producer or country of the filming locations, that yields to different mapping specifications.

The literature presents some approaches for data-driven mapping discovery intended for expert users (e.g. [2,14]) while [4] is the only mapping specification approach that bootstraps with exemplar tuples provided by non-expert users. As to our knowledge, this is the first paper that addresses the problem of mapping discovery for non-expert users and proposes a solution that starts from their information needs. This paper lays the basis for this novel mapping specification paradigm by putting forth a general framework made up of: (1) a theoretical foundation for the dataspace model, the query answering model, and the notion of candidate mapping set, i.e. a set of mappings that can be used to answer a given query; (2) an interactive and incremental algorithm for user-driven mapping discovery that delivers a candidate mapping set that fits the user information need expressed by the issued query. For ease of presentation, we assume that the user selects a data source to issue a query. The algorithm starts from the partial answers that can be computed over the data source and incrementally build such candidate mapping set by searching in the dataspace data samples and deriving mapping lattices that are explored to deliver mappings for user feedback. At each step, the user is challenged to retain the mappings that fits his/her information need. The algorithm provides for a multi-criteria selection strategy for candidate mapping sets that aims at a rapid convergence towards user satisfaction by taking into account key features of possible candidate mapping sets, user requirements and context, and user feedback. The paper also includes a proof of algorithm correctness.

Thanks to this result, we perform an important step towards realizing the vision of a fully fledged pay-as-you-go information integration approach for analytics on dataspaces. The impact of this work to the state of the art is twofold. From an effectiveness point of view, mapping specifications meet the information needs of the actual mapping consumers, i.e. end users, that can evolve over time and can be influenced by the context. From an efficiency point of view, Pareto's principle tells us that (1) there are many widely shared commonalities in information needs, thereby increasing the chances of mapping reuse, and (2) information needs are almost always highly focused and topical, not requiring mapping discovery at the level of traditional data integration systems.

The rest of the paper is organized as follows: Sect. 2 formalizes the framework; Sect. 3 introduces the notion of candidate mapping set; Sect. 4 presents the overall algorithm whose details are provided in Sects. 5 and 6; in Sect. 7 we prove algorithm correctness; Sect. 8 discusses the state of the art and presents concluding remarks.

2   The Dataspace and Query Answering Model

In this Section, we formally introduce the models underlying the user-driven mapping discovery approach.

2.1   The Dataspace Model

The data modeling abstraction we adopt to represent a dataspace is as light-weight as possible, for the goals of our study. To this end, it is (a) *fully decentralized*; (b) *schema-flexible*, in the sense that data sources can be schemaless and users do not need to know the complete and exact structure of the data to query it; (c) *based on data exchange*, in that when answering user queries the data source includes data objects that reflects a given set of mappings between the data source and the dataspace [7].

Let C, V, and N be infinite pairwise disjoint sets of *constants* (i.e., atomic data objects), *variables*, and marked *null values*, resp.

Definition 1. *A* data source *is a finite set of* statements *modeled as triples* (s, p, o), *for* $s, p, o \in C \cup N$.

Given a set of statements G, in the following $\mathsf{constants}(G)$ and $\mathsf{nulls}(G)$ denote the sets of all constant and null values, respectively, appearing in the triples of G, and $\mathsf{adom}(G) = \mathsf{constants}(G) \cup \mathsf{nulls}(G)$.

A dataspace is a collection of autonomous data sources locally connected through semantic mappings used to answer past user requests. For ease of presentation, we assume that data sources are aligned in their objects.

Definition 2. *A* dataspace *is a finite set of data sources, each labeled with a distinct identifier in* I: $D = \{i_1 : G_1, \ldots, i_n : G_n\}$. *If the statement* (s, p, o) *belongs to* G_i *then it is univocally identified as* $i : (s, p, o)$.

We introduce a query language where queries have a conjunctive pattern matching form, closed to our data model.

Definition 3. *A* statement pattern *is a triple* (x, y, z), *where* $x, y, z \in C \cup V$. *Given a set* S *of statement patterns*, $\mathsf{vars}(S)$ *denotes the set of all variables occurring in* S.

The query language is used to define both queries on dataspaces and mappings between data sources in a dataspace.

Definition 4. *A* query *is a pair* (H, B) *where* H *and* B *are finite sets of statement patterns such that* $\mathsf{vars}(H) \subseteq \mathsf{vars}(B)$.

Let D *be a dataspace and* $G_j \in D$ *a data source. A* mapping φ *from* D *to* G_j *is a pair* (H, B) *where* H *and* B *are finite sets of statement patterns, denoted as* destination *and* source, *respectively, and the statement patterns in* B *are possibly prefixed by a data source identifier in* D. *In the following,* $H \leftarrow B$ *will be used to denote* (H, B).

Example 1. Figure 1a depicts our very simple running example in a PIM scenario consisting of a dataspace $D = \{G_1, G_2, G_3\}$ and a query Q. G_1 is a portion of a social graphs containing information about what user's friends like, G_2 contains information about movies (e.g. IMDB), and G_3 is a knowledge graph like DBPedia. The user query Q is issued over G_1 and asks for the countries related to the movies friends like. G_1 has not enough information to resolve Q. This is the triggering condition for the proposed approach for user-driven mapping specification that aims at discovering mappings that resolve Q and the delivered answers satisfy the user.

It is worth noting that mappings are a special type of the so-called source-to-target tuple-generating dependencies (s-t tgds) [7]. As such, they have the potential to introduce labeled-null values when there are one or more variables in the head of the mapping which do not appear in its body. The *semantics* of φ on D, denoted as $\varphi(D)$, is D extended with the statements of the unique (up to isomorphism) core universal solution for D w.r.t. φ[7]. Informally speaking, the core is the best solution to be materialized for the data exchange problem. This notion can be easily extended to a set of mappings M and will be denoted as $M(D)$.

In this paper, we are interested in reasoning about query-driven mapping discovery in dataspaces. For ease of presentation, we assume that (some of) the existing mappings are already resolved. Hence, we consider query evaluation on a single fixed data source, potentially already "extended" with triples.

2.2   Query Semantics

The query semantics relies on the notion of homomorphism and, unlike standard query evaluation semantics, permits "partial" evaluation of user queries.

Definition 5. *Let G be a data source and B be a finite set of statement patterns. A V-homomorphism from B to G is a function $h : (\mathsf{C} \cup \mathsf{V}) \to (\mathsf{C} \cup \mathsf{N})$ such that h is the identity on C and $h(B) \subseteq G$.[1]*

Definition 6. *Let G be a data source, $Q = (H, B)$ be a query on G, and $B' \subseteq B$.*

We say that the V-homomorphism h from B' to G is maximal on B' in G if h cannot be extended to a V-homomorphism from $B' \cup \{b\}$ to G for any $b \in B - B'$.

We denote as $H(B')$ the set of all V-homomorphism h to G that are maximal on B' and $max_G(B)$ the set of all subsets B' of B in G for which $H(B') \neq \emptyset$.

For given query $Q = (H, B)$, we say that Q is total on G if $B \in max_G(B)$; Q is partial on G, otherwise.

The embedding-set semantics of Q on G is represented by the set

$$Q^e(G) = \bigcup_{B' \in max_G(B)} H(B').$$

---

[1] $h(B)$ means the point-wise application of h to the variables and constants in B.

$G_1 = \{(John, likes, The\ name\ of\ the\ rose), (John, likes, Sagrada\ familia),$
$\qquad (Sagrada\ familia, country, Spain), (likes, type, vote), \ldots\}$
$G_2 = \{(The\ name\ of\ the\ rose, type, Movie), (The\ name\ of\ the\ rose, country, Italy),$
$\qquad (The\ name\ of\ the\ rose, country, West\ German), (The\ name\ of\ the\ rose, location, Campo\ Imperatore), \ldots\}$
$G_3 = \{(Campo\ Imperatore, province, L'\ Aquila), (L'\ Aquila, country, Italy), \ldots\}$
$Q = (?y, movie\_place, ?z) \leftarrow (?x, likes, ?y), (?y, type, Movie), (?y, country, ?z)$

(a) The dataspace and the query of the reference example.

$\varphi_1 = (?a, country, ?b) \leftarrow 2 : (?a, country, ?b)$
$\varphi_2 = (?a, type, Movie) \leftarrow 2 : (?a, type, Movie)$
$\varphi_1' = (?a, country, ?d) \leftarrow 2 : (?a, location, ?b), 3 : (?b, province, ?c), 3 : (?c, country, ?d)$
$\varphi_3 = (The\ name\ of\ the\ rose, country, ?z) \leftarrow 2 : (The\ name\ of\ the\ rose, location, ?w), 3 : (?w, province, ?z)$
$\varphi_4 = (Sliding\ doors, country, ?z) \leftarrow 2 : (Sliding\ doors, location, ?z)$

(b) Some candidate mappings.

Fig. 1. The running example

Example 2. Following the running example, we recall G_1 has not enough information to resolve Q. In other words, the body of Q is not maximal in G_1. Indeed, the two V-homomorphism for the body of Q, $h_1 = \{?x \mapsto John, ?y \mapsto nor^2\}$ and $h_2 = \{?x \mapsto John, ?y \mapsto Sagrada\ familia, ?z \mapsto Spain\}$, are maximal on $body_1 = \{(?x, likes, ?y)\}$ and $body_2 = \{(?x, likes, ?y), (?y, country, ?z)\}$, respectively. Therefore, $max_G(B) = \{body_1, body_2\}$ and $Q^e(G) = \{h, h'\}$.

3   On the Notion of Candidate Mapping Sets

This section introduces the notion of candidate mappings set. Given a user query that is partial on the queried data source, a candidate mapping set is a set of mappings that can be used to resolve the query, that is that makes the body of the query total. We first exemplify this notion through simple examples then we introduce the formal definition.

Example 3. Following Example 2, let us consider the maximal subset $body_1$ and the corresponding V-homomorphism $H(body_1) = \{h_1\}$.

The missing statement pattern set $\{(?y, country, ?z), (?y, type, Movie)\}$ in Q is explicit in the dataspace. Hence, what we need are the two mappings φ_1 and φ_2 of Fig. 1b which moves the necessary information from G_2 to G_1. Specifically, φ_1 and φ_2 add the tuples $(nor, country, Italy)$, $(nor, country, West\ German)$, and $(nor, type, Movie)$ to G_1 and Q becomes total.

The answer set is $\{(nor, movie\_place, Italy), (nor, movie\_place, West\ German)\}$. The set $\{\varphi_1, \varphi_2\}$ is therefore a candidate mapping set.

Next, suppose that the user is not satisfied with the query result. The user has in mind the country of the filming location. One mapping that can be

---

[2] For compactness reasons, in the following *The name of the rose* will be abbreviated in the text with *nor*.

defined to this purpose is the mapping φ'_1 of Fig. 1b. The set $\{\varphi'_1, \varphi_2\}$ is therefore another candidate mapping set and the query answer this time will be $\{(nor, movie\_place, Italy)\}$.

The definition of candidate mapping set we are going to introduce aims at characterizing mappings from a structural point of view. No assumption is made on the mapping meaningfulness.

Definition 7. *Let $Q = (H_q, B_q)$ be a query on a datasource G of a data space D such that Q is partial on G. A set of s-t tgds M is a candidate mapping set for Q on G if:*

1. *M is the core [7] in the sense that the canonical solution for M on D, $M(D)$, coincide with the core of M on D. In the following $M(D)|_G$ denotes the update of G in $M(D)$;*
2. *there exists at least one V-homomorphism h from B_q to $M(D)|_G$, i.e. Q is total on $M(D)|_G$;*
3. *each h extends at least one homomorphism in the embedded-set semantics of Q on G, $Q^e(G)$;*
4. *for every mapping $\varphi = (H, B) \in M$, H and B are "connected" in the sense that their join graphs has one connected component.*

It is worth noting that the last condition capture the notion of "useful" schema mapping. Indeed, if H is not connected then φ can be split into two mappings while B is usually one connected component in schema-mapping languages [6].

4   A General Algorithm for User-Driven Mapping Discovery

The **goal of user-driven mapping discovery** can be formulated as: given a query $Q = (H, B)$, a dataspace D and a datasource $G \in D$ such that Q is partial on G, find a candidate mapping set M such that the user is satisfied with M, that is that brings answers that satisfy the user information need expressed through Q.

This goal raises two challenging issues: how to discover candidate mapping sets M for Q and how to deliver the candidate mapping set M that best fit the user information need in a limited number of interactions.

As to the first issue, we observe that the definition of $Q^e(G)$ is such that we are given a set of matches on G for the subsets of the body in $max_G(B)$. Each match is "partial" by virtue of missing data in G. Therefore, the algorithm for mapping discovery targets the missing data by relying on two elements that can be derived from each body subset $B' \in max_G(B)$. As we aim at a non-empty answer, it uses the V-homomorphism images (i.e., the values of the bounded query variables) in $H(B')$ to extract examples from the dataspace. Moreover, it uses the missing predicate set in $B - B'$ to generate the target of the missing mappings.

Example 4. Continuing from Example 2, let us consider $body_1$ and $H(body_1) = \{h_1\}$. Then, the missing statement pattern set is $\{(?y, country, ?z), (?x, type, Movie)\}$ and the answer is not empty for Q if we discover some mappings that extends G_1 with statements matching the above pattern set.

For instance, let us consider the statement pattern $(?y, country, ?z)$. We do not know the statement patterns involved in the missing mappings but we know the value that must be searched in the dataspace in order to extend h to $?z$: $h(?y) = nor$. More precisely, we search for one possible mapping in the dataspace that adds to G_1 statements of the form $(nor, country, o)$ where o is a data value in $\mathsf{C} \cup \mathsf{N}$ that will be associated with $?z$. This is the case, for instance, of mappings φ_1 and φ'_1 defined in Fig. 1b.

As to the second issue, the algorithm provides for a multi-criteria selection strategy for candidate mapping sets that aims at a rapid convergence towards user satisfaction by taking into account key features of possible candidate mapping sets, user requirements and context, and user feedback.

4.1   The Algorithm

Algorithm 1 shows the pseudo-code of the general algorithm that recursively discovers mappings in Depth First Search (DFS) fashion. Fixed a subset $\widehat{B}$ of the query body B that belongs to $max_G(B)$, the algorithm incrementally builds a candidate mapping set M for the missing statement patterns in $B - \widehat{B}$ that meets the user information need. In other words, M extends the queried graph source G with data that can be used to resolve the query Q and the delivered answers satisfy the user.

The algorithm starts with the empty mapping set, i.e. $M = \emptyset$, and incrementally extends some of the V-homomorphisms in $H(\widehat{B})$ to B. At each call, it selects a maximum connected component b of $B - \widehat{B}$ (step 2)[3]. Then, it explores the space of the sets of mappings M_b over D that make some of the V-homomorphisms in $H(\widehat{B})$ maximal on $\widehat{B} \cup \{b\}$ (steps 4–11). The search space is scanned in sequential order according to a multi-criteria exploration strategy that aims at rapidly converging towards mapping set M_b the user is satisfied with (steps 5 and 7).

Each selected set of mappings M_b is submitted to user feedback (steps 8–9): (a) the queried data source is virtually extended with the statements from M_b, $M_b(D)$, (b) the partial answers to Q that can be computed from $M_b(D)\,|_G$ are shown to the user, (c) user feedback is collected and instantiated in F. This paper does not address the problem of feedback collection and analysis. Among the possible feedback variables, we assume that flags $F.CONTINUE\_M$ and $F.CONTINUE\_S$ can be instantiated stating whether to continue to explore the search space or not. If the user is satisfied with the obtained answers (i.e. $F.CONTINUE\_M$ and $F.CONTINUE\_S$ have $FALSE$ value) then the algorithm is recursively called (step 12), otherwise the user feedback is exploited

---

[3] This is always possible since we assume that B is connected for all queries.

Algorithm 1. Query-driven mapping discovery

Input: D: dataspace, G: queried data source, $Q = (H, B)$: query, $\widehat{B}$, M
Output: A sequence of candidate mapping set M for Q and D
1: **if** $B - \widehat{B} \neq \emptyset$ **then**
2:    Let $b \subseteq B - \widehat{B}$ be the next maximum connected component of $B - \widehat{B}$ such that
        b shares at least one variable with $\widehat{B}$
3:    $V = \mathsf{vars}(b) \cap \mathsf{vars}(\widehat{B})$
4:    **repeat**
5:        $Samp \leftarrow SelectDataSample(V, H(\widehat{B}), D)$
6:        **repeat**
7:            $M_b \leftarrow SelectMappingSet(Samp)$
8:            Show $Q(M_b(D) \mid_G)$
9:            $F \leftarrow collectFeedback()$
10:        **until** $F.CONTINUE\_M$
11:    **until** $F.CONTINUE\_S$
12:    mapping-discovery($M_b(D), \widehat{B} \cup \{b\}, Q, M \cup M_b$)
13: **else**
14:    output M
15: **end if**

---

together with the other criteria to select the next set M_b. While the discovery process can obviously led to termination with failure, for the sake of simplicity, the algorithm assumes that a satisfactory mapping can be always found.

The recursive call at step 12 relies on $M_b(D)$, the updated subset $\widehat{B} \cup \{b\}$ of B and the mapping set that extends M with M_b. When all the connected components of B have been explored, the algorithm outputs the discovered set M of candidate mappings (step 14).

Example 5. Let us consider Example 1 extended with the tuples:

$$(John, likes, Sliding\ doors) \in G_1$$
$$(Sliding\ doors, type, Movie) \in G_2$$
$$(Sliding\ doors, country, UK)(Sliding\ doors, location, London) \in G_2$$

Thus, $h_3 = \{?x \mapsto John, ?y \mapsto Sliding\ doors\}$ is added to $H(body_1)$.

Given $\widehat{B} = body_1$ there are two connected components, i.e. $(?y, country, ?z)$ and $(?x, type, Movie)$.

When $b = \{(?y, country, ?z)\}$ is selected at step 2, some solutions that can be delivered at step 11 are $M_b = \{\varphi_1\}$ and $M_b = \{\varphi_1'\}$, but also $M_b = \{\varphi_3, \varphi_4\}$ (see Fig. 1b). Each of these solutions extend the V-homomorphisms in $H(body_1)$ to variable $?z$. Let us assume that the user is satisfied with the last one. Then, the algorithm will focus on $b = \{(?y, type, Movie)\}$ and will be able to discover $M_b = \{\varphi_2\}$ that satisfies the user. Then $\{\varphi_2, \varphi_3, \varphi_4\}$ is the candidate mapping set delivered at step 14.

The core steps of the algorithm are steps 5 and 7 where a set of data samples is selected and the semi-lattices of the mappings derivable from the data sample set is explored to select M_b. For instance, one data sample for the reference example is $(nor, country, Italy)$ and $(?y, country, ?z) \leftarrow 2 : (?y, country, ?z)$ is one mapping derivable from this data sample. In the following, we introduce the key notions of data sample and data witness that are at the basis of these two steps. Details about the two steps are given in the next sections.

4.2   Data Samples

A data sample is a connected component in D that can be exploited to specify the mappings for b, i.e. M_b. M_b must extend the queried graph source $G \in D$ with tuples that make some of the V-homomorphisms in $H(\widehat{B})$ maximal on $\widehat{B} \cup \{b\}$. Therefore, data samples must contain the values in the image of the variables in $V = \text{vars}(b) \cap \text{vars}(\widehat{B})$ under the V-homomorphism $H(\widehat{B})$.

More precisely, given a V-homomorphism $h \in H(\widehat{B})$ and the dataspace D, a *data sample* for h and V in D is a connected component of statements in D that contains the values in $h(V)$.

Each data sample s of length i, $s = \{(p_0, p_1, p_2), \ldots, (p_{3i-3}, p_{3i-2}, p_{3i-1})\}$, is associated with a (V, i)-assignment function $pos_s : V \rightarrow 2^{[0, 3i-1]} - \{\emptyset\}$ that records the positions where $h(V)$ occur in ds, i.e. for each $v \in V$, for each $k \in pos_s(v)$: $p_k = h(v)$.

Example 6. Following Example 5, when $b = \{(?y, country, ?z)\}$ is selected at step 2, $V = \text{vars}(b) \cap \text{vars}(\widehat{body_1}) = \{?y\}$. The following three statement sets are data samples for $h_1 \in H(body_1)$ and V in D because they are connected components in D containing $h_1(?y) = nor$:

$s_1 : \{2 : (nor, type, Movie)\}$
$s_2 : \{2 : (nor, country, Italy)\}$
$s_3 : \{2 : (nor, location, Campo\ Imperatore),$
$\qquad 3 : (Campo\ Imperatore, province, L'Aquila)\}$

s_1 and s_2 have length 1 and share the same $(V, 1)$-assignment function: $?y \mapsto 0$. Whereas s_2 has length 2 and its $(V, 2)$-assignment function has the same signature as above.

The set of all data samples for h and V in D is denoted as $S(h, V, D)$, the set of all data samples for h and V in D having the same length i and sharing the same (V, i)-assignment function pos is denoted as $S(h, V, D, i, pos)$. Therefore, $S(h, V, D) = \bigcup S(H, V, D, i, pos)$.

4.3   Data Witnesses

A data witness is a baseline schema mapping the algorithm derives from a data sample s and a maximum connected component b.

For instance, the straightforward data witness for the data sample s_2 of Example 6 and $b = \{(?y, country, ?z)\}$ is the mapping

$$(nor, country, ?z) \leftarrow 2 : (nor, country, Italy)$$

where s_2 is the source and the V-homomorphism h_1 is applied to b in the destination. As this mapping extends the queried data source with tuples matching b, it can be easily shown that h_1 becomes maximal on $\widehat{B} \cup b$. The following definition introduces this kind of data witnesses.

Definition 8. *Let $s \in S(h, V, D)$ be a data sample for the V-homomorphism h and b a maximum connected component. The mapping $\varphi_{s,h,b} = (h(b), s)$ is a data witness for s.*

When $h(b)$ contains free variables, as in the case above, we obtain a data witness $(h(b), B)$ that essentially introduces novel values on the target, i.e. values that are not related to the source. Beside this data witness, we want to derive the s-t tgds that bound the free variables in $h(b)$ to the source of the mapping in order to specify, for instance, the following mapping: $(nor, country, ?z) \leftarrow 2 : (nor, country, ?z)$. To this end, the following definition introduces the second kind of data witnesses.

Definition 9. *Let $\varphi = (h(b), s)$ be a data witness for $s \in S(h, V, D, i, pos)$ and b, where $s = \{(p_0, p_1, p_3), \ldots, (p_{3i-3}, p_{3i-2}, p_{3i-1})\}$.*
The mapping $(h(b), s')$ where $s' = \{(p'_0, p'_1, p'_3), \ldots, (p'_{3i-3}, p'_{3i-2}, p'_{3i-1})\}$ is a data witness that extends φ if

- *$p'_k = p_k$, for all $k \in pos(V)$;*
- *$(\mathsf{vars}(s') \neq \emptyset) \subseteq \mathsf{vars}(h(b))$;*
- *there exists a homomorphism h from $\mathsf{adom}(s') \cup \mathsf{vars}(s')$ to $\mathsf{adom}(s)$ such that h is the identity on $\mathsf{constants}(s')$.*

The set of data witnesses that extends φ is denoted as $ext(\varphi)$.

5   The Space of Data Sample Sets and Selection Strategies

In this section, we define the space of data sample sets that is explored in step 5 and the multi-criteria selection strategy that is used for exploration.

We recall that a data sample set is used to derive the set of mappings to submit to user. At each iteration, one of the aims of the algorithm is to try to extend to b as many V-homomorphism in $H(\widehat{B})$ as possible. Therefore, a data sample set should contain data samples from all the V-homomorphism in $H(\widehat{B})$.

Definition 10. *A data sample set $Samp$ for V, $H(\widehat{B})$ and D selected at step 4, is defined as the union of one data sample set for each $h \in H(\widehat{B})$:*

$$Samp = \bigcup_{h \in H(\widehat{B})} S(h, V, D, Samp)$$

where $S(h, V, D, Samp) \subseteq S(h, V, D, i_h, pos_h) \subseteq S(h, V, D)$.

Example 7. Following Example 6, we now consider the two V-homomorphism in $H(body_1) = \{h_1, h_3\}$, and the data sample sets:[4]

$$
\begin{aligned}
S(h_1, \{?y\}, D, 1, \{?y \mapsto 0\}) = \{ & \{2 : (nor, type, Movie)\}, \\
& \{2 : (nor, country, Italy)\}, \\
& \{2 : (nor, country, West\ German), \\
& \{2 : (nor, location, Campo\ Imperatore)\}\}
\end{aligned}
$$

$$
\begin{aligned}
S(h_3, \{?y\}, D, 1, \{?y \mapsto 0\}) = \{ & 2 : (sd, type, Movie) \\
& 2 : (sd, country, UK) \\
& 2 : (sd, location, London)\}
\end{aligned}
$$

Then $S(h_1, \{?y\}, D, 1, \{?y \mapsto 0\}) \cup S(h_3, \{?y\}, D, 1, \{?y \mapsto 0\})$ is an example of data sample set for $H(body_1)$.

The space of data sample sets is therefore derivable from the data sample sets of the V-homomorphism in $H(\widehat{B})$ and can be very large. At each call, the *SelectDataSample* function at step 5 selects a data sample set *Samp* to deliver. To this end, different selection criteria are considered to rapidly converge towards mappings that satisfy the user information needs.
SelectDataSample follows two phases:

1. $Samp' \leftarrow BuildDataSample(V, H(\widehat{B}), D)$;
2. $Samp \leftarrow SubsetSelect(Samp')$.

In the first phase, it builds the next data sample set $Samp'$ to be explored by relying on the whole data sample sets $S(h, V, D, i, pos)$ for each V-homomorphism $h \in H(\widehat{B})$ and according to a mapping quality criteria, namely, mapping compactness. In the second phase, it selects at each call the next subset *Samp* of $Samp'$ that best meets the user information need according to finer-grain selection criteria defined for single data samples.

5.1 Mapping Compactness for the Generation of Data Sample Sets

In the first phase, *BuildDataSample* relies on the whole data sample sets $S(h, V, D, i, pos)$. This means that the space of data sample sets to be explored is made up of sets of the following form:

$$
Samp' = \bigcup_{h \in H(\widehat{B})} S(h, V, D, i_h, pos_h), \text{ where } S(h, V, D, i_h, pos_h) \subseteq S(h, V, D).
$$

The exploration is driven by a total ordering of the data sample sets in the space that founds on a key feature of the mappings derivable from each data sample set, namely, *mapping compactness*. The main idea is that more compact

---

[4] For compactness reasons, in the following *Sliding doors* will be abbreviated in the text with sd^4, .

mappings, i.e. mappings with less statement patterns on the source side, are preferable to less compact ones. When more than one variable are involved, for instance, this means that the corresponding values are close together rather than sparse in distant statement patterns.

The value of mapping sparsity (the opposite of compactness) of any data sample set $S(h, V, D, i_h, pos_h)$ is essentially i_h, that is the length of the data samples. Indeed, mappings are derived from data witnesses and, as shown in Definition 8, the length of the source side of each data witness is the length of the data sample. Thus, the value of sparsity of any data sample set $Samp'$ can be straightforwardly computed by summing up all the i_h values and the space of all the data sample sets $Samp'$ can be totally ordered at increasing sparsity values.

In this way, starting from the top, function $BuildDataSample$ delivers at each call the next data sample set $Samp$ of the ordering. In this context, it is worth noting that it is not necessary to generate the whole search space for ordering. Indeed, data sample sets can be generated at increasing sparsity values when needed. At the first call, the starting value for i is 1 and the function tries to generate $S(h, V, D, 1, pos_h)$, for each $h \in H(\widehat{B})$. When the computation fails or all alternatives have been delivered, the next value for i is considered.

Example 8. Following Example 7, $BuildDataSample$ will first deliver the data sample set $S(h_1, \{?y\}, D, 1, \{?y \mapsto 0\}) \cup S(h_3, \{?y\}, D, 1, \{?y \mapsto 0\})$ (the total sparsity score is 2), then $S(h_1, \{?y\}, D, 2, \{?y \mapsto 0\}) \cup S(h_3, \{?y\}, D, 1, \{?y \mapsto 0\})$ (the total sparsity score is 3), and so forth.

Notice that mapping φ_1' derives from the data samples for h_1 of length 3, $S(h_1, \{?y\}, D, 3, \{?y \mapsto 0\})$; φ_3 from $S(h_1, \{?y\}, D, 2, \{?y \mapsto 0\})$; φ_4 from $S(h_3, \{?y\}, D, 1, \{?y \mapsto 0\})$; φ_1 and φ_2 from $S(h_1, \{?y\}, D, 1, \{?y \mapsto 0\}) \cup S(h_3, \{?y\}, D, 1, \{?y \mapsto 0\})$. In the following, we will show how φ_1 can be derived. All the other mappings can be discovered in the same way.

5.2   A Multi-criteria Selection Strategy for Data Sample Set Refinement

In the second phase, $SelectSubset$ receives in input a data sample set $Samp'$ and sequentially selects $Samp'$ subsets according to a multi-criteria selection strategy that assess the relevance of data samples with respect to the user information needs.

As in [13], this selection problem is modeled as a multi-objective optimization problem. Let MT be the set of metrics under consideration and $B_m(\cdot)$ be an oracle computing the user benefit with respect to the metric $m \in MT$ of any subset of $Samp'$. The total user benefit is computed as a weighted linear combination of the individual benefit of each metric, i.e. $B(\cdot) = \sum_{m \in MT} w_m \cdot B_m(\cdot)$. Therefore, a total ordering of the subsets of $Samp'$ is possible and $SelectSubset$ sequentially scans this ordered set to deliver, at each step, the next best subset of $Samp'$.

As to the metrics, our main aim is to rapidly converge towards the set of mappings M_b that meet the user information needs expressed through Q when the partial query body $\widehat{B} \cup \{b\}$ is considered. To this end, we propose two kinds of metrics: mapping meaningfulness and data source metrics.

Mapping meaningfulness is another key feature of the mappings that can be derived from each data sample set. Essentially, any discovered mapping can be interpreted a solution to the problem of approximating a query over a data source by using the data available in dataspace [11]. In this light, some mappings, and the data samples that originate them, are more meaningful than others. For instance when $b = \{(?y, country, ?z)\}$ is considered, the data sample $2 : (nor, country, Italy)$ is certainly more meaningful than the data sample $2 : (nor, type, Movie)$. Various relatedness functions have been proposed in the literature and can be used to compare the literals in b with those in the data samples. When the metric $m \in MT$ is the mapping meaningfulness metric, $B_m(Samp)$, with $Samp \subseteq Samp'$, is an aggregation function that receive in input the relatedness score between each data sample in $Samp$ and b.

Data source metrics are metrics that concern the quality of data sources. Their definitions depends on the application scenario and the user task and, as also evidenced in [13], it is more natural, useful, and accurate to talk about the quality of a source with respect to some specific context. For instance, in the application scenario introduced in Sect. 1 where a data enthusiast interacts with an exploratory data analysis tool, possible metrics of interest are accuracy, freshness, and the level of detail. All these metrics can be measured for the topics of interests, such as sports, medicine, and politics, and used in accordance with the user context and background. For instance, depending on the level of user knowledge on a specific topic, jargon-free sources will be preferred to specialized sources or vice-versa, and, the kind of query and the involved topics could suggest if up-to-date sources are better than stable sources. In dynamic scenarios like, for instance, the case of a mobile user that issues requests related to his/her current position, the context will explicitly include the user position and other metrics of interest could be related to the source position. In this paper, we present a general algorithm for user-driven mapping discovery and data source metrics are intentionally left unspecified. Any instantiation of the algorithm in a specific application scenario will detail them according to the application requirements. As to the application of data source metrics to data samples, each statement in a data sample inherits the metric scores of the source it belongs to and the data sample score can be computed by aggregating the scores of all the statements in the data sample.

Finally, this phase can also benefit from user feedback. For instance, when the user confirms a portion of a query answer, this means that the data samples that originate it must belong to any data sample set delivered in the following steps. On the contrary, if a query answer does not meet the user information need, the data samples that originate it must be excluded.

Example 9. Following Example 8, given $S(h_1, \{?y\}, D, 1, \{?y \mapsto 0\}) \cup S(h_2,$ $\{?y\}, D, 1, \{?y \mapsto 0\})$, function *BuildDataSample* according to mapping meaningfulness could deliver at the first call the data sample set $Samp_1 = \{2 : (nor, country, Italy), 2 : (nor, country, West German), 2 : (sd, country, UK)\}$. Then, let us assume that the user is not satisfied with the delivered answers that essentially interpret the movie place as the production country, then another possible set that can be delivered is $Samp_2 = \{2 : (nor, location, Campo\ Imperatore), 2 : (sd, location, London)\}$.

6   Mapping Set Selection

Once a data sample set *Samp* has been selected, the algorithm in steps 6–10 iteratively deliver the sets of mapping specifications M_b that can be derived from such data samples.

To this end, it starts from the data witnesses that originates from *Samp*. In the following $DW(Samp)$ will denote the set of data witnesses that can be computed from the data samples in *Samp* according to Definition 8 and $EDW(Samp)$ the set of data witnesses that extend the data witnesses in *Samp* according to Definition 9, i.e. $EDW(Samp) = \bigcup_{\varphi \in DW(Samp)} ext(\varphi)$.

6.1   The Mapping Semilattice

Notice that each data witness can be used as a seed to derive other mappings. The partial order covering relationship is introduced to this purpose.

Definition 11. *Let* $\varphi_1 = (H_1, B_1)$ *and* $\varphi_2 = (H_2, B_2)$ *be s-t tgds. We say* φ_1 *covers* φ_2, *denoted* $\varphi_2 \sqsubseteq \varphi_1$, *if:*

- φ_1 *does not add existential variables w.r.t.* φ_2, *i.e.* $\mathsf{vars}(H_1) \setminus \mathsf{vars}(B_1) = \mathsf{vars}(H_2) \setminus \mathsf{vars}(B_2)$;
- *there exists a homomorphism* h *from* $\mathsf{constants}(\varphi_1) \cup \mathsf{vars}(\varphi_1)$ *to* $\mathsf{constants}(\varphi_2) \cup \mathsf{vars}(\varphi_2)$ *such that* h *is the identity on* $\mathsf{constants}(B_1)$.

For instance the mapping $(?y, country, ?z) \leftarrow 2 : (?y, country, ?z)$ covers the mapping $(nor, country, ?z) \leftarrow 2 : (nor, country, ?z)$.

Given a set of data witnesses DW, we denote as $cov(DW)$ the set of mappings that cover DW, i.e. $cov(DW) = \{\varphi' \mid \varphi \sqsubseteq \varphi' \text{ and } \varphi \in DW\}$.

$(cov(DW(Samp)), \sqsubseteq)$ and $(cov(EDW(Samp)), \sqsubseteq)$ are therefore semi-lattices the algorithm explores to deliver alternative mapping sets M_b for the maximum connected component b. For example, Fig. 2 shows a small portion of semi-lattices in $(cov(EDW(Samp_1)), \sqsubseteq)$ that can be derived from $Samp_1$ of Example 9.

6.2   Lattice Exploration for the Delivery of Solutions M_b

Each mapping set M_b delivered at step 7 is a set of mappings selected from the semilattices $(cov(DW(Samp)), \sqsubseteq)$ and $(cov(EDW(Samp)), \sqsubseteq)$ that satisfies the two properties shown in the following definition.

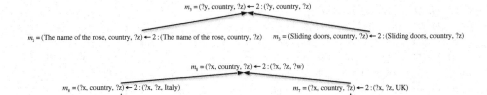

Fig. 2. A small portion of the DAG derivable from the running example.

Definition 12. *Given* $(cov(DW(Samp)), \subseteq)$ *and* $(cov(EDW(Samp)), \subseteq)$, *a solution for b is a set of mappings* M_b *such that*

- $M_b \subseteq (cov(DW(Samp)), \subseteq) \cup (cov(EDW(Samp)), \subseteq)$;
- *each mapping* $\varphi \in M_b$ *is not covered by any other mapping in* M_b, *i.e. it is* $\subseteq$*-maximal in* M_b;
- *each* V*-homomorphism* $h \in H(\widehat{B})$ *can be "covered" either in* $(cov(DW(Samp)), \subseteq)$ *or in* $(cov(EDW(Samp)), \subseteq)$ *but not in both, i.e., all mappings* $\varphi \in M_b$ *that covers data witnesses originating from data samples in* $S(h, V, D)$ *either belong to* $(cov(DW(Samp)), \subseteq)$ *or to* $(cov(EDW(Samp)), \subseteq)$.

Example 10. Let us consider the lattices shown in Fig. 2. According to Definition 12 $\{m_1, m_3\}$ is not a solution because m_3 covers m_1 while $\{m_1\}$ and $\{m_3\}$ are solutions.

Function *SelectMappingSet* at step 7 iteratively explores the lattices in $(cov(DW(Samp)), \subseteq)$ and $(cov(EDW(Samp)), \subseteq)$ to deliver alternative solutions M_b. As *Samp* has been selected with the aim of fitting the user information need, lattice exploration aims at rapidly understanding if the mappings that can be derived from *Samp* are actually of interest for the user.

In this paper, we only give an intuition of how this could be performed. Different exploration strategies will be tested in our future work. First, it is worth noting that existential variables introduces null values and are thus less informative than universal variables [12]. Then, the process can start by exploring $(cov(EDW(Samp)), \subseteq)$ only. Moreover, the notion of mapping meaningfulness introduced in Subsect. 5.2 can be exploited in this context too. For instance, with reference to Fig. 2, mapping m_1 is preferred to m_4. Generally speaking, the function can start by selecting some meaningful data witnesses and deliver them as M_b. At steps 8–9, the algorithm shows to the user the answers that can be obtained from $M_b(D)|_G$. In this phase, as in [4], the user can be challenged with simple questions to understanding if it is interested in exploring the available semi-lattices or not. For instance, if (s)he says that (s)he is interested in solutions like $(nor, movie\_place, Italy)$ that stems from m_1, then in the next step *SelectMappingSet* can select the mapping that covers m_1, i.e. m_3. Otherwise, all mappings that cover m_1 can be safely excluded from further exploration.

7   Proof of Algorithm Correctness

In this section we prove that the algorithm is correct, that is that M is a candidate mapping set. Before enunciating the main theorem, we introduce the following proposition. Proofs can be found in the full version of the paper.

Proposition 1. *Let M_b be a solution delivered at step 7. Then each $h \in H(\widehat{B})$ that is covered in M_b can be extended to a $\vee$-homomorphism from $\widehat{B} \cup b$ to $M_b(D)|_G$.*

The proof explicitly defines the image of the $\vee$-homomorphism that extend h.

Theorem 1. *Given a query $Q = (H_q, B_q)$ on a data source G of a dataspace D, then the set of mappings M delivered at step 14 is a candidate mapping set.*

The proof shows that M satisfies all the conditions stated in Definition 7.

8   Related Works and Concluding Remarks

Up to now, the discovery of mappings has been addressed by leveraging on data examples (see e.g. [2,4,14]). Among the state-of-the-art approaches, the approaches that are closest to our proposal are [2,4], the former intended for expert users while the latter for end users. In [2], the mapping designer provides universal data samples that represent a partial specification of the semantics of the desired schema mapping. Actually, the data witnesses the algorithm discovers can be interpreted as data samples and the top mappings in the semi-lattices are the canonical mappings of the covered data examples in [2]. In [4], they lift the universality assumption arguing that universal data examples are hard to be produced by non-expert users. The contexts that originate the examples in our approach and, consequently, the algorithms are completely different. In particular, the query-driven mapping discovery algorithm we propose is able to deliver candidate mapping sets, i.e. sets of mappings that can be used to solve the user query. Moreover, our approach supports a multi-criteria selection strategy that aims at rapidly converging towards the mappings that best fit the user information need.

The interaction with end users is the distinctive element of some pay-as-you-go approaches for dataspace systems. User feedback has been exploited to annotate, select, and refine schema mappings [3] and to confirm matches [9]. As to our knowledge, this is the first paper that proposes an algorithm for the discovery of mappings starting from user information needs.

Our user-driven mapping discovery algorithm can be interpreted as an algorithm for graph-based query approximation (see e.g. [11,16]). Indeed, in both cases, the input to the algorithm is a user query that cannot be solved over the queried data source. However, in graph-based query approximation, mappings are given and exploited to solve the query. In our case, instead, the partial answers that can be computed over the data source are the input to the process of mapping discovery that aims at introducing new mapping specifications.

Finally, the problem of queries that cannot be rewritten as such due to insufficient mappings has been addressed also in [10]. However, the proposed solution is completely different as they remove parts from the original query to create a query approximation and try to rewrite the latter with possible success.

This paper addresses the problem of mapping specification for non-expert users in data analysis contexts and lays the basis for a novel paradigm that interacts with the information seeking activities of users. Much work is left to be done. As a fist step, we will focus our research efforts in implementing the algorithm in a reference application scenario. This step will lead to the study and experimental evaluation of different approaches for the exploration of the space of data sample sets and the mapping lattices.

References

1. Abiteboul, S., Marian, A.: Personal information management systems. In: EDBT Tutorial (2015)
2. Alexe, B., ten Cate, B., Kolaitis, P.G., Tan, W.C.: Designing and refining schema mappings via data examples. In: Proceedings of ACM SIGMOD, pp. 133–144 (2011)
3. Belhajjame, K., Paton, N.W., Embury, S.M., Fernandes, A.A.A., Hedeler, C.: Incrementally improving dataspaces based on user feedback. Inf. Syst. 38(5), 656–687 (2013)
4. Bonifati, A., Comignani, U., Coquery, E., Thion, R.: Interactive mapping specification with exemplar tuples. In: Proceedings of SIGMOD, pp. 667–682 (2017)
5. Buoncristiano, M., et al.: Database challenges for exploratory computing. SIGMOD Rec. 44(2), 17–22 (2015)
6. ten Cate, B., Kolaitis, P.G.: Structural characterizations of schema-mapping languages. Commun. ACM 53(1), 101–110 (2010)
7. Fagin, R., Kolaitis, P.G., Popa, L.: Data exchange: getting to the core. ACM TODS 30(1), 174–210 (2005)
8. Heath, T., Bizer, C.: Linked Data: Evolving the Web into a Global Data Space. Synthesis Lectures on the Semantic Web. Morgan & Claypool Publishers, San Rafael (2011)
9. Jeffery, S.R., Franklin, M.J., Halevy, A.Y.: Pay-as-you-go user feedback for dataspace systems. In: Proceedings of ACM SIGMOD, pp. 847–860 (2008)
10. Kantere, V., Orfanoudakis, G., Kementsietsidis, A., Sellis, T.K.: Query relaxation across heterogeneous data sources. In: Proceedings of ACM CIKM, pp. 473–482 (2015)
11. Mandreoli, F., Martoglia, R., Penzo, W.: Approximating expressive queries on graph-modeled data: The GeX approach. Elservier JSS 109, 106–123 (2015)
12. Mecca, G., Papotti, P., Raunich, S.: Core schema mappings. In: Proceedings of ACM SIGMOD, pp. 655–668 (2009)
13. Rekatsinas, T., Deshpande, A., Dong, X.L., Getoor, L., Srivastava, D.: Finding quality in quantity: the challenge of discovering valuable sources for integration. In: CIDR (2015)

14. Rivero, C.R., Hernández, I., Ruiz, D., Corchuelo, R.: Mostodex: a tool to exchange rdf data using exchange samples. Elsevier JSS **100**, 67–79 (2015)

15. Shvaiko, P., Euzenat, J.: A survey of schema-based matching approaches. JDS **4**, 146–171 (2005)

16. Torre-Bastida, A.I., Bermúdez, J., Illarramendi, A., Mena, E., González, M.: Query rewriting for an incremental search in heterogeneous linked data sources. In: Larsen, H.L., Martin-Bautista, M.J., Vila, M.A., Andreasen, T., Christiansen, H. (eds.) FQAS 2013. LNCS, vol. 8132, pp. 13–24. Springer, Heidelberg (2013). doi:10. 1007/978-3-642-40769-7_2

Ontologies and Human Users: A Systematic Analysis of the Influence of Ontology Documentation on Community Agreement About Type Membership

Francesca Zarl[1], Martin Hepp[2(✉)], Alex Stolz[2], and Walter Gerbino[1]

[1] Department of Psychology, Università degli Studi di Trieste, Trieste, Italy
francesca.zarl@phd.units.it, gerbino@units.it
[2] E-Business and Web Science Research Group,
Universität der Bundeswehr München, Neubiberg, Germany
{martin.hepp,alex.stolz}@unibw.de

Abstract. In this paper, we study the impact of the human-readable documentation of Web ontologies on the ability of human users to agree on the membership of instances according to a given ontology. We first introduce a model of the problem and then present a user study, in which we measured the impact of documentation features in schema.org on the quality of annotations with n = 73 study participants. The paper concludes with a discussion of implications for ontology design in the context of the Semantic Web.

Keywords: Ontology engineering · Schema.org · Data quality · Human factors · Ontology documentation · Web ontologies

1 Introduction

Ontologies in computer science are conceptual models that are shared by a community and which are specified using one or more modalities, namely human-readable documentation and formal axioms [1–4]. They aim at a shared specification of conceptual elements, like classes, properties, or enumerations, i.e. types, for data interoperability and other purposes. Historically, the design decisions of such ontologies have been mostly based on (1) the usefulness of the conceptual distinctions for the consumption and processing of respective data and knowledge bases by computers [cf. 5, 6], and by (2) the consistency and normalization of the representation [cf. 7].

In recent years, there is a growing understanding that cognitive aspects at the interface between human users and ontologies are an important variable in ontology engineering [8–16]. The conceptual modeling community has discussed similar issues since the 1990 s, for an overview, see e.g. [17].

These advances, however, have so far limited influence on the design and documentation of actual Web ontologies. This is unfortunate, because the amount and data quality of the growing body of structured data available from the Web will critically depend on the ability of, and effort for, humans to publish data according to the conceptual elements specified in the ontologies. This will depend on at least two tasks that

H. Panetto et al. (Eds.): OTM 2017 Conferences, Part II, LNCS 10574, pp. 418–426, 2017.
https://doi.org/10.1007/978-3-319-69459-7_28

include human cognition, namely (1) how well human users can grasp the intended meaning of the ontology elements ("ontology perspicuity", [18]), and (2) how well they are able to decide whether a given entity instance or relationship instance is a member of an entity type (class) or relationship type (property), i.e. the user's performance in the classification task. Our working assumption is that cognitive barriers between the creators of ontologies and the other stakeholders in the entire lifecycle of an ontology are significant sources of data quality problems for the Web of Linked Data.

In this paper, we introduce a model of the problem and present a user study, in which we measured the impact of documentation features in schema.org[1] on the quality of annotations with n = 73 study participants. We then discuss implications for the design and presentation of Web ontologies.

1.1 Problem Statement and Relevance

Ontologies are artifacts that aim at establishing and maintaining type systems in data interchange operations so that the *overlap of the extension* (i.e. the set of individual phenomena, like entities, relationships, or attributes) of a type between (1) the data source (e.g. the creator of markup or data related to an ontology) and (2) the consuming application or human is maximized. An ideal ontology will have a perfect match of type agreement between all parties and systems involved. For the sake of simplicity, we hereby do not distinguish entity types ("classes") from relationship types ("properties"), albeit there might be relevant differences in the general case.

When the owner of data wants to publish it in a form suitable for consumption by Semantic Web applications, he or she has to master the following tasks that include cognitive challenges:

1. **Ontology search,** i.e. finding a suitable ontology for the data, based on domain coverage and support by relevant target applications (e.g. Web search engines), and
2. **Typing of entities and relationships,** i.e. finding the most suitable ontology element (like class or property) for entities and relationships in the data, and deciding whether a particular entity or relationship falls within the intension of the ontology element.

The second task can occur either at the level of individuals, i.e. the owner of data has to decide on the basis of instance data, or at the level of conceptual data models, like database schemas. The latter is typical for database-driven Web applications. In theory, owners of data might also have to import and extend existing ontologies or align two or more ontologies. We do not further consider these cases in this paper, because they are not relevant for mainstream consumers of data like Web search engines, since those would typically not understand data based on user-specific extensions of the official ontology.

Both tasks have already been studied in the context of tool development for ontology search, like e.g. LOV [19] and Watson [20]. We argue, however, that the cognitive challenges should be considered independently of possible tool support,

---

[1] http://schema.org

because they are influenced by multiple aspects [cf. 15, 21]: (1) conceptual modeling decisions (some modeling patters are more easily understood than others), (2) mental models of the human users (how does the model fit to existing conceptualizations in the users' minds), (3) naming and terminology, (4) presentation of the ontology, namely features of the ontology documentation, like textual descriptions, examples, counter-examples, concept hierarchy, list of applicable properties, etc.

1.2   Organization of the Paper

In Sect. 2, we define our research design and method. In Sect. 3, we present the findings from the online experiment in the context of schema.org. In Sect. 4, we evaluate and discuss the results, and identify threats to validity. Section 5 concludes the work and points to future research.

2   Research Design

In this section, we describe our research questions, method, and data.

2.1   Research Questions

In our study, we want to analyze how the richness of the presentation of an ontology influences the ability of human users to apply the ontology correctly. This leads to the following research questions:

RQ 1: How does the number of features in the presentation of an ontology influence the ability of human users to apply the ontology correctly?
RQ 2: Which types of features are more effective than others?
RQ 3: Are the effects of the number and types of features uniform across different types in the ontology or do the effects differ very much by the type?
RQ 4: Are the effects of features different for experts vs. less experienced users?
RQ 5: Are users able to assess their degree of understanding correctly?

2.2   Method and Data

We first drew a random sample of $n = 12$ classes from the most popular classes in schema.org. As a metric for popularity, we chose the information provided in the Github source code of schema.org[2]. We manually reviewed the sample and removed one class (SearchAction), because it was not suited for the structure of our experiment. The resulting 11 classes are WebSite, WebPage, Product, PostalAddress, Person, Organization, Offer, ImageObject, BlogPosting, Blog, and Article.

We then manually created a set of $n = 88$ objects that were a mix of (1) clear members, (2) borderline members, (3) borderline non-members, and (4) clear

---

[2] https://github.com/schemaorg/schemaorg/blob/sdo-callisto/data/2015-04-vocab_counts.txt

non-members for a given type. The exemplars are either textual descriptions or images, and were created with a single of the types from the sample in mind and assigned for testing to one single type. For each object, we added a gold standard assessment based on multiple rounds of discussions among experts.

Next, we populated the following features for the classes from schema.org if available:

1. Hierarchy position (e.g. "Thing ->Intangible ->Offer")
2. Example as text
3. Counter example as text
4. Textual description of pitfalls and subtleties ("Watch out")
5. More specific types (Subclasses)
6. Position in the hierarchy (Superclasses)
7. Description as text

The human-readable name of the type was our baseline feature. We manually added those features of a given class that are not currently supported by schema.org, or were empty for the particular type.

The data was then used to conduct an online experiment in which users were shown a type with a randomly selected set of documentation features. They were then asked to assess whether a given exemplar was a member of that type. In addition, we asked for a difficulty rating on a 4-point Likert scale. The set of features shown remained the same for the same user and the same type in order to avoid memorization effects (if a user had seen a feature for the same type before, the effects of the features would be blurred). Each participant had to make judgments for all $n = 11$ types and $n = 4$ exemplars per type, totaling to 44 judgments per participant.

The final survey was then completed by $n = 73$ participants from varying backgrounds.

3   Results

In the following, we summarize the findings from the experiment. Due to space constraints in this paper and because some of them are less useful in a greyscale version, we do not include the visualizations here. They are available from the companion website http://www.heppnetz.de/publications/cognition-odbase2017.

3.1   Raw Impact of Features

We first defined eight different categories of outcomes by the Cartesian product of

```
{True Positive (TP); True Negative (TN); False Positive (FP); False
          Negative (FN)} X {easy, difficult}
```

and plotted bar charts of the respective data. Note that at this level, we are aggregating over all types and all groups of users and just considered the number of features but not their kind. We observed that there is no uniform effect of the number of features on task performance if we look at the aggregate over all types and all users. This could be explained by the fact that differences among types are stronger than differences between the features, or the fact that very intuitive type names are sufficient to activate existing, intersubjectively consistent notions of category.

We then compared task perfomance with the data from the self-assessment of task difficulty. Compatible with our intuition, a higher confidence usually correlates with a better task performance. In our case, this observation at least holds true for (objectively) easy examples where the ratio of correct judgments was highest for examples considered the most easy ones, and vice versa. For (objectively) more difficult examples, correct judgements were higher for those examples assessed more difficult, whereas incorrect judgments were still lower for examples deemed easier.

3.2   Heatmaps

We then created heatmaps that break down the effect of ontology features by individual schema.org types and individual features rather than the number of features. The colors in the respective visualizations indicate the type performance, defined as the ratio of correct assessments as compared to the gold standard. We observe that the absolute task performance varies much stronger across types than is the influence of the features. We can also see that features can indeed *reduce* the user's understanding of the intension of a type.

The task performance (averaged for each participant) varies more clearly between types than between types of features. For example, the task performance for article and organization is much better than for blog or blog posting. Without any additional information (no ontology documentation features given), participants face greater difficulties categorizing products. Conversely, for blog, postal address, offer and person, they do better.

There are more correct rejections for blog and organization if no features are shown than if features were present. Similarly, we see most true positives for WebSite if no features were given, while ImageObject, WebPage and Offer benefit from the availability of features. What is striking is that more misses and false alarms were produced in general when more features became available, only for Article it was the opposite. This might indicate information overload, or that users are more hesitant to assume class membership for more types that are more fully specified. This effect could be tested in future work by augmenting type descriptions by redundant information. This would reveal whether such causes a bias.

By comparison of the means of the response ratings, an independent-samples t-test between the modalities "text" and "image" was not significant. Grouping the dataset by types, there are some cases where the null hypothesis of equal means can be rejected. For BlogPosting, images lead to significantly higher confidence on average than text. Interestingly, for ImageObject, text causes a significantly higher confidence on average than images. For Person, text also leads to significantly higher response ratings on average than images.

4   Evaluation and Discussion

We have presented a research design for measuring the effects of ontology documentation features on the task performance of human ontology users of classifying given exemplars according to the types of an ontology. While we have focused on testing entity types, the same design can also be used for relationship types and equivalence of enumerated values.

4.1   Comparison with Research Questions

Contrasting our findings with the original research questions, we come to the following assessment:

RQ 1: How does the number of features in the presentation of an ontology influence the ability of human users to apply the ontology correctly

There is a complex interplay between the types themselves and the effects of features in ontology documentation. The pure number of features is not correlated with task performance and we observe several cases where more features actually reduce task performance. We have to keep in mind that the types are (1) the most popular ones from (2) the most widely used ontology on the entire Semantic Web. From that popularity, we can assume that none of the 11 types is entirely novel for any of the participant, and that there are existing associations with the names of the type alone in the minds of the participants. In fact, we can see that the differences across types are more substantial than the effects of the features alone. This can be an indication that we have to pay more attention to existing conceptualizations in the target users' minds, i.e. the category systems in the minds of typical users, and the terminology that will activate those when engineering Web ontologies.

We also analyzed the increase of categorization sensitivity as an effect of the amount of information available from the ontology documentation and could find a significant correlation between doc gain and criterion shift, as well as a large variability of those metrics by type. The types in our sample that benefit from more features (like Organization) also suffer from making users more conservative to accept type membership.

RQ 2: Which types of features are more effective than others?

We can see that there is a huge variation by the type. For instance, the feature "position in the hierarchy" has a positive impact on five of the 11 types, and a negative impact on the other six. Thus, we can infer that the same documentation feature has very different effects for different types. This could partly be explained by variation in quality of the content of the feature, but more likely indicates the need to customize the selection of features by type and user group. In the concrete context of schema.org in clearly indicates that the features of the documentation must be revisited, since several of the features of 11 of the most popular types actually reduce the task performance.

RQ 3: Are the effects of the number and types of features uniform across different types in the ontology or do the effects differ very much by the type?

As we have already seen, the types show a much greater variation than the features. For us, it indicates that drawing upon types and names for types that activate established, shared conceptualizations should be an important requirement for ontology engineering. We have to build ontologies that cling to the category systems in the minds of the user community much more than align them with the computational processing of resulting instance data. We may have to rethink our understanding of how much power the ontology engineer actually has when it comes to shaping the intension of the conceptual elements.

RQ 4: *Are the effects of features different for experts vs. less experienced users?*

The differences between the two clusters are much smaller than the differences inside the clusters caused by the different types.

RQ 5: *Are users able to assess their degree of understanding correctly?*

In general, task performance and self-reported confidence do not correlate. One may find additional insight when looking into individual effects within the different types, but we did not do this in our current analysis.

4.2   Threats to Validity

Our work is based on popular types in a popular Web ontology, i.e. schema.org. While this increases the practical relevance of the data, it introduces a several potential threats to validity:

1. Users might have been exposed to the full documentation at schema.org before, and have memorized features or shaped their understanding of the type from previous experiences. In our future work, we will repeat our experiments with artificial types from a domain unknown to the participants (e.g. categories from science fiction or specific branches from biology or the arts and sciences). We would argue, however, that since we use only very popular types from schema.org, the effect should be similar for all types.
2. Since schema.org itself suffers from quality issues, the actual of the content of the features could be a source of error. However, we can see that even features that are pretty objective (like the position in the hierarchy) show the same effects as we observe for other features.

5   Conclusion and Outlook

We have shown that features in the presentation of ontologies to human users play an important role in establishing shared type systems for data interoperability in the form of ontologies. Typical features in Web ontologies have very different effects on different types, and can actually negatively influence the performance of users of the ontology. We have traced this back to the interplay of the sensitivity gain and the criterion change, both due to more information available in the documentation.

References

1. Gruber, T.R.: A translation approach to portable ontology specifications. Knowl. Acquis. **5**, 199–220 (1993)
2. Gruber, T.: Every ontology is a treaty - a social agreement - among people with some common motive in sharing. AIS SIGSEMIS Bull. **1**, 4–8 (2005)
3. Guarino, N., Giaretta, P.: Ontologies and knowledge bases: towards a terminological clarification. In: Mars, N. (ed.) Towards Very Large Knowledge Bases: Knowledge Building and Knowledge Sharing, pp. 25–32. IOS Press, Amsterdam (1995)
4. Hepp, M.: Ontologies: state of the art, business potential, and grand challenges. In: Hepp, M., de Leenheer, P., de Moor, A., Sure, Y. (eds.) Ontology Management: Semantic Web, Semantic Web Services, and Business Applications, pp. 3–22. Springer, Boston (2007). doi:10.1007/978-0-387-69900-4_1
5. Uschold, M., Grüninger, M.: Ontologies: principles, methods, and applications. Knowl. Eng. Rev. **11**, 93–155 (1996)
6. Gómez-Pérez, A., Fernández-López, M., Corcho, O.: Ontological Engineering. Springer, London (2004)
7. Guarino, N., Welty, C.A.: Evaluating ontological decisions with ontoclean. Commun. ACM **45**, 61–65 (2002)
8. Stark, J., Esswein, W.: Rules from Cognition for Conceptual Modelling. In: Atzeni, P., Cheung, D., Ram, S. (eds.) ER 2012. LNCS, vol. 7532, pp. 78–87. Springer, Heidelberg (2012). doi:10.1007/978-3-642-34002-4_6
9. Chiew, V., Wang, Y.: From cognitive psychology to cognitive informatics. In: Proceedings of the 2nd IEEE International Conference on Cognitive Informatics, p. 114. IEEE Computer Society (2003)
10. Kotis, K., Vouros, A.: Human-centered ontology engineering: the HCOME methodology. Knowl. Inf. Syst. **10**, 109–131 (2006)
11. Warren, P., Mulholland, P., Collins, T., Motta, E.: The Usability of Description Logics. In: Presutti, V., d'Amato, C., Gandon, F., d'Aquin, M., Staab, S., Tordai, A. (eds.) ESWC 2014. LNCS, vol. 8465, pp. 550–564. Springer, Cham (2014). doi:10.1007/978-3-319-07443-6_37
12. Wilmont, I., Hengeveld, S., Barendsen, E., Hoppenbrouwers, S.: Cognitive Mechanisms of Conceptual Modelling. In: Ng, W., Storey, Veda C., Trujillo, Juan C. (eds.) ER 2013. LNCS, vol. 8217, pp. 74–87. Springer, Heidelberg (2013). doi:10.1007/978-3-642-41924-9_7
13. Peroni, S., Motta, E., d'Aquin, M.: Identifying Key Concepts in an Ontology, through the Integration of Cognitive Principles with Statistical and Topological Measures. In: Domingue, J., Anutariya, C. (eds.) ASWC 2008. LNCS, vol. 5367, pp. 242–256. Springer, Heidelberg (2008). doi:10.1007/978-3-540-89704-0_17
14. Ernst, N.A., Storey, M.-A., Allen, P.: Cognitive support for ontology modeling. Int. J. Hum Comput Stud. **62**, 553–577 (2005)
15. Engelbrecht, P.C., Dror, I.E.: How psychology and cognition can inform the creation of ontologies in semantic technologies. In: Proceedings of the 2009 conference on Information Modelling and Knowledge Bases XX, pp. 340–347. IOS Press (2009)
16. Evermann, J., Fang, J.: Evaluating ontologies: towards a cognitive measure of quality. Inf. Syst. **35**, 391–403 (2010)
17. Ramesh, V., Parsons, J., Browne, Glenn J.: What Is the Role of Cognition in Conceptual Modeling? A Report on the First Workshop on Cognition and Conceptual Modeling. In: Goos, G., Hartmanis, J., van Leeuwen, J., Chen, Peter P., Akoka, J., Kangassalu, H., Thalheim, B. (eds.) Conceptual Modeling: Current Issues and Future Directions. LNCS, vol. 1565, pp. 272–280. Springer, Heidelberg (1999). doi:10.1007/3-540-48854-5_21

18. Fox, M.S., Gruninger, M.: On ontologies and enterprise modelling. In: Kosanke, K., Nell, J.G. (eds.) International Conference on Enterprise Integration Modelling Technology, pp. 190–200. Springer, Heidelberg (1997). doi:10.1007/978-3-642-60889-6_22
19. Baker, T., Vandenbussche, P.-Y., Vatant, B.: Requirements for vocabulary preservation and governance. Libr. Hi Tech **31**, 657–668 (2013)
20. d'Aquin, M., Motta, E.: Watson, more than a Semantic Web search engine. Semant. Web J. **2**, 55–63 (2011)
21. Ehrlich, K.: The essential role of mental models in HCI: Card, Moran and Newell. In: Erikson, T., Mc.Donald, D.W. (eds.) HCI Remixed: Reflections on Works That Have Influenced the HCI Community, pp. 281–284. MIT Press (2008)
22. Macmillan, N.A., Creelman, C.D.: Detection Theory: A User's Guide. Lawrence Erlbaum Associates, Mahwah (2005)

DLUBM: A Benchmark for Distributed Linked Data Knowledge Base Systems

Felix Leif Keppmann$^{(\boxtimes)}$, Maria Maleshkova, and Andreas Harth

Karlsruhe Institute of Technology, Karlsruhe, Germany
{felix.leif.keppmann,maria.maleshkova,andreas.harth}@kit.edu

Abstract. Linked Data is becoming a stable technology alternative and is no longer only an innovation trend. More and more companies are looking into adapting Linked Data as part of the new data economy. Driven by the growing availability of data sources, solutions are constantly being newly developed or improved in order to support the necessity for data exchange both in web and enterprise settings. Unfortunately, currently the choice whether to use Linked Data is more an educated guess than a fact-based decision. Therefore, the provisioning of open benchmarking tools and reports, which allow developers to assess the fitness of existing solutions, is key for pushing the development of better Linked Data-based approaches and solutions. To this end we introduce a novel Linked Data benchmark – Distributed LUBM, which enables the reproducible creation and deployment of distributed interlinked LUBM datasets. We provide a system architecture for distributed Linked Data benchmark environments, accompanied by guiding design requirements. We instantiate the architecture with the actual DLUBM implementation and evaluate a Linked Data query engine via DLUBM.

Keywords: Linked Data · Linked Data benchmarking · Distributed benchmarking · LUBM · DLUBM

1 Introduction

Linked Data (LD) has witnessed a rapid evolution and growth during the past years. Driven by the growing availability of data sources, solutions are constantly being newly developed or improved in order to support data exchange both in web and enterprise settings. Unfortunately, currently the choice whether to use Linked Data is more an educated guess than a fact-based decision. Therefore, the provisioning of benchmarking tools and evaluation reports, which allow developers to assess the fitness of existing solutions, is key for pushing the development of better Linked Data-based approaches and solutions. In addition, to be able to evaluate different alternatives, we require benchmarks that are capable of providing truly distributed Linked Data settings in a reproducible way.

The Linked Data setting is, on the one hand, characterized by datasets that consist of multiple distinct graphs represented in the Resource Description Framework (RDF). Each graph is addressed by a different Uniform Resource

© Springer International Publishing AG 2017
H. Panetto et al. (Eds.): OTM 2017 Conferences, Part II, LNCS 10574, pp. 427–444, 2017.
https://doi.org/10.1007/978-3-319-69459-7_29

Identifier (URI) and is accessible via the Hypertext Transfer Protocol (HTTP). The documents returned by hosts during requests at these HTTP URIs represent the graphs in one of the RDF serialization formats, e.g., N-Triples, Turtle, or RDF/XML. Graphs may contain references to other graphs at different URIs. On the other hand, clients in the Linked Data setting must be capable of handling remote documents. While RDF query engines enable the answering of queries against a local graph, Linked Data query engines must retrieve graphs by requesting documents that contain graph representations via HTTP. Furthermore, to leverage the full potential of Linked Data, these query engines must support resolving of links within received RDF graphs [16,17], i.e., query engines must support link following in the overall dataset of interlinked graphs during query answering.

The characteristics of such distributed Linked Data settings pose new challenges to the design of adequate Linked Data benchmarks, which are ignored or only partially covered by existing approaches. In particular, we identify the following four main problem areas:

Data Generation. Existing benchmarks [9,10,14,15,20,22] for RDF, including [14,20] that are only partially but name-wise related to Linked Data, lack certain capabilities to cover major aspects of Linked Data settings. In particular, benchmark data generators must be capable of generating datasets that consist of distinct but interlinked graphs. These graphs must represent a subset of the overall generated information of the dataset and must not prevent virtual (and physical) separation, but contain valid links to other graphs in the dataset. Existing benchmarks for RDF generate monolithic graphs and not separated and interlinked ones.

Dataset Distribution. Distinct characteristics of Linked Data are the distribution of datasets in a network, c.f., the Linking Open Data Cloud (LODC) [1], and accessibility via HTTP. Therefore clients, e.g., Linked Data query engines, must be capable of not only evaluating but also retrieving documents from hosts and following links to related graphs. Existing benchmarks for RDF focus on single evaluation engines but do not provide support for distribution, as common for Linked Data settings.

Deployment Complexity. In addition to the lack of certain capabilities and due to the distributed nature of Linked Data, we face complex deployment requirements for the simulation of Linked Data settings. For example, one challenge is the setup of multiple hosts, with the distribution to and interlinking of graphs at these hosts, and with the resolvability of links in certain network environments.

Reproducibility. The complexity of the deployments, which extends beyond the reliable generation of datasets, leads to difficult reproducibility. We need to be able to reliably reproduce not only RDF graphs with certain characteristics but also the distribution of the data, including all aforementioned challenges. The setup of a benchmark with the same parameters at different locations or times must result in an equal setting with the same characteristics.

With respect to these challenges, we make the following contributions:

1. Our approach on reproducible distributed benchmark environments for Linked Data, including requirements derived from the problem areas and complemented by our architecture designed to cope with these requirements.
2. The Distributed LUBM (DLUBM) as a specific implementation of our architecture based on the Lehigh University Benchmark (LUBM) [15]. DLUBM includes (1) a Linked Data generator, (2) integration with container-based virtualization and distribution technologies, (3) supporting tools for automatic composition and deployment, and (4) reproducibility by pervasive parametrization and declaration.
3. The evaluation of a Linked Data query engine by utilizing DLUBM instances of different scale and granularity. These experiments have been conducted for comparability on computing resources of broadly available platform providers, in particular, in the Elastic Compute Cloud (EC2) provided by Amazon Web Services (AWS).

In the following, after giving an overview of related work in Sect. 2, we present in Sect. 3 our approach on creating distributed benchmark environments for Linked Data. In Sect. 4, we describe DLUBM as a specific implementation of our approach and evaluate the implementation in Sect. 5 by testing a Linked Data query engine via DLUBM. We conclude our work in Sect. 6.

2   Related Work

With the growing proliferation of graph-based data in general and Linked Data in particular, benchmarking has become a prominent topic in the Semantic Web community. Many benchmarks and data generators already exist, including for instance LUBM [15], Berlin SPARQL Benchmark (BSBM) [9], SPARQL Performance Benchmark (SP2Bench) [22], gMark [6], Generating Random RDF (Grr) [10], or Waterloo SPARQL Diversity Test Suite (WatDiv) [2]. Still, the main focus of these solutions is on covering the main features of the SPARQL Protocol and RDF Query Language (SPARQL), on query optimisation, and not specifically on exploring the Linked Data setting. One important requirement is the ability to generate distinct but interlinked datasets. In this context, existing benchmarks for RDF focus on monolithic graphs but not on separated graphs in interlinked RDF datasets.

In general, the desirable characteristics of benchmarks [18] (e.g., repeatable, fair, verifiable, and economical) are not always taken into consideration. Moreover, sometimes the specifics and requirements in RDF and graph data management are neglected as pointed out in a series of benchmark surveys [11,12,24]. In this context, the Linked Data Benchmark Council (LDBC) [3] aims to establish benchmarks, and benchmarking practices for evaluating graph data management systems. It proposes a "choke-point"-driven design of graph database benchmarks, which combines user input with input from expert systems architects.

We are also witnessing the development of benchmarks that cover specific domains. For instance, related to social network benchmarking, LDBC developed the Social Network Benchmark (SNB) [14], which introduces a synthetic social network graph with three workloads: SNB-Interactive, SNB-BI, and SNB Algorithms. Similarly, Facebook features LinkBench [4], a benchmark targeting the workload of Online Transaction Processing (OLTP) on the Facebook graph. LinkBench focuses only on transactions and uses a synthetic graph generator, which is unfortunately capable of reproducing very little of the structure or value correlations found in real networks. Finally, the BG benchmark [7] proposes to evaluate simple social networking actions under different Service Level Agreements (SLAs). Naturally, domain-specific benchmarks put emphasis on particular characteristics such as network structure and node correlations.

In summary, while there are a number of benchmarks supporting SPARQL features, including query optimisation, and while there are also some benchmarks with focus on certain domains, what is still missing is a benchmark that targets specifically Linked Data and not RDF in general. This is especially true for providing options for having virtually or physically distributed datasets, having valid links between graphs in these datasets, and at the same time being able to manage the deployment complexity of such a distributed setting.

3   Distributed Benchmark Environments for Linked Data

Our approach on realizing truly distributed benchmark environments for Linked Data is guided by a set of requirements derived from the problem areas introduced in Sect. 1. With these requirements, we ensure the broad applicability of our architecture and abstract away from specific implementation characteristics. Thereby, the approach itself is independent from domains and technologies but exposes explicit requirements to dataset generation, virtualization, and distribution technologies. Our DLUBM benchmark environment, described in Sect. 4, is one specific implementation. In the following, we present first the guiding requirements and second our architecture.

3.1   Requirements

The requirements that guide our approach are divided in four main groups, each covering one challenge: deployment-aligned data generation, network layer distribution, deployment automation, and pervasive declaration.

Deployment-aligned Data Generation. Our goal is the creation of a truly distributed benchmarking environment not for arbitrary data, but for Linked Data. Therefore, and with respect to the first problem area "data generation" introduced in Sect. 1, the first main requirement for our approach is the generation of RDF graphs that are distinct but interlinked and are suited to settings that adhere to the well-known Linked Data [8] principles:

1. "Use URIs as names for things"
2. "Use HTTP URIs so that people can look up those names"
3. "When someone looks up a URI, provide useful information, using the standards (RDF, SPARQL)"
4. "Include links to other URIs, so that they can discover more things"

These Linked Data principles imply requirements on the generation of fitting graphs that can be used in a distributed environment. First, the data generator must support the generation of completely separate RDF graphs, e.g., split into different files. Depending on the deployment process of the distributed benchmark environment, the generation of a relevant subset of the graphs may be an advantage. Second, the data generator must support the correct interlinking of these graphs with respect to the actual deployment of the distributed environment, i.e., the HTTP URIs within generated graphs must point to the correct related graphs, including their correct hostnames. Third, if SPARQL queries are dynamically generated along with the datasets, the generator must support the correct use of HTTP URIs in these queries with respect to the actual deployment of graphs in the distributed benchmark environment.

Network Layer Distribution. An implicit requirement to a distributed benchmarking environment for Linked Data, introduced by the second problem area "dataset distribution" in Sect. 1 and imposed by the second and fourth Linked Data principles, is the interlinking and lookup in distributed datasets, i.e., HTTP URIs in graphs of the datasets can be looked up to discover related graphs. While this can be achieved with a single host through distribution on the application layer (i.e., graphs identified and looked up by different HTTP URIs hosted at one server) simulating a realistic setting in the context of scenarios in the LODC, the Internet of Things (IoT) [5], the Web of Things (WoT) [13], or the Semantic Web of Things (SWoT) [19], can only be achieved if these graphs are distributed at least at the level of different hosts. Therefore, the second main requirement guiding our approach is the distribution of interlinked RDF graphs at the network layer of the Open Systems Interconnection (OSI) model, in particular using different Internet Protocol (IP) addresses or Domain Name System (DNS) records. Graphs must be provided by distinct hosts and be resolvable via HTTP. With respect to established cloud and emerging container-solutions, we neither require nor prohibit physically separate hosts.

Deployment Automation. With the third problem area "deployment complexity" in Sect. 1, we focused on the complexity of deployments that accompanies distributed applications. As examples, we pointed out: the setup and networking of multiple hosts, the distribution to and interlinking of graphs at these hosts, and the resolvability of links in certain network environments. In addition, benchmark environments are not setup for long-term provisioning, but for experiments with different parametrizations and computing resources. Furthermore, benchmark environments must be flexible to be deployed, at appropriate scale: on limited development resources or large scale testing facilities. Our third main requirement on distributed benchmarking environments for Linked Data is, therefore, the automation of the deployments, thus reducing the complexity to a sufficient degree that renders the benchmark environments usable in short-term and enables temporary experimental settings of different scales.

Pervasive Declaration. Comparing different solution alternatives for the same problem area is one of the key goals of benchmarks. The same set of configuration parameters leads to the same set of resources, which are used to evaluate and compare different solutions for the same problem. Therefore, and with respect to the fourth problem area "reproducibility" introduced in Sect. 1, the fourth and final main requirement guiding our approach is the reproducible creation of the distributed benchmarking environment. The declaration of the same set of parameters must lead to the same generated datasets, distribution, and interlinking. In addition, the computing resources available for the hosts may be covered for greater comparability. The interweaving of Linked Data generation and the actual distributed deployment, under different conditions, is especially challenging for reproducibility. In order to be able to handle this, for our approach, we distinguish between different types of parameters. On the one hand, we introduce parameters that influence the scale and granularity of the distributed benchmarking environment. These may be global (e.g., the number of graphs and the size of overall datasets for scaling) or local, i.e., host-specific (e.g., the type of data to be generated in case of different granularities within the overall dataset). On the other hand, we introduce parameters that influence the interlinking of graphs. These are deployment-specific and enable the adjustment of generated data to the environment, in which the distributed benchmark is deployed (e.g., URI templates to generate correct links, according to the IP or DNS schema of the environment).

3.2   Architecture

The architecture of our distributed benchmarking environment was especially designed to fulfil the aforementioned requirements. In particular, the requirement of pervasive declaration to ensure reproducible creation of the benchmark environment is challenging, since a solution must also cover implications from all other main requirements. While the scope of reproducibility in pure RDF/ SPARQL benchmarks is limited to the generation of equal graphs and queries, it is extended to the generation of distributed environments with equal characteristics in the case of Linked Data.

For the requirement of deployment-aligned data generation, we built on Linked Data-capable data generators or on established RDF benchmarks. These are extended to generate graphs for Linked Data, e.g., in our implementation in Sect. 4. In general, data generators are exchangeable in our approach as long as they support the parametrized generation of datasets that contain distinct interlinked graphs, including the adjustment of links to the actual deployment. For the requirements of network layer distribution and deployment automation, we built on container-based virtualization technologies that, independently of the specific deployment platforms, allow the definition of hosts, networks, and other deployment-specific tasks. For the requirement of pervasive declaration, we built on the parametrization of the data generator, the declarative definition of containers based on container blueprints for the setup of the included systems and software, and the declarative definition of compositions of containers. Taking, in addition, the computing resources into account that are provided for the deployment of a composition, we are able to reach reproducible creation for the overall distributed benchmark environments.

Figure 1 shows an overview of our architecture. On the one hand, we describe the lifecycle, similar to a state machine, as a set of states and transitions that a benchmark environment passes trough from initialization to termination. On the other hand, we describe the artefacts that are created during the lifecycle by users or the benchmark tooling and describe the technologies that accompany each state. Beside the technologies required for the environment, the transitions between states can and partially should also be supported by technology in order to enable the increased deployment automation. With every state, the set of required technologies is extended and, thereby, details of the implementations are narrowed down.

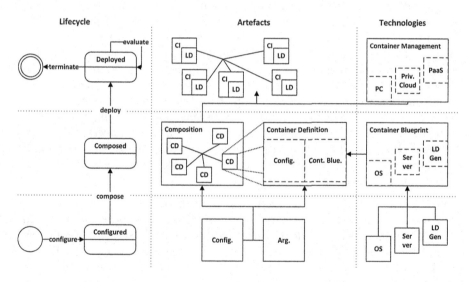

Fig. 1. Architecture – Lifecycle, Artefacts, and Technologies

Configuration. The *configure* transition leads from the initial state to the *Configured* state. In this state, an implementation of the architecture is still independent from any specific virtualization technology or deployment platform. The *technologies* accompanying this state are technologies that are indirectly configured, e.g., the operating system, servers, or the Linked Data generator. These technologies are setup in a declarative way by container blueprints and optionally pre-assembled as container images, that again reduce the potential variables in the overall system. The *artefacts* generated or determined during the *configure* transition are limited to simple configuration files or program arguments. These configurations contain the essence of parameters that are required to define the characteristics of the distributed benchmark environment, which is then derived from these parameters in the following transitions.

Composition. The *compose* transition leads from the *Configured* state to the *Composed* state. In this state, an implementation of the architecture is specific to one virtualization technology but independent from deployment platforms. The *technologies* accompanying this state are container-based virtualization technologies. These enable us to define in a declarative way: (1) the detailed setups of systems in the form of container blueprints, including their operating systems, applications, processes, and optional parametrization; and (2) the detailed setups of distributed applications in the form of compositions, including the containers that should be part of the application, their parameters, and the network connections between containers and to external networks. An implementation of the architecture must provide container blueprints for the Linked Data generation and provisioning. In addition, the *compose* transition should be supported by composition generators to expand configurations to compositions. During the *compose* transition, we expand the benchmark-specific parameters of the configuration and create as *artefact* a virtualization technology-specific composition, based on the provided container blueprints. By utilizing the container blueprints, in combination with the composition generator, we reduce the set of parameters in the overall benchmark environment to the aforementioned set of configuration parameters. These parametrize the composition generator and, thereby, the container instances and the data generators as well. For instance, during this transition we determine the correct amount of containers and the size of generated datasets to fit scaling parameters, or container-specific parameters to generate correct graphs and generate correct links to other graphs based on the host configurations in the composition.

Deployment. The *deploy* transition leads from the *Composed* to the *Deployed* state. In this state, an implementation of the architecture is specific to one virtualization technology and one deployment platform. The *technologies* accompanying this state are the container management of the container-based virtualization and the computing resources provided to this management. An important characteristic of these virtualization technologies is the platform-agnostic handling of standalone containers and compositions, i.e., they can be deployed on local computers, private clouds, or on Platform as a Service (PaaS) solutions. Our strategy behind decoupling the composition and the deployment platform is as follows: the more broadly available the deployment platforms used for the benchmarking are, the more comparable the benchmarking results will be. While custom local computing power or private clouds may be sufficient for development and test purposes, comparative benchmarking should be executed on detailed specified hardware settings and, ideally, run on broadly available computing resources. In this way, the benchmarking results become not only more comparable but also more reproducible. The *artefacts* generated during the *deploy* transition are container instances, setup according to the composition, which run on the chosen deployment platform, are connected by a network, provide correctly interlinked graphs, and expose this Linked Data to clients. At this state, the benchmark environment is ready for use.

Evaluation. Having reached the *Deployed* state, we indicate that the benchmark environment is in use via the *evaluate* transition. Experiments can repeatedly be performed to conduct evaluations by utilizing the deployed environment. After the completion of the experiments, the benchmark environment is obsolete and can be stopped (as indicated by the *terminate* transition to the final state). Furthermore, any reserved computing resources may be released.

4 Distributed LUBM

With the Distributed LUBM (DLUBM), we provide a specific implementation of the architecture introduced in Sect. 3 that realises a truly distributed and interlinked Linked Data benchmarking environment. The implementation of the data generator component, which we introduce below in more detail, is derived from the well-known LUBM [15] benchmark. We support the same scaling and semantically same data but provide new contributions in the form of new functionalities. In particular, we enable the generation of datasets of different size and granularity as well as the correct interlinking of contained graphs with respect to the actual deployment. Furthermore, we support virtualization based on the Docker[1] ecosystem by providing DLUBM container blueprints, container images, and the generation of compositions that enable the deployment of distributed multi-container DLUBM instances on standalone Docker or computing clusters managed by Docker Swarm on all supported platforms. Our DLUBM implementation is configurable by a small set of parameters, deployable with a high degree of automation on local computers, private clouds, or PaaS computing resources, and creates reproducible and truly distributed Linked Data environments for benchmarking.

4.1 Structure and Parameters

Our DLUBM benchmark environment is based on the original LUBM scenario, which consists of information about universities, departments of universities, and detailed information about entities involved in departments such as professors, students, or courses, as well as interrelations between these entities. We support different granularity levels to denote the structural level of detail at which graphs about entities are provided by separate hosts. For example, the granularity level *department*, visualized in Fig. 2, will lead to separate hosts that provide one or more department graphs as well as hosts for one or more university graphs and a global graph. The number of universities and departments per host is parametrized. In contrast, the granularity level *global* will lead to one host that provides

---

[1] https://www.docker.com/.

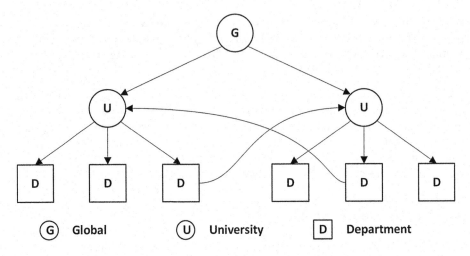

Fig. 2. DLUBM – Host Structure and Interlinking of Granularity "department" [21]

all graphs. Currently, we support the levels *global, university,* and *department*. The global graph contains links to all universities, the university graphs contain links to all departments of the corresponding university, and the department graphs contain information about all other entities that partially contain again links to other universities.

Our DLUBM benchmark environment can be configured with a small set of parameters. We give in Fig. 3 an overview of all important parameters. In particular, we can describe the characteristics of a DLUBM instance as function of the global parameters:

Parameter	Scope	Description
Seed	global	Seed of the data generator
Granularity	global	Structural level with distinct hosts
University Offset	global	Offset of the first university to generate
University Amount	global	Amount of universities to generate
University Limit	global	Limit of universities per host
Department Limit	global	Limit of departments per host
Host Depth	local	Structural level of the host
Host University Offset	local	Offset of the first university to generated
Host University Amount	local	Amount of universities to generate
Host Department Offset	local	Offset of the first department to generated
Host Department Amount	local	Amount of departments to generate
Ontology	deploy	URI for links to the ontology
University Template	deploy	URI template for links to universities
Department Template	deploy	URI template for links to departments

Fig. 3. DLUBM – Parameters

```
DLUBM Instance = DLUBM(seed, granularity, university offset,
university amount, university limit, department limit)
```

These parameters must be provided to the composition generator that derives the local parameters (Fig. 3) for every container definition during the compose transition (c.f., Sect. 3.2). The generator adds a container definition for a fixed ontology URI to the composition and a single global container definition, or global and university container definitions, or container definitions at all three levels, depending on the global granularity level parameter. Every container is parametrized by the local parameters to generate and provide its part of the overall dataset. The last type of parameters is relevant for deployment, but not for the characteristics of the DLUBM instance. In particular, the ontology URI and templates for URIs to universities and departments must be provided in the container definition. During initialization of a container the Linked Data generator replaces template variables, e.g., path parameters for university and department indices, and, thereby, generates resolvable links. The creation of these templates is taken over by the composition generator.

4.2   Software Components

We developed different software components[2], which, integrated with existing technology, implement our architecture. We provide a data generator, a container blueprint for data provisioning containers, a container image based on the blueprint to ease deployment, a composition and query generator, and utilize different deployment helpers, as described in the following.

Linked Data Generator. We provide the DLUBM data generator for generating distinct but interlinked graphs. The data generator is an extension of the LUBM artificial data generator[3] by Vesse and, originally, by Guo [15]. Our extension enables the data generator to (1) generate a subset of graphs of the overall datasets on different hosts, (2) generate only a specific granularity level, and (3) generate correct links to other graphs in the dataset by utilizing URI templates. The generator still supports the generation of original LUBM data.

Container Blueprint. We provide a Dockerfile, i.e., the aforementioned container blueprints, as well as pre-build Docker images[4] for the DLUBM data generator. The Dockerfile defines in a declarative way the setup of an operating system, the generation of documents with graph representations, and a web server to provide these documents. With the Docker image a pre-build version of the defined system exists, ready to use for parametrized instantiation of containers in the Docker ecosystem.

Composition Generator. We provide a generator for Docker compose files, i.e., the aforementioned compositions, that is capable to expand DLUBM configurations to compositions of containers that represent the complete derived

---

[2] https://github.com/fekepp/dlubm.
[3] https://github.com/rvesse/lubm-uba.
[4] https://hub.docker.com/r/fekepp/dlubm.

```
...
u1_dl0_ul1_uo0_ua5_gu_s0:
  deploy:
    labels: [traefik.docker.network=traefik_proxy,
      traefik.port=80, 'traefik.frontend.rule=Host:u1.dlubm.ddns.me']
  environment: ['DLUBM_ONTOLOGY=http://o.dlubm.ddns.me/univ-bench.owl', DLUBM_SEED=0,
    DLUBM_GRANULARITY=UNIVERSITY, DLUBM_UNIVERSITY_AMOUNT=5, DLUBM_UNIVERSITY_OFFSET=0,
    DLUBM_UNIVERSITY_LIMIT=1, 'DLUBM_UNIVERSITY_TEMPLATE=http://
    u{UNIVERSITY_INDEX}.dlubm.ddns.me/u#',
    DLUBM_DEPARTMENT_LIMIT=0, 'DLUBM_DEPARTMENT_TEMPLATE=http://
    u{UNIVERSITY_INDEX}.dlubm.ddns.me/d{DEPARTMENT_INDEX}#',
    DLUBM_HOST_DEPTH=UNIVERSITY, DLUBM_HOST_UNIVERSITY_AMOUNT=1,
    DLUBM_HOST_UNIVERSITY_OFFSET=1,
    DLUBM_HOST_DEPARTMENT_AMOUNT=0, DLUBM_HOST_DEPARTMENT_OFFSET=0]
  image: fekepp/dlubm:latest
  networks: [proxy]
...
```

Fig. 4. DLUBM Composition – Example of a Parametrized Container Declaration

DLUBM instance. In Fig. 4, we show an excerpt of a DLUBM Docker compose file that contains as an example the container definition of one DLUBM container instance. The definition includes deployment-specific parameters, e.g., the redirection rules for a reverse proxy, as well as previous presented DLUBM-specific parameters.

Query Generator. We also support the generation of LUBM SPARQL queries that are aligned with the compositions. Since the URIs in generated graphs are deployment-specific, queries must be adjusted to the deployment. Generated queries provide the same semantics as the original LUBM queries but are tailored to the actual DLUBM instance in terms of URIs.

Deployment Helpers. The Docker ecosystem provides several components that ease the management and deployment of large-scale container environments, which we mention but refer to their documentation for further details. We use Docker Compose[5] for deployment of compositions to local Docker instances. We use Docker Stacks for deployment of compositions to multiple Docker instances via a cluster managed in a Docker Swarm[6]. In the latter case, the distribution and lifecycle of containers on multiple computing resources is dynamically managed. We use Traefik[7], a reverse proxy with support for local Docker instances as well as Docker Swarm environments. Hereby, the assignment of DNS records to containers and the corresponding routing of requests to these records is part of the composition and thus is automatically managed.

5   Evaluation

We evaluate our approach by utilizing the architecture and the DLUBM implementation for measuring the performance of a Linked Data query engine.

---

[5] https://docs.docker.com/compose/.

[6] https://docs.docker.com/engine/swarm/.

[7] https://traefik.io/.

In the following, we first introduce the query engine, then describe the experimental setup, and finally discuss the results.

5.1 Linked Data Query Engine

For our experiments, we evaluate Linked Data-Fu (LD-Fu) [23], a combination of SPARQL query evaluation engine and interpreter of Notation3 (N3) rule programs. LD-Fu is accompanied by the corresponding semantics for the interpretation of rules and supports separate handling of specific ontologies, e.g., the HTTP vocabulary[8] for execution of HTTP requests, or the Math ontology for providing built-in mathematical functions. LD-Fu maintains during interpreter runs an internal RDF graph that is enriched by RDF triples given as part of rule programs, derived by the interpretation of rules, or requested via HTTP. The internal RDF graph can be queried via SPARQL. Thereby, LD-Fu enables requests to resources, link following, as well as querying of requested, aggregated, and derived data.

5.2 Experiments

A documentation of our experiments, including guidlines, highly-automated deployment, experiments, and results, is available online[9].

Deployment Platform. Our experiments have been conducted on computing instances in the Elastic Compute Cloud (EC2)[10] of Amazon Web Services (AWS)[11]. In particular, we used the following setup of 20 instances[12]:

While the instance for experiments is provisioned with Ubuntu, all Docker Swarm instances use the Docker-centric operating system RancherOS. The master instance and all worker instances form a Docker Swarm that manages the provisioning of containers. At the master instance, a container including a reverse proxy handles the mapping of incoming requests at a dynamically allocated DNS entry to containers managed by the Docker Swarm. Thereby, the assignment of containers to hostnames can be declared in the composition. For the experiments, we restrict DLUBM containers to run only at worker instances (Fig. 5).

Linked Data Environment. For comparability with the original LUBM benchmark results, we use two instances of our DLUBM environment with similar characteristics: DLUBM(0, DEPARTMENT, 0, 1, 1, 1) leading to 18 hosts and DLUBM(0, DEPARTMENT, 0, 5, 1, 1) leading to 100 hosts. Both instances are configured at department granularity for the generation of multiple hosts, i.e., with one host per university and per department. Thereby, we maintain comparability of the query results with the results of the work on the non-distributed

---

[8] https://www.w3.org/TR/HTTP-in-RDF10.

[9] https://github.com/fekepp/dlubm-ldfu-eval.

[10] https://aws.amazon.com/ec2.

[11] https://aws.amazon.com.

[12] The amount of 20 instances still allows usage of our documented experiments without requesting a limit increase for running more instances on AWS EC2.

Amount	Type	Operating System	Description
1	m4.xlarge	Ubuntu 16.04	Experiments instance
1	m4.xlarge	RancherOS 1.0.1	Docker Swarm master instance
18	t2.small	RancherOS 1.0.1	Docker Swarm worker instances

Fig. 5. Experiments – Deployment Platform at AWS EC2

LUBM [15], that have been based among others on querying one and five universities.

Queries. For our measurements, we use the original 14 SPARQL queries of the LUBM benchmark, adjusted by our composition generator with the correct URIs for our DLUBM environment. In addition, we added the "DISTINCT" modifier to eliminate duplicate results.

Rules. The LD-Fu interpreter is capable of evaluating N3 rules but without advanced built-in inferencing for the Resource Description Framework Schema (RDFS) or the OWL Web Ontology Language (OWL). We, therefore, add in separate experiments entailment rule sets with RDFS and OWL-LD[13] entailment rules. In addition, and to emphasize the modular approach of LD-Fu, we add two custom rules in an extra run to the OWL-LD rule set, that derive most of the still missing results for the LUBM queries.

5.3 Measurements

In the following, we illustrate the results of the query evaluation as well as the different time and inferencing metrics, measured during the evaluation of LUBM queries on both DLUBM configurations with LD-Fu.

Query Results. We measured the evaluation of DLUBM queries without rule set, with every rule set, and with the combinations of rule sets. Due to space constrains, we show in Fig. 6 only the query results for DLUBM(0, DEPARTMENT, 0, 1, 1, 1) without rule sets, with RDFS, OWL-LD, and with OWL-LD + C (extended by the two custom rules). Query 1, 3, and 14, contain simple selections and can be handled without entailment rules. For all other queries, we see an improvement with RDFS, and further for query 11 with OWL-LD, which supersedes the RDFS result. Finally, our two custom rules, that handle inferencing related to students and head of departments, complement the OWL-LD rule set in most cases.

Times. During the experiments, we measured the different times for both configurations per query. In Fig. 7, we visualize from top to bottom – the overall runtime for query evaluation, the average request time for successful requests, and the average request time for failing requests; on the left side – for the configuration with 18 hosts; on the right side – for the configuration with 100 hosts.

---

[13] http://semanticweb.org/OWLLD/.

Q	A	None		RDFS		OWL-LD		OWL-LD + C	
1	4	4	100%	4	100%	4	100%	4	100%
2	0	0	100%	0	100%	0	100%	0	100%
3	6	6	100%	6	100%	6	100%	6	100%
4	34	0	0%	34	100%	34	100%	34	100%
5	719	0	0%	719	100%	719	100%	719	100%
6	7790	0	0%	5916	76%	5916	76%	7790	100%
7	67	0	0%	59	88%	59	88%	67	100%
8	7790	0	0%	5916	76%	5916	76%	7790	100%
9	208	0	0%	103	50%	103	50%	208	100%
10	4	0	0%	0	0%	0	0%	4	100%
11	224	0	0%	0	0%	224	100%	224	100%
12	15	0	0%	0	0%	0	0%	15	100%
13	1	0	0%	0	0%	0	0%	0	0%
14	5916	5916	100%	5916	100%	5916	100%	5916	100%

Fig. 6. Experiments – Query Results and Completeness

Failing requests appear due to a hard-coded limitation within the LUBM data generator, that generates links to universities with an index between 0 and 1000. These universities must not necessary exist, in particular not for low university numbers as in our evaluation. Therefore, diagrams 5 and 6 represent the average request times for several failing requests.

The diagrams show in essence two noticeable details. First, the performance of a Linked Data query engine is not only dependent on the performance of the engine itself, but has also to cope with a potential unpredictable network environment, which is out of its control. This is visible in the failing requests (diagrams 5 and 6). HTTP requests are, as far as possible, executed by LD-Fu in parallel and are, therefore, considerably faster than in serial execution. However, if some requests are significantly delayed, e.g., indicated by the increased average request times, for example, during evaluation of query 1, 13, and 14 in diagram 5, the engine has to wait for these requests to fail or to return with valuable content. The delay caused by these requests is reflected in the overall runtime for the respective queries in the runtime diagrams 1 and 2.

Second, in runtime diagram 2, showing query evaluation on the configuration with 100 hosts, we can clearly see the impact of entailment rules. Already with RDFS entailment rules, the runtime is significantly higher, but increases for OWL-LD entailment rules many times over. Unclear, however, is the correlation with the average times for successful requests (diagram 4). Either the connection handling is slowed down by the processing power consumed for evaluation of entailment rules, or the processing of returned payload is slowed down for the same reason.

Derived Triples. The triples, which have been derived by the interpreter in the different settings, are listed in the following. For DLUBM(0, DEPARTMENT, 0, 1, 1, 1), the LD-Fu engine derived 0 triples (No Rule Set), 115860 triples

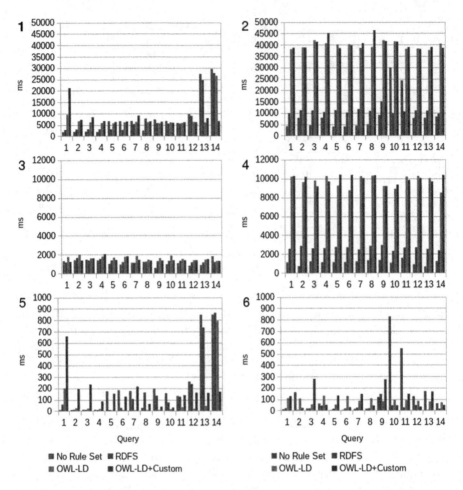

Fig. 7. L/R: 18/100 Hosts; T/B: Runtime, Avg. Req. Time Success/Failure

(RDFS), 182949 triples (OWL-LD), and 184939 triples (OWL-LD + Custom). For DLUBM(0, DEPARTMENT, 0, 5, 1, 1), the LD-Fu engine derived 0 triples (No Rule Set), 700552 triples (RDFS), 1115161 triples (OWL-LD), and 1127154 triples (OWL-LD + Custom).

6   Conclusion

High-quality benchmarks and evaluation results are essential not only for measuring the suitability of a certain technology or setup but also for pushing the development of better approaches. We contribute to the evolution of Linked Data technologies by introducing the Distributed LUBM for creating reproducible and distributed benchmark environments for Linked Data. While we extend the LUBM benchmark with features for generating Linked Data, the approach itself is independent from a specific data generator, as long as the generator provides means for generating distinct graphs and provides means for adapting the links between these graphs to interlink the overall dataset. Substitution of components, e.g., the data generator, is possible due to our architectural approach as well as the introduced design requirements, which clearly separate the involved artefacts, the used technologies, and the current state of the system. We demonstrate the practical use of DLUBM by evaluating a Linked Data query engine via DLUBM with the original LUBM queries and different entailment rule sets.

References

1. Abele, A., McCrae, J.P., Buitelaar, P., Jentzsch, A., Cyganiak, R.: Linking Open Data cloud diagram, March 2017, http://lod-cloud.net/
2. Aluç, G., Hartig, O., Özsu, M.T., Daudjee, K.: Diversified stress testing of RDF data management systems. In: Mika, P., Tudorache, T., Bernstein, A., Welty, C., Knoblock, C., Vrandečić, D., Groth, P., Noy, N., Janowicz, K., Goble, C. (eds.) ISWC 2014. LNCS, vol. 8796, pp. 197–212. Springer, Cham (2014). doi:10.1007/978-3-319-11964-9_13
3. Angles, R., Boncz, P.A., Larriba-Pey, J.L., Fundulaki, I., Neumann, T., Erling, O., Neubauer, P., Martinez-Bazan, N., Kotsev, V., Toma, I: The Linked data benchmark council: a graph and RDF industry benchmarking effort. SIGMOD Rec. (2014)
4. Armstrong, T.G., Ponnekanti, V., Borthakur, D., Callaghan, M.: LinkBench: a database benchmark based on the facebook social graph. In: Proceedings of the SIGMOD International Conference on Management of Data (2013)
5. Atzori, L., Iera, A., Morabito, G.: The internet of things: a survey. Comput. Netw. (2010)
6. Bagan, G., Bonifati, A., Ciucanu, R., Fletcher, G.H.L., Lemay, A., Advokaat, N.: gMark: schema-driven generation of graphs and queries. IEEE Trans. Knowl. Data Eng. (2016)
7. Barahmand, S., Ghandeharizadeh, S.: BG: a benchmark to evaluate interactive social networking actions. In: Proceedings of the Conference on Innovative Data Systems Research (2013)
8. Bizer, C., Heath, T., Berners-Lee, T.: Linked data - the story so far. Int. J. Semant. Web Inf. Syst. (2009)
9. Bizer, C., Schultz, A.: The Berlin SPARQL benchmark. Int. J. Semant. Web Inf. Syst. (2009)
10. Blum, D., Cohen, S.: Grr: generating random RDF. In: Antoniou, G., Grobelnik, M., Simperl, E., Parsia, B., Plexousakis, D., De Leenheer, P., Pan, J. (eds.) ESWC 2011. LNCS, vol. 6644, pp. 16–30. Springer, Heidelberg (2011). doi:10.1007/978-3-642-21064-8_2
11. Dominguez-Sal, D., Martinez-Bazan, N., Muntes-Mulero, V., Baleta, P., Larriba-Pey, J.L.: A discussion on the design of graph database benchmarks. In: Nambiar, R., Poess, M. (eds.) TPCTC 2010. LNCS, vol. 6417, pp. 25–40. Springer, Heidelberg (2010). doi:10.1007/978-3-642-18206-8_3
12. Duan, S., Kementsietsidis, A., Srinivas, K., Udrea, O.: Apples and oranges: a comparison of RDF benchmarks and real RDF datasets. In: Proceedings of the SIGMOD International Conference on Management of Data. ACM (2011)

13. Duquennoy, S., Grimaud, G., Vandewalle, J.J.: The Web of Things: interconnecting devices with high usability and performance. In: Proceedings of the International Conference on Embedded Software and Systems (2009)
14. Erling, O., Averbuch, A., Larriba-Pey, J., Chafi, H., Gubichev, A., Prat, A., Pham, M.D., Boncz, P.: The LDBC social network benchmark: Interactive workload. In: Proceedings of the SIGMOD International Conference on Management of Data (2015)
15. Guo, Y., Pan, Z., Heflin, J.: LUBM: A benchmark for OWL knowledge base systems. Web Semantics: Science, Services and Agents on the World Wide Web (2005)
16. Harth, A., Hose, K., Karnstedt, M., Polleres, A., Sattler, K.U., Umbrich, J.: Data summaries for on-demand queries over Linked Data. In: Proceedings of the International Conference on World Wide Web (2010)
17. Hartig, O., Bizer, C., Freytag, J.-C.: Executing SPARQL queries over the web of linked data. In: Bernstein, A., Karger, D.R., Heath, T., Feigenbaum, L., Maynard, D., Motta, E., Thirunarayan, K. (eds.) ISWC 2009. LNCS, vol. 5823, pp. 293–309. Springer, Heidelberg (2009). doi:10.1007/978-3-642-04930-9_19
18. Huppler, K.: The art of building a good benchmark. In: Proceedings of the TPC Technology Conference on Performance Evaluation & Benchmarking (2009)
19. Jara, A.J., Olivieri, A.C., Bocchi, Y., Jung, M., Kastner, W., Skarmeta, A.F.: Semantic web of things: an analysis of the application semantics for the IoT moving towards the IoT convergence. Int. J. Web Grid Serv. (2014)
20. Joshi, A.K., Hitzler, P., Dong, G.: LinkGen: multipurpose linked data generator. In: Groth, P., Simperl, E., Gray, A., Sabou, M., Krötzsch, M., Lecue, F., Flöck, F., Gil, Y. (eds.) ISWC 2016. LNCS, vol. 9982, pp. 113–121. Springer, Cham (2016). doi:10.1007/978-3-319-46547-0_12
21. Keppmann, F.L., Harth, A.: Adaptable interfaces, interactions, and processing for linked data platform components. In: Proceedings of the SEMANTiCS Conference (2017)
22. Schmidt, M., Hornung, T., Lausen, G., Pinkel, C.: SP2Bench: a SPARQL performance benchmark. In: Proceedings of the International Conference on Data Engineering (2009)
23. Stadtmüller, S., Speiser, S., Harth, A., Studer, R.: Data-fu: a language and an interpreter for interaction with read/write Linked Data. In: Proceedings of the International World Wide Web Conference (2013)
24. Weithöner, T., Liebig, T., Luther, M., Böhm, S.: What's wrong with OWL benchmarks? In: Proceedings of the International Workshop on Scalable Semantic Web Knowledge Base Systems (2006)

Norwegian State of Estate Report as Linked Open Data

Ling Shi[1(✉)], Dina Sukhobok[2], Nikolay Nikolov[2],
and Dumitru Roman[2]

[1] Statsbygg, Pb. 8106 Dep, 0032 Oslo, Norway
ling.shi@statsbygg.no
[2] SINTEF, Pb. 124 Blindern, 0314 Oslo, Norway
{dina.sukhobok,nikolay.nikolov,
dumitru.roman}@sintef.no

Abstract. This paper presents the Norwegian State of Estate (SoE) dataset containing data about real estates owned by the central government in Norway. The dataset is produced by integrating cross-domain government datasets including data from sources such as the Norwegian business entity register, cadastral system, building accessibility register and the previous SoE report. The dataset is made available as Linked Data. The Linked Data generation process includes data acquisition, cleaning, transformation, annotation, publishing, augmentation and interlinking the annotated data as well as quality assessment of the interlinked datasets. The dataset is published under the Norwegian License for Open Government Data (NLOD) and serves as a reference point for applications using data on central government real estates, such as generation of the SoE report, searching properties suitable for asylum reception centres, risk assessment for state-owned buildings or a public building application for visitors.

Keywords: State-owned real estates · Linked Data · Open Government Data · RDF

1 Introduction

One significant part of public spending is on buildings and properties needed by public administrations. A State of Estate (SoE) report[1] – a report containing integrated data on state-owned real estates[2] can help the governments use them more effectively[3]. In Norway, such a report is published as an attachment[4] to the proposed parliamentary resolution No. 1 every four years by Statsbygg[5] on behalf of the Ministry of Local

---

[1] An example of such a report from the UK government can be found at: https://www.gov.uk/government/uploads/system/uploads/attachment_data/file/200448/SOFTE2012_final.pdf.
[2] Real estates can also be called real properties, properties or cadastral parcels if the properties are registered at the national cadastral system.
[3] https://www.theguardian.com/news/datablog/2013/may/21/downsizing-government-estate.
[4] https://www.regjeringen.no/contentassets/f4346335264c4f8495bc559482428908/no/sved/stateigedom.pdf.
[5] http://www.statsbygg.no/Om-Statsbygg/About-Statsbygg/.

© Springer International Publishing AG 2017
H. Panetto et al. (Eds.): OTM 2017 Conferences, Part II, LNCS 10574, pp. 445–462, 2017.
https://doi.org/10.1007/978-3-319-69459-7_30

Government and Modernization[6]. The data collection and quality control process has historically been resource demanding and error prone and the result was static and did not reflect the changes after the report was published. Though the report is available online as a PDF file, the data from the report is not easily reusable because of the data format, quality, and lack of semantic descriptions of the complex real property domain. A State of Estate (SoE) business case was introduced in [1] to carry out the reporting task in a more effective way by publishing and integrating government data from both open and proprietary sources: the Norwegian business entity register, cadastral system, building accessibility register and the previous SoE report. Sharing the SoE dataset in a Linked Data format enables data reuse, opens up possibilities for using the SoE data in innovative ways, and helps increase transparency in the government administration.

Our contribution in this paper is the SoE dataset together with the publication process of state-owned real estates in Norway as Linked Open Data – a result of publishing and integrating several cross-domain government datasets in RDF[7]. Complex queries can be run on multiple interlinked source datasets to generate lists of inconsistencies between them. The lists are used to improve the data quality in the source systems and, afterwards, the data quality in the resulting SoE dataset is also improved when the source datasets are updated and republished. Publishing the resulting SoE dataset as Linked Data avoids manual data collection, simplifies the process of SoE report generation, and also promotes innovative services such as risk assessment of state-owned buildings by integrating the dataset with natural hazards datasets.

The structure of the paper is as follows. Section 2 introduces the source datasets. The SoE Linked Data generation process is described in Sect. 3. Section 4 provides a description of the resulting SoE dataset. Section 5 presents the usage examples of the published dataset. Related work is discussed in Sect. 6. Section 7 summarizes the paper and outlines further work.

2 Source Datasets

This section presents the details on the source datasets and the major challenges on working with and processing the datasets. The following datasets from different sources serve as input for the generation of the SoE dataset:

- The central government organization dataset – a subset of data from the Norwegian Business Entity Register administrated by the Brønnøysund Register Centre[8];
- The cadastral datasets – a subset of data from the Norwegian Cadastral System[9] administrated by the Norwegian Mapping Authority;

---

[6] https://www.regjeringen.no/en/dep/kmd/id504/.

[7] https://www.w3.org/RDF/.

[8] https://www.brreg.no/home/.

[9] http://www.kartverket.no/en/Land-Registry-and-Cadastre/.

- The building accessibility dataset from the Building Accessibility Register[10] administrated by Statsbygg;
- The previous SoE dataset administrated by Statsbygg;
- The municipality boundaries dataset administrated by the Norwegian Mapping Authority.

The non-geospatial datasets are prepared by dataset providers in tabular format and the geospatial datasets as shape files[11]. Both tabular format and shape files are supported by the RDF conversion tool used for the generation of the SoE data.

2.1 The Central Government Organizations Dataset

The Norwegian Business Entity Register provides a complete dataset that covers all the public and private business entities in Norway. The dataset is assigned the Norwegian License for Open Government Data (NLOD)[12] and available through a Web API on the Brønnøysund Register Centre's website[13] and the Norwegian government's open data sharing platform[14].

The central government organizations dataset is a subset of the complete list of organizations and, because of the structure of the data, there is no single attribute that can be used to extract all central government organizations. The top level central government organizations (e.g., the parliament, ministries) can be identified using a filter on organizational format that equals to "STAT". The subordinate organizations need to be extracted for each of the top level organizations by using the parent organization attribute. There exist some organizations that are not covered by the above iterative method and the organizations are added from a manual exception list which is a result of domain experts' evaluation of organizations from the previous SoE report compared to the current organization list. Moreover, the central government organizations dataset is subject to changes, though the change frequency is not often. Organizational change after a general election, fusions between organizations and privatization of government organizations and other reasons can have effect on the ownership relationship, which causes further difficulties when reconciling the data.

2.2 The Cadastral Datasets

The Norwegian cadastral data (including land ownership) are partially open and available as a map-based Web application[15] that presents detailed information of only one cadastral parcel at a time. This distribution of the data does not support programmatic access to cadastral data for all the state-owned real estates and it is not suitable for the purposes of the SoE report generation. The cadastral data are also

---

[10] https://byggforalle.no/uu/sok.html?&locale=en.

[11] https://en.wikipedia.org/wiki/Shapefile.

[12] https://data.norge.no/nlod/en/1.0.

[13] http://data.brreg.no/oppslag/enhetsregisteret/enheter.xhtml.

[14] http://hotell.difi.no/?dataset=brreg/enhetsregisteret.

[15] http://seeiendom.no.

available through a Web service or a database dump under subscription-based licensing. Thereby, the cadastral datasets suitable for programmatic access are essentially proprietary and closed for public access. The current data export approach is based on a cadastre database dump and returns four sub-datasets:

- The cadastral parcel ownership dataset for state-owned or state-leased cadastral parcels;
- The cadastral parcel geospatial dataset;
- The building dataset of buildings built on state-owned or state-leased cadastral parcels;
- The building geospatial dataset.

Table 1 contains sample records from the cadastral parcel ownership dataset. The organization number (*Org. No.*) is an attribute shared by the central government organization dataset. The *Municipality* is an attribute shared by the municipality boundaries dataset. The *Cadastral ID* is an attribute shared by the other cadastral sub-datasets.

The four cadastral sub-datasets are subject to change due to ownership changes and other kinds of changes related to cadastral parcels and buildings.

2.3   Building Accessibility Dataset

The Building Accessibility Register covers many aspects of accessibility[16] of public buildings at different levels of a construction: building, floor and room. The building accessibility data is open and available through a Web application[17] which only returns the accessibility data of one building by each search. The dataset is a simplified subset of the building accessibility data and provides primarily five indicators at the building level chosen by building accessibility experts. These indicators are *hasStepFreeMainAccess*, *hasStepFreeSideAccess*, *hasElevator*, *hasHandicapToilet* and *hasHandicapParking*. The dataset is generated through an API-based export from the building accessibility database.

2.4   The Previous State of Estate Report Dataset

The previous State of Estate report dataset was dated 2013–2014 and is open and available as an online PDF file with limited possibility of data manipulation and further processing. The data are also stored in a proprietary relational system at Statsbygg, the access to which was granted for the purposes of the SoE report. Only Statsbygg's properties and buildings are updated after the report was generated and Statsbygg administrates less than 40% of the state-owned buildings. Therefore, this dataset is only used as an alternative reference data source for quality check in this business case as exemplified in Sect. 3.6. No further updates need to be provided and supported for this dataset.

---

[16] https://en.wikipedia.org/wiki/Accessibility.

[17] https://byggforalle.no/uu/sok.html?&locale=en.

Table 1. Example records from the cadastral parcel ownership dataset

Name	ROLE	Cadastral ID	Org. no.	Municipality	Date From	Date To
Statens Vegvesen	H- Hjemmelshaver	0214/107/402/0	971032081	0214	22.10.2014	01.01.1753
Statens Vegvesen	H- Hjemmelshaver	0214/112/2/0	971032081	0214	23.02.2006	01.01.1753
Statens Vegvesen	H- Hjemmelshaver	0214/121/9/0	971032081	0214	23.02.2006	01.01.1753
Norsk Institutt for Skog Og Landsk	F - Fester	0214/42/1/78	970167641	0214	05.02.1953	01.01.1753

2.5 The Municipality Boundaries Dataset

The Municipality Boundaries geospatial dataset is one of the open datasets[18] provided by Norwegian Mapping Authority and is downloadable[19] in SOSI[20] format. The downloaded dataset is then converted to a shape format. The dataset covers the whole Norway and contains a national identifier and name of each municipality in addition to the geometry of the municipality's boundary as polygons. In comparison to the other input datasets, the municipality boundaries are relatively stable, but the municipalities could be changed due to administrative reforms.

2.6 Challenges in Integrating the Source Datasets

Open government data's burdens are inherent to large-scale distributed data integration, collective data manipulation and transparent data consumption [2]. Though the data providers are the most authorized actors in their domains in the public sector, none of the datasets are 100% accurate and consistent. The method to extract central government organizations described in Sect. 2.1 does not cover the whole scope of central government organizations because some organizations are not required to be registered using the specified method and need to be identified and added to the list manually.

There also exist inconsistencies between the source systems which increase the complexity in data integration and impact the quality of the data in the resulting SoE dataset. A common reason for inconsistencies is due to different domain focus. Propman's (Statsbygg's property management system) domain focus is real estate and building management for Statsbygg while the cadastral system focuses on legal rights and obligations on cadastral parcels and buildings in Norway. State-owned properties abroad are registered in Propman, but they are not available in the Norwegian cadastral system. Buildings with areas of less than 15 square meters are registered in Propman, but it is not mandatory to register them in the Norwegian cadastral system and they do not have corresponding national building identifiers. Furthermore, a physical building can be extended several times and the extensions are assigned new cadastral building numbers and therefore it can be connected to more than one national cadastral building number though Propman treats it as only one building.

---

[18] http://www.kartverket.no/en/data/Open-and-Free-geospatial-data-from-Norway/.

[19] http://data.kartverket.no/download/content/geodataprodukter.

[20] http://www.kartverket.no/en/geodataarbeid/SOSI-Standard-in-English/SOSI-Standard-in-English/.

Additionally, inconsistencies between systems are often caused by delayed or missing registrations or updates. The ownership change between organizations in the public sector is not always officially registered either to save the registration cost or due to lack of compulsory reporting routines. After a fusion between organizations the business entity register is updated while the previous organization number and name may still be registered in the cadastral system as a real rights holder. There are also examples of old ministries or organizations from earlier government periods registered as owners in the cadastre. Though all cadastral parcels in the previous SoE report have registered national cadastral parcel identifiers, the registration is manual and sometimes includes invalid values. More than 70% of buildings in the previous SoE report lack of national cadastral building identifiers because this attribute was not mandatory.

The above challenges will be addressed and discussed in Sects. 3.1 and 3.6 to improve the data quality of both the source datasets and the resulting dataset.

3   Linked Data Generation Process

The process of generating Linked Data for state-owned properties is shown in Fig. 1. The data acquisition step collects and prepares the datasets from multiple cross-domain sources as described in Sect. 2. Data cleaning is introduced in Sect. 3.1, conversion to RDF is described in Sect. 3.2, data augmentation using SPARQL CONSTRUCT[21] queries is detailed in Sect. 3.3, data interlinking is explained in Sect. 3.4, dataset publishing for the source datasets and result dataset is presented in Sect. 3.5, and data quality assessment on the result dataset is described in Sect. 3.6. DataGraft[22] [3–5] – a cloud-based platform for data cleaning and Linked Data generation – has been used for generating the Norwegian SoE dataset.

3.1   Data Cleaning

The data cleaning step aims to remove syntactic and/or semantic errors in the source datasets by following the tabular data cleaning approach described in [6]. Examples of data cleaning and preparation tasks include:

- The source datasets inherits the number and time-date formatting from their original systems which are Norwegian in this case. The Norwegian decimal separator is comma and it is replaced by point so that the dataset conforms to the decimal formatting used in the RDF conversion tool in DataGraft (e.g., the decimal 300,5 is replaced by 300.5).
- The cleaning process also recognizes and unifies null values for attributes with null values or similar (e.g., converting date values from "", "0", "101", etc., to the unified conventional null date value "17530101").
- Not all records in the previous SoE report dataset have valid cadastral parcel identifier values and many records lack cadastral building identifiers (ref. Sect. 2.6).

---

[21] https://www.w3.org/TR/rdf-sparql-query/#construct.

[22] https://datagraft.io/.

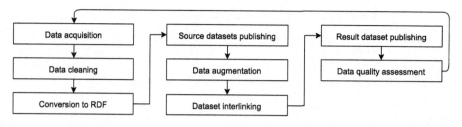

Fig. 1. SoE Linked Data generation process

A multistep procedure was developed to assign unique identifiers for cadastral parcels and buildings to handle the situation of missing or non-valid national identifiers. The national cadastral parcel identifier is assigned as unique identifier in the first place if it is available; otherwise a unique identifier is generated by concatenating municipality number, cadastral unit number, property unit number and leaseholder number.

- The default separator semicolon may also occur as part of the text in some of the text columns, which will cause wrong mapping of columns. The text columns containing semicolons are first identified and then cleaned by removing the extra semicolons.

3.2   Conversion to RDF

This section describes the process of converting the source datasets to RDF format. In addition to standard and established ontologies such as DBpedia-owl and schema.org, the proDataMarket ontology [7][23] is used as a central reference model for RDF transformation. The proDataMarket ontology reuses the Land Administration Domain Model (LADM) defined in ISO 19152:2012[24] standard and cadastral parcel concept specified by the European Union's INSPIRE data specifications[25].

The transformation scripts for the datasets are freely available as public transformations on DataGraft for registered users[26]. Table 2 shows the latest transformation scripts for each dataset with links to the actual transformation scripts provided as footnotes.

The transformation scripts are also published at a third party site[27]. The scripts are coded in Clojure[28] and include two parts: a data cleaning pipeline for data cleaning and preparation tasks; and a part for RDF mapping.

The central government organizations dataset reuses the established vocabulary from *schema.org* and *DBpedia-owl* to map and convert data to RDF. Attributes like the

---

[23] Available at http://vocabs.datagraft.net/ and the names start with proDataMarket.

[24] http://www.iso.org/iso/catalogue_detail.htm?csnumber=51206.

[25] http://inspire.ec.europa.eu/data-model/approved/r4618-ir/html/.

[26] The scripts are currently best visible using Chrome.

[27] https://zenodo.org/record/834300/files/SPARQLQueriesForLinkedDataGeneartion.pdf.

[28] https://clojure.org/.

Table 2. Transformation script files for the involved datasets

Input dataset	Source format	Size	Transformation scripts	Output, triples	Dataset is public
Central government organizations dataset	CSV	46 KB (217 records)	Central government organization transformation[a]	5049	Yes
Cadastral parcel ownership dataset	CSV	3.83 MB (18105 records)	Cadastral parcel ownership transformation[b]	559743	No
Cadastral parcel geospatial dataset	Shape file	4 GB (3673391 records)	Cadastral parcel geospatial dataset transformation[c]	~118 M	No
Building dataset	CSV	2.59 MB (23651 records)	Buildings built on the state-owned or state-leased cadastral parcels transformation[d]	454192	No
Building point geo-dataset	Shape file	288 MB (4637654 records)	The building geospatial dataset transformation[e]	23188270	No
Building accessibility dataset	CSV	81 KB (665 records)	Building accessibility dataset transformation[f]	19285	Yes
Previous SoE dataset	CSV	1,91 MB (11241 records)	The previous SoE dataset transformation[g]	663219	Yes
Municipality boundaries dataset	Shape file	24,1 MB (428 records)	The municipality boundaries dataset transformation[h]	3424	Yes

[a] https://datagraft.io/prodatamarket_publisher/transformations/the-central-government-organization-transformation-1010c106-2254-4c8b-9480-b88d47a41323

[b] https://datagraft.io/prodatamarket_publisher/transformations/the-cadastral-parcel-ownership-transformation

[c] https://datagraft.io/prodatamarket_publisher/transformations/the-cadastral-parcel-geospatial-dataset-transformation

[d] https://datagraft.io/prodatamarket_publisher/transformations/buildings-built-on-the-state-owned-or-state-leased-cadastral-parcels-transformation

[e] https://datagraft.io/prodatamarket_publisher/transformations/the-building-geospatial-dataset-transformation

[f] https://datagraft.io/prodatamarket_publisher/transformations/the-building-accessibility-dataset-transformation

[g] https://datagraft.io/prodatamarket_publisher/transformations/the-historical-soe-dataset-transformation

[h] https://datagraft.io/prodatamarket_publisher/transformations/the-munincipalitiy-boundaries-dataset-trans-formation

organization number, name, type, founding date and parent organization are mapped to *schema:leiCode, schema:legalName, dbpedia-owl:type, schema:foundingDate* and *schema:parentOrgnization* respectively.

The ProDataMarket ontology is the core ontology used to map and transform the four cadastral datasets, the building accessibility dataset and the previous SoE dataset. A cadastral parcel is mapped to *prodm-cad:CadastralParcel* and a building is mapped to *prodm-cad:Building*. Ownership relationships are mapped to *prodm-cad:RealRights* which connect a *prodm-cad:RightsHolder* to *prodm-cad:CadastralParcel* or *prodm-cad:Building*. The *prodm-cad:RightsHolderOrganization* subsumes both *prodm-cad: RightsHolder* and *schema:Organization*, which links automatically the rights holders to the organizations converted using schema.org vocabulary. The geospatial information of cadastral parcels and buildings are mapped to *gsp:asWKT* either as *sf:MultiPolygon* or *sf:Point*. The building accessibility is modelled as indicators on buildings by connecting *prodm-com:Building* to *prodm-com:Indicator* through object property *prodm-com:hasIndicator*.

The geospatial information of municipality boundaries are also mapped to *gsp: asWKT* as *sf:MultiPolygon*. The municipality code is mapped to *au:nationalCode*, and the municipality name is mapped to *rdfs:label*.

3.3   Data Augmentation Using SPARQL CONSTRUCT Queries

The source cadastral datasets cover both the state-owned and non-state-owned properties and buildings. In the data augmentation process SPARQL CONSTRUCT[29] queries are executed on the datasets from RDF conversions in Sect. 3.2 to generate subsets of state-owned properties and buildings. Table 3 lists the SPARQL CONSTRUCT queries that generate new datasets by only selecting triples related to the state-owned cadastral parcels or buildings. For example, the state-owned or state-leased cadastral parcel geospatial dataset is generated by first executing a CONSTRUCT query to indicate all the state-owned cadastral parcels by integrating the cadastral parcel geospatial dataset with the central government organization dataset. Afterwards, a second CONSTRUCT query is executed to select out triples related to ownership information about the state-owned or state-leased cadastral parcels. CONSTRUCT query can also be used to do calculations such as calculating the area summary for a cadastral parcel that can include one or more land parcels.

SPARQL CONSTRUCT queries are also used to infer triples based on known business rules. For example, the ownership of buildings does not exist in the original datasets, but it can be inferred by using this rule: *The owner or lessor of a cadastral parcel owns the buildings on the cadastral parcel.*

3.4   Interlinking with Other Datasets

Datasets interlinking is a fundamental prerequisite of the semantic Web [8]. The resulting SoE dataset is linked with several central Linked Open datasets in order to

---

[29] https://www.w3.org/TR/rdf-sparql-query/#construct.

Table 3. SPARQL CONSTRUCT queries

Input datasets as RDF		CONSTRUCT queries	Output datasets as RDF		
Name	Triples		Name	Triples	Public
Cadastral parcel geospatial dataset; Central government organization dataset	~118 M	1. CONSTRUCT query #1[a] to indicate state-owned or state-leased cadastral parcels 2. CONSTRUCT query #3[b] to select geospatial information about state-owned or state-leased cadastral parcels	The state-owned or state-leased cadastral parcel geospatial dataset	416604	Yes
Cadastral parcel ownership dataset	559743	1. CONSTRUCT query #1 2. CONSTRUCT query #2[c] to select ownership information about state-owned or state-leased cadastral parcels	The state-owned or state-leased cadastral parcel ownership dataset	206962	Yes
State-owned or state-leased cadastral parcel geospatial dataset	416604	CONSTRUCT query #4[d] to calculate the area of each cadastral parcel as summary of the belonging land parcels	The state-owned or state-leased cadastral parcel areas	40476	Yes
Building dataset	454192	1. CONSTRUCT query #1 2. CONSTRUCT query #5[e] to select information about buildings built on state-owned or state-leased cadastral parcels	The state-owned building	267862	Yes
Building dataset; State-owned or state-leased cadastral parcel ownership dataset	~1 M triples	CONSTRUCT query #6[f] to generate ownership dataset for state-owned buildings	The state-owned building ownership dataset	15380	Yes
Building geospatial dataset	23188270	CONSTRUCT query #7[g] to generate geospatial dataset of state-owned buildings	The state-owned building geospatial dataset	42295	Yes

[a] https://datagraft.io/prodatamarket_publisher/queries/soe-construct-query-1
[b] https://datagraft.io/prodatamarket_publisher/queries/soe-construct-query-3
[c] https://datagraft.io/prodatamarket_publisher/queries/soe-construct-query-2
[d] https://datagraft.io/prodatamarket_publisher/queries/soe-construct-query-4
[e] https://datagraft.io/prodatamarket_publisher/queries/soe-construct-query-5
[f] https://datagraft.io/prodatamarket_publisher/queries/soe-construct-query-6
[g] https://datagraft.io/prodatamarket_publisher/queries/soe-construct-query-7

increase its reusability to support queries on cross-domain distributed datasets. Norway as a country is linked to http://sws.geonames.org/3144096/ from the GeoNames dataset. The municipalities are modelled as administrative units and they have *owl: same* links to both the DBpedia[30], GeoNames[31] and lenka.no (the Norwegian RDF-resources for geographical breakdown)[32]. For example, the municipality of Oslo is linked to http://dbpedia.org/page/Oslo at DBpedia on municipality names, and is also linked to http://data.lenka.no/geo/inndeling/03/0301 at *lenka.no* on the national municipality identifier. The linking triples are produced using SPARQL CONSTRUCT queries. In the result dataset 405 triples are connected to DBpedia, 422 triples are connected to GeoNames and 404 triples are connected to Lenka.no.

3.5 Datasets Publishing

Both the RDF conversion results of source datasets from Sect. 3.2 and the data augmentation results of SPARQL CONSTRUCT queries from Sect. 3.3 are published via the DataGraft platform. Published data can be queried through the generated SPARQL endpoint or accessed via APIs. The resulting SoE dataset is described in more details in Sect. 4.

3.6 Data Quality Assessment Using the Interlinked Datasets

Challenges related to data quality are mentioned in Sect. 2.6 and a rule-based approach for data quality assessment and improvement has been introduced in [9]. Table 4 shows some examples of quality check scenarios where SPARQL queries are executed on interlinked datasets to identify inconsistencies between source systems. All the queries are shared and freely available. The table also associates possible reasons to the inconsistencies. The results from the SPARQL queries can help the responsible staff control and improve the data quality in the source systems by following the suggested quality improvement strategies. The updated source datasets with better data quality will then be reloaded to the Linked Data generation process to produce an updated resulting SoE dataset with improved quality.

---

[30] http://wiki.dbpedia.org/wiktionary-rdf-extraction .

[31] http://www.geonames.org/ .

[32] http://data.lenka.no/ .

Table 4. Data quality assessment using interlinked source datasets

No.	The SPARQL query identifies	# of triples	Possible reasons	Suggested quality improvement strategy
1	The owner name difference between cadastral system and business entity register[a]	146	Delayed or missing updates of owner names in the cadastre	Update the owner names in the cadastre
2	The state-owned properties that are missing in the previous SoE report[b]	6880	The properties were acquired after the previous report was made	No actions needed though it reflects partially the quality of the previous SoE report
			The properties were forgotten to be registered in the previous SoE report	
3	The state-owned properties from the previous SoE report that are missing in the resulting SoE dataset[c]	2857	The properties were sold to a non-central government organization after the previous report was made	No actions needed
			The properties are abroad	
			There has been organization change with the owner and the owner's organization number is no longer valid in the business entity register	Update the owner's organization number and name in the cadastre
			The ownership change between organizations in the public sector is not always officially registered in the cadastre	Inform the current owner organization to update the ownership in the cadastre
			The owner's organization is not officially registered as central government organization in the business entity register	Update the organization in the business entity register if it is applicably or add it to the manual exception list of the central government organization dataset

[a] https://datagraft.io/prodatamarket_publisher/queries/soe-query1-the-owner-name-difference
[b] https://datagraft.io/prodatamarket_publisher/queries/soe-query2-missing-soe-records
[c] https://datagraft.io/prodatamarket_publisher/queries/soe-query3-missing-result-soe-records

Table 5. Technical details of the resulting SoE dataset

Name	Norwegian state of estate report dataset
URL	https://datagraft.io/prodatamarket_publisher/sparql_endpoints/norwegian-state-of-estate-report-04693e1f-4060-48c1-8ab9-888a6c95f6d6
VOID file	https://datahub.io/dataset/norwegiansoe/resource/f5a83fbb-4324-43c2-a3da-5865b1f2d44e
Data download	https://rdf.datagraft.net/4035596353/db/repositories/norwegian-state-of-estate-report-6/statements
Ontology	http://vocabs.datagraft.net/
Version	1.0
Version date	27 June 2017
License	Norwegian Licence for Open Government Data (NLOD)

Table 6. RDF data sample on cadastral ownership dataset

@prefix prodm-cad: <http://vocabs.datagraft.net/proDataMarket/0.1/Cadastre#>
@prefix dbo: <http://dbpedia.org/ontology/>
@prefix dul: <http://www.ontologydesignpatterns.org/ont/dul/DUL.owl#>
@prefix dc: <http://purl.org/dc/terms/>
@prefix xsd: <http://www.w3.org/2001/XMLSchema#>
@prefix rdfs: <http://www.w3.org/2000/01/rdf-schema#>
@prefix dc11: <http://purl.org/dc/elements/1.1/>
@prefix schema: <https://schema.org/>
<http://vocabs.datagraft.net/proDataMarket/0.1/Cadastre#CadastralParcel/0214121900>
prodm-cad:hasCadastralID "214/121/9/0"
<http://vocabs.datagraft.net/proDataMarket/0.1/Cadastre#RealRights/0214121900971032081>
a prodm-cad:RealRights;
dbo:type "HJEMMELSHAVER"@no, "OWNER"@en;
dul:defines <http://vocabs.datagraft.net/proDataMarket/0.1/Cadastre#CadastralParcel/
0214121900>, <http://vocabs.datagraft.net/proDataMarket/0.1/
Cadastre#RightsHolderOrganization/971032081>;
dc:source "cadaster" ;
prodm-cad:hasStartDate "2006-02-23T01:00:00.000+01:00"^^xsd:dateTime ;
prodm-cad:hasEndDate "1753-01-01T01:00:00.000+01:00"^^xsd:dateTime
prodm-cad:RealRights rdfs:subClassOf dc11:Rights
<http://vocabs.datagraft.net/proDataMarket/0.1/Cadastre#RightsHolderOrganization/
971032081>
a prodm-cad:RightsHolderOrganization;
schema:leiCode "971032081"

458 L. Shi et al.

4 SoE Dataset Overview

Table 5 lists the technical details of the resulting SoE dataset. The result is a new dataset of state-owned properties and buildings which contains all the publicly available data published in Tables 2 and 3. There are a total of 1,223,208 unique triples in the dataset. In addition to DataGraft, the dataset dump and documentation are also published at a third-party site Zenodo[33]. The dataset is registered in the datahub.io data catalogue[34]. The resources defined in datahub.io for the Norwegian SoE dataset include the Norwegian SoE SPARQL endpoint, RDF dump, proDataMarket vocabulary, VOID file, Linked Data Generation SPARQL queries, and example SPARQL queries to help users understand and use the dataset.

Table 6 shows RDF data example that models cadastral ownership described in Table 1. The current plan to update and maintain the dataset is scheduled every 6 months on DatarGraft.io though patch releases are also supported when necessary. Data quality assessment and evaluation are compulsory steps in the publishing process to improve the data quality. In addition to datahub.io as a dataset sharing and user feedback channel, a data marketplace for property data is under development which focuses on selling, sharing and maintaining the datasets.

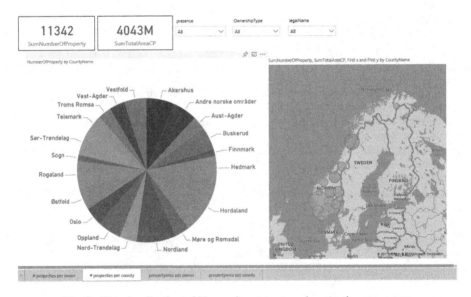

Fig. 2. The visualization of Norwegian state-owned properties per county

---

[33] https://zenodo.org/.

[34] Available at https://datahub.io/dataset/norwegiansoe.

5 Application Scenarios and Use Cases

Generating the SoE report is the main application scenario of the resulting SoE dataset. The report includes a table-based list of all state-owned properties and buildings and also as aggregated list grouped by municipalities and organizations – both lists can be generated by SPARQL queries. Figure 2 presents an example of visualization of Norwegian state-owned properties per county, both as a pie chart and on the map. There are 11342 state-owned properties in Norway and the total area is 4043 million square meters as shown in the figure.

In addition, the dataset can also be integrated with other contextual datasets to generate added value as follows.

Risk and vulnerability analysis of state-owned buildings. The dataset can be integrated with natural hazard datasets such as flood continuance and storm mean hours. Flood continuance map presents the areas in Norway that can possibly flood and it reveals areas where the danger of flooding needs to be further assessed. The geospatial dataset is provided by The Norwegian Water Resources and Energy Directorate as an open source dataset[35]. It is converted to RDF and published at DataGraft and accessible via the data endpoint[36]. Examples of queries that are enabled by such integration between the SoE dataset and the flood continuance dataset include:

- How many buildings owned by Statsbygg have flood risk in Norway?[37]
- Which state-owned buildings have the flood risk in Ås municipality?[38]

The result of the risk and vulnerability analysis helps the property owners to take proactive actions in maintenance and thereby reduce the damage and cost when natural hazard occurs.

Searching properties suitable for asylum reception centres. This is a demanding task that has significant effect on local communities. The demographical statistics and relevant geographical data can be integrated with the Norwegian SoE dataset to help identifying state-owned properties suitable for asylum reception centres.

Visitor application for public buildings. A subset of the Norwegian SoE dataset can be integrated with cultural heritage data, traffic data, weather data and other contextual data to provide input to a public building application for visitors.

6 Related Work

State-owned properties and buildings are shared as a downloadable text reports in the U.K. and Norway and similar reports are available for other countries. However, no systematic transformations of such reports to Linked Open Data exist to date.

---

[35] http://nedlasting.nve.no/gis/.

[36] https://rdf.datagraft.net/4035596353/db/repositories/statsbygg_data-2.

[37] https://datagraft.io/prodatamarket_publisher/queries/soe-query4-floodriskstatsbygg.

[38] https://datagraft.io/prodatamarket_publisher/queries/soe-query5-floodrisk-as.

Cadastral data including land ownership data is one of the core datasets used in the LOD generation of the SoE dataset. Openness of land ownership data is presented at the OKFN's website[39]. The statistics from 2015 show that the land ownership data is 100% open in Denmark and Uruguay, and only partially open in several other countries including, for example, Norway with 45% openness and Spain with 5% openness. There is no known effort to publish the land ownership data as Linked Data in Norway. There are few known efforts in Europe or worldwide to transform and publish cadastral data as Linked Data, and even fewer as Linked Open Data. One related research is [10] that developed a process to generate, integrate and publish geospatial Linked Data from several Spanish national datasets including the administrative units from the Spanish cadastre, and this process methodology has been applied in [11] to integrate two cadastre datasets for a city in Colombia. Neither of the Spanish and Colombian datasets from the research is open to the public.

The Norwegian government implemented PSI-directive in the Norwegian law on 1st January 2009 [12]. One of the central Norwegian LOD projects was Semicolon[40] which aimed to improve interactions in the Norwegian public sector. The project presented a status report on open and linked data in Norway. Examples of datasets available in LOD formats included business entity register, registry of municipalities and counties, central register of parties and their income, travel information for the public transportation system in Oslo, city bike stand status, etc., as described in [13]. The PlanetData project[41] made a report on Norwegian LOD extensions [14], which included two business cases and six updated or new datasets in RDF. Though the business entity register has been published as RDF in the aforementioned projects, the endpoint is not stable enough to be reused for central government organizations dataset in the SoE case.

7   Summary and Outlook

This paper introduced the SoE dataset (containing information about state-owned properties integrated from a variety of relevant sources), together with the process of generating a Linked Data representation of the dataset. Several government datasets over multiple systems/databases including the cadastral system and business entity register of Norway are integrated. The Linked Data generation process includes data cleaning, source dataset publishing, conversion to RDF, data augmentation using SPARQL CONSTRUCT queries and dataset publishing. The technical details of the resulting SoE dataset were presented. The data quality challenges were discussed in detail and SPARQL queries on interlinked source datasets were proposed to improve data quality. New services can also be generated by integrating the resulting SoE dataset with other contextual data.

In terms of future work, the preparation of source datasets currently includes several manual steps and it can be automatized to a certain level. The data cleaning process can

---

[39] http://global.census.okfn.org/dataset/land accessed 20 March 2017.

[40] http://www.semicolon.no/.

[41] http://www.planet-data.eu/.

be more user friendly by introducing visualization tools for semantic and syntactic error checks. The experience and process methods can further be tested with similar datasets from other countries.

Acknowledgements. The work has been funded by the European Commission under the proDataMarket project (H2020-ICT-2014-1 644497). The authors would like to thank Bjørg Pettersen and the proDataMarket consortium for additional contributions.

References

1. Shi, L., Pettersen, B.E., Østhassel, I., Nikolov, N., Khorramhonarnama, A., Berre, A.J., Roman, D.: Norwegian state of estate: a reporting service for the state-owned properties in Norway. In: Bassiliades, N., Gottlob, G., Sadri, F., Paschke, A., Roman, D. (eds.) RuleML 2015. LNCS, vol. 9202, pp. 456–464. Springer, Cham (2015). doi:10.1007/978-3-319-21542-6_30

2. Ding, L., Lebo, T., Erickson, J.S., DiFranzo, D., Williams, G.T., Li, X., Michaelis, J., Graves, A., Zheng, J.G., Shangguan, Z., Flores, J.: TWC LOGD: a portal for linked open government data ecosystems. Web Semant. Sci. Serv. Agents World Wide Web **9**(3), 325–333 (2011)

3. Roman, D., Nikolov, N., Putlier, A., Sukhobok, D., Elvesæter, B., Berre, A., Ye, X., Dimitrov, M., Simov, A., Zarev, M., Moynihan, R.: DataGraft: one-stop-shop for open data management. In: The Semantic Web Journal (SWJ) – Interoperability, Usability, Applicability (published and printed by IOS Press, ISSN: 1570-0844) (2017, to appear). doi:10.3233/SW-170263

4. Roman, D., Dimitrov, M., Nikolov, N., Putlier, A., Sukhobok, D., Elvesæter, B., Berre, A., Ye, X., Simov, A., Petkov, Y.: DataGraft: simplifying open data publishing. In: Sack, H., Rizzo, G., Steinmetz, N., Mladenić, D., Auer, S., Lange, C. (eds.) ESWC 2016. LNCS, vol. 9989, pp. 101–106. Springer, Cham (2016). doi:10.1007/978-3-319-47602-5_21

5. Roman, D., Dimitrov, M., Nikolov, N., Putlier, A., Elvesæter, B., Simov, A., Petkov, Y.: DataGraft: a platform for open data publishing. In: The Joint Proceedings of the 4th International Workshop on Linked Media and the 3rd Developers Hackshop. (LIME/SemDev@ESWC 2016) (2011)

6. Sukhobok, D., Nikolov, N., Pultier, A., Ye, X., Berre, A., Moynihan, R., Roberts, B., Elvesæter, B., Mahasivam, N., Roman, D.: Tabular data cleaning and linked data generation with Grafterizer. In: Sack, H., Rizzo, G., Steinmetz, N., Mladenić, D., Auer, S., Lange, C. (eds.) ESWC 2016. LNCS, vol. 9989, pp. 134–139. Springer, Cham (2016). doi:10.1007/978-3-319-47602-5_27

7. Shi, L., Nikolov, N., Sukhobokb, D., Tarasovac, T., Roman, D.: The proDataMarket ontology for publishing and integrating cross-domain real property data. In: The Journal "Territorio Italia. Land Administration, Cadastre and Real Estate", vol. 2 (2017, to appear)

8. Bizer, C., Heath, T., Ayers, D., Raimond, Y.: Interlinking open data on the web. In: Demonstrations Track, 4th European Semantic Web Conference, Innsbruck, Austria, June 2007

9. Shi, L., Roman, D.: Using rules for assessing and improving data quality: a case study for the Norwegian State of Estate report. In: The Proceedings of the Doctoral Consortium, Challenge, Industry Track, Tutorials and Posters @ RuleML+RR 2017, hosted by International Joint Conference on Rules and Reasoning 2017 (RuleML+RR 2017), London, UK, 11–15 July 2017

10. Vilches-Blázquez, L.M., Villazón-Terrazas, B., Corcho, O., Gómez-Pérez, A.: Integrating geographical information in the Linked Digital Earth. Int. J. Digital Earth **7**(7), 554–575 (2014)
11. Saavedra, J., Vilches-Blázquez, L.M., Boada, A.: Cadastral data integration through Linked Data (2014)
12. Lov om rett til innsyn i dokument i offentleg verksemd. §9 (2006)
13. Roman, D., Norheim, D.: An overview of Norwegian linked open data. In: proceedings of the Fourth International Conference on Information, Process, and Knowledge Management (eKNOW), pp. 93–96 (2012)
14. Roman, D., Mjelva, J.K., Norheim, D., Grønmo, R.: NorthPole Report on Norwegian LOD extensions. http://www.planet-data.eu/sites/default/files/D12.2.2.pdf. Accessed 23 July 2017

The InfraRisk Ontology: Enabling Semantic Interoperability for Critical Infrastructures at Risk from Natural Hazards

Dumitru Roman[⊠], Dina Sukhobok, Nikolay Nikolov,
Brian Elvesæter, and Antoine Pultier

SINTEF, Pb. 124 Blindern, 0314 Oslo, Norway
{dumitru.roman, dina.sukhobok, nikolay.nikolov,
brian.elvesater, antoine.pultier}@sintef.no

Abstract. Earthquakes, landslides, and other natural hazard events have severe negative socio-economic impacts. Among other consequences, those events can cause damage to infrastructure networks such as roads and railways. Novel methodologies and tools are needed to analyse the potential impacts of extreme natural hazard events and aid in the decision-making process regarding the protection of existing critical road and rail infrastructure as well as the development of new infrastructure. Enabling uniform, integrated, and reliable access to data on historical failures of critical transport infrastructure can help infrastructure managers and scientist from various related areas to better understand, prevent, and mitigate the impact of natural hazards on critical infrastructures. This paper describes the construction of the InfraRisk ontology for representing relevant information about natural hazard events and their impact on infrastructure components. Furthermore, we present a software prototype that visualizes data published using the proposed ontology.

Keywords: Ontology · Infrastructure components · Natural hazards · Events

1 Introduction

A natural hazard can be defined as a natural process that poses a threat to human life or property [1]. Extreme natural hazard events have the potential to cause devastating impacts to infrastructure networks, resulting in significant economic losses. In Europe, the number of disasters due to natural hazards increased in recent decades due to a combination of climate change effects, and changes in physical and social systems. For the period between 1998 and 2009, natural hazards and technological accidents caused nearly 100,000 fatalities and affected more than 11 million people, at the same time resulting in overall economical impact of about 200 billion euro [2]. These natural hazards have included hydrometeorological hazards (e.g. storms, floods) and geophysical hazards (e.g. landslides, earthquakes). Thus, floods, along with storms, are natural hazards that cause the highest economic losses in Europe. The flood-related losses in the EEA member countries over the period from 1998 to 2009 amounted to more than

© Springer International Publishing AG 2017
H. Panetto et al. (Eds.): OTM 2017 Conferences, Part II, LNCS 10574, pp. 463–479, 2017.
https://doi.org/10.1007/978-3-319-69459-7_31

60 billion euro [3]. In [4], the authors estimated that the expected annual damage (EAD) from flooding events in Europe may increase to 23.5 billion euro by 2050.

Given the potential economic losses caused by natural hazards, it is necessary to analyse the effects of natural hazards on the infrastructure, in particular critical infrastructure. Reliable transport infrastructure is of a great value to society as it facilitates the effective transportation of people and goods. The EU transport network has over 4.5 million km of paved roads and 212,500 km of rail lines[1]. Transport infrastructure plays a fundamental role in the EU and the ability to transport goods safely, quickly and cost-efficiently is highly important for international trade and economic development [5]. The complex interdependency of European infrastructure networks results in spreading the interruptions in infrastructure networks to many parts of Europe.

In this context, novel methodologies and tools are needed to analyse the potential impacts of extreme natural hazard events and aid in the decision-making process regarding the protection of existing critical road and rail infrastructure as well as the development of new infrastructure. One example of initiative addressing such aspects was the InfraRisk project.[2] The project aimed to develop reliable stress tests on European critical infrastructure using integrated modelling tools for decision-support. An important aspect of the project was to set the foundations for the development of a Geographical Information System (GIS) knowledge base of major global infrastructure failures, enabling users uniform, integrated, and reliable access to data on historical failures of critical transport infrastructure. The potential users of such a knowledge base are infrastructure managers, but also researchers (risk management, transportation, civil engineering, natural sciences, etc.). The knowledge base can serve as a case study for the events an infrastructure manager might consider important, and provide them with data of good/bad practices of managing solutions during and after the event.

In order to share common understanding of the data structure among the knowledge base users and enable semantic interoperability of infrastructure failure related data, the InfraRisk ontology was developed. Although the ontology was primarily developed to support data sharing and data usage within the scope of the InfraRisk project, it is generic and can be used for publishing and integrating various kinds of infrastructure components and natural hazards data. In this paper we present the design and implementation of the InfraRisk ontology for describing infrastructure failures due to the natural hazard events. Furthermore, we present a software prototype developed to consume data using the proposed ontology and interactively visualize information about various infrastructure components and natural hazards. The contributions of this paper are thereby two-fold:

1. First, we describe the InfraRisk ontology (design, implementation) for enabling semantic interoperability for critical infrastructures at risk from natural hazards;
2. Second, we propose a software prototype to visualize infrastructure components and natural hazards data made available using the developed ontology.

---

[1] http://ec.europa.eu/transport/infrastructure/tentec/tentec-portal/site/en/facts_and_figures.html.

[2] https://www.infrarisk-fp7.eu/.

The rest of the paper is organized as follows. Section 2 discusses related work. Section 3 describes the development process of the InfraRisk ontology. Section 4 illustrates an example of using the ontology for data publishing and integration, and presents the software prototype developed to visualize the data made available using the developed ontology. Section 5 summarizes the paper and outlines directions for future work.

2   Related Work

Defining and modeling natural hazards and their consequences is inconsistent across various natural hazard studies, databases and vocabularies. We analyzed the most common natural-hazard and infrastructure related terminologies and vocabularies in order to use available ontological knowledge in the ontology development process.

The terminology used by UNISDR (United Nations office for Disaster risk reduction)[3] defines *natural hazard* as a process, phenomenon or human activity associated with natural processes and phenomena that may cause loss of life, injury or other health impacts, property damage, social and economic disruption or environmental degradation. Natural hazards, as well as any other hazardous event can cause a *disaster* – a serious disruption of the functioning of a community or a society involving widespread human, material, economic or environmental *losses* and impacts, which exceeds the ability of the affected community or society to cope using its own resources. It is further commented that disasters are often described as a result of the combination of: the exposure to a hazard; the conditions of vulnerability that are present; and insufficient capacity or measures to reduce or cope with the potential negative consequences. Disaster impacts may include loss of life, injury, disease and other negative effects on human physical, mental and social well-being, together with damage to property, destruction of assets, loss of services, social and economic disruption and environmental degradation.

The International Federation of Red Cross and Red Crescent Centres (IFRC)[4] defines *disaster* as a sudden, calamitous event that seriously disrupts the functioning of a community or society and causes human, material, and economic or environmental *losses* that exceed the community's or society's ability to cope using its own resources. Though often caused by nature, disasters can have human origins. *Natural hazards* are considered as types of disasters and are defined as naturally occurring physical phenomena caused either by rapid or slow onset events.

The concepts *Natural Hazard*, *Loss* and *Event* as a generalized concept representing occurrence of a particular set of circumstances, are key concepts used in various terminologies and vocabularies in this field. The above mentioned concepts were also taken as the most general concepts in the domain and were used as a basic classes in the developed ontology.

---

[3] https://www.unisdr.org/.

[4] http://www.ifrc.org/en/.

A few linked open vocabularies have emerged to capture natural hazard consequence data. For example, the Management of a Crisis Vocabulary (MOAC)[5] provides
a minimum set of classes and properties for describing crisis management activities.
The vocabulary is specifically designed to aid the disaster information managers to
carry out activities in response to a disaster, but doesn't describe losses related to
natural hazard events.

None of existing terminologies and vocabularies cover aspects related to how
natural hazard events affect infrastructure components and therefore they were found
unsuitable for a direct use in the InfraRisk project. Nevertheless, the analysis of
existing terminologies and vocabularies has provided us a good baseline for ontological
knowledge to be built upon in the development of the InfraRisk ontology.

3   InfraRisk Ontology Development Process

The InfraRisk ontology was developed in accordance with existing guidelines and
methodologies for ontology development process, in particular the one proposed by [6].
As a first step in the ontology development process we defined the ontology domain,
scope and purpose, and requirements. After that we analyzed existing ontologies in
order to find a way to refine and extend them for our particular domain and task. The
next step was to collect domain knowledge to determine important terms in the ontology
and build and refine a conceptual model using Object-Role Modelling (ORM). Finally,
the conceptual model was realized in a concrete language (RDFS/OWL). Defining the
domain and scope of the ontology is described in Sect. 3.1. The resulting conceptual
model is discussed in Sect. 3.2. The process of the ontology implementation is discussed
in the Sect. 3.3.

3.1   Defining the Scope of the Ontology

The main purpose of the ontology creation was to relate global major infrastructure
failures with natural hazard events. Although the ontology was primarily developed to
follow the InfrarRisk project's focus on critical transport infrastructure (more specifically, European Ten-T core network[6]) and high-impact natural events, one can expect
the developed ontology to be applicable in a wider critical infrastructure context.

In order to define the scope of the ontology, a set of competency questions were
developed with infrastructure components and natural hazards experts in the project.
The followings are samples of competency questions used in the process:

- Which tunnels/bridges are located in country X?
- Which road bridges have collapsed between 1990 and 2014 in region Y and were
  triggered by floods?
- Which infrastructure failures were triggered by the 2003 flood of Danube river?
- What were the consequences (monetary loss) of the collapse of bridge Y in 2010?

---

[5] http://www.observedchange.com/moac/ns/.

[6] https://ec.europa.eu/transport/themes/infrastructure/ten-t-guidelines/maps_en.

- How many tunnels collapsed in region Y due to floods during 2001-2011 and incurred monetary losses more than X amount euro?
- What were the events that triggered the collapse of bridge Y in 2002?
- What were the infrastructure failure events that were triggered by no distinguishable natural hazards or by low/moderate/high/"black swan" natural hazards?
- Which original natural hazard events caused cascading hazard events (and subsequently caused infrastructure failure)?
- Which type of infrastructure failure causes the biggest losses relatively to other types?
- Due to the flood event in country X, which bridges had to be closed to traffic?
- What were the casualties due to the collapse of bridge Y in 2010?
- Which highway segment on infrastructure X could not be accessed due the earthquake in 2010?

The scope of transport infrastructure therefore covers road and rail transport infrastructures and their elements. The scope of natural hazards covers disasters affecting road and rail transport infrastructures and their elements. This includes components such as bridges (single or viaducts), tunnels, off ramps, embankments and slopes, and road and rail surface segments. In terms of natural hazards we consider events such as floods, earthquakes, landslides and any cascading hazards.

3.2 Building the Conceptual Model

Conceptual modeling methodologies have proven to be very effective for building information systems in a graphical interface at the high level of abstraction. Conceptual data schemes and ontologies have a lot of similarities, as both model concept relations and rules (constraints) [7]. The idea of reusing conceptual modeling techniques for ontology development is proposed by several authors (e.g. [7–9]) and provides a lot of advantages such as ability to use numerous existing conceptual modeling tools and methods.

A conceptual model for the InfraRisk ontology was developed with the help of the ORM (Object Role Modeling) data modeling approach. ORM models consist of objects (mapped to classes in the ontology) playing roles (relations) [10]. One advantage of using this technique is that ORM diagrams can be translated into pseudo natural language statements. This enables non-computer scientists (e.g. infrastructure experts) to evaluate the developed model. The ORM model for the InfraRisk ontology defines a conceptual model that relates major global infrastructure failures with natural hazard events. In the following we provide a brief description of key aspects covered by the conceptual model: infrastructure components and events (consequences and natural hazards).

Infrastructure components

An *Infrastructure* represents a transport mode, e.g., *Road* or *Rail* in our context. It has a *name, description* and a *geographical feature*. An infrastructure consists of one or more *Infrastructure Components*, e.g. *Bridge, Tunnel*, etc. (see Fig. 1). Each component has a *name, description*, a number of *lanes*, and a *geographical feature*. An Infrastructure Component can be connected to other Infrastructure Components. As mentioned above,

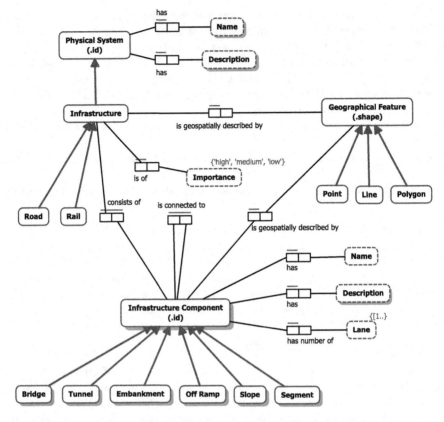

Fig. 1. ORM model for Infrastructure Component

the scope of this ontology is limited to components such as *bridges, tunnels, embank-ments, off ramps, slopes* and *segments* (e.g. of a road or rail line). Each of these infrastructure component types has its own set of properties as shown in the ORM models in Appendix A.

Events

An *Event* represents an incident where a *Natural Hazard* or *Infrastructure Component Failure* has occurred. It has a *name, description, location, date* and *consequence* (see Fig. 2). An Infrastructure Component Failure concerns the *full* or *partial collapse* of an Infrastructure Component.

A *Consequence* represents the expected losses in a specific location as a result of a given event. The Consequence can be a *Monetary Loss, Societal Loss* or *Usability Problem* concerning closure of or reduced traffic on an Infrastructure Component (see Fig. 3). The conceptual model distinguishes between three types of *Natural Hazard* events, namely *Earthquakes, Floods* and *Landslides*. The conceptual models for natural hazards are detailed in Appendix B.

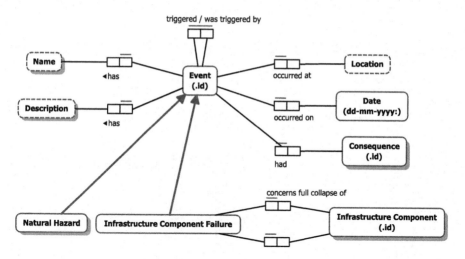

Fig. 2. ORM model for Event and Infrastructure Component Failure

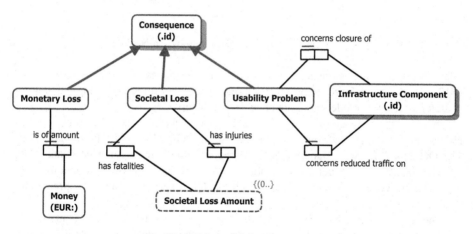

Fig. 3. ORM model for Consequence

3.3 Realizing the Conceptual Model in RDFS/OWL

The next step in the development of the InfraRisk ontology was its realization in a concrete language that can be used for publishing data. The InfraRisk conceptual model was specified using class hierarchy in RDFS[7]/OWL[8] and implemented in the Neologism vocabulary publishing platform[9] (see Fig. 4). RDFS is the most basic schema language commonly used in the semantic Web to model concepts, properties and their

---

[7] https://www.w3.org/TR/rdf-schema/.

[8] https://www.w3.org/TR/webont-req/.

[9] http://neologism.deri.ie/.

All terms at a glance

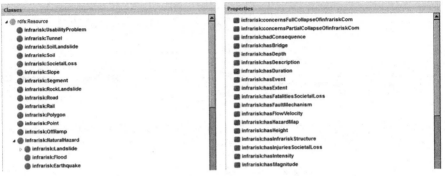

Fig. 4. Screenshot of the InfraRisk ontology in Neologism

relationships and characteristics (such as domains and ranges of properties). In its turn OWL is built upon RDFS and provides a larger vocabulary for web ontology modelling and can be used to model more advanced relationships.

In order to cover geospatial attributes of natural hazard and infrastructure component data, the InfraRisk ontology makes use of OGC GeoSPARQL standard [11]. The OGC GeoSPARQL standard supports representing and querying geospatial data on the semantic Web. It defines a vocabulary for representing geospatial data in RDF, and also provides an extension to the SPARQL query language for processing geospatial data.

The InfraRisk RDFS/OWL vocabulary is openly available[10] and contains 28 classes and 48 properties. The vocabulary is licensed under Creative Commons Attribution (CC BY)[11] and is available as an endpoint[12] via the DataGraft platform[13].

4   Software Prototype

Five datasets have been published on the DataGraft [12, 13] platform during the InfraRisk project using the ontology as a central reference model. The source datasets included data about natural hazard events (floods and landslides) in Europe that

---

[10] http://vocabs.datagraft.net/infrarisk.

[11] https://creativecommons.org/licenses/by/3.0/.

[12] https://rdf.datagraft.net/4037543173/db/repositories/infrarisk-vocabulary-1.

[13] https://datagraft.io/.

resulted in failures of critical transport infrastructure during the period 1972-2016. Data was obtained from InfraRisk project partners.

The DataGraft platform's warehouse for the RDF data is realized through the Semantic Graph Database-as-a-Service (DBaaS) component – a fully managed, cloud-based version of GraphDB[14] semantic graph database (triplestore). To meet the requirements of working with geospatial data and linked data, the DBaaS component introduces support for GeoSPARQL. The implementation of the GeoSPARQL specification in DBaaS is delivered as an additional plug-in for the GraphDB engine.

After the data has been published, it was possible to perform queries on data related to historical failures of critical transport infrastructure using the SPARQL query language. In order to query geographic information, GeoSPARQL extension functions for spatial computations are used. The following shows a SPARQL query retrieving infrastructure failures that occurred in 2015.

```
PREFIX geo: <http://www.opengis.net/ont/geosparql#>
PREFIX geof: <http://www.opengis.net/def/function/geosparql/>
PREFIX : <https://www.infrarisk-fp7.eu/vocabs/#>
PREFIX rdf: <http://www.w3.org/1999/02/22-rdf-syntax-ns#>
PREFIX rdfs: <http://www.w3.org/2000/01/rdf-schema#>
PREFIX xs: http://www.w3.org/2001/XMLSchema#

SELECT distinct ?eventCause ?eventDescription ?coordinates
?hasFatalities ?hasInjuries

WHERE {
    [] :hasinfrariskCom ?infrastructureComponent;
       :isGeospatiallyDescribedBy ?point.
    ?infrastructureComponent rdf:type :InfrastructureComponent .
    ?point geo:asWKT ?coordinates .
    ?infrastructureComponentFailure a :InfrastructureComponentFailure .
    ?infrastructureComponentFailure ?icAssociation ?infrastructureComponent .
    ?event rdf:type ?eventClass .
    ?eventClass rdfs:subClassOf :Event .
    ?event ?eventAssociation ?infrastructureComponent .
        ?causedBy :hasEvent ?event .
        ?causedBy :hasName ?eventCause .
    ?event :hadConsequence ?societalLoss .
    ?societalLoss rdf:type :SocietalLoss .
```

Table 1 shows a result sample of the above SPARQL query.

Thus, using the InfraRisk ontology one can represent and query integrated data from originally heterogeneous data sources. The results of the queries can further be

---

[14] http://graphdb.ontotext.com/.

Table 1. Result excerpt of the SPARQL query

Event cause	Event description	Coordinates	Has fatalities	Has injuries
Storm Frank	Bridge subsided and partially collapsed	POINT (53.688716 - 1.840771)	no	no
Rockfall	Rockfall blocked & damaged tracks	POINT (46.749723 8.642357)	no	no
Mudflow due to heavy rain	Mudflow caused by thunderstorm covered road and trapped several cars	POINT (34.886522 - 118.904150)	no	yes

visualized in various tools. A graphical user interface (GUI) application prototype was developed to visualize the data published using the ontology on a map using various interaction mechanisms. The prototype's GUI is based on the open source MASTER application[15]. It is an HTML5 application which can be used on smartphones, tablets and desktop computers. In addition to the map view, the application was integrated with the Google Street View technology. It allows the user to navigate along the roads photographed by Google. This mode provides an interesting alternative for viewing of hazard events (see Fig. 5).

Data about infrastructure components and events published using the InfraRisk ontology are retrieved via SPARQL queries and the results are presented in the GUI application (see Fig. 6). The output data is formatted using JSON.

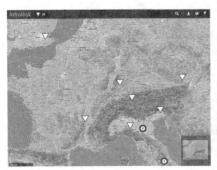

Fig. 5. Visualizing infrastructure events in the software prototype

---

[15] https://github.com/SINTEF-9012/mobileMaster.

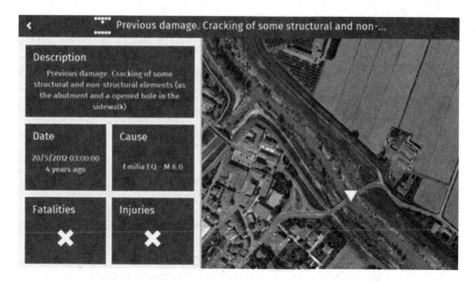

Fig. 6. Example of event details visualized in the software prototype

5 Summary and Outlook

This paper provided an overview of the InfraRisk ontology developed in order to assist publishing and integration of data about infrastructure failures due to natural hazard events. The ontology was developed in accordance with well-known ontology development guidelines. The ontology aimed to relate critical transport infrastructure with high-impact natural hazard events. The ontology was used to integrate and publish datasets about natural hazard events resulted in failures of critical transport infrastructure as Linked Open Data.

Furthermore, we developed a software prototype that visualizes data about infrastructure components and natural hazards published using the proposed ontology.

The ontology and the software prototype were developed based on the InfraRisk project's focus on European critical transport infrastructure, however they are suitable for use with infrastructure components failure data from other sources than those defined in the project. Future activities will be related to publication of data from various sources using the ontology, standardization of the ontology, and further improvements to the developed prototype.

Acknowledgements. The work in this paper is partly supported by the EC funded projects InfraRisk (Grant number: 603960) and proDataMarket (Grant number: 644497). The authors would like to thank Pierre Gehl (UCL), Khaled Taalab (UCL), Pieter van Gelder (PSCT), Yuliya Avdeeva (PSCT), Maria Jose Jimenez (CSIC), Mariano Garcia-Fernandez (CSIC), Bryan T. Adey (ETH), Miguel José Segarra Martínez (Dragados) and Mark Tucker (ROD), as well as other InfraRisk project partners for their involvement and contributions to the development of the InfraRisk ontology.

Appendix A. ORM Models for Various Infrastructure Components

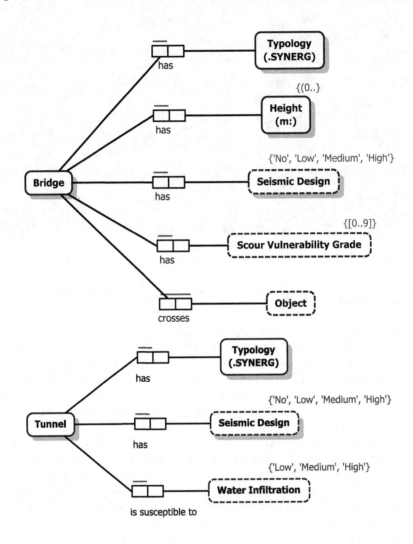

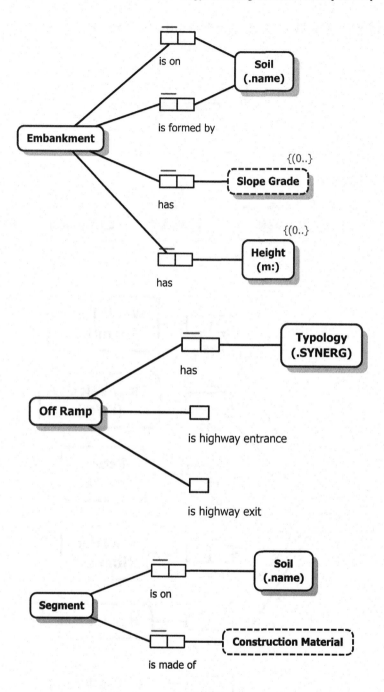

Appendix B. ORM Models for Various Natural Hazards

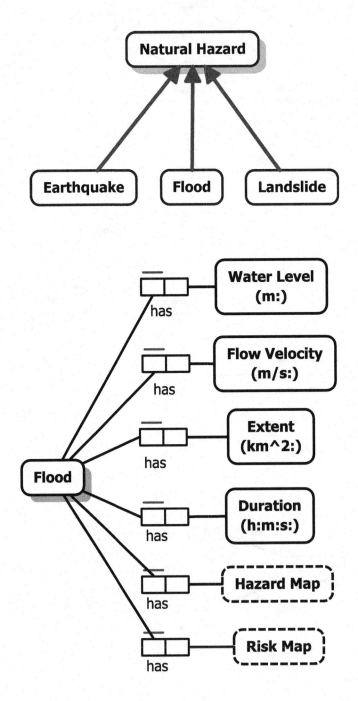

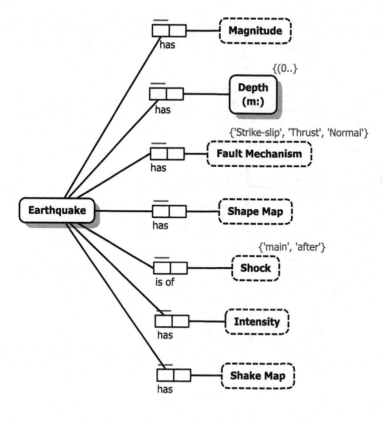

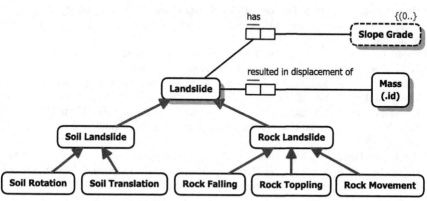

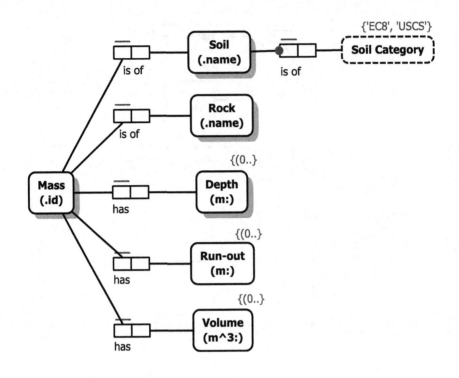

References

1. Hyndman, D., Hyndman, D.: Natural hazards and disasters. Cengage Learn. (2016)
2. Mapping the impacts of natural hazards and technological accidents in Europe. An overview of the last decade (2010)
3. Exploring nature-based solutions. The role of green infrastructure in mitigating the impacts of weather- and climate change-related natural hazards (2015)
4. Brenden, J., et al.: Increasing stress on disaster-risk finance due to large floods. Nat. Clim. Change 4(4), 264–268 (2014)
5. Eurostat regional yearbook, Luxembourg: Publications office of the European Union (2016)
6. Noy, N.F., McGuinness, D.L.: Ontology development 101: A guide to creating your first ontology (2001)
7. Jarrar, M., Demey, J., Meersman, R.: On using conceptual data modeling for ontology engineering. In: Spaccapietra, S., March, S., Aberer, K. (eds.) Journal on Data Semantics I. LNCS, vol. 2800, pp. 185–207. Springer, Heidelberg (2003). doi:10.1007/978-3-540-39733-5_8
8. Cranefield, S., Haustein, S., Purvis, M.: UML-based ontology modelling for software agents. In: Proceedings of the Workshop on Ontologies in Agent Systems, 5th International Conference on Autonomous Agents, Montreal, Canada (2001)
9. Baclawski, K., Kokar, Mieczyslaw K., Kogut, Paul A., Hart, L., Smith, J., Holmes, William S., Letkowski, J., Aronson, Michael L.: Extending UML to Support ontology engineering for the semantic web. In: Gogolla, M., Kobryn, C. (eds.) UML 2001. LNCS, vol. 2185, pp. 342–360. Springer, Heidelberg (2001). doi:10.1007/3-540-45441-1_26

10. Halpin, T.: Object-role modeling: Principles and benefits. Int. J. Inf. Syst. Model. Des. (IJISMD) 1(1), 33–57 (2010)

11. OGC GeoSPARQL - A Geographic Query Language for RDF Data (2012)

12. Roman, D., Nikolov, N., Putlier, A., Sukhobok, D., Elvesæter, B., Berre, A., Ye, X., Dimitrov, M., Simov, A., Zarev, M., Moynihan, R.: DataGraft: one-stop-shop for open data management. To appear in the Semantic Web Journal (SWJ) – Interoperability, Usability, Applicability (2017). IOS Press, ISSN: 1570–0844, doi:10.3233/SW-170263

13. Roman, D., Dimitrov, M., Nikolov, N., Putlier, A., Sukhobok, D., Elvesæter, B., Berre, A., Ye, X., Simov, A., Petkov, Y.: DataGraft: simplifying open data publishing. In: Sack, H., Rizzo, G., Steinmetz, N., Mladenić, D., Auer, S., Lange, C. (eds.) ESWC 2016. LNCS, vol. 9989, pp. 101–106. Springer, Cham (2016). doi:10.1007/978-3-319-47602-5_21

Usability of Visual Data Profiling in Data Cleaning and Transformation

Bjørn Marius von Zernichow[✉] and Dumitru Roman

SINTEF, Pb. 124 Blindern, 0314 Oslo, Norway
{BjornMarius.vonZernichow, Dumitru.Roman}@sintef.no

Abstract. This paper proposes an approach for using visual data profiling in tabular data cleaning and transformation processes. Visual data profiling is the statistical assessment of datasets to identify and visualize potential quality issues. The proposed approach was implemented in a software prototype and empirically validated in a usability study to determine to what extent visual data profiling is useful and how easy it is to use by data scientists. The study involved 24 users in a comparative usability test and 4 expert reviewers in cognitive walkthroughs. The evaluation results show that users find visual data profiling capabilities to be useful and easy to use in the process of data cleaning and transformation.

Keywords: Data preparation · Visual data profiling · Usability testing · Interactive data cleaning and transformation

1 Introduction

Data collection has become a necessary function in most large organizations both for record keeping and in support of different data analysis activities that are strategically and operationally critical [1]. In this context, proper data quality is a crucial aspect of extracting accurate information from data sources. Hence, incorrect or inconsistent data may distort analysis and compromise the benefits of any data-driven approaches. Examples of data quality issues, also labelled anomalies, include occurrences of missing, extreme, erroneous or duplicate values [2].

To illustrate the impact of poor quality data, IBM has estimated the yearly cost of inadequate data quality to be $3.1 trillion in US in 2016 [3]. Further, a recent survey [4] shows that data scientists spend 60% of their time on cleaning and organizing data, and 57% ranked this as a repetitive and tedious activity.

Considering the potential negative impact of poor data quality, there has been considerable research during the last decades, and different methods and tools have been proposed to cope with data cleaning [1]. Data cleaning is the process and technique of identifying and resolving missing values, outliers, inconsistencies, and noisy data, to improve data quality [5]. Closely related to data cleaning processes, additional data transformation procedures, i.e. changing the data format while preserving the original meaning, are often required to improve data quality [5].

In the context of data cleaning, data profiling is the statistical assessment of datasets to identify quality issues such as potential outliers or missing values with the goal of

© Springer International Publishing AG 2017
H. Panetto et al. (Eds.): OTM 2017 Conferences, Part II, LNCS 10574, pp. 480–496, 2017.
https://doi.org/10.1007/978-3-319-69459-7_32

achieving improved data quality [2]. Since determining what defines an error is context-dependent, human judgment is usually involved to determine whether the issues are actual errors and how the issues should be treated. The data quality assessment can be facilitated by a data profiling tool that performs statistical analysis [2, 5].

Visual data profiling is an extension of data profiling, achieved by supplementing statistical assessment of datasets with adequate data visualizations [2]. The integration of statistical analysis and visual analysis can reduce the time users spend on exploring and assessing data quality issues by providing constant real-time feedback on content and structure of datasets. Considering that data scientists use more than half their time cleaning and organizing data, and often find this activity tedious, visual data profiling approaches should be considered more often as it reduces the time and cost data scientists spend when addressing data quality issues.

The basic principle behind visual data profiling is to let the visual data profiling system perform the review of data quality and identification of data quality issues. The system collects statistics and information about the data, and then returns metadata that describes the quality of the data. Based on this information, the data scientist can make an informed decision about how any issues should be treated.

A recent example of a data cleaning and transformation framework is Grafterizer [6, 7], part of the cloud-based DataGraft[1] [8–10] platform. Grafterizer represents the state-of-the-art within data preparation research, supporting a wide range of cleaning and transformation operations. The framework provides an interactive user interface, and detailed specification and customization of transformation steps. Still, Grafterizer does not yet support visual data profiling that can ease the process of data cleaning, transformation, and improving data quality, for data scientists. Grafterizer provides research opportunities for evaluating usability of visual data profiling since the existing version serves as a benchmark in a comparison with the proposed prototype.

To address the problems with data quality, and time/cost consuming data preparation activities, we propose an approach that simplifies the data cleaning and transformation processes in Grafterizer, and reduces the effort spent on preparing data for analysis. We present a software prototype of the visual data profiling approach that features an interactive spreadsheet table view, suggestions for relevant data cleaning and transformation operations, and data quality feedback from a visual data profiling system. The goal was not only to create a prototype featuring the enumerated capabilities, but also to extensively evaluate it. To evaluate the usability of the approach and the prototype, a study was carried out that involved 24 users in a comparative usability test, and 4 expert reviewers in streamlined cognitive walkthrough sessions.

Key contributions of this paper include:

1. An *approach* to using visual data profiling in tabular data cleaning and transformation processes to improve data quality. The visual data profiling approach is realized by means of a prototype that includes features for identifying and visualizing data quality issues, i.e. missing values and outliers.

---

[1] https://datagraft.io.

2. An *evaluation of the usability* of the visual data profiling approach by empirical validation of the prototype. A comparative usability study and expert reviews have been conducted to evaluate the usefulness and ease of use.

The remainder of this paper is organized as follows. Related work is presented in Sect. 2. Section 3 introduces the proposed visual data profiling approach. The implementation of the approach in a software prototype is presented in Sect. 4, and the evaluation of the approach and prototype is discussed in Sect. 5. Finally, Sect. 6 summarizes this paper and outlines avenues for future work.

2   Related Work

The development of the visual data profiling approach draws upon current research, and is inspired by existing solutions within the areas of data profiling technologies, visual analysis systems, and tabular data preparation approaches.

Profiler [2] is an example of a system for data quality analysis that includes data mining and anomaly detection techniques in addition to visualizations of relevant data summaries that can be used to evaluate data quality issues and possible causes. Profiler integrates statistical and visual analysis to reduce the time spent on data cleaning activities. The Profiler architecture and framework were developed by the former Stanford Visualization Group, now UW Interactive Data Lab. This team also developed Polaris [11] that evolved into the commercialized business and analytics software Tableau[2], and Data Wrangler [12] that together with Profiler merged into the commercialized data preparation solution, Trifacta[3].

The above-mentioned profiling solutions all originated in research environments, are well documented in research literature, and represent effective and user-friendly approaches to data profiling. Moreover, Talend[4] uses similar visual profiling techniques as Trifacta to automatically explore data characteristics and data quality issues. Talend focuses on ease of use and an intuitive user-interface.

In terms of usability testing of our visual data profiling approach, it would be challenging to use Trifacta or Talend as the system under test. First, it is difficult to isolate the data profiling capabilities from the data cleaning and transformation functionality. Hence, it would be problematic to know what is really evaluated. Second, the solutions are not open-source, and cannot be further developed to extend the existing version of Grafterizer.

Generating visualizations from large data sets requires an understanding of users' needs and preferences along with knowledge of visual encoding rules and perception guidelines [13]. There are two general approaches to building a visual analysis system. First, considering visual encoding only will generate all possible valid visualizations without acknowledging the specific needs and preferences of users [14–16]. Second, introducing a visualization recommender system in a visualization pipeline [14–16]

---

[2] https://www.tableau.com.

[3] https://www.trifacta.com.

[4] https://www.talend.com/products/data-preparation.

will potentially reduce the information overload of presenting all available visualizations. Tracking and storing information provided by the recommender system enables adaptation of the visualization system due to an evolving knowledge about which visualizations are valid and preferred by users [14].

Voyager [16] is an exploratory data analysis tool that is open-source, originated in research, and provides state of the art within open source data exploration. Voyager specifies visualizations through Vega-Lite [17], a high-level declarative JSON specification language based on Wilkinson's Grammar of Graphics [18], ggplot2 [19] and Tableau VizQL [11, 20]. Vega [21] is the underlying formal model for rendering Vega-Lite specifications. Our visual data profiling approach is inspired by Voyager, Vega, and Tableau, and implements a high-level declarative language to specify visualizations.

Microsoft Excel is a widely-used tool to prepare data for analysis and gaining insight into data. A central feature of Excel is the direct manipulation interface [22] where users can interact with the table to manipulate the dataset (e.g. selecting columns and/ or rows, right-clicking for options). The advantage of a direct manipulation interface, is that many users are already familiar with this interface, and less time is required to learn to use the tool. The proposed visual data profiling approach relies on the implementation of a spreadsheet-like table view for direct manipulation of data.

The proposed visual data profiling approach draws upon existing research to include capabilities for statistical profiling, suggestions for data cleaning and transformation operations, and a direct manipulation table. The approach differs from existing solutions by expanding the profiling capabilities with more relevant data quality feedback and visualizations for missing values and data distribution.

3   Proposed Approach

The requirements of the proposed approach emerge from needs of existing users of Grafterizer (Fig. 1), and as a research opportunity to propose an approach that will contribute to improving data quality in this context. Grafterizer provides state of the art functionality within data cleaning and transformation capabilities, but there is still a need for improving user experience by providing approaches that assist the users in achieving their goals of cleaning and transforming data. User feedback shows that Grafterizer has a steep learning curve, and is rather complex to use. Hence, novel approaches should be considered to provide useful functionality for improving data quality, and a user interface that is easy to learn and easy to use. Based on this exiting situation, our visual data profiling approach should provide the necessary statistical profiling capabilities that are needed to assist the user in identifying data quality issues, and ease the process of improving quality. The visual data profiling capabilities should be integrated with a table view interface that lets the user manipulate columns and rows directly. Furthermore, the user interface should provide data cleaning and transformation functionality that is relevant to the user, and appropriately addresses the goals that the user tries to achieve. The applied data cleaning and transformation sequences should finally be reflected in a stepwise pipeline.

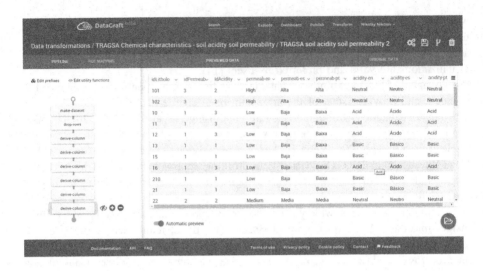

Fig. 1. Grafterizer user interface without visual data profiling capabilities

To facilitate the requirements process, a wireframe was created to describe the user interface and functionality, and the needs of users that led to a set of requirements. The wireframe outlined the basic graphical user interface components and functionality to resemble the final version of the application [23]. The wireframe was the first step to realizing the visual data profiling approach. Wireframes can be directly used in the implementation of the user interface of a prototype that supports the visual data profiling approach.

The user interface of the visual data profiling approach illustrated in the wireframe in Fig. 2, consists of the following main components and capabilities:

- A visual data profiling component (Fig. 2, component 1).
- A tabular table view that provides data cleaning and transformation functionality (Fig. 2, component 2).
- A sidebar that suggests relevant data cleaning and transformation actions to the user (Fig. 2, component 3).
- A steps pipeline that reflects applied data cleaning and transformation steps (Fig. 2, component 4).

The profiling assisted data cleaning and transformation process involves the following sequence of steps [2, 24, 25]:

1. **Discovery:** The user starts the data cleaning and transformation process by discovering the content, structure, and quality of the dataset. The visual data profiling system performs statistical assessment of data quality and returns the summarized feedback to the user.

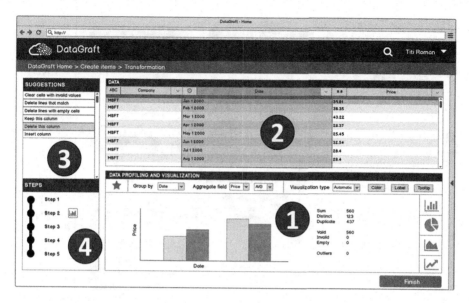

Fig. 2. Visual data profiling approach wireframe

2. **Cleaning and transformation:** Based on the statistical assessment of data quality, the user applies the appropriate procedures to clean the dataset, e.g. by correcting missing values. The dataset is further transformed to change shape into a desired format, e.g. by deleting a column.
3. **Validation:** Assisted by the data profiling system, the user validates the result of the applied cleaning and transformation procedures to ensure the output dataset has the intended content and structure.

4   Realization of the Proposed Approach in a Software Prototype

Prototyping is applied as an iterative design and development process to realize the concepts and requirements that are defined in the proposed visual data profiling approach [26, 27]. By prototyping, we always had something functional to test with users, collect feedback, implement changes, and then iterate. The prototype adds interactivity to the user interface wireframe, and provides functionality needed to demonstrate and validate the visual data profiling approach.

4.1   System Architecture

The high-level system architecture (Fig. 3) is based on a microservice architecture, and implements the design principles of Separation of Concerns (SoC) [28]. SoC is traditionally achieved in layered architectures, e.g. in a 3-Tier architecture, by defining

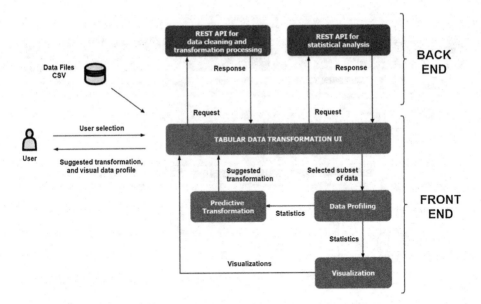

Fig. 3. Visual data profiling microservice architecture

interfaces and encapsulating information. A 3-Tier architecture would separate concerns into a presentation layer, an application tier, and a data layer.

A microservice architecture would take the SoC one step further by dividing the application tier and data layer into separate, domain-driven services that would operate autonomously from other services. A network-protocol would provide secure end point access to the services. While the SoC in a layered architecture is horizontal, the SoC in a microservice architecture would be both horizontal and vertical.

4.2 Data Cleaning and Transformation Functions

The key functionality that was needed to evaluate usability of the visual data profiling approach is implemented in the prototype. The functionality is based on which data cleaning and transformation steps are needed to demonstrate and validate the visual data profiling approach in a user scenario developed by Statsbygg[5] and SINTEF[6]. The user scenario is named 'State of Estate', and is based on a dataset (reporting state-owned properties in Norway) that is cleaned and transformed by utilizing Grafterizer [8]. Statsbygg is the Norwegian government's advisor in construction and property affairs, and serves as a building commissioner, property manager and developer. One of the purposes of cleaning and transforming the State of Estate dataset is to integrate information about public buildings in Norway with for example accessibility in buildings [8].

---

[5] http://www.statsbygg.no.

[6] https://www.sintef.no.

In total 14 data cleaning functions were defined and implemented in the prototype. Examples of functions include setting first row as header, replacing values, setting text to uppercase, concatenating values, and filling empty cells with a given value.

4.3   Implementation of the Software Prototype

The visual data profiling approach was implemented in a software prototype in four iterative stages. The final iteration of the prototype reflects the proposed functionality of the initial wireframe, and desired functionality to evaluate the usability of visual data profiling.

The UI of the prototype depicted in Fig. 4 implements basic functionality of the following components: *Component 1*, the file import, is implemented for prototype development purposes only; *Component 2*, the table view, is a direct-manipulation table with Excel-like features such as right-clicking functionality (e.g. copy/paste, insert column/row); *Component 3*, the transformations sidebar, implements a rule-based system that suggests relevant data cleaning and transformation procedures; *Component 4*, the steps pipeline, displays a functioning pipeline that reflects all steps applied; *Component 5*, the visual data profiling service, features (from left to right) a data distribution chart, a chart that displays number of missing values, and basic measures of central tendency. The leftmost data profiling chart represents missing values and valid (non-null) values for the currently selected column. The three remaining charts (from left to right) represent the distribution of the currently selected column.

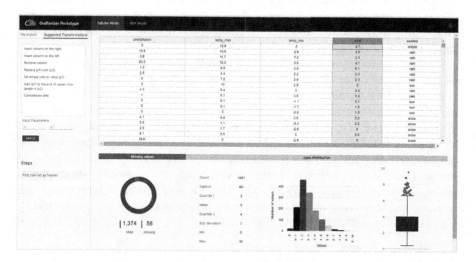

Fig. 4. Implementation of prototype, final iteration

5 Evaluation: Usability Testing of the Software Prototype

5.1 Evaluation Methodology and Setup

In terms of usability and user acceptance of a system, it is essential that users believe that the system is useful and easy to use in order to adopt the technology [29, 30]. A user will consider a system to be useful if it enhances his or her work performance, and a system is easy to use if a user thinks that learning and using the system requires an acceptable amount of effort in terms of time and cost. Hence, a visual data profiling extension should not only provide the capabilities that the user needs, but the solution should also be considered useful in data scientists' work activities, and be easy to use [29]. We refer to these qualities as the usability of the visual data profiling system. The usability study addressed the following research questions:

- RQ1: How *useful* is the visual data profiling approach for users of tabular data cleaning and transformation tools?
- RQ2: How *easy to use* is the visual data profiling approach for users of tabular data cleaning and transformation tools?
- RQ3: Will the visual data profiling approach introduce usability issues in tabular data cleaning and transformation applications, and if so; which types of usability issues occur and how can they be corrected?

To understand users' experience with visual data profiling approaches, we have defined the typical users as data consumers, more specifically data scientists, that use data for data-driven decision making. The data scientist is an analytical expert that explores and analyses large volumes of data to solve complex problems and reveal business insights. Dedicated solutions for cleaning and transforming tabular data, e.g. Grafterizer, are often part of data scientist's toolbox.

We used two complementary methods of usability testing to evaluate whether users find the visual data profiling approach to be useful and easy to use.

A *comparative usability test*, survey based, was used to collect statistics and attitudinal data from users through an online questionnaire [31] which contains Likert-type rating scales. The test compared the prototype against the existing version of Grafterizer in terms of usefulness and ease of use. The survey was anonymized and voluntary, and only non-sensitive information was collected. A representative group of users was selected to participate in the survey. Voluntary participants from project meetings in current research initiatives with SINTEF were invited to participate in the comparative usability test, respond to the survey questionnaire, and provide qualitative feedback on the visual data profiling approach:

- EW-Shopp[7] (project meeting February 2017)
- proDataMarket[8] (project meeting March 2017)
- euBusinessGraph[9] (project meeting May 2017)

---

[7] http://www.ew-shopp.eu.

[8] https://prodatamarket.eu.

[9] http://eubusinessgraph.eu.

We also used *streamlined cognitive walkthrough* as a usability inspection method where evaluators inspected the user interface by completing a set of tasks to simulate users' problem solving approaches [32–35]. The aim of this process was to identify usability issues introduced by the visual data profiling approach in data cleaning and transformation processes. In total four expert reviewers were selected to participate in the sessions. Users were divided in two subgroups and two corresponding sessions:

- Session 1: Two Human-Computer Interaction (HCI) experts from SINTEF Digital[10];
- Session 2: Two expert reviewers from the Logic and Intelligent Data (LogID) group at University of Oslo[11].

5.2   Analysis of Findings from the Comparative Usability Test

In total 24 participants responded to the survey questionnaire. The same users evaluated both the existing version of Grafterizer and the visual data profiling prototype, which defines the test setup as a within-subjects design [31]. The advantage of using this type of test design is that it removes some sources of variation in the datasets, as compared to between-subjects design where different users test each version of the application.

The online survey questionnaire[12] asked respondents to rate each application on the dimensions of *usefulness* and *ease of use*, respectively.

The summarized results from all respondents are illustrated in Figs. 5 and 6 below. The figures indicate the mean value of each question asked, e.g. the rating score of question Q1 in Fig. 5 shows the average of all 24 respondents' rating score on that specific question. High rating scores indicate that users perceive the application to be highly useful and easy to use, while the opposite is true for low scores.

The results that are illustrated in Figs. 5 and 6 indicate that the visual data profiling approach consistently is rated higher than the existing version of Grafterizer on both dimensions of usefulness and ease of use. Still, it is insufficient to draw such conclusions based only on the kind of descriptive statistics [31] we find in Figs. 5 and 6. We need to determine if this difference between the applications is statistically significant, and if it is larger than we would expect from pure chance.

Since the usability test is a within-subject comparison of two applications, and the survey test results are continuous values, a paired t-test can be applied to appropriately determine if there is a significant difference between the mean ratings of the two applications [31]. The approach suggested by Sauro and Lewis [31] has been applied to compare the mean rating between the prototype and the existing version of Grafterizer.

---

[10] http://www.sintef.no/en/information-and-communication-technology-ict/departments/networked-systems-and-services/human-computer-interaction-hci.

[11] http://www.mn.uio.no/ifi/english/research/groups/logid.

[12] https://goo.gl/forms/P3pD8zVPOj3uOSLT2.

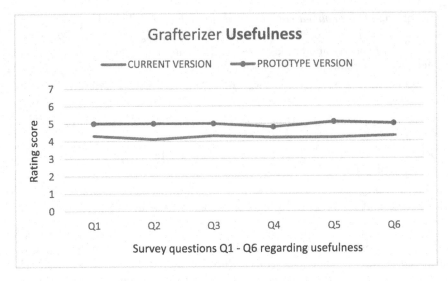

Fig. 5. Comparative usability test results (usefulness)

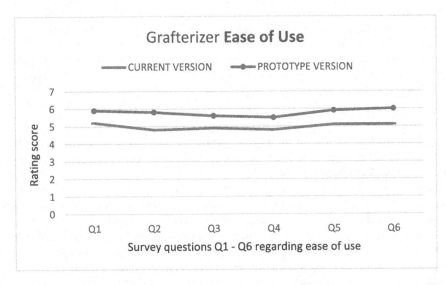

Fig. 6. Comparative usability test results (ease of use)

We used a paired t-test to determine statistical significance of survey results for the *usefulness* dimension:

Paired t-test:

$$t = \frac{\widehat{D}}{\frac{s_D}{\sqrt{n}}}$$

(1)

where

$\widehat{D}$ is the mean of the difference between the scores
s_D is the standard deviation of the difference between the scores
n is the sample size, i.e. the number of survey respondents
t is the test statistic

Using the t-test to calculate the test statistic t of the values in Table 1 below, we get the following t value:

$$t = \frac{4.63}{\frac{4.48}{\sqrt{24}}} = 5.09$$

Table 1. Survey rating scores, and difference, in terms of usefulness

Respondent	Prototype	Existing version	Difference
1	6	6	0
2	10	11	−1
3	24	24	0
4	34	35	−1
5	31	24	7
6	26	26	0
7	32	24	8
8	38	36	2
9	19	18	1
10	34	27	7
11	34	32	2
12	26	14	12
13	28	25	3
14	11	10	1
15	30	19	11
16	39	24	15
17	36	30	6
18	35	29	6
19	38	37	1
20	40	32	8
21	42	34	8
22	33	31	2
23	36	32	4
24	38	29	9
Mean	**30**	**25.4**	**4.63**

To determine whether the *t* value is significant, we use the TDIST function in Excel:

TDIST function:

$$TDIST(t \text{ value}, \text{degrees of freedom}, \text{one}-\text{sided}=1/\text{two}-\text{sided}=2) \quad (2)$$

The degrees of freedom are equal to n − 1, and we use a two-sided test in the comparison. n = 24, which leads to the following calculation:

$$TDIST(5.09, 23, 2) = 0,000037$$

The calculations of statistical significance indicate that we can be approximately 99.999% sure that the prototype and the existing version have different scores, i.e. the difference is not due to chance. Hence, the prototype's rating score of 30 is statistically significantly higher than the existing version's score of 25.4. We can conclude that the users perceive that the prototype is more useful than the existing version of Grafterizer.

In terms of the *ease of use* dimension, the mean rating score for the prototype is 34.6, while the rating score of the existing version of Grafterizer is 30. The difference in rating scores is then 4.58, and applying the paired t-test leads to the conclusion that the rating score of the prototype is significantly higher than the existing version's score. The calculations of statistical significance indicate that we can be approximately 99.999% sure that the prototype and the existing version have different scores.

Based on the above analysis, we can conclude that the users perceive that the prototype is both more useful and easier to use compared to the existing version of Grafterizer.

5.3   Analysis of Findings from the Cognitive Walkthrough

The two groups of expert reviewers went through user scenarios that were divided into tasks of the following format:

Task 1
I want to set first row as header.
Expert evaluation (questions answered by reviewers):

a. Will the user know what to do next?
b. Will the user get appropriate feedback if the correct action is taken?

The sessions resulted in an eight pages long document that describes the responses from the reviewers, and includes a discussion of the findings. To categorize and analyse the findings from the streamlined cognitive walkthrough sessions, a bottom-up approach [30] was used to organize and analyse the findings from the sessions. By using this method, we emphasize the advantage it provides by keeping the researcher open to the results the process will reveal. The method requires more time to organize and analyse than would a top-down approach that starts with predefined concepts, but this disadvantage is outweighed by the potential of identifying more usability issues.

Table 2. Identified usability issues and suggestions for further research

Category	Usability issues	Suggestions for further research
Visual data profiling	• Some of the charts are not domain specific enough • The functionality and purpose of each visual data profiling chart are not clear • Outlier detection and correction of missing values are too generic	• Explore visual recommender system approaches to suggest relevant and domain specific charts to the user • Explore approaches that include multivariate data profiling (i.e. by profiling two or more columns to reveal relevant information related to data cleaning and transformation)
'Excel' table view	• Missing information about data type of selected values. Lack of possibility to specify parameters directly in the table view	• Explore direct table manipulation approaches to data cleaning and transformation to extend capabilities of the tabular table view
'Suggested transformations' sidebar	• The sidebar is overlooked/ignored in several cases because the suggested transformations are too generic and not specifically aimed at the current dataset • Users also prefer to use the right-clicking functionality of the Excel-like table view	• Explore approaches within predictive data cleaning and transformation, based on machine learning techniques, to provide more intelligent and relevant suggestions

The main findings from the reviews are summarized and categorized in Table 2 above. With each type of usability issue follows a suggestion on how the issue could be corrected.

In terms of *learnability* of the visual data profiling approach, the expert reviews show that the system needs to recommend charts that are domain specific and relevant to the user. This improvement will probably increase the speed, and ease of use, of learning new and basic functionality to perform the specific data cleaning and transformation tasks. Advanced capabilities (i.e. clicking and zooming charts to display detailed information) are not intuitive, and should be considered moved up one level in the user interface hierarchy to be visible always (e.g. by providing access to detailed information in a drop-down menu). The expert reviews also identified a need for a more consistent pattern of visual data profiling sequences (e.g. every time a user clicks a table column, he or she would know what happens next in the visual data profiling view).

Furthermore, the table view and 'Suggested transformation sidebar' need to be consistent by displaying the exact same range of data cleaning and transformation options. Users were confused when only a subset of options were available when right-clicking the table view. The approach should also consider including a mode where the sidebar 'Suggested transformations' can be hidden on demand by the user to free up more space for the table view.

In general, the expert reviews indicate that users were satisfied with the immediate feedback that the visual data profiling approach provided. Feedback included information such as status of missing values, potential extreme values, and number of distinct values. Still, the partial lack of explicit feedback after clicking columns and rows of the table view, resulted in uncertainty about which parts of the dataset had been profiled. Hence, the visual data profiling approach should provide immediate feedback to the user by indicating which columns or rows have been selected, and indicate the data type of the values.

6 Summary and Outlook

With the increasing amounts of data in today's organizations and businesses, proper data quality has become essential to extract and analyse content from large volume data sources. Incorrect or inconsistent data may distort the results of analysis processes, and reduce the potential benefits of applying data-driven approaches in organizations. Furthermore, data scientists spend more than half their time on preparing data for analysis. Hence, there are considerable research opportunities to ease the process of data cleaning and transformation, and improve data quality.

As a response to the demand for solutions that improve data quality and reduce time spent on cleaning and transforming data, this paper proposes a visual data profiling approach that implements powerful visual data profiling capabilities. The visual data profiling approach has been evaluated in terms of usability, and found to be perceived useful and easy to use by users. Furthermore, critical usability issues have been identified and proposed as further work in future iterations of the prototype. We have also contributed to proposing a visual data profiling approach that can be further researched and implemented on the DataGraft platform to extend, or replace, the existing version of Grafterizer.

Future work includes the implementation of a visual recommender system for data profiling that can recommend relevant, personalized and domain specific visualizations to the user. Furthermore, the visual data profiling approach would benefit from combining a visual recommender system and an intelligent approach to the domain-specific data cleaning and transformation problem. Such a framework could relieve the burden of technical specification in a domain specific language, and guide the user through an incremental process of cleaning and transforming data.

Acknowledgements. The work in this paper is partly supported by the EC funded projects proDataMarket (Grant number: 644497), euBusinessGraph (Grant number: 732003), and EW-Shopp (Grant number: 732590).

References

1. Hellerstein, J.M.: Quantitative data cleaning for large databases. United Nations Economic Commission for Europe (UNECE), February 2008
2. Kandel, S., Parikh, R., Paepcke, A., Hellerstein, J.M., Heer, J.: Profiler: integrated statistical analysis and visualization for data quality assessment. In: Proceedings of the International Working Conference on Advanced Visual Interfaces, New York, NY, USA, pp. 547–554 (2012)
3. Redman, T.C.: Bad Data Costs the U.S. $3 Trillion Per Year. Harvard Business Review, 22 September 2016. https://hbr.org/2016/09/bad-data-costs-the-u-s-3-trillion-per-year. Accessed 18 Mar 2017
4. CrowdFlower|2016 Data Science Report. https://visit.crowdflower.com/data-science-report. Accessed 19 Mar 2017
5. Han, J., Pei, J., Kamber, M.: Data Mining: Concepts And Techniques. Elsevier, Amsterdam (2011)
6. Sukhobok, D., et al.: Tabular Data Cleaning and Linked Data Generation with Grafterizer. ESWC (Satellite Events), pp. 134–139 (2016)
7. Sukhobok, D., Nikolov, N., Roman, D.: Tabular data anomaly patterns. In: Proceedings of the 3rd International Conference on Big Data Innovations and Applications (Innovate-Data 2017), 21–23 August 2017, to appear
8. Roman, D., et al.: DataGraft: One-Stop-Shop for Open Data Management. In: The Semantic Web Journal (SWJ) – Interoperability, Usability, Applicability. IOS Press (2017, to appear). ISSN 1570-0844
9. Roman, D., et al.: Datagraft: Simplifying open data publishing. In: ESWC (Satellite Events), pp. 101–106 (2016)
10. Roman, D., et al.: DataGraft: a platform for open data publishing. In: Joint Proceedings of the 4th International Workshop on Linked Media and the 3rd Developers Hackshop. (LIME/SemDev@ESWC 2016) (2016)
11. Stolte, C., Tang, D., Hanrahan, P.: Polaris: a system for query, analysis, and visualization of multidimensional relational databases. IEEE Trans. Visual Comput. Graphics 8(1), 52–65 (2002)
12. Kandel, S., Paepcke, A., Hellerstein, J., Heer, J.: Wrangler: interactive visual specification of data transformation scripts. In: Proceedings of the SIGCHI Conference on Human Factors in Computing Systems, pp. 3363–3372 (2011)
13. Mutlu, B., Veas, E., Trattner, C., Sabol, V.: VizRec: a two-stage recommender system for personalized visualizations. In: Proceedings of the 20th International Conference on Intelligent User Interfaces Companion, New York, NY, USA, pp. 49–52 (2015)
14. Voigt, M., Franke, M., Meissner, K.: Using expert and empirical knowledge for context-aware recommendation of visualization components. Int. J. Adv. Life Sci 5, 27–41 (2013)
15. Mutlu, B., Veas, E., Trattner, C., Sabol, V.: Towards a recommender engine for personalized visualizations. In: International Conference on User Modeling, Adaptation, and Personalization, pp. 169–182 (2015)
16. Wongsuphasawat, K., Moritz, D., Anand, A., Mackinlay, J., Howe, B., Heer, J.: Voyager: exploratory analysis via faceted browsing of visualization recommendations. IEEE Trans. Visual Comput. Graphics 22(1), 649–658 (2016)
17. Vega-Lite. https://vega.github.io/vega-lite/. Accessed 19 Mar 2017
18. Wilkinson, L.: The Grammar of Graphics. Springer Science & Business Media, New York (2006)

19. Wickham, H.: ggplot2: Elegant Graphics for Data Analysis. Springer, New York (2016)
20. Mackinlay, J., Hanrahan, P., Stolte, C.: Show me: automatic presentation for visual analysis. IEEE Trans. Visual Comput. Graphics 13(6), 1137–1144 (2007)
21. Satyanarayan, A., Russell, R., Hoffswell, J., Heer, J.: Reactive vega: a streaming dataflow architecture for declarative interactive visualization. IEEE Trans. Visual Comput. Graphics 22(1), 659–668 (2016)
22. Bakke, E., Karger, D.R.: Expressive query construction through direct manipulation of nested relational results. In: Proceedings of the 2016 International Conference on Management of Data, pp. 1377–1392 (2016)
23. The Guide to Prototyping Process & Fidelity. Studio by UXPin. https://www.uxpin.com/studio/ebooks/prototyping-process-fidelity-guide/. Accessed 13 Apr 2017
24. Heer, J., Hellerstein, J.M., Kandel, S.: Predictive interaction for data transformation. In: CIDR (2015)
25. Chen, S.: Six Core Data Wrangling Activities eBook. Trifacta, 23 November 2015
26. Hanington, B., Martin, B.: Universal Methods of Design: 100 Ways to Research Complex Problems, Develop Innovative Ideas, and Design Effective Solutions. Rockport Publishers, Gloucester (2012)
27. The ultimate guide to prototyping. Studio by UXPin. https://www.uxpin.com/studio/ebooks/guide-to-prototyping/. Accessed 13 Apr 2017
28. Familiar, B.: Microservices, IoT and Azure: Leveraging DevOps and Microservice Architecture to deliver SaaS Solutions. Apress, New York (2015)
29. Davis, F.D.: Perceived usefulness, perceived ease of use, and user acceptance of information technology. MIS Q. 13, 319–340 (1989)
30. Barnum, C.M.: Usability Testing Essentials: ready, set… Test! Elsevier, Amsterdam (2010)
31. Sauro, J., Lewis, J.R.: Quantifying the User Experience: Practical Statistics for User Research. Morgan Kaufmann, Burlington (2016)
32. Nielsen, J.: Usability inspection methods. In: Conference Companion on Human Factors in Computing Systems, pp. 413–414 (1994)
33. Spencer, R.: The streamlined cognitive walkthrough method, working around social constraints encountered in a software development company, pp. 353–359 (2000)
34. Mahatody, T., Sagar, M., Kolski, C.: State of the art on the cognitive walkthrough method, its variants and evolutions. Intl. J. Hum.-Comput. Interact. 26(8), 741–785 (2010)
35. Cognitive Walkthrough|Usability Body of Knowledge. http://www.usabilitybok.org/cognitive-walkthrough. Accessed 10 May 2017

Semantic-Based Approach for Low-Effort Engineering of Automation Systems

Aparna Saisree Thuluva[1,2]($\boxtimes$) (iD), Kirill Dorofeev[3], Monika Wenger[3],
Darko Anicic[1]($\boxtimes$), and Sebastian Rudolph[2]($\boxtimes$) (iD)

[1] Siemens AG - Corporate Technology, Munich, Germany
{aparna.thuluva,darko.anicic}@siemens.com
[2] TU Dresden, Dresden, Germany
sebastian.rudolph@tu-dresden.de
[3] Fortiss An-Institut Technische Universität München, Munich, Germany
{dorofeev,wenger}@fortiss.org

Abstract. Industry 4.0, also referred to as the fourth industrial revolution aims at mass customized production with low-cost and shorter production time. Automation Systems (ASs) used in the manufacturing processes should be flexible to meet the constantly changing needs of mass customized production. Low-effort engineering of an Automation System (AS) is an important requirement towards this goal. Secondly, transparency and interoperability of ASs across different domains open a new class of applications. In order to address these challenges we propose a low-effort approach to engineer, configure and re-engineer an AS by employing Web of Things and Semantic Web Technologies. The approach allows for creating semantic specification for a new functionality or an application. It automatically checks whether a target AS can run a new functionality. We developed an engineering tool with a graphical user interface for our approach that enables an engineer to easily interact with an AS when discovering its functionality, engineering, configuring and deploying new functionality on it.

Keywords: Low-effort engineering · Web of Things · Semantic Web of Things · Industry 4.0 · Mass customized production · Semantic-based engineering · Semantic matchmaking

1 Introduction

Mass Customized Production aims at producing individualized products to meet the needs of every customer. It requires constantly changing settings to manufacture customized products, which increases the production cost and time [1]. Industry 4.0 aims to reduce the cost and shorten the production time[1] [2] by maximizing the digitalization and automation in manufacturing processes using

---

[1] https://ukmanufacturing2015.eng.cam.ac.uk/proceedings/Industry4.0AN10715.pdf.

© Springer International Publishing AG 2017
H. Panetto et al. (Eds.): OTM 2017 Conferences, Part II, LNCS 10574, pp. 497–512, 2017.
https://doi.org/10.1007/978-3-319-69459-7_33

Automation Systems (ASs). In order to fulfill the requirements of mass customized production, ASs used in a manufacturing process should be flexible. A task in a manufacturing process is usually achieved through ASs by communicating, co-operating and exchanging information with each other.[2] Therefore, ASs should be flexible in order to easily engineer and configure the interactions between them. Moreover, typically an Automation System (AS) possesses under-used equipment, which if used efficiently can reduce the cost and time for manufacturing products. In order to achieve this, an AS should be flexible to update its functionality. That is, an AS equipment must be transparent to an engineer when creating a new functionality or re-configuring an existing one. Further on, interoperability between ASs that belong to diverse domains is another challenge that should be addressed[3]. These ASs use different communication protocols such as Profinet[4], Modbus [3], OPC UA[5] and so on, which restrains the interoperability between them. In the direction of addressing these challenges, Web of Things (WoT) and Semantic Web Technologies (SWT) are good candidates. SWT enable interoperability across domains by formalizing the information models using ontologies, which enable interoperability by providing un-ambiguous and machine-readable descriptions.

Recently ASs in the manufacturing process are becoming part of Internet of Things (IoT) [4]. In IoT, physical devices are embedded into electronic systems which can connect to the Internet, where they can be discovered, monitored, controlled and interacted with each other over various network interfaces. But IoT lacks a universal application protocol that can work across many networking interfaces, which restrains the integration of devices from diverse domains into a single application [5]. This is addressed by Web of Things (WoT) by leveraging the Web standards to physical devices that enable devices from diverse domains to be integrated into Web applications with minimal effort [6]. Initial standards are being developed for WoT by the W3C WoT working group[6] to create an abstract layer for inter-operable IoT applications.

Therefore, we apply WoT and SWT to address the above challenges and propose an approach to engineer, configure and re-engineer an AS with minimum effort. Our main contributions are the following: (1) we enable an AS infrastructure to be transparent for rapid application development, (2) we develop an approach, which allows to create semantic specification for a functionality or an application, (3) the approach allows to automatically checks whether a target AS has the capability to run a new application (4) we developed an engineering tool with graphical user interface called Semantic Web of Things for Automation Systems (SWAS) tool, to support an AS engineer in tasks (1), (2), (3).

---

[2] https://www.plattform-i40.de/I40/Redaktion/EN/Downloads/Publikation/interaction-model-I40-components.pdf.

[3] https://ukmanufacturing2015.eng.cam.ac.uk/proceedings/Industry4.0AN10715.pdf.

[4] http://us.profinet.com/technology/profinet/.

[5] https://opcfoundation.org/about/opc-technologies/opc-ua/.

[6] https://www.w3.org/WoT/WG/.

2   State of the Art

The state of the art in engineering of ASs is based on model-driven software design. It is divided into five phases: 1. Design phase 2. Development phase 3. Engineering phase 4. Commissioning phase and 5. Operation phase [7]. According to the model-driven design, an engineer specifies a field function or a data point in a model in the Design phase. The code generation is run to produce a skeleton of a service that is supposed to implement the function or data point in the Development phase. Finally, in the Engineering phase, the engineer implements the service skeleton, deploys and configures it on an AS in the Commissioning phase and starts it.

In contrast to this, we propose an engineering approach where semantics is used for multiple purposes. First, it enables transparency of an AS equipment for rapid application development by providing semantic descriptions for it (semantic description for device functionalities, configurations and relations to other devices). Second, the same semantic-based approach can be applied for cross-domain interoperability between ASs. Further on, in the design phase the approach allows an engineer to create semantic specification for a new functionality. Interactions between functionalities specified in an application are established using semantics (e.g. to match whether the output of one function matches with the input of other function) in the engineering phase. Semantics is further used to discover the existing functionality on the target AS and to check if the target AS has the functionality to run an application. In commissioning phase, parameterization of AS equipment is provided based on semantic constraints of that equipment. Finally, it also facilitates deployment of an application on an AS, thanks to the proposed run-time (which is introduced in the next section) for device level semantic processing.

State of the art in engineering of ASs also includes, the Totally Integrated Automation Portal (TIA Portal).[7] It is a software framework for controller programming, HMI development and drives parameterizing. TIA Portal, among other tasks, is responsible for allowing an engineer to configure and program Siemens Programmable Logic Controllers (PLC), allowing a programmer to create control applications that complete various manufacturing tasks.

3   Use Cases

In this section we describe two industrial use cases for engineering and re-engineering of an AS on the FESTO[8] Process Automation workstation shown in Fig. 1. It consists of two tanks. There are various sensors and actuators attached to the tanks such as the following: an ultrasonic sensor, float sensors, proximity

---

[7] https://www.siemens.com/global/en/home/products/automation/
industry-software/automation-software/tia-portal.html.

[8] http://www.festo-didactic.com/int-en/learning-systems/process-automation/
compact-workstation/mps-pa-compact-workstation-with-level,flow-rate,
pressure-and-temperature-controlled-systems.htm.

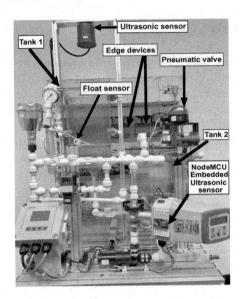

Fig. 1. FESTO process automation workstation

sensors, temperature sensor, heater, overflow sensor, pump, a pneumatic valve and so on. The workstation captures the process of steering liquid, measuring level of liquid, protecting overflow of liquid, protecting a pump from dry run, measuring flow of liquid, keeping liquid within certain temperature range and within certain level range. Therefore, the workstation is already complex as found in real production environment.

The ultrasonic sensor and a float sensor are deployed on Tank 1. The float sensor is a binary sensor, which detects the overflow of liquid in Tank 1. The ultrasonic sensor measures the level of liquid in Tank 1. A pneumatic valve controls liquid flow from Tank 1 to Tank 2. In order to automate the process of ensuring overflow protection on Tank 1, the devices on the workstation can be engineered as follows:

- **Engineering use case:** An engineer can design, engineer and configure the interactions between AS devices with minimum effort in order to fulfill a task specified by an automation process.
  Example: In order to ensure overflow protection on Tank 1, the devices on the workstation should be engineered and configured in such a way that, when overflow is detected on Tank 1, then liquid in Tank 1 should be pumped out to Tank 2.
- **Re-engineering use case:** An engineer can update the functionality on an AS with minimum effort simply by installing a new functionality on it.
  Example: If any of the device on the FESTO workstation used in the above engineering process is malfunctioning, then the devices should be re-engineered to ensure overflow protection on Tank 1.

We use these two as the running examples to describe our methodology for engineering, configuration and re-engineering of an AS, in a step-by-step manner in Sect. 4.

4 Methodology for Semantic-Based Engineering of Automation Systems

In this section we describe our approach for semantic-based engineering of an AS. The basic setup for this approach is that an AS is WoT-enabled by embedding it into an electronic system that can connect to the Web and interact with other devices using existing Web standards[9] [8–10].

4.1 Building Blocks

Figure 2a shows the device building blocks of our approach, which are embedded into a WoT-enabled AS. They enable an AS to interpret semantic models locally and to control the operation of AS devices. The WoT-enabled ASs under our consideration are resource-constrained devices that have limited memory and processing power. The PC-based semantic reasoning techniques are not feasible to be applied on these devices [11]. Therefore, we use an embedded **Micro-reasoner**, which in the scope of this work is deployed on an embedded Linux system. It consists of two parts: **Micro Event Engine** (MEE) implemented in C and a **Datalog reasoner**, which is an open source C and LUA[10] implementation.[11] MEE is based on the work from [12]. MEE uses Event rules to do Complex Event Processing (CEP) of the events coming from AS devices. Datalog reasoner facilitates a device with rule-based reasoning in datalog [13]. Both the components are integrated into Micro-reasoner, which enables it to process

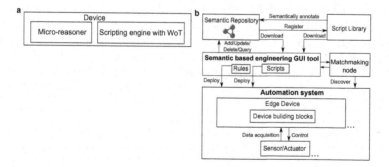

Fig. 2. System architecture of Semantic-based low-effort engineering of an AS (a) Device building blocks (b) Integrated system architecture

---

[9] http://mqtt.org/documentation.
[10] https://www.lua.org/.
[11] http://www.ccs.neu.edu/home/ramsdell/tools/datalog/datalog.html.

events with MEE in the context evaluated by Datalog reasoner. If needed, they can also be used separately. Micro-reasoner is offered as a CoAP RESTful web service, which provides an easy way to interact with Micro-reasoner to add or delete event rules and add, query and delete datalog facts and rules over a RESTful API. In addition to Micro-reasoner, a scripting engine with WoT Scripting API[12] implemented in LUA is also embedded into an AS as shown in Fig. 2a. The scripting engine interprets the scripts sent to the AS. In our approach the Micro-reasoner runs on an edge device (we used SIMATIC IOT2040[13]) which is a part of an AS as shown in Fig. 2b. The system architecture of our approach will be further extended in the next sections.

Example: Consider the engineering use case example presented in the Sect. 3. Here we show how the engineering of the AS devices can be done (to ensure overflow protection on Tank 1) by using the device building blocks on edge devices.

The FESTO workstation shown in Fig. 1 is equipped with three edge devices. All sensors and actuators on the workstation are embedded with NodeMCUs[14] and then they are connected to the edge devices. Suppose that an engineer wants to prevent the overflow on Tank 1 by putting into interaction the float sensor on Tank 1 and the pneumatic valve. When float sensor raises an overflow event, then the pneumatic valve opens and lets the liquid flow from Tank 1 to Tank 2. Here we show how this can be done by deploying a simple event rule and a WoT script (which are shown below). The rule and script are deployed on the edge device which controls float sensor and pneumatic valve.

The rule head "valveOpenAction(Y)" is triggered when the rule body is executed. The rule body consists of an event, "overflow(X)". It is an event raised by the float sensor when the overflow occurs in Tank 1. In such a case, the where clause of the rule body is executed. The where clause in the rule below consists of "valveOpen(X)" predicate. The engine invokes an external function which in our case is the WoT script shown below. The script is used to open the valve on a device with provided URI. The function "invokeAction" is defined in WoT scripting API, it takes "action name" and its input parameters as input and invokes the action. In our example, the name of an action on the pneumatic valve is "open", it takes a boolean value as input to control the valve operation. Therefore, when the event "overflow(true)" occurs, then the script is executed and the valve gets opened. In this way the overflow protection on Tank 1 is ensured.

$$
\begin{aligned}
\texttt{valveOpenAction[\_,\_](Y) :- overflow[\_,\_](X) where} \\
\texttt{(X = true, Y := valveOpen(X)) [count 1].}
\end{aligned}
\tag{1}
$$

---

[12] https://w3c.github.io/wot-scripting-api/.

[13] http://docs-europe.electrocomponents.com/webdocs/1536/0900766b815365c3.pdf.

[14] https://nodemcu.readthedocs.io/en/master/.

Listing 1.1. WoT script to invoke action on a pneumatic valve

```
function valveOpen(X) wot = require("wot")
  valve = wot.consumeDescriptionUri("coap
    ↪ ://192.168.2.60:5683/Valve/")
  valve:invokeAction("open", "{\"value\":"..X..""}")
  return true
end
```

In this example an event rule and a script are used to engineer an AS. That is, to establish interactions between AS devices (e.g., float sensor and pneumatic valve). In the similar manner an event rule and/or a script can be used to define a new functionality, which in turn can be used to update the functionality of an AS. In more complex cases an action can be triggered on detection of a complex event. For example, instead of the single overflow event, a rule can trigger an action upon detection of multiple events with certain temporal and semantic constraints. This rule language is based on the work from [12].

4.2   System Architecture

Figure 2b presents the system architecture of our approach integrated with device building blocks. Firstly, the semantic repository is used to store the semantic descriptions of new functionalities and applications, which are used to engineer or update the functionality of an AS. For this purpose, we extended the thingweb repository developed by the W3C WoT Working Group. It is an Apache Jena TDB[15] with a RESTful interface.[16] The second component is the script library, which is used to store scripts for engineering an AS. These scripts are semantically annotated and stored in the repository in order to facilitate their discovery and re-use. The next component in the system architecture is the "semantic-based engineering tool". It provides a graphical user interface to do end-to-end engineering in accordance to our approach. The tool provides an interface to the semantic repository in order to add, delete, update application semantic descriptions to the repository and to download them from the repository. The tool also provides an easy access to ASs by deploying event rules or scripts on an AS. The tool interacts with Matchmaking node in order to discover functionality on an AS. The tool and Matchmaking node connect to an edge device which is the part of an AS. Each of these components will be explained in detail in the following sections.

4.3   Modeling

Device Semantic Modeling. Semantic modeling of an AS and its devices plays an important role to address the challenges of cross-domain interoperability and transparency of ASs. A semantic description of an AS acts as an

---

[15] https://jena.apache.org/documentation/tdb/.
[16] https://github.com/thingweb/thingweb-repository.

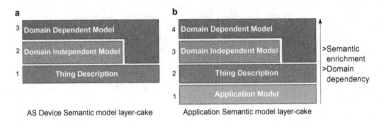

Fig. 3. Semantic layer-cake approach (a) Device semantic modeling approach (b) Application semantic modeling approach

interface to it. In order to model the WoT-enabled AS, their services and applications semantically, we developed a semantic-layering approach as shown in Fig. 3a. According to this approach, semantic modeling of an AS is done in 3 layers. In the first layer an AS is described in terms of its functionalities and configurations. We use W3C WoT Thing Description (TD) in this layer. The motivation for WoT is to access physical devices on the Web in a similar fashion as accesing Web pages. In order to facilitate this, TD of a device is stored on the device itself. This enables the transparency of an AS infrastructure for rapid application development. TD also enables cross-domain interoperability between ASs from diverse domains such as energy, manufacturing and so on, by providing an abstract description of their functionalities in terms of properties, events and actions. A TD is serialized in JSON-LD[17] format. In the second layer, TD is enriched with domain-independent vocabulary to model contextual information and non-functional properties of an AS device. For this purpose, we reuse the widely accepted existing vocabularies such as, Semantic Sensor Networks ontology (SSN) [14]. It models sensors, actuators and their contextual information semantically. QUDT[18] is used to model the quantities and their units of measurement. The schema.org[19] ontology is used to model the non-functional properties. Common set of domain-independent vocabularies are used to describe ASs, their services and applications which creates interoperability between ASs. The third layer is called domain-dependent layer. This layer models the domain specific features of an AS. For example, eCl@ssOWL [15] which is an OWL ontology that models Industrial devices semantically is used to model the AS and its devices that belong to industrial domain. Figure 4a shows the TD of the ultrasonic sensor from the FESTO workstation enriched with domain-independent and domain-dependent semantic annotations.

Application Semantic Modeling. In our approach, an application is a semantic description of an event rule and/or a WoT script (see Sect. 4.1), which can be used either to **engineer** an AS to establish the interaction between AS devices

---

[17] https://www.w3.org/TR/json-ld/.
[18] http://www.qudt.org/.
[19] http://schema.org/.

as shown in Sect. 4.1 or to **re-engineer** an AS by updating its functionality. The application semantic modeling is done using semantic-layering approach as shown in Fig. 3b. It differs from device semantic modeling only in the first layer. It has an additional layer called "Application model", which defines the vocabulary to model the logic of an application and its requirements. The Application model is shown in Fig. 5. It consists of two parts:

a. Ultrasonic Sensor Thing Description

```
{ "@context" :
    ["https://w3c.github.io/wot/w3c-td-
        context.jsonld",
    "http://SWAS/interactions/interactions-
        context.jsonld"],
    "qu":
        "http://purl.oclc.org/NET/ssnx/qu/qu#",
    "ssn":http://purl.oclc.org/NET/ssnx/ssn#,
    "schema": "http://schema.org/",
    "eclass": http://www.ebusiness-
        unibw.org/ontologies/eclass/5.1.4/#,
    "name": "MyUltrasonicSensor",
    "@type": ["ssn:Sensor",
        "eclass:C_AKE655002-tax"],
    "uris" :
        ["coap://192.168.2.82:5683/ultrasonicSe
            nsor",
        "http://192.168.2.82:8080/ultrasonicSen
            sor"],
    "encodings":["JSON"],
    "ssn:onPlatform":"Tank1",
    "ureasoner" : "true",
```

```
    "properties":[ {
        "@id" : "level",
        "@type":
    http://SWAS/interactions/liquidLevelProperty,
        "name" : "liquidLevel",
        "valueType" : {"type" : "float"},  .
        "writable" : "false",
        "qu:unit" : {"@type"  : "qu:millimetre"},
        "schema:minValue" : "0.2",
        "schema:maxValue" : "800",
        "hrefs" : ["liquidLevel"] } ]}
```

b. Ultrasonic Sensor Datalog facts

```
name("td","MyUltrasonicSensor").
ureasoner("td","true").
uris("td","coap://192.168.2.82:5683/ultrasonicSensor").
onPlatform("td","Tank1").
properties("td","propetery1")
name("propetery1","liquidLevel").
hrefs("propetery1","liquidLevel").
@type("property1",http://SWAS/interactions/liqui
    dLevelPropetry").
```

Fig. 4. (a) Thing Description of an ultrasonic sensor (b) Datalog facts generated from the ultrasonic sensor Thing Description

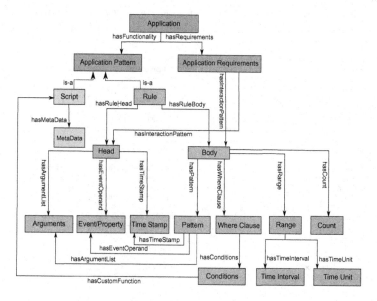

Fig. 5. The Application semantic model

1. **Application pattern:** describes the logic of an application. The Application model is an OWL ontology, which defines the schema to model an event rule or a script. It models a script in terms of its meta-data or a rule in terms of rule head, rule body, rule pattern, the list of arguments, where clause and the range or count of a rule and so forth. The Property or Event from the TD model are used as event operands in this model.
2. **Application requirements:** models the requirements that should be fulfilled by a target AS in order to install an application. The class "Application Requirements" is used to model these requirements. The typical requirements are, a target device or devices should match the functionalities modeled in the rule body and/or rule head. It should also match the additional requirements specified in the "hasRequirements" part of an application.

The applications are serialized in JSON-LD format in the similar manner as TD. They are stored in the semantic repository as shown in Fig. 2. Figure 6 shows an application semantic description, which defines a new functionality for the re-engineering example described in Sect. 3. The logic of the application is the following: the rule body observes the liquid level in a tank and raise an overflow event when the level is above certain threshold. In order to install this application a target AS should have the functionality to measure the level of liquid in a tank, which gives liquid measurements in floating point data-type. Lastly, Micro-reasoner should be running on the target AS. These requirements are modeled in the "hasRequirements" part of the application.

An application semantic modeling has the following benefits: (1) the model is directly install-able on an AS with Micro-reasoner, which can interpret it and run it without any code generation or implementation. Therefore, it lowers the effort during engineering process, (2) the model can be discovered, shared, extended and documented for later use (3) modeling the application requirements

```
{"@context":
["https://w3c.github.io/wot/w3c-wot-td-context.jsonld",
"http://SWAS/applicationvocabulary.jsonld"
"name" : "overflowEventApplication",
"@type" : "application",
"description" : "Raises an overflow event when the
liquid level is above certain threshold in a tank"
"ruleId":
["coap://SWAS/applications//overflowAlarmApp"],
"hasRuleBody":{
"hasPattern":{
"hasEventOperand":{
 "Property":{
  "@type":
  ["http://SWAS/interactions/liquidLevelProperty"]},
"hasEventArguments":{
 "ArgumentList" : ["X"]}}},
"hasWhereClause" : {
 "ConditionsList" : [{
 "Condition" : {
  "RelativeOperator":
{"@type" : "GreaterThanOrEqualTo"},

"hasLeftWhereConditionArgument" : {
 "Argument"      :"X"},
"hasRightWhereConditionArgument" : {
 "Argument"      :"50"}}}]},
"hasCount" : {
 "Count" : "1"}},
"hasRuleHead" : {
"Event" : {
 "name" : "overflowEvent",
      "@type" :
["http://SWAS/interactions/overflowEvent"],
 "valueType" : {"type" : "boolean"},
 "hrefs" : ["overflowEvent"] },
"hasEventArguments" : {
 "Arguments" : ["X"] }},
"hasRequiements" : {
 "hasInteractionPattern" :{
 "Property" : {
  "@type"
["http://SWAS/interactions/liquidLevelProperty"],
  "valueType" : {"type" : "float"} },
  "ureasoner" : "true"} }
```

Fig. 6. Semantic description of Overflow Event Application serialized in JSON-LD format

semantically enables matchmaking between the application requirements and a target AS functionality to be done automatically, therefore, it reduces the human intervention during the engineering process.

4.4 De-centralized Device Discovery and Matchmaking

In our approach TD of an AS is stored locally on the AS itself to enhance transparency of AS equipment as described in Sect. 4.3. On one hand, it optimizes maintainability of AS functionality as TD can be updated easily whenever the functionality of the AS is updated. Further on, it simplifies Plug and Play of an AS (new devices are added to the system on the fly). In order to discover functionality of an AS during an engineering process, we propose a de-centralized approach, which enables local discovery of AS functionality. To implement this, firstly all AS involved in an engineering process should be discovered. An AS has an embedded edge device, which in turn has Micro-reasoner running on it. Micro-reasoner is offered as a CoAP server in our implementation. Therefore, we implemented CoAP multicasting to discover all the edge devices running on AS as explained below:

Device Discovery. The Constrained Application Protocol (CoAP) is a specialized web transfer protocol for use with constrained nodes and constrained (e.g.,low-power, lossy) networks [16]. One of the discovery mechanism specified by this protocol is a *service discovery* that allows to discover all CoAP servers in the network, which support the discovery functionality. Using the service discovery mechanism, we implemented a CoAP-discovery in Eclipse Californium[20] - a Java implementation of CoAP protocol. CoAP supports requests to a multicast group. Therefore, a CoAP-server with enabled discovery functionality, being plugged into the network, should also join a multicast group to be able to get CoAP packets sent not only to the server directly, but also to the predefined "All CoAP Nodes" address group. Internet Assigned Numbers Authority (IANA) assigned the following addresses for the use by CoAP nodes as "All CoAP Nodes" address: 224.0.1.187[21] and FF0X:FD15 for IPv4 and IPv6 respectively. A GET-request to this multicast-group address results in a series of CoAP-responses from all CoAP-servers that previously joined the "All CoAP Nodes" group. In our implementation a device that has a CoAP-server running after start-up iterates through all its IPv4 interfaces, excluding the loop-back interfaces, and, if set in the server configuration file, each socket, created for the corresponding interface joins the CoAP multicast group. Then any CoAP-client can send a multicast request, getting a corresponding address to connect to the discovered server. Note, that in case if client and server are connected to the network using multiple interfaces, this procedure allows us to find a way to establish the connection, if there exists any. The CoAP multicasting is implemented in Matchmaking node shown in Fig. 2. Moreover, all the edge devices are enabled with discovery

---

[20] https://eclipse.org/californium/.

[21] https://www.iana.org/assignments/multicast-addresses/multicast-addresses.xhtml.

functionality so that they can be discovered when a multicast request is sent by Matchmaking node.

Discovery of AS Functionality and Matchmaking. The local discovery of AS functionality and matchmaking on an AS is enabled by Datalog reasoner embedded into an edge device. We developed an algorithm to convert TDs on an edge device into datalog facts. These facts along with some datalog rules are then stored into Datalog reasoner to enable query answering about AS functionality and to do matchmaking between its functionality and an application requirements automatically using datalog rule-based reasoning. As Micro-reasoner is offered as a CoAP server, an engineer or another AS can connect to it and query about its functionality.

During an engineering process, an engineer first discovers all the edge devices on an AS using CoAP multicasting and then sends datalog queries to each edge device to discover their functionalities. Figure 4b shows some of the datalog facts generated from the ultrasonic sensor TD. These facts are added to Datalog reasoner. Additionally, the datalog rules, which are shown below are also added to the reasoner, which enables the AS to answer queries about its functionality.

$$\text{hasInteraction(ThingUri, IType) :-} \atop \text{uris(T, ThingUri), properties(T, P), @type(P, IType).} \tag{2}$$

$$\text{thingType(Name, Type) :- @type(T, Type), name(T, Name), uris(T, U).} \tag{3}$$

$$\text{thingType(Name,SType) :-} \atop \text{subClassOf(Type, SType),thingType(Name, Type).} \tag{4}$$

$$\text{hasInteractionOnPlatform(ThingUri, Platform) :-} \atop \text{uris(T, ThingUri), ureasoner(T, UR), properties(T, X),} \atop \text{@type(X, IType), valueType(X, Y), type(Y, Z),} \atop \text{onPlatform(X, Platform).} \tag{5}$$

If the AS has the capability to run a new application, then it sends its URI and the platform it is deployed on, as a response to the datalog query sent by Matchmaking node. Therefore, the application can be installed on the AS which fulfills its requirements. Below we present the datalog query generated by Matchmaking node from the overflow event application description shown in Fig. 6.

$$\text{hasInteractionOnPlatform(ThingUri, Platform) :-} \atop \text{uris(T, ThingUri), ureasoner(T,"true"), properties(T, X),} \atop \text{@type(X, "http://SWAS/interactions/liquidLevelProperty"),} \atop \text{valueType(X, Y), type(Y, "float"), onPlatform(X, Platform)?} \tag{6}$$

4.5   Semantic-Based End-to-End Engineering Tool

We developed an engineering tool with a graphical user interface called SWAS tool to enable engineering and re-engineering of an AS with our semantic-based approach. A screen-shot of the tool is shown in Fig. 7. It is an Eclipse application implemented in Java. It supports an AS engineer in all phases of an engineering process. The design goal of the tool is not creation of semantic models. Semantic models are created using ontology engineering tools and stored in a semantic repository. Instead, the tool enables an engineer to easily interact with an AS. The interaction with semantic repository is also provided, as well as interaction with matchmaking node in device discovery phase.

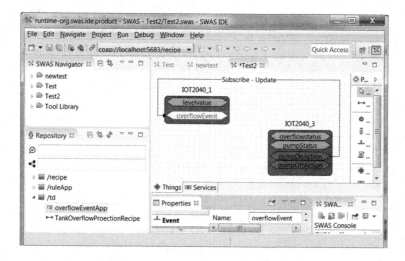

Fig. 7. Semantic-based engineering tool with graphical user interface

The engineering process is done with SWAS tool in a step-by-step manner as described below: (1) an AS engineer first discovers the required application from the semantic repository using a SPARQL BGP or Text-based query, (2) then he selects the required application and presses the "Discover" button to discover the matching devices on an AS which can install the application, (3) the list of matching devices will be displayed in the Search Results. The engineer can select one of the matching devices where he wants to deploy the application, (4) then he can simply deploy the application on the device with the click of a button. Apart from this, scripts can also be deployed on the devices from the tool. After deployment of an application on a device, the tool can also be used to update the TD of an AS with the newly installed functionality.

5   Discussion

The goal of our approach is to make an AS flexible and transparent to meet the constantly changing settings of mass customized production. For this purpose, we proposed a semantic-based approach to enhance the transparency of an AS by semantically modeling an AS equipment's functionalities, capabilities and configurations. This reduces the reliability on domain knowledge of the experts and makes the knowledge available also to the Web developers. Therefore, even the Web developers can engineer the ASs and it also reduces the errors during the engineering process. This enhances the flexibility in engineering of ASs. On the other hand, semantic modeling of an AS information also enables interoperability between ASs from heterogeneous domains. In comparison to this, in state of the art engineering approaches a gateway has to be built between multiple domains in order to make them inter-operable which is significantly more complex than achieving interoperability by semantic modeling of AS with TD.

Moreover, engineering of an AS using state of the art engineering approaches is more complex and involves significant effort than the semantic-based approach proposed in this paper. Firstly, it is very complex to discover a library function that can be reused to engineer a new application. For this purpose only a string matching search is available, which makes the process extremely difficult. Typically, there exists numerous libraries and an engineer should know the name of the library required for the specific use-case. In contrast to this, in our approach, semantic modeling of an application enables semantic-based discovery of an application which is more efficient.

Secondly, the library function block should be then configured to map the program to the physical hardware, which, depending on the setup, can be seen as from easy to very complicated and error-prone process. In comparison to this, the semantic-based approach makes an application configuration and prameterization process simpler as the device configurations are semantically described and semantic constraints can be defined to guide the configuration process. For example: it can be checked whether configured parameters satisfy defined ranges (e.g., minimum, maximum values), units of measurement (e.g., degree Celsius) and so forth. Furthermore, if the interaction between devices is needed each of those interactions should be configured and programmed separately. But in our approach the interactions are semantically described. These semantic descriptions serve as a transparent documentation which can be used in later applications. Moreover, semantically described interactions can be configured and implemented with less effort. Lastly, matchmaking of application requirements and AS functionality is done automatically in the semantic-based approach, which cannot be done in state of the art engineering approaches.

SWAS tool is designed based on already established open source engineering tool called 4diac[22], taking into account the success of 4diac with respect to the expectations of engineers. However, our additional design decision was to develop an engineering tool, which would be easy to use for Web developers. The goal of our approach is not only to help engineers to lower the engineering effort but also to enable Web developers to become engineers of industrial ASs.

---

[22] https://eclipse.org/4diac/.

We proved that this new approach fulfills all these requirements by using semantics for multiple purposes. We conducted two experiments on the FESTO process automation workstation shown in Fig. 1. On this FESTO workstation, we attached a micro controller to each sensor and actuator of the workstation which provides basic web interface and semantic description of this interface using TD to interact with a sensor or actuator. Each edge device, that is, IoT2040 plays the role of a controller for sensors and actuators. It is equipped with Micro-reasoner, which enables it to interpret semantic descriptions of AS devices and applications (coming from SWAS tool). Our deployment is general in sense that can be applied to any sensor or actuator, and can be used when implementing other scenarios too. Moreover the procedure is simple and can be scaled to many ASs. On this setup we conducted experiments for engineering and re-engineering use cases using the examples described in Sect. 3. The experiments proved that it is feasible to do de-centralized discovery and automated matchmaking locally on an AS. This feature can be useful in an environment where new devices are plugged on the fly and they need to be instantly discoverable. Lastly, our experiments showed that SWAS tool simplifies the engineering process with its integrated user interface, which supports end-to-end engineering. Upon first interactions with engineers they stated that the tool was easy to use. We got positive and encouraging feedback about the UI. These experiments proved that our approach is feasible for engineering and re-engineering of ASs in real-world scenarios.

Our experiments were performed on process automation use cases. But in general the devices from diverse domains could be used. Even in this case, the interoperability can be enabled by describing them with TD in terms of properties, events and actions, and engineering can be done easily using our approach. In the re-engineering scenario, when a new change is required for manufacturing a product variant, a functionality can be added, updated or re-configured on an AS similarly using this approach.

6   Conclusion and Future Work

We proposed a semantic-based approach for low-effort engineering of an Automation System, making it flexible for mass customized production. It enables horizontal integration of systems by providing cross-domain interoperability between them, which opens up possibilities for new applications. The approach makes an automation equipment transparent in the sense that the equipment exposes its functionality via a semantic interface. We implemented local discovery and matchmaking on an Automation System to enable local decision support, and to enhance the autonomy of the device. Furthermore, we implemented a tool that provides an integrated user interface and simplifies the usability of our semantic-based engineering approach. We deployed our implementation on a FESTO process automation workstation to test the proposed engineering and re-engineering approach. The experience showed that our approach is feasible in the real-world scenarios.

Future steps will involve improvements on the matchmaking procedure for more complex use cases. We will perform extensive experiments on more complex test beds to test scalability of the implementation. Depending on results of the experiments our intention is to deploy it on the real industrial manufacturing plants.

References

1. Da Silveira, G., Borenstein, D., Fogliatto, F.S.: Mass customization: literature review and research directions. Int. J. Prod. Econ. **72**(1), 1–13 (2001)
2. Siemens: Modeling new perspectives: digitalization - the key to increased productivity, efficiency and flexibility (white paper). In: DER SPIEGEL (2015)
3. Modbus, I.D.A.: Modbus application protocol specification v1.1a. North Grafton, Massachusetts. 4 June 2004. www.modbus.org/specs.php
4. Shrouf, F., Ordieres, J., Miragliotta, G.: Smart factories in industry 4.0: a review of the concept and of energy management approached in production based on the internet of things paradigm. In: 2014 IEEE International Conference on Industrial Engineering and Engineering Management (IEEM), pp. 697–701. IEEE (2014)
5. Dominique, G., Vlad, T., Friedemann, M., Erik, W.: From the internet of things to the web of things: resource oriented architecture and best practices. In: Uckelmann, D., Harrison, M., Michahelles, F. (eds.) Architecting the Internet of Things, pp. 97–129. Springer, Heidelberg (2011). doi:10.1007/978-3-642-19157-2_5
6. Dominique, G., Vlad, T.: Towards the web of things: web mashups for embedded devices. In: Workshop on Mashups, Enterprise Mashups and Lightweight Composition on theWeb (MEM 2009), Proceedings of WWW (International World Wide Web Conferences), Madrid, Spain (2009)
7. Butzin, B., Golatowski, F., Niedermeier, C., Vicari, N., Wuchner, E.: A model based development approach for building automation systems. In: 2014 IEEE Emerging Technology and Factory Automation (ETFA), pp. 1–6. IEEE (2014)
8. Fette, I., Melnikov, A.: The websocket protocol, RFC 6455 (2011)
9. Belshe, M., Peon, R., Thomson, M., Melnikov, A.: Hypertext transfer protocol version 2.0. internet draft (2013)
10. Kovatsch, M., Duquennoy, S., Dunkels, A.: A low-power COAP for contiki. In: Proceedings of the 8th IEEE International Conference on Mobile Ad-hoc and Sensor Systems (2011)
11. Seitz, C., Schönfelder, R.: Rule-based OWL reasoning for specific embedded devices. In: Aroyo, L., Welty, C., Alani, H., Taylor, J., Bernstein, A., Kagal, L., Noy, N., Blomqvist, E. (eds.) ISWC 2011. LNCS, vol. 7032, pp. 237–252. Springer, Heidelberg (2011). doi:10.1007/978-3-642-25093-4_16
12. Anicic, D., Rudolph, S., Fodor, P., Stojanovic, N.: Stream reasoning and complex event processing in etalis. Semant. Web **3**(4), 397–407 (2012)
13. Ceri, S., Gottlob, G., Tanca, L.: Logic Programming and Databases. Springer Science & Business Media, Heidelberg (2012)
14. Compton, M., Barnaghi, P., Bermudez, L., Garca-Castro, R., Corcho, O., Cox, S., Graybeal, J., Hauswirth, M., Henson, C., Herzog, A., Huang, V., Janowicz, K., Kelsey, W.D., Phuoc, D.L., Lefort, L., Leggieri, M., Neuhaus, H., Nikolov, A., Page, K., Passant, A., Sheth, A., Taylor, K.: The SSN ontology of the W3C semantic sensor network incubator group. Web Semant. Sci. Serv. Agents World Wide Web **17**, 25–32 (2012)
15. Hepp, M.: eClassOwl: a fully-fledged products and services ontology in OWL. In: Poster Proceedings of ISWC, Galway (2005)
16. Shelby, Z.: Constrained RESTful environments (CoRE) link format (2012)

Author Index

Printed in the United States
By Bookmasters